Soil Biology

Volume 23

Series Editor
Ajit Varma, Amity Institute of Microbial Technology,
Amity University Uttar Pradesh, Noida, UP, India

For further volumes:
http://www.springer.com/series/5138

Günther Witzany
Editor

Biocommunication in Soil Microorganisms

Springer

Editor
Günther Witzany
Telos-Philosophische Praxis
Vogelsangstr. 18c
5111-Bürmoos
Austria
witzany@sbg.at

ISSN 1613-3382
ISBN 978-3-642-26572-3 ISBN 978-3-642-14512-4 (eBook)
DOI 10.1007/978-3-642-14512-4
Springer Heidelberg Dordrecht London New York

Cover design: SPi Publisher Services

Printed on acid-free paper

Springer is part of Springer Science+Business Media (www.springer.com)

Preface

Why Biocommunication of Soil Microorganisms?

Although research on signal-mediated interactions of microorganisms – especially prokaryotes – in the ocean is a broad field of investigation, research on communication between soil microorganisms is not as well developed in comparison. This contradicts the importance of the roles of soil bacteria on parasitic and symbiotic interactions with plants, animals, and fungi in dry, wetland, and wasteland ecologies and in flood waters and their consequences on terrestrial life. Additionally, it is equally important to investigate the main sources of genetic innovation, exchange, and storage of soil bacteria (such as decomposers, nitrogen fixers, disease suppressors, aerobes, anaerobes, actinobacteria, and sulfur oxidizers), i.e., the roles of phages, plasmids, and related genetic parasites. Focusing on these viral colonizers and viral-derived regulatory elements of all prokaryotic life is important because they determine the interactional competences of soil bacteria and their group identity, i.e., their competence in producing and emitting shared signal molecules, interpreting incoming messages via appropriate receptors, measuring them, and generating appropriate response behaviors. Although these factors have been investigated by physiological, chemical, and mechanistic perspectives, it has become increasingly clear that signal-mediated interactions, i.e., biocommunication processes, additionally rely on semiotic rules that have to be correctly followed or biocommunication will not occur. This means that the rules of signal use are not strict (natural) laws: although very conservative, rules of signal use may be changed or even generated de novo for adaptational purposes.

If we speak of biocommunication of soil microorganisms, we first must become clear with the up-to-date terms of communication (and with the signaling system, which is used to communicate what we call language). Therefore, we should rely on the results of the pragmatic turn in the philosophy of science discourse in the seventies and eighties of the last century, which was the result of a discourse between 1920 and 1980 to clarify the conditions for generating correct sentences in science.

Communication is defined as an interaction between at least two living agents, which share a repertoire of signs (which represents a kind of natural language) that are combined (according to syntactic rules) in varying contexts (according to pragmatic rules) to transport content (according to semantic rules).

These three levels of semiotic rules are complementary parts of any natural language or code. If one level is missing, according to Charles Morris (see Morris 1946), we cannot seriously speak of language or communication. So the most recent definition of communication is: sign-mediated and rule-governed interactions, i.e., interactions that depend on a commonly shared repertoire of signs and rules of sign-use. However, these features are lacking in abiotic interactions. Additionally, we know that mathematical and mechanistic theories of language are less helpful in investigations on natural languages and real-life communication processes because such theories cannot explain typical features of living agents that communicate, which are not formalizable, i.e., for which no algorithm is available, such as the de novo-generation (innovation) of sentences/sequences. This means that no natural language or code speaks or codes itself but needs living agents that are competent in using such languages or codes (Witzany 2010).

In the biology of the twentieth century, the physiology of all kinds of cells, tissues, organs, and organisms of all organismic kingdoms was the mainstream direction in biological research and experiments. In the 1970s, an increasing use of "communication" as a metaphor also occurred in biology. During the last decade of this period, interest in communication (no longer being used as a metaphor) within and between organisms overtook that of the pure physiological understanding of organisms. Cell-to-cell communication now dominates contemporary cell biology, including an enormous knowledge about a great variety of signaling pathways serving for both organization and coordination of production, release, uptake, and processing of "information" within and between cells.

In parallel, the use of "language" as a metaphor increased from the middle of the twentieth century with growing knowledge about the genetic code. Most of the processes that evolve, constitute, conserve, and rearrange the genetic storage medium DNA are terms that were originally used in linguistics, such as nucleic acid language, genetic code, "codes without commas" (F. Crick), coding, copying, translation, transcription, sequence homology, etc. Meanwhile, the linguistic approach also lost its metaphorical character and the similarity between natural languages/codes, and the genetic storage medium DNA are not only accepted but are adapted in epigenetics, bioinformatics, biolinguistics, protein linguistics, and biosemiotics. The advantage of methodical adaptation of communication and linguistic terminology is in having appropriate tools for differentiation at specific levels, which is otherwise difficult to describe nonreductively by pure physiology.

This means that language-like structures and communication processes occur at the bottom of living nature. Language and communication are not at all evolutionary inventions of humans, nor are they anthropomorphous adaptations to describe nonhuman living nature. It simply became, and still becomes, obvious that every coordination and organization within and between cells, tissues, organs, and organisms needs signs, i.e., chemical molecules that serve as signals or symbols in messages or serve as vital indicators of environmental conditions. Because no code codes itself, as no language speaks itself, these signs need to be sensed and interpreted in a correct way by biological agents, i.e., there must be subjects/representatives of sign production and sign interpretation. This means that sensing, as well as interpretation, may fail with the result of nonappropriate behavior and even fatal consequences for cells, tissues, organs, and organisms.

The method of analyzing any part of a machine in detail to get a picture of its whole functional blueprint, which can then be used to reproduce or manipulate it, or to produce an even more perfect one (taking genetic engineering as an example), is still useful if we are dealing with machines. However, growing evidence of the aims of several biological processes makes it doubtful now whether investigating organisms with this mechanistic attitude will still be useful in the future:

On the Interorganismic Level (Between Same and Related Organisms)

Communication between cells, cellular parts, tissues, organs, and organisms is far from being a procedure, which can be reduced to mechanistic input/output or cause/reaction descriptions. It is evident today that communication processes between living organisms include a variety of circumstances and competences that must be fulfilled in parallel if communicative acts are to have successful consequences, such as common coordination.

First of all, no single organism is able to communicate as an emerging property. It must be a community, a society, or a swarm of organisms that each share an identity (group) and a competence to sense others as being part of this identity or not (self/nonself differentiation competence), even if this competence is shared genetically solely. To communicate, it is necessary that an organism has some skills that serve as signs (signals, symbols), such as chemical molecules either produced directly by itself or as secondary metabolites or even molecules in the surroundings that are not produced by the organism but can still be manipulated, according to the organismic needs.

Secondly, organisms must share a competence to use these signs in a coherent manner, which means using these signs in a strict temporal and spatial context. In most cases, it is not just one signaling molecule but several that are combined in a certain manner to transport messages (information). This represents a common feature of sign-use in communication processes, which is called their correct combination or syntax.

Thirdly, organisms are part of a habitat in which they live together with similar organisms of the same or related species, and in some cases, with an abundance of nonrelated organisms of other kingdoms. This context exactly represents the natural history of organismic swarms or communities in which they – and this is only a recently experienced feature – evolved and developed certain abilities to appropriate response behaviors according to their survival. These include sensing, learning, and memory, which are the preconditions for faster adaptations.

Finally, the signaling molecules, which serve as signs, transport messages with meanings (semantics). The informational (semantic) content, which is transported, triggers certain response behaviors by the same or related, or even unrelated, organisms. Interestingly, the signal sequence or signal content does not necessarily depict a single meaning, i.e., function but can vary according to different situational contexts. This means that identical signs can transport a variety of different messages according to different contextual needs. This is important in very dense ecological habitats (as demonstrated below), for example, in the oral cavity of humans where communication of up to 500 different microbial species must function in order to prevent oral diseases. The different uses of identical signs (sequences) enable the generation of dialects within same species that can transport messages, which are microecosphere-specific. These include a very sensitive self/nonself recognition between slightly differently adapted populations of the same species in the same ecological habitat.

Although sign-mediated interactions (i.e., communication processes) are very reliable in most cases, they do not function mechanistically in a strict sense. Syntax (combination), pragmatics (context), and semantics (content) must function in parallel to ensure and optimize coordination and thus survival of group members. These semiotic rules do not function mechanistically but may be varied, deleted, or, in certain circumstances, generated de novo. Additionally, semiotic rules do not function by themselves but need semiotic subjects, i.e., living organisms that use such rules. If no living organism is present, semiotic rules, signs, and communication are absent. Although highly conserved semiotic rules are modifiable, environmental circumstances, such as stress, trigger adaptational responses. In such cases, signals may transport new messages, which previously did not exist, broadening the communicative competences of organisms, i.e., broadening evolutionary capabilities. This is different in the case of abiotic processes, where semiotic (syntactic, pragmatic, semantic) rules of sign-use are unnecessary as natural laws are sufficient alone. No semiotic rules are used or necessary for water molecules to freeze into ice.

On the Intraorganismic Level

During the last 2 decades, more and more indicators replaced the mechanics of intracellular generation and use of signaling molecules by interactive information processing between parts of the cellular organism. Natural genetic engineering (Shapiro 2009) identified the whole processing of most steps and substeps of genetic content arrangement and rearrangement necessary for adaptational purposes in the fields of replication, transcription, translation, repair, marking (epigenetics), and immune functions. In contrast to former assumptions of DNA as a stable genetic information storage medium, which can be altered only by errors (mutations) or damage, it became increasingly clear that the driving force of evolutionary novelty is a vast abundance of highly dynamic, mobile networking agents. These agents are active shortly after transcription out of the DNA storage medium and include RNAs such as the prominent tRNA, mRNA, and rRNA agents and a great variety of regulatory RNAs, most of them small, noncoding RNAs before translational processes into proteins (Witzany 2009). Also, transposons and/or retroposons, which represent mobile genetic elements, serve as competent genetic content operators.

Recent research indicates that all of these noncoding RNAs with its higher order regulatory functions, as well as the three prominent RNA agents mentioned above, and the whole range of transposable elements are remnants of former viral or viral-like agents (Villarreal 2005, 2009; Witzany 2010). It also became clear that the evolutionary role of viruses is not a derivative one as suggested by models that interpreted viruses (1) to be escaped as transcripts out of cellular organisms or (2) as descended from free-living bacteria and having lost their cellular functions, as with the regressive hypothesis. In contrast to this virus-first hypothesis is the identification of a high abundance of viral genetic sequences that are not found in any cellular DNA content (Forterre and Prangishvili 2009, Koonin 2009, Villarreal and Witzany 2010). This is in agreement with the early RNA world theory, where cellular life evolved after viruses. In this new perspective, cellular DNA is the preferred habitat for persistent viral settlers, which not only integrate but rearrange and transfer viral competences to the cellular host and therefore broaden cellular evolutive and developmental potentials.

The interrelation between nucleic acid language and linguistics is predominant in the field of bioinformatics, which is a successful tool in genetic comparison techniques such as phylogenetic analyses and comparative genomics. For several decades, it was assumed that the molecular syntax of genetic sequences determined the meaning (semantics) of these sequences according to Manfred Eigen. With the rise of epigenetics, it became clear that different marking (methylation) patterns of an identical genetic sequence can lead to different reading patterns and, consequently, to the production of different products from this genetic data set. Changing environmental circumstances such as stress or nutrient availability may alter these markings (histone modifications, methylation patterns), which may lead (not necessarily) to inheritable features (Jirtle and Skinner 2007). The evolution of epigenetic marking remained a mystery for a long time. According to the virus-first hypothesis, epigenetic marking is a viral competence. All viruses mark their genome in order to be able to differentiate self from nonself agents. If we assume that viruses are evolutionarily older than cellular life, epigenetic marking is a viral competence transferred to cellular life to broaden host informational content and evolutionary as well as developmental capabilities.

The integration of viral features to cellular hosts is not a rare event. Considering that viruses are ten times more abundant in the environment than cellular micro-organisms, which all are infected by phages and plasmids, it seems rather doubtful that this rare habitat of cellular genomes contains free sequence space that is not subjected to competing viral settlers. Therefore, I predict that future investigations will show a much higher level of these persistent viral agents.

In Vitro Analyses Lack Context-Dependent Behaviors of Real Life Habitats

In vitro investigations focus on ecological setups, which do not represent the entire interactional context in which an organism is involved *in vivo*. The evolution and development of each organism depends on the *in vivo* habitat with its inter-, intra-, and transorganismic triggers on genetic reading patterns, which are absent from *in vitro* setups. Therefore, it is likely that isolated organisms in laboratory setups lack a variety of features, which would be triggered in *in vivo* habitats by natural circumstances such as symbiotic and parasitic microorganisms. This may lead to restricted conclusions on their intra- and interorganismic biocommunicative capabilities.

Biocommunication of Soil Microorganisms

As a consequence of these findings, I tried to integrate this biolinguistic and biocommunicative features into a uniform description of all key levels of communication within the organismic kingdoms of plants, fungi, animals, and bacteria, based on recent empirical data (Witzany 2010). Accordingly, biocommunication occurs on three levels: (1) intraorganismic, within an organism, (2) interorganismic, between the same or related species and (3) transorganismic, between organisms that are not related.

The biocommunicative approach demonstrates that cells, tissues, organs, and organisms coordinate and organize by communication processes and that genetic nucleotide sequence orders in cellular and noncellular genomes are structured like language, i.e., they follow combinatorial (syntactic), context-sensitive (pragmatic), and content-specific (semantic) rules. Without sign-mediated interactions, no vital functions within and between organisms can be coordinated. This feature is absent in nonliving matter. Additionally, the biocommunicative approach investigates natural genome editing competences of viruses and viral-like agents. Natural genome editing from a biocommunicative perspective is competent agent-driven generation and integration of meaningful nucleotide sequences into preexisting genomic content arrangements and the ability to (re)combine and (re)regulate them according to context-dependent (i.e., adaptational) purposes of the host organism.

It became increasingly clear that, particularly in these investigations, in contrast to the relatively rich database concerning the role of bacteria and their obligate viral settlers in the oceans, the basic knowledge of biocommunication in soil microorganisms is far from satisfactory. Although soil habitats are equally as important to terrestrial organisms as the sea is to aqueous organisms, main focus was not on soil organismic life. To give a recent overview on the biocommunication of soil microorganisms, Ajit Varma encouraged me to edit this book.

Contributions to the biocommunication of soil microorganisms

After the introduction in which a general overview on the key levels of communication of bacteria is given in the first section on intracellular biocommunication of soil bacteria, Stephen Abedon begins with the various levels of biocommunication of phages with soil microorganisms. Robert Armon gives a systematic overview on the interactional patterns of soil bacteria and their bacteriophages. Kurt Williamson reports on advances of our understanding of soil viral ecology. K.V. Srividhya and S. Krishnaswamy describe identification methods of persistent viral agents and their defectives (prophages and phage remnants) within soil microbials. Omar Bagasra and Gene Pace report on the important role of transposable elements in genome formatting of soil microbes. Makoto Kimura, Guanghua Wang, Natsuko Nakayama, and Susumu Asakawa investigate the role of bacteriophages on soil bacteria in rice paddies. Antonet Svircev, Susan Lehman, Peter Sholberg, Dwayne Roach, and Alan Castle investigate the phage mediated genetic exchange between soil microorganisms with their hosts. Brian Cheetham, Gabrielle Whittle, Michael Tang, and Margaret Katz identify a series of genetic elements within the *Dichelobacter nodosus* genome, which modulate expression of virulence determinants and are themselves controlled in a coordinated manner. Dalit Roth, Asaf Madi, Dror Y. Kenett, and Eshel Ben-Jacob introduce the Genome Holography method for the analysis of gene expression data. Margaret Riley reflects on the bacterial species concept in light of the large scope for horizontal gene transfer and on relevant methods for identification and measurement.

The second section on intercellular and transorganismic biocommunication of soil microorganisms starts with Katherine Pappas and Miguel Cevallos reporting on the plasmids of Rhizobiaceae, which receive, integrate, and release signals that profoundly determine bacterial–host cohabitation. Yves Dessaux, Emilie Chapelle, and Denis Faure differentiate two biocommunicative strategies in soil ecosystems such as quorum sensing and quorum quenching. Max Teplitski, Massimo Merighi, Mengsheng Gao, and Jayne Robinson report on the role of plasmids, transposons, and other viral defectives as effective tools for signal production for biocommunicative needs. Paul Paré, Huiming Zhang, Mina Aziz, Xitao Xie, Mi-Seong Kim, Xin Shen, and Jinlin Zhang investigate biocommunication of beneficial microbes that drive growth and development of plants. Max Teplitski and Sathish Rajamani investigate biocommunication between soil algae and bacteria. Ilona Pfeiffer reports on possible communication strategies between bacteria and fungi that cause antagonistic or symbiotic interactions between them. Ralf Oelmüeller, Neeraj Shrivastava, Meghna Pohani, Kailash Upadhyaya, Irena Sherameti, Paul Murugan, Shashibala Singh, and Ajit Varma investigate symbiotic and transkingdom signaling of bacteria, plants, and fungi in the rhizosphere.

Bürmoos, Austria Günther Witzany

References

Forterre P, Prangishvili D (2009) The great billion-year war between ribosome- and capsid-encoding organisms (cells and viruses) as the major source of evolutionary novelties. Ann N Y Acad Sci 1178: 65–77

Jirtle RL, Skinner MK (2007) Environmental epigenomics and disease susceptibility. Nat Rev Genet 8:253–262

Koonin EV (2009) On the origin of cells and viruses: primordial virus world scenario. Ann NY Acad Sci 1178:47–64

Morris C (1946) Signs, language, and behavior. Prentice Hall, New York

Shapiro J (2009) Revisiting the central dogma in the 21st century. Ann N Y Acad Sci 1178:6–28

Villarreal LP (2005) Viruses and the evolution of life. ASM Press, New York

Villarreal LP (2009) Origin of group identity. Viruses, addiction and cooperation. Springer, New York

Villarreal LP, Witzany G (2010) Viruses are essential agents within the roots and stem of the tree of life. J Theor Biol 262: 698–710

Witzany G (2009) Noncoding RNAs: Persistent viral agents as modular tools for cellular needs. Ann N Y Acad Sci 1178:244–267

Witzany G (2010) Biocommunication and natural genome editing. Springer, Dordrecht

Contents

Contributors

Stephen T. Abedon Department of Microbiology, The Ohio State University, 1680 University Dr., Mansfield, OH 44906, USA, abedon.1@osu.edu

Robert Armon Division of Environmental, Water & Agricultural Eng., Technion-Israel Institute of Technology, Haifa, 32000, Israel, cvrrobi@tx.technion.ac.il

Susumu Asakawa Graduate School of Bioagricultural Sciences, Nagoya University, Furocho, Chikusa, Nagoya, 464-8601, Japan

Mina Aziz Department of Chemistry and Biochemistry, Texas Tech University, Lubbock, TX 79409, USA

Omar Bagasra Department of Biology, South Carolina Center for Biotechnology, Claflin University, 400 Magnolia Street, Orangeburg, SC 29115, USA, obagasra@claflin.edu

Eshel Ben-Jacob School of Physics and Astronomy, Tel Aviv University, 69978 Tel Aviv, Israel, eshel@tamar.tau.ac.il, eshelbj@gmail.com

Emilie Chapelle Institut des sciences du vegetal, UPR 2355, CNRS, avenue de la terrasse, 91198, Gif-sur-Yvette CEDEX, France

Alan J. Castle Department of Biological Science, Brock University, 500 Glenridge Avenue, St. Catharines, ON, Canada L2S 3A1

Miguel A. Cevallos Programa de Genómica Evolutiva, Centro de Ciencias Genómicas, Universidad Nacional Autónoma de México, Apartado Postal 565-A, Cuernavaca, Morelos, Mexico

Brian Cheetham Molecular and Cellular Biology, University of New England, Armidale, NSW 2351, Australia, bcheetham51@gmail.com

Yves Dessaux Institut des sciences du végétal, CNRS UPR 2355, Bâtiment 23, Avenue de la terrasse, 91198 Gif sur Yvette CEDEX, France, dessaux@isv.cnrs-gif.fr

Denis Faure Institut des sciences du vegetal, UPR 2355, CNRS, avenue de la terrasse, 91198 Gif-sur-Yvette CEDEX, France

Mengsheng Gao Soil and Water Sciences Department, Institute of Food and Agricultural Sciences (IFAS), Genetics Institute, University of Florida, 1376 Mowry Rd, Gainesville, FL, USA

Margaret E. Katz Molecular and Cellular Biology, University of New England, Armidale, NSW 2351, Australia

Dror Y. Kenett School of Physics and Astronomy, Tel Aviv University, 69978 Tel Aviv, Israel

Mi-Seong Kim Department of Chemistry and Biochemistry, Texas Tech University, Lubbock, TX 79409, USA

Makoto Kimura Graduate School of Bioagricultural Sciences, Nagoya University, Furocho, Chikusa, Nagoya, 464-8601, Japan, kimuram@agr.nagoya-u.ac.jp

S. Krishnaswamy Centre of Excellence in Bioinformatics, School of Biotechnology, Madurai Kamaraj University, Madurai, 625021, India, krishna@mrna.tn.nic.in

Susan M. Lehman Division of Healthcare Quality Promotion, Centers for Disease Control and Prevention, Mail Stop C-16,1600 Clifton Rd NE, Atlanta, GA 30333, USA

Asaf Madi Faculty of Medicine, Tel Aviv University, 69978 Tel Aviv, Israel

Massimo Merighi Glycosyn Inc, Boston, MA, USA

Paul Murugan Defence Institute of High Altitude Research (DIHAR), Defence R & D Organisation, Leh, Ladakh, 194101, India

Natsuko Nakayama Graduate School of Bioagricultural Sciences, Nagoya University, Furocho, Chikusa, Nagoya, 464-8601, Japan

Ralf Oelmüller Institute of General Botany and Plant Physiology, Friedrich-Schiller University of Jena, Dornburger Street 159, 07743 Jena, Germany

Katherine M. Pappas Department of Genetics & Biotechnology, University of Athens, Panepistimiopolis, Ilissia, Athens, TK 15701, Greece, kmpappas@biol.uoa.gr

D. Gene Pace Department of English and Foreign Languages, South Carolina Center for Biotechnology, Claflin University, 400 Magnolia Street, Orangeburg, SC 29115, USA

Paul W. Paré Department of Chemistry and Biochemistry, Texas Tech University, Lubbock, TX 79409, USA, Paul.Pare@ttu.edu

Ilona Pfeiffer Department of Microbiology, University of Szeged, Kozep fasor 52, 6726 Szeged, Hungary, pfeiffer@bio.u-szeged.hu

Meghna Pohani Amity Institute of Microbial Technology, Amity University Uttar Pradesh, Sector-125, Noida, Uttar Pradesh, India

Sathish Rajamani Department of Microbiology and Immunology, Dartmouth Medical School, 208 Vail building, Hanover, NH 03755, USA

Margaret A. Riley University of Massachusetts, 611 N. Pleasant Street, Amherst, MA 01003, USA, riley@bio.umass.edu

Dwayne Roach Department of Biological Science, Brock University, 500 Glenridge Avenue, St. Catharines, ON Canada L2S 3A1

Jayne Robinson Department of Biology, University of Dayton, Dayton, OH USA

Dalit Roth Faculty of Medicine, Tel Aviv University, 69978, Tel Aviv, Israel

Xin Shen Department of Chemistry and Biochemistry, Texas Tech University, Lubbock, TX 79409, USA

Irena Sherameti Institute of General Botany and Plant Physiology, Friedrich-Schiller University of Jena, Dornburger Street 159, 07743 Jena, Germany

Peter Sholberg Agriculture and Agri-Food Canada, Pacific Agri-Food Research Centre, Summerland, BC, Canada V0H 1Z0

Neeraj Shrivastava Amity Institute of Microbial Technology, Amity University Uttar Pradesh, Sector-125, Noida, Uttar Pradesh, India

Shashibala Singh Defence Institute of High Altitude Research (DIHAR), Defence R & D Organisation, Leh, Ladakh, 194101, India

K.V. Srividhya Centre of Excellence in Bioinformatics, School of Biotechnology, Madurai Kamaraj University, Madurai 625021, India

Antonet M. Svircev Agriculture and Agri-Food, 4902 Victoria Ave. North, Vineland Station, ON Canada L0R 2G0, Antonet.Svircev@agr.gc.ca

Max Teplitski Soil and Water Sciences Department, University of Florida – Institute of Food and Agricultural Sciences (IFAS), 1376 Mowry Rd, Gainesville, FL, USA, maxtep@ufl.edu

Michael Ting Molecular and Cellular Biology, University of New England, Armidale, NSW 2351, Australia

Kailash C. Upadhayaya School of Life Sciences, Jawahar Lal Nehru University, New Delhi, India

Ajit Varma Amity Institute of Microbial Technology, Amity University Uttar Pradesh, Sector-125, Noida, Uttar Pradesh India, ajitvarma@aihmr.amity.edu

Guanghua Wang Northeast Institute of Geography and Agroecology, Chinese Academy of Sciences, Harbin 150081, China

Kurt E. Williamson Department of Biology, College of William and Mary, Williamsburg, VA 23187, USA, kewilliamson@wm.edu

Gabrielle Whittle Molecular and Cellular Biology, University of New England, Armidale, NSW 2351, Australia

Guenther Witzany Telos-Philosophische Praxis, Vogelsangstr. 18c, 5111, Buermoos, Austria, Witzany@sbg.at

Xitao Xie Department of Chemistry and Biochemistry, Texas Tech University, Lubbock, TX 79409, USA

Huiming Zhang Department of Chemistry and Biochemistry, Texas Tech University, Lubbock, TX 79409, USA

Jinlin Zhang Department of Chemistry and Biochemistry, Texas Tech University, Lubbock, TX 79409, USA

Chapter 1
Introduction: Key Levels of Biocommunication of Bacteria

Guenther Witzany

1.1 Introduction: Communicative Competences of Bacteria

Bacteria (prokaryotes) communicate and therefore are able to organize and coordinate their behavior similar to a multicellular organism (Losick and Kaiser 1997; Kaiser and Losick 1993; Ben Jacob and Levine 2006; Bassler and Losick 2006). But what does communication mean? In contrast to older concepts which summarize communication processes as information exchange only, nowadays communication processes are investigated as sign-mediated *interactions*, i.e., the informational content which is transported with signs triggers all kinds of different behavior. Also such different processes as production, release, uptake, and interpretation of signal molecules represent behavioral patterns. Signs are, in most cases, chemical molecules, in some cases also tactile interactions, which serve as signals both within and between prokaryotic organisms.

Bacteria are symbiotic organisms covering the whole range from mutualism to parasitism. They may be beneficial for their (eukaryotic) hosts and without them host survival would not function. Others are neutral, i.e., they do not harm the host. Many of them also cause diseases, with sometimes epidemic characteristics and, often, lethal consequences.

Bacteria represent one of the main success stories of evolution. They originated at the early beginning of life similarly to archaea which represent a different organismic kingdom (Woese et al. 1990; Koonin et al. 2006; Koonin and Wolf 2008; Koonin 2009). Bacteria are found in all ecological niches and share a common flux of their gene pool with a high rate of gene order recombination for adaptational purposes of great diversity (Pal et al. 2005). More than in any other organismic kingdom it is in common use to speak about the languages and even

G. Witzany
Telos-Philosophische Praxis Vogelsangstraße, 18c, A-5111-Buermoos, Salzburg, Austria
e-mail: witzany@sbg.at

G. Witzany (ed.), *Biocommunication in Soil Microorganisms*, Soil Biology 23,
DOI 10.1007/978-3-642-14512-4_1, © Springer-Verlag Berlin Heidelberg 2011

dialects of bacteria (Kaiser and Losick 1993; Swift et al. 1994; Bassler 1999, 2002; Schauder and Bassler 2001; Ben Jacob et al. 2004).

Quorum sensing is the term of description for sign-mediated interactions in which chemical molecules are produced, secreted, and uptaken by bacteria (Crespi 2001; Manefield and Turner 2002; Greenberg 2003; Tu and Bassler 2007). They are recognized by the bacterial community dependent on (a) a critical concentration and in a special ratio to (b) the population density (Daniels et al. 2004; Lerat and Moran 2004; Waters and Bassler 2005). These molecules trigger the expression of a great variety of gene transcriptions. Many bacteria use multiple quorum sensing codes; each may be modulated by posttranscriptional or other regulatory engineering (Loh et al. 2002).

There are also communication processes between different species of bacteria and between bacteria and nonbacterial life such as eukaryotic hosts (Konaklieva and Plotkin 2006). Beneath the semiochemicals (Gr.: Semeion = sign) necessary for developmental processes of bacterial communities such as division, sporulation, and synthesis of secondary metabolites, there are physical contact-mediated behavioral patterns which are important in biofilm organization (Davis et al. 1998; Fuqua and Greenberg 2002; Voloshin and Kaprelyants 2004; Parsek and Greenberg 2005). Also, abiotic influences serve as signs which indicate specific nutrients or other environmental circumstances such as dehydration or hydrodynamic changes or changes of ph-level in soil and floodwaters.

As communities of bacteria species, which are able to coordinate their behavior and have advantages over single bacteria organisms, are much more common, it is not surprising that the evolutionary drive went into rising communicative complexity (Ben Jacob 2003). We should not forget that in comparison to the first two billion years of life on earth with closed prokaryotic symbiology the rise and growth of the multicellular eukaryotes (animals, fungi, plants) was a *crucial advantage* for bacterial lifestyle to colonize vertical hosts with their great spatial and motility resources.

In general in biocommunication, we can differentiate three classes of signaling molecules for different purposes, i.e., signaling within the organism to coordinate gene expressions to generate adequate response behavior, signaling between the same or related and different species. With a limited number of molecules and a limited number of combinatorial rules, they generate quite different interactions for different purposes all mediated by signs. As in every sign-mediated interaction sign users share a common set of syntactic rules, i.e., how signs may be combined; of pragmatic rules which determine a great variety of interactional contexts, e.g., development, growth, mating, virulence, attack, or defense. The situational context of these complex interactional processes determines the meaning of signs, i.e., semantics of signals. Independent of organismic complexity, the complementarity of these three levels of semiotic rules can be identified, in principle, in every sign-mediated interaction within and between organisms (Witzany 2010). This leads to the generation of intra- and intercellular processes which enable bacterial communities to generate memory which may be inheritable but can alter epigenetically, i.e., different reading/meaning patterns of the same genetic data set with differences at the phenotypic level without altering the genetic data set.

The link between linguistics and genetics has been obvious since the detection of the universal grammar and the structural code of DNA. Chomsky's meaning-independent syntax approach lead to the broad acceptance and usage of bioinformatic methods and systems biology. Researchers in bacteria communication like (Ben Jacob et al. 2004) suggested with good reason that this approach reduces linguistic competences found in bacterial communication and has to be satisfied by both semantic aspects, i.e., the meaning of signals which serve as signs, and pragmatic aspects, which focus on the context-dependent variety and differences of the behavioral patterns in common-goal coordination, shared knowledge, memory, and mutual targets. Apart from that, it is coherent with the presupposition by Charles Morris of any nonreductionistic analysis of language-like structures, the obligate complementarity of syntax, semantics, and pragmatics (Morris 1946).

1.2 Semiochemical Vocabulary and Communicative Goals of Bacteria

The semiochemical vocabulary used by bacteria is of great variety, especially because some signaling molecules are multiple reusable components (Henke and Bassler 2004). Acyl homoserine lactones (AHLs) and linear oligopeptides are used as signs in diverse processes. Cyclized oligopeptides function as virulence genes. γ-Butryolactones (GBLs) are used as antibiotics and in sporulation processes. Furanosyl diester (AI-2) is used in a variety of processes (Sun et al. 2004) and in luminescence. cis-11-Methyl-2-dodecenoic acid (DSF) serves in virulence and pigmentation. 4-Hydroxy-2-alkyl quinolines (PQS, HAQs) are important in whole regulation processes and for virulence as are palmic acid methyl esters (PAME). Putrescine is important in swarming motility like biofilm organization. A-signal is used in early developmental processes and aggregation. C-signal is a cell surface-associated protein and serves to coordinate motility and the developmental process of building a fruiting body. Cyclic dipeptide is a secondary metabolite (Shapiro 1998; Visick and Fuqua 2005).

Gram-negative bacteria use homoserine lactones (LuxR/LuxI) as signs in communication processes (Swift et al. 1994; Schauder and Bassler 2001; Lenz et al. 2004), whereas Gram-positive bacteria use oligopeptides in quorum sensing communication. As in all organisms, noncoding RNAs are important in higher order regulatory pathways such as small RNAs and microRNAs are used by bacteria to regulate special genetic expression patterns, which play an important role as appropriate response behavior to stress or nutrient availability (Teplitski et al. 2000; Masse and Gottesman 2002; Wassarman 2002; Vogel and Sharma 2005; Majdalani et al. 2005), e.g., in controlling the quorum sensing pathways (Bauer and Robinson 2002).

At present, three kinds of communicative goals are distinguished: (1) reciprocal communication, i.e., active sign-mediated interactions, which are beneficial for both interacting parts such as decision-making processes (Brockhurst et al. 2008); (2) messages which are produced as response on a triggering event, which may be an indicator for a receiver which was not specially targeted by the producer. A coincidental event which is neutral – except for the energy costs of production – to the producer but beneficial for the receiver; (3) signaling to manipulate the receiver, i.e., to cause a response behavior which is one-sided – beneficial to the producer and harmful to the receiver (Visick and Fuqua 2005), often in that they behave against their normal goals (Keller and Surette 2006).

The three classes of intra-, inter-, and transorganismic (trans-specific) communication enable bacteria to generate and coordinate different behavioral patterns: self and nonself identification, i.e., "recognition" and identification of self and other colonies and measurement of their size, pheromone-based courtship for mating, alteration of colony structure in formatting of fruiting bodies, initiation of developmental and growth processes, e.g., sporulation.

In receiving signals from same or related species or nonbacterial organisms the signaling molecules bind to specialized sensor proteins which function as receptors. They transmit the message to an intracellular regulator (Fuqua et al. 1996; Visick and Fuqua 2005), i.e., the signal molecule transits the cell membrane through diffusion or by specific transport pathways. Inside the cell the signaling molecule, in most cases, binds to a cytoplasmic target protein. It may be that a diffusible molecule is chemically engineered to an active signal after entering the target cell (Visick and Fuqua 2005). Organization of cellular production of response molecules leads to signal-dependent transcription control of DNA.

Bacteria have to distinguish between species-specific signaling and signaling which is able to modulate behaviors interspecifically (Bassler 1999; Federle and Bassler 2003; Waters and Bassler 2006). With these communicative competences, they are able to coordinate species-specific behavioral patterns as well as to coordinate behaviors between diverse species (Hughes and Sperandio 2008).

1.3 Transorganismic Communication of Soil Bacteria

If we look at beneficial symbioses between bacteria and plants, we refer to the complex communication networks between soil bacteria, mychorrizal fungi, and plant roots (Hayashi 2005; Imaizumi-Anraku et al. 2005). Mychorrizal fungi secret molecules in the surrounding environment which serve as nutrients for soil bacteria and trigger their activation to degrade special nutrients which are then available for mychorrizal fungi (Bonfante 2003; Bonfante and Anca 2009). Their hyphal growth serves as the developmental and growth area of plant roots, themselves being dependent on nutrients which are prepared by the mychorrizal fungi. Plant roots can also mimic bacterial signaling molecules, either to trigger bacterial production of special molecules or to disturb bacterial communication pathways (Teplitski et al. 2000; Bauer and Robinson 2002; Daniels et al. 2004).

Rhizobia bacteria are integrated into plant cells by phagocytosis when they interact symbiotically with the plant roots (Samaj et al. 2004). In other cases where rhizobia fail to fix nitrogen inside the root nodules because they are being deceptive, plants are sanctioning these rhizobia (Kiers et al. 2003) and prevent their spread to stabilize mutualistic symbioses with bacterial colonies (Keller and Surette 2006). Root exudates of different kinds regulate plant and microbial communities in the rhizosphere. This is necessary to stabilize equilibrium and inhibit the continuity of attacks by pathogenic bacteria in the soil (Walker et al. 2003; Bais et al. 2003). The full range of trans-specific communication processes between bacteria and plant roots is important for developmental and growth processes in the entire plant kingdom (Manefield and Turner 2002; Kent and Triplett 2002; Sharma et al. 2003).

Chemical molecules, which serve as signs in intercellular communication processes of bacteria, are similar to pheromones in social insects and animals. This may be an indicator of evolutionary lineages that evolved in the bacterial "chatter" (Velicer 2003). Interbacterial communication uses hormone-like signaling to sense specific host locations such as intestinal habitat. In this specialized ecosphere, a bacteria–host communication occurs which means the host cells and bacterial cells share a common meaning function for the same signaling molecules (Sperandio et al. 2003).

Living as endosymbionts as potential candidates for symbiogenesis (Margulis 1996, 1999, 2004; Margulis and Sagan 2002), as documented in the origin of eukaryotic endosomes like mitochondria, indicates the important role of bacteria for the entire history of evolution (Witzany 2005). The interactions may be pericellular colonization events but also an intracellular lifestyle. These different symbiotic interactions range from acquisition of novel genetic material to reduction in size and content connected with gene loss (Batut et al. 2004). Successful living processes of higher eukaryotes would not be viable without beneficial symbiosis with bacteria. The cell mass of an adult human assembles 20% of human origin and up to 80% of exogenic settlers (Blech 2000), most of them bacteria.

1.4 Interorganismic Communication

Interorganismic Communication is the sign-mediated interactions (coordination) between the same and the related species and includes the ability to sense self and nonself members. For a long time, it was assumed that bacteria live predominantly as monads. However, it has been recognized that this is a very rare exception (Federle and Bassler 2003; Dunn and Handelsman 2002). Bacterial colonies live, in almost all cases, not alone but in coexistence with other bacterial species self-coordinated by a diversity of sign-mediated interactions (Gray 1997; McNab and Lamont 2003; Ben Jacob and Levine 2006). Bacteria use intraspecific and interspecific signaling in all ecological in vivo situations (Keller and Surette 2006). This also implies a broad variety of conflicts within and between species (Velicer et al. 2000;

Xavier and Foster 2007). The mutual, neutral, and manipulative aims of communication processes are special kinds of response behavior to certain degrees of beneficial up to conflictual relationships (Keller and Surette 2006).

Dependent on the availability of nutrients, some bacteria suppress normal cell development which leads to the development of a different cell type, which is better suited for adequate response behavior for this situational context. It means that different environmental conditions can lead to different gene expressions within the same genetic data set. It has been shown that if the same colony is exposed several times to these changing contexts they react more immediately. This indicates that bacterial communities are able to develop collective memory and learn from the experience (Ben Jacob et al. 2004; McNab and Lamont 2003). In the case of changing environmental conditions, the suppression of cell division may lead to cell elongation which enables cell colonies to change the modus of motility. This is an important feature of socio-bacterial behavior, e.g., swarming coordination and organization for surface colonization (Shapiro 1998, 2007).

Some authors have documented altruistic strategies in mixed colony formations, which seems to be an advantage to the mixing among microcolonies. Altruistic behavioral strategies enable strengthened self-identity and a sustainable equilibrium in multilevel colonized ecological niches (Velicer and Yu 2003; Kreft 2004; West et al. 2007).

Interestingly, bacteria use a common contextual interpretation of incoming signals by each member of the colony. The response behavior is appropriate to the majority vote (Ben Jacob et al. 2004; Ben Jacob 2009) in a context-dependent decision.

The identification of nonself species is a competence which is possible through species-specific and group-specific quorum sensing and is coherent with the assumption that smaller groups of the same bacterial species are able to built types of quorum-sensing "dialects." These are important in the high density of coexistent bacterial life habitats to prevent confusion and enable more complex coordination (Taga and Bassler 2003) such as in the oral cavety of humans (see below). Interestingly, the prokaryotic cell–cell communication has structural analogs to cross-kingdom signaling between bacteria and fungi (Wang et al. 2004).

1.4.1 Interpretation and Coordination

Bacteria have profound effects on human health, agriculture, industry, and other ecospheres. Therefore, they target the multiple drugs which fight them (Camara et al. 2002). They develop drug resistance by coordination of special defensive behavior called biofilm organization (Sutherland et al. 2004; Burmølle et al. 2007; Danhorn and Fuqua 2007). Biofilm organization is a special kind of coordination with a high density of physical contact and contact-specific signaling (Bassler and Losick 2006) between members of a bacterial identity group. Biofilmorganisation in most cases depends on coordination of group members which share self (group identity) and nonself (Bacteria which are recognized to be not part of the group). In most habitats, there is also an organization with one or more nonself groups concerning group density, i.e., symbiotic signaling. This includes the release and uptake of molecules, which serve as indicators that, e.g., signalize to certain group members to undergo apoptotic processes if group density is too high to continue population survival concerning nutrient or even living space availability. If bacteria realize a critical mass via quorum sensing, they organize a high density of communal body by moving their flagellas which may resist even strong antibiotics (Wadhams and Armitage 2004; Diggle et al. 2007a, b). Biofilms are constructed on abiotic surfaces, e.g., on stones in rivers and other aqueous surfaces, as well as biotic ones, e.g., in the respiratory track of animals. Each human who had a strong cold remembers like persistent the mucus in the bronchial tube remained.

Nutrient availability also regulates the structure of biofilm organization (Stanley and Lazazzera 2004) as hydrodynamic forces (Wuertz et al. 2004). Interestingly, it has been found that biofilm organization is linked with coordinated DNA release, which is integrated in the biofilm (Spoering and Gilmore 2006).

Bacteria decide, in special cases – to mention another coordinative pattern – , to form fruiting bodies of different types and shapes for sporulation (Ben Jacob 2009). This enables bacterial communities to more efficiently disseminate the spores. The fruiting body building is governed by context-specific rules with different roles for different subgroups of bacterial communities for coordination (Kaiser and Welch 2004). Some have to serve for motility to density, followed by direction decision and decision of cell types, cell growth and developmental stages in all the different steps until the fruiting body is ready for the sporulation event. Without communicative hierarchical organization, this would not be possible. If communication is disturbed body building is not assured, so bacterial communities have developed special strategies to single out so-called "cheaters" (Velicer et al. 2000; Ben Jacob et al. 2004; Sandoz et al. 2007), which do not follow the rules for coordinating this special behavior.

As documented into Fig. 1.1 (Kohlenbrander et al. 2002), one of the most interesting and best investigated phenomena of bacterial communication is the *symbiology* of multiple colonies coexistent in the human oral cavity (Kohlenbrander et al. 2002, 2005; Rickard et al. 2006). Bacteria on human teeth and oral mucosa establish a homeostasis of pathogenic and mutualistic bacteria by a complex system of sign-mediated interactions both species-specific and trans-specific. The dental plaque in the oral cavity of humans is a unique habitat, which is not found in any other species (Sahasrabudhe and Dewhirst 2001). The homeostasis is not static but is the result of a continous and dynamic relationship between different species-colonies dependent on intervals of daily hygiene. The interacting species number approximately 500 different species (Moore and Moore 1994; Kroes et al. 1999; Paster et al. 2001).

Each member of these communities must be capable of self and nonself distinction and be able to distinguish between species-specific signaling and trans-specific signaling or even "noise" (same molecule but without informational content). As a community they must be able to measure their own colony size and the size of the

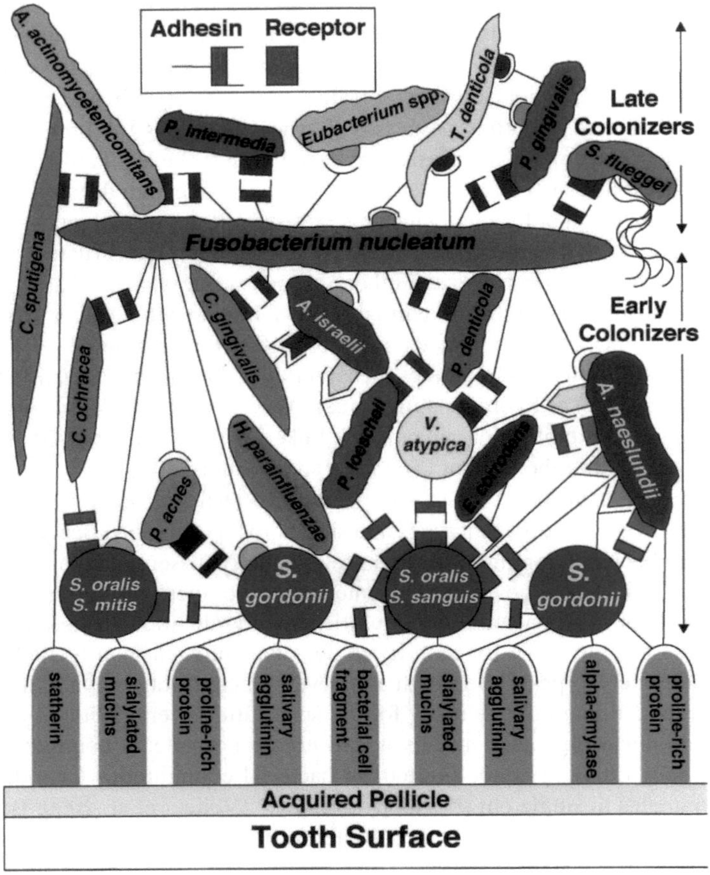

Fig. 1.1 Spatiotemporal model of oral bacterial colonization, showing recognition of salivary pellicle receptors by early colonizing bacteria and coaggregations between early colonizers, fusobacteria, and late colonizers of the tooth surface. Each coaggregation depicted is known to occur in a pairwise test. Collectively, these interactions are proposed to represent development of dental plaque and are redrawn from Kolenbrander and London (79). Starting at the bottom, primary colonizers bind via adhesins (round-tipped black line symbols) to complementary salivary receptors (vertical round-topped columns) in the acquired pellicle coating the tooth surface. Secondary colonizers bind to previously bound bacteria. Sequential binding results in the appearance of nascent surfaces that bridge with the next coaggregating partner cell. Several kinds of coaggregations are shown as complementary sets of symbols of different shapes. One set is depicted in the box at the top. Proposed adhesins (symbols with a stem) represent cell surface components that are heat inactivated (cell suspension heated to 85°C for 30 min) and protease sensitive; their complementary receptors (symbols without a stem) are unaffected by heat or protease. Identical symbols represent components that are functionally similar but may not be structurally identical. Rectangular symbols represent lactose-inhibitable coaggregations. Other symbols represent components that have no known inhibitor. The bacterial strains shown are *Actinobacillus actinomycetemcomitans*, *Actinomyces israelii*, *Actinomyces naeslundii*, *Capnocytophaga gingivalis*, *Capnocytophaga ochracea*, *Capnocytophaga sputigena*, *Eikenella corrodens*,

other colonies and molecules that have the same chemical structure but are not part of a biotic message. Special communication patterns with detailed hierarchical steps of signal production and transmission include (1) metabolite exchange, (2) cell–cell recognition, (3) genetic exchange, (4) host signal recognition and signal recognition of same or related species. Owing to the high number of competing and cooperating species, there is a special short- and long-term community architecture established according different spatial and temporal conditions. If the communication on the intra-, inter-, and metaorganismic level is successful, i.e., the signal transmission and reception enables colonies to live in a dynamic homeostasis, then the human oral cavity will avoid cavity diseases (Kohlenbrander et al. 2002, 2005).

1.5 Intraorganismic Communication

In contrast to transorganismic (trans-species) and interorganismic (same and related) biocommunication of bacteria, we term intraorganismic communication those sign-mediated interactions within the bacterial organisms, i.e., the signaling, regulation, coordination of all processes within prokaryotic cells including all genetic and epigenetic processes.

Only some higher order regulations (operons) that code for physically interacting proteins are found in almost all bacterial (and archaeal) genomes. Recent research indicates high dynamics of new gene orders as documented in the horizontal gene transfer events with their intensive intragenomic recombination (Imaizumi-Anraku et al. 2005; Xie et al. 2004). This exchange of whole genes or gene-blocks enables bacterial lifestyles to combine several bacterial competences, i.e., phenotypes. The transformation process includes the release of naked DNA, followed by the uptake and recombination, i.e., the integration, with 17 steps identified to date (Thomas and Nielsen 2005) (see Fig. 1.2). Thus, we can recognize the outcomes of a diversity of mobile DNA contents (Bordenstein and Reznikoff 2005), not a mass of individualized genetic texts, but a bacterial gene pool as a genetic text repertoire which is available for each individual bacteria and the resource for bacterial genome innovation and evolution (Gogarten and Townsend 2005; Olendzenski and Gogarten 2009). Horizontal gene transfer is a main resource for integrating newly evolved genes into existing genomes and does not need the slow steps of chance mutations to alter the genomes but accelerated genome innovations in both bacteria and archaea (Jain et al. 1999, 2003; Brown 2003). Important in this context of genomic innovation is not the sequence acquisition alone but also the contextualization (Solomon and Grossman 1996); it means also their loss (Berg and Kurland 2002). It seems now that the

Fig. 1.1 (Continued) Eubacterium spp., *Fusobacterium nucleatum*, *Haemophilus parainfluenzae*, *Porphyromonas gingivalis*, *Prevotella denticola*, *Prevotella intermedia*, *Prevotella loescheii*, *Propionibacterium acnes*, *Selenomonas flueggei*, *Streptococcus gordonii*, *Streptococcus mitis*, *Streptococcus oralis*, *Streptococcus sanguis*, *Treponema spp.*, and *Veillonella atypical* (with permission by Kohlenbrander et al. 2002)

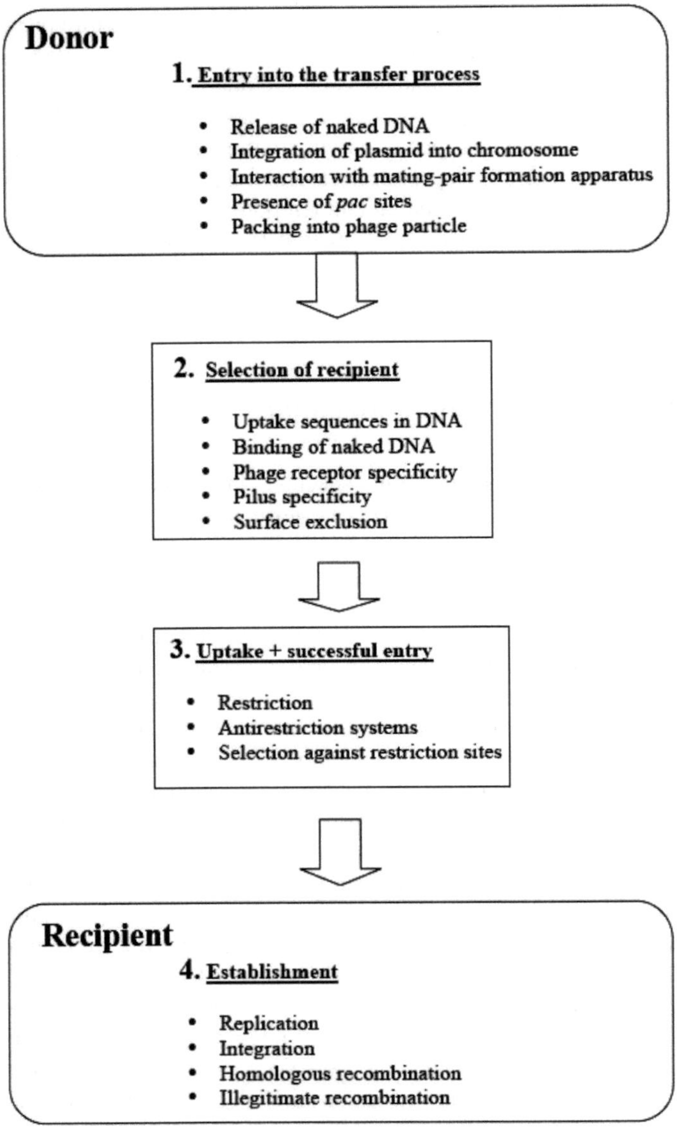

Fig. 1.2 The process of horizontal gene transfer. A schematic outlining the stages through which DNA must go on its journey from donor to recipient bacteria. The process begins with DNA in a potential donor cell becoming available and ends when this DNA becomes a functional part of a recipient cell's genome (Fig. 2 accordingly Thomas and Nielsen 2005)

phylogeny of microbial species is not a tree of life, but an evolutionary network or a ring of life, mediated by genetic exchange, i.e., acquisition and loss of genetic data sets (Rivera and Lake 2004; Kunin et al. 2005) (Fig. 1.3).

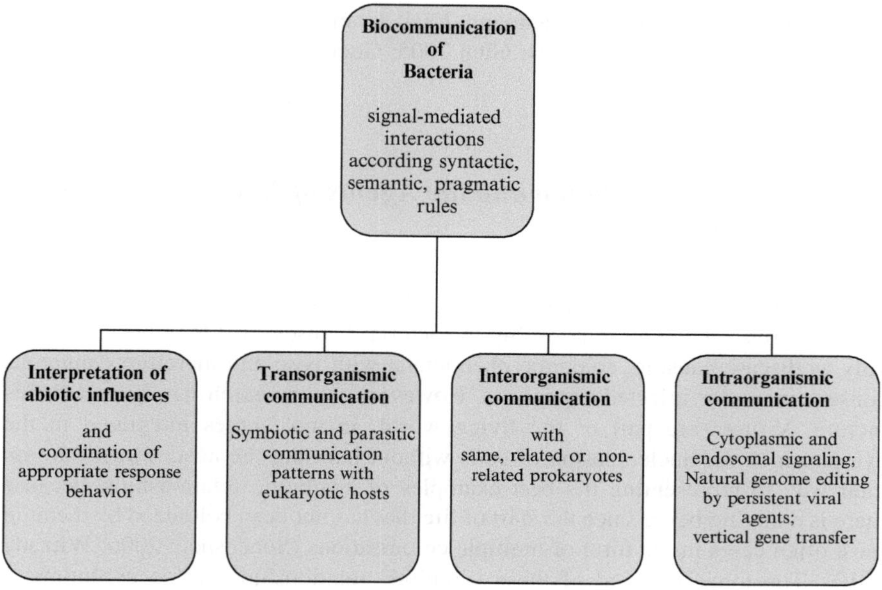

Fig. 1.3 Levels of biocommunication of bacteria

1.5.1 Intracellular Communication

Signal-dependent transcription regulation of the DNA serves for a great variety of response behavior. One of the most interesting phenomena is the fact that in the first two billion years of life on planet earth, the immense density of bacterial life has not been an event of the mass of individual organisms but their commonly shared gene pool which was in constant flux, as we now know, through investigations on horizontal gene transfer.

It means that the evolution of bacteria was not a random event of chance mutations and their selection but transfer of whole genes and gene-blocks representing real phenotypes that were transferred. This leads to different combinatorial patterns of genetic encoded phenotypes and the rise of bacterial diversity. It also enables bacterial pathogens to optimize their disease-causing coordination and is therefore targeted to special kinds of drug developments for medical purposes (Tettelin et al. 2005). New empirical data seem to suggest that the phenomenon of horizontal gene transfer is driven by viral competences inherent in bacterial settlers such as phages, plasmids, retroplasmids, and transposons (Villarreal 2005; see also Chiang and Lambowitz 1997).

For a long time, it has been proposed that tubulin plays an important role in cytoskeletal functions of eukaryotes whereas prokaryotes lack this system. Recent research has shown that tubulin is a very ancient system for genetic data set segregation also in bacteria which plays important roles in filament formation, movement, and orientation (Graumann 2004; Graumann and Defeu-Soufo 2004; Defeu-Soufo and Graumann 2004; Gitai 2005; Guerrero and Berlanga 2007).

1.5.2 Bacterial Evolution and the Agents of Natural Genome Editing

To elucidate communicative competences of bacteria, we also have to look at the roles of viruses and their relationship to bacteria. Viruses have long been accepted only as disease causing, epidemic phenomena with lytic and therefore dangerous consequences for infected organisms. However, new research has corrected this picture. Viruses are part of the living world, in most cases integrated in the cytoplasm or the nucleoplasma of cells without harming the host. Viruses are on their way to representing the best examples of symbiotic relationships, because there is no living being since the start of life that has not been colonized by them, in most often cases in the form of multiple colonisations (Sonenshein 2006; Witzany 2010). The longest period of these symbiotic relationships during evolutionary history share viruses, archaea, and bacteria. As viruses are extremely biosphere specific, i.e., they adapt to special host tissues, the identification of various forms of, e.g., bacteria is to identify primarily the viruses that colonize them (Rohwer and Edwards 2002; Rohwer 2003). This is also the concept of "bacteriophages," in that bacteria are identified best by identifying the viruses that are associated with them. Host identification in this way is a special method called phage typing (Goyal et al. 1987; Abedon et al. 2009; Ackermann 2007, 2009).

1.5.3 Lytic vs. Persistent Viral Life Strategies

As mentioned in recent years, the lytic consequences of viral infection are a special case if viruses are not able to develop a sessile lifestyle without harming the host. In most cases, viruses living within organisms help to ward off competing parasites from the host and becoming part of its evolutionary history. Persistent viruses are decisive for species diversity and host genome editing. Nearly all natural cellular competences such as expression, transcription, translation, and recombination with all their detailed steps derived from viral competences. Even the DNA replication pathways, after a period of early RNA influence (Forterre 2002, 2005, 2006), seem to be a special viral strategy for the conservation of coded phenotypes by warding off RNA parasites (Villarreal and DeFilippis 2000; Villarreal 2005).

Since observations have become more evident that viruses are able to integrate genetic material into the host genome, it has become clear that viruses have infection lifestyles and endosymbiotic and even symbiogenetic lifestyles. They transfer phenotypic capabilities on the host, which noninfected hosts from the same species do not possess. As endosymbiotic viruses, which are dependent on the host's replication, they are part of the host history in that they are inheritable and part of the genomic identity of the host as documented in some several 10,000 infection events in the human genome by endogenous retroviruses (Villarreal 2004).

The two viral lifestyles are not in strict opposition but, in most cases, are part of a symbiotic process. It starts with an infection by a virus. In the infected host, it arrives at an equilibrial status where the immune system does not eliminate the virus but controls its replication without fatal consequences for the host organism. The persistent status lasts during most phases of the host's life, but may return to the lytic lifestyle if the host-immune system is under stress (Villarreal et al. 2000). Most often the integration occurs by mutual neutralization of toxic capabilities by an antitoxin of a competing genetic settler (Pandey and Gerdes 2005). The whole range of toxin/antitoxin addiction modules we can find throughout all genetic contents in living nature most likely is of viral origin (Lehnherr et al. 1993; Lehnherr and Yarmolinsky 1995; Yarmolinsky 1995; Gerdes and Wagner 2007). Therefore, the persistence is sometimes called temperate lifestyle.

Also bacteria may be infected by viruses without being harmed. If infected bacteria meet noninfected bacteria, it may be that the noninfected acquires lysis; the lysogenic strain does not lyse itself, but is lethal to the noninfected one. The colonized bacteria has a virus-derived molecular genetic identity, which has an advantage against the noninfected one through an acquired ability. This lysogenic bacteria, termed prophage, has an immunity function for the bacteria which the noninfected bacteria lack. Prophage is a virus that is integrated into the bacterial host genome. Both the acute lytic phages and the persistent prophages such as T4 and lamda are highly abundant in oceans and in the soil and seem to be the most dynamic life form on the entire planet (Hendrix et al. 2000; Hendrix 2002; Chibani-Chennoufi et al. 2004; Hendrix 2008). Some viruses are not integrated in the host genome but persist as plasmids and replicate independently from the host genome (Villarreal 2005).

When we speak about the relationship of bacteria and viruses in most cases we speak about phage ecology (Abedon et al. 2009). Most prokaryotic viruses are double-stranded DNA viruses with either linear or circular genome morphology and are packaged in an icosahedral capsid. Whereas acute viruses in most cases code for their own replication, recombination, and repair proteins, the persistent phages lack such genes and use the host-cellular replication. This involves a totally different gene word order (Villarreal 2005) in acute lytic and in persistent phages. This is documented in the very different nucleotide words (di-, tri-, and tetranucleotides). Nucleotide word frequencies of acute phages are very dissimilar to those of their hosts while persistent or temperate phages share nucleotide word frequencies with the host. This means the molecular syntax from acute and persistent phages is constructed totally differently according to the different strategies. Different life strategies with different behavioral patterns need a completely different semantic content in the genome expressed in a different syntactic arrangement of nucleotides (Witzany 2010).

As the bacterial cell walls differ substantially between different types of bacteria, a different behavior is necessary for viruses for recognition, attachment, and penetration. Owing to these diverse barriers of the bacterial cell walls, the prokaryotic viruses do not enter the host cells physically but attach to the cell surface and inject their genomes through contractile tails or pilot proteins. Also, the progeny of the virus has to deal with this barrier (Villarreal 2005).

Bacterial DNA does not have highly stable structures as do eukaryotes and in most cases, it is circular with a unique origin of replication. In contrast to that viral double-stranded DNA is a linear DNA with integrated short terminal repeats. Since bacterial viruses do not use a transport technique as they need in eukaryotes to be transported out of the nucleus, bacterial viruses differ a great deal from eukaryotic viruses.

All bacteria have a restriction/modification system which is a connected form of two viral competences. Only the descendants of mitochondria lack this system which causes them not to be exposed to viral selection. It may be that they have transposed their ability to the eukaryotic nucleus which cares in a more efficient way for cell immunity (Villarreal 2005).

1.5.4 Bacteria as Global Habitat for Viruses

Horizontal gene transfer between bacteria as being responsible for genetic plasticity in prokaryotes may be a capability which is acquired by viral infections. Then, viral genetic inventions are transferred to bacteria via persistent lifestyles of viruses and are not an exchange phenomenon performed by bacteria. In contrast to the "gene-shuttle" hypothesis of HGT, i.e., the horizontal transfer of genes between bacteria via plasmids and similar transfer techniques, the infectious perspective focuzises on a vertical gene transfer (VGT) by viral swarms into the life habitat of cellular organisms such as bacteria. The advantage of the latter one is that the origin and recombination of genetic contents has its roots in viral competence of natural genome editing, whereas in HGT hypothesis, the de novo-generation and recombination of new genetic contents remains randomly derived or even unclear.

As new research indicates, the agents of gene transfer are plasmids, retroplasmids, bacteriophages, and transposons. They effect DNA movements and act in all prokaryotes. DNA movement is achieved through transformation, conjugation, and transduction. Transformation is the transfer of DNA between related bacteria mediated by encoded proteins. Conjugation is performed by conjugative plasmids, which are independently replicating genetic elements. These elements code for proteins which facilitate their own transfer (Frost et al. 2005). Transduction is a DNA transfer mediated by phages which can package host DNA in their capsid and inject it into a new host followed by integration into the host genome (Holmes et al. 2003). Phages, plasmids, retroplasmids, and transposons, therefore, played a crucial role in bacteria evolution (Chen et al. 2005). Bacteria are the most genetically adaptable organisms with enormous capabilities to react appropriately to extreme changes of their ecological habitats. This does not stem from their high reproductive rates but from their great ability to acquire DNA segments by plasmids, bacteriophages, and transposons which transport complete and complex sets of genes from external sources (Shapiro 2007).

When we consider the age of the ocean and the dense abundance of bacterial and viral life in it, then we can say that the possibility of genetic arrangements, rearrangements, and exchange does not need long time periods to create the basics of the complexity of life, because the exchange rate is of astronomical order. If we imagine that 1 ml of seawater contains one million bacteria and ten times more viral sequences, it can be determined that 10^{31} bacteriophages infect 10^{24} bacteria per second (Tettelin et al. 2005). Since the beginning of life, this behavioral pattern has been an ongoing process. The enormous viral genetic diversity in the ocean seems to have established pathways for the integration of complete and complex genetic data sets into host genomes, e.g., acquisition of complex new phenotypes via a prophage can include the acquisition of more than 100 new genes in a single genome editing event (Ryan 2006). Similar interactive patterns are estimated to occur in soil habitats.

Owing to the virus-induced genomic plasticity of bacteria, they are an ideal global biotic matrix to evolve and develop varieties in genome editing, i.e., competent content arrangement of bacterial gene word order coherent with its regulation network. Bacteria are the smallest living organisms with relatively simple genomic structures where the competitive situation between an abundance of viral infective elements leads to the adaptation of lytic viruses to temperate viruses integrated as plasmids in cytoplasma and even persistent viruses integrated in the host genome. The viral competences can develop in this global bacterial habitat as the bacterial species due to their immense genetic flux between viral colonization events and immunity reactions such as restriction/modification (Kulakauskas et al. 1995; Hambly and Suttle 2005).

The highly conserved genome edited functions such as replication, transcription, translation, recombination, and all the substeps evolved primarily in the competitive situation between viral competences to colonize a host and to ward off competing parasites. This includes that biotic self and nonself recognition functions as we know it from diverse immunity systems are also of viral origin, i.e., the integration and all genetic/genomic modification steps that what we call natural genome editing are of viral origin. Therefore, the immense importance of horizontal gene transfer for bacterial species evolution, diversity, and competences is derived from viral genome editing competences and is, in most cases, infection induced by persistent nonlytic viruses (Villarreal 1999; Frost et al. 2005). As phylogenetic analyses demonstrate, the main protein enzymes for natural genome editing are viral inventions and not of cellular origin (Villarreal 2004, 2005). Also, the origin of eukaryotic nucleus was thought to be an ancient prokaryote but phylogenetic analyses show that its ancestor most likely was a large DNA virus (Takemura 2001; Bell 2001, 2006). Interestingly, the early genetic invention of capsid proteins detected in viruses infecting archaea seems also to be of viral origin and of common ancestry to eukaryotic and bacterial viruses (Nandhagopal et al. 2002; Rice et al. 2004;

Khayat et al. 2005). These clearly indicates that the escape theory which assumes viruses to be rements of cellular host genomes cannot be substantiated, because most of viral genes have not counterpart in cellular life (Villarreal and Witzany 2010).

1.6 The Origins of Bacterial Group Identity

1.6.1 Obligate Viral Settlers of Bacteria

In comparison to investigations on bacterial life in the ocean, our knowledge about soil bacteria is rudimentary. The number of bacterial species in soil and its dry, wet, and floating ecospheres is just an estimate, but it may be of a similar magnitude to that in the ocean (Fierer and Jackson 2006; Fierer et al. 2007). The abundance of phages infecting soil bacteria is also just an estimate. The authors in this book will give an overview on the current knowledge on the various levels of sign-mediated intra- and interorganism interactions in which soil bacteria are involved. Additionally, it is important to be aware of soil-related ecospheres such as the enteric bacterial habitat in animals and, of particular importance, the rhizosphere of plants with its unique symbiotic relationships with organisms such as rhizobial bacteria, plant roots, fungi, protozoa, insects, nematodes, and both exogenous and endogenous persistent as well as lytic viruses and their satellites and hyperparasites that are linked to all these organisms. This means that the variety of viruses which are specialized to the various organismic kingdoms, such as ssRNA, dsRNA, dsDNA, ssDNA, and retroviruses, are in constant interaction which, from today's view, is an optimal resource for evolutionary novelty and adaptation.

Soil is in most cases colonized by fungi and bacteria. Soil is a resource which is needed by all terrestrial eukaryotes, which represents a highly complex network of dependencies of different ecospheres such as flood waters, lakes, forests, dry lands, etc. In this respect, soil has a similar function to water in the ocean, in that it is the basis for all terrestrial life forms. At the basics of organic soil are bacteria with their obligate settlers, phages, and plasmids, and then the single-celled protozoa and fungi, all of which are settled by bacterial colonizers. Higher eukaryotes like animals and plants depend vitally on these settlers because, as in the case of plants, rhizosphere ecology depends on biocommunication between plant root cells (of three different types), mycorrhizal fungi, and rhizobial bacteria. Digestive processes in animals depend on intestinal bacterial colonizers, without which animals could not survive. Additionally, the oral cavity of terrestrial animals such as humans is a complex bacterial ecosphere with up to 500 different species in a symbiotic lifestyle as described above (see Fig. 1.1). One of the obligate symbiotic settlers of animal intestinal tracts is E. coli, with its obligate lytic T4 phage parasite. T4 phage shows a similarity to some eukaryotic DNA viruses (e.g., herpesvirus). Additionally, T4 phage shows genetic similarities not only to other viruses, but also to eukaryotic cells such as T4 DNA polymerase, lysozyme and, in particular, group

I self-splicing introns, which can be found in mitochondria of fungi, nuclei of protists and chloroplasts, as well as in mitochondria of plants, but not in many prokaryotes (Villarreal 2005).

Interestingly, the identity of bacterial strains is determined by their colonizing viruses, that is, their phages or plasmids, and therefore the identification of bacterial strains is called "phage typing." The knowledge that viruses colonize bacterial genomes was discovered just 60 years ago, when Salvator Luria termed this a "molecular genetic parasite," whilst Lwoff termed this hereditary virus as having a "temperate" lifestyle or called it a "prophage" (Villarreal 2005). With these, it was clear that there are two different lifestyles of viruses: a lytic acute one and a silent one (lysogenic). Later, a third was detected: a continuous and chronic virus production without lysis of the host cell and without silent persistence.

Another interesting observation was that if lysogenic bacterial strains and non-lysogenic strains were mixed, the noninfected were lysed but the lysogenic were not. This seemed to be an advantage in that the infected bacteria were protected by their prophage, whereas the noninfected were not. The identity of the infected bacterial strains is therefore another one as those of the noninfected. In competitive situations, infected bacteria have an immunity advantage in comparison to noninfected bacteria. The genetic identity is virus-derived (Villarreal 2005). With this identity, the bacteria host is able to identify and preclude other competing genetic parasites. Defective phages can also be a part of this identity. Although they are not able to produce infective viruses, they can effectively preclude infection from the host by related parasites.

The most abundant bacteria-infecting viruses are the T4 phage and the lambda phage (Tetart et al. 2001; Desplats and Krisch 2003). If the bacterium is lysogenic, they integrate in the host chromosome. When viruses integrate in the host cytoplasma, rather than in the host chromosome, they are known as episomes. Episomes are exogenous genetic elements, such as plasmids, which replicate independently of the host genome and are derived from transposable elements, i.e., viruses. The overwhelming majority of bacterial phages are dsDNA viruses of linear or circular genomes. Of these, 96% are tailed and the remaining 4% are isometric. Whereas lytic phages such as T4 code for their own replication, recombination and repair, persistent pages use the host system instead. The next most common bacterial phages are the ssDNA viruses with rolling circle replicons, whereas dsRNA viruses and ssRNA viruses are found very rarely. Whereas bacterial DNA is circular with a unique origin of replication, and most dsDNA genes from viruses are linear with terminal repeats, it has been found that these repeats facilitate replication via circular theta forms and RCR intermediates (Villarreal 2005).

As an immunity function, it is common for all prokaryotes to use a restriction/modification system. The restriction system acts as endonuclease which degrades unmodified DNA and its counterpart modification, i.e., methylase which covalently modifies DNA protecting it from degrading endonuclease. Prokaryotes that undergo symbiogenesis, such as mitochondria in eukaryotic cells, lack the restriction enzyme, which seems to be an indicator for lacking phage-selective pressure within the new host habitat.

In vivo bacteria are the most genetically adaptable organisms in that they can change their genetic molecular syntax very rapidly due to their high clonal reproductivity. The ability to acquire complete and complex genetic datasets from the outside also enables genetic adaptation. This has been termed horizontal gene transfer in that it is assumed to be an exchange and transfer system of genetic elements within microbial ecospheres. In reality, this seems to be virus-infection driven and leads to an altered genetic (molecular) syntax of the host which does not originally stem from the genetic lineage of the bacterial population. In this respect, we should speak of a vertical gene transfer (VGT). This is a very important fact as phage–phage interactions can lead to phage immunity for the host and rapidly infected bacteria may acquire multiple drug resistances and also virulence.

There are several examples of phages that can be colonized by other phages which are phage-specific colonizers rather than host-specific. This means a different lifestyle. In some cases, it is well documented that these phage–phage interactions are complementary and provide features which are absent in phages that are not part of these interactions. This means that a bacterial host which has been colonized multiple times will have a variety of features not available to a less infected bacterial host of the same species. If, for example, a bacterium acquires a virulence-associated prophage, this means that it will be able to acquire more than 100 new genes in one infection event (Villarreal 2005).

1.6.2 The Role of Persistent Viruses in Gene Word Order of Bacteria

The ability of bacteria to communicate and coordinate via quorum sensing processes depends on a variety of features which assemble group identity, that is, the ability to sense group members as part of self and nongroup members as nonself agents. A coherent exchange of signaling molecules ensures the ongoing processes. Each of these capabilities must be genetic because they are determined by the production, release, uptake, and interpretation of protein products which are genetically encoded. In order to understand this, there are questions that need to be answered, such as: what is the genetic make-up of a bacterial population, what are the adaptational purposes for changing these genetic datasets and finally, what agents are involved in the natural genome editing of bacteria, i.e., determine their genetic content arrangement?

The behavioral patterns demonstrate an intense relationship between bacteria and viruses in that the viruses colonize bacterial cells as exogenetic (plasmids) or endogenetic (phages) parasites. Episomal (plasmid) persistent viruses replicate independently of the host genome whereas the replication of endogenous parasites depends on host-genome replication. Parasitism reaches from lytic to nonlytic lifestyles. The nonlytic persistent lifestyle of phages determines the gene word order (molecular syntax) of the bacterial genetic text. Additionally, it is necessary to look at a great variety of phage–phage interactions which can result in multiple alterations within the host genome. Also, parasites of parasites of parasites (hyperparasites) are not rare but common, so that the interrelationships between several viral strains within the bacterial cell have to be mentioned. In this respect, the traditional process described as horizontal gene transfer – a gene transfer from one bacterium to another – is not the only source of change of genetic content arrangements. More important seems to be the content arrangements generated by phages/plasmids and their interrelation partners of competing/colonizing parasites or their defectives. As mentioned above, the gene transfer can assemble up to 100 new genes within one single event, which is a completely different contribution to evolutive novelties than single randomly derived point mutations could cause. Because of the abundance of phages/plasmids in the bacterial world, this could be seen as the rule more than the exception and has important consequences on previous thoughts on prokaryote evolution.

Bacteria are undoubtedly the best adaptational organisms on earth. An abundance of fast-arranged genetic variations together with genetic adaptations occurs and can resist any environmental circumstances, for example, intense heat and radiation. The acquisition of complex genetic datasets by virally derived infection events seems to be the main source of evolutionary adaptational processes, and this was not in the focus of bacterial research within the last few decades. This means that the main genetic resources did not derive from direct cellular predecessors or from the genetic lineage of the bacterial population horizontally, but is the result of natural genome-editing activities of viruses.

1.6.3 Infection-Driven Group Identity and Group Immunity

If a bacterial strain is persistently infected by a phage and brought into a competitive situation with a bacterial strain which is not infected by the same virus, the noninfected strain will undergo lysis. This means that infection and colonization of bacteria are interconnected with the acquisition of an immunity function which does not destroy the infected one by the noninfected one. Infected bacteria share a common immunity which noninfected bacteria do not have. Phage colonization in a nonlytic persistent lifestyle has a symbiotic function which protects host cells and host strains. Colonized bacteria now have a virus-derived genetic identity (Villarreal 2005), which is dissimilar to that of noncolonized bacteria of the same strain. Members of virally colonized bacterial colonies are able to sense the different identity of bacteria that are colonized by different viruses and can also identify those that are noncolonized. Colonized bacteria can also sense other genetic parasites. If we look at the interactions of viruses and bacteria, we therefore have to look at three different but interconnected levels of relationships:

- Acute lytic and persistent viruses to the prokaryotic hosts
- Acute lytic and persistent viruses to other acute host viruses

- Acute lytic and persistent viruses to other persistent host viruses.

These interrelated processes are highly dynamic and constantly changing processes because the competition between viruses with their high mutation rate (gene word-order-plasticity) to reach a persistent status, which includes the exclusion of related parasites, is an ongoing process.

Persistent endogenous viruses do not need genes for their own replication, recombination, and repair, whereas acute lytic viruses show a strong tendency to encode these features by themselves.

1.7 Transfer of Viral Competences as Modular Tools

1.7.1 Molecular Identity Markers

One of the important features of lytic T4 phages is their ability to modify nucleotide sequences at a high frequency. Interestingly, this ability to modify molecular syntax, such as the exchange of hydroxymethylcytosine for cytidine, serves to mark the molecular identity of the phage genome distinctly from the host genome. Additionally, this marked phage DNA is protected from phage-encoded restriction endonucleases II and IV that degrade unmodified host DNA. Marking is an identity sharing action which serves for both the self/nonself identification/differentiation and as a kind of immune function. As a third advantage, the modification prevents restriction by Mcr endonucleases. Lytic phages that mark their genome through modification are able to distinguish self from nonself (host) DNA (Villarreal 2005).

1.7.2 Persistent Phages Determine Bacterial Identity

In contrast to the T4 phage which is a well investigated acute (lytic) bacterial virus, phage lambda is a typical persistent virus. Persistent viruses have enormous impact on host genomes including recombination, immunity, and identity, including identification competence of (nonself) competitors. This identification competence is important for any population determining the variety of coordination processes within habitats with multiple bacterial populations, such as the human oral cavity (see above). The phage family of lambda seems to represent a large genetic pool which interacts continuously in the exchange and deletion, as well as in the assembly, of a variety of genes. For example, lambda and related P2 as well as P22 can recombine with each other although the core genes of lambda viruses are not conserved. However, they differ completely in the area of immunity. As a common feature of phage parasites, the persistent colonization of bacteria by phages prevents infection by phage-related parasites. In this respect, the phage colonization of bacteria results in a symbiotic interaction with the immune function of the host that clearly determines host identity (Villarreal 2005). Persistent viruses have a bistable genetic switch that only allows the expression of genes which are associated with immune functions. With the expression of only one gene (cI), a lambda lysogen is immune to super-infection by phages related to lambda. Interestingly, other related persistent phages (P2, P22) differ in mechanisms of gene expression associated with immune function (Villarreal 2005). It seems likely that the most selective pressure on a persistent phage is that of resisting super-infection by related phages. In this respect, it can be understood why 12% of the P22 genome is dedicated to preventing growth of competing related phages.

Also P2 is a prevalent colonizer of E.coli and is much more prevalent than the lambda phage (Bertani and Deho 2001). Like other prophages, P2 integrates near to various tRNA sites (7bp anticodon loop) in a site-specific manner. Although P2 can be reactivated and is therefore not seen as defective, its lifestyle is clearly symbiotic in that it protects host bacteria from infection by competing (related) viruses. The fitness advantage for persistent P2 is that the host is not damaged by infection of lytic viruses. Interestingly, the feature of host immunity against competing viruses is derived from a persistent retrovirus in P2, whereby the reverse transcriptase coding element disrupts competitor genomes. This clearly represents a genetic agent that identifies and destroys nonself genetic competitors. P2 is activated to produce virions only if another infection by a satellite P4 family occurs, which as a defective phage propagates P2 activity (Villarreal 2005). P4 can be parasitized by a retron (a defective retrovirus). The result of this hyper-hyper-parasitation is phage R73, which is nearly identical to P4 but contains a retron (Villarreal 2005). It is important to notice, that this is only one example of an abundance of highly complex networks of interconnected relationships and interactions between various phages and their parasites. This is the real precondition for the evolution of higher order regulatory networks as it is found in eukaryotic life cycles. Also, the retron of P4 depends in its persistence on addiction modules, such as homing endonuclease genes and introns and inteins. Only the presence of the mobile intron element prevents homing endonuclease genes and therefore protects the host genome from P4 infection (Villarreal 2005).

The lysogenic lifestyle (nonlytic but persistent) of bacterial viruses is the result of a counterbalanced effect of two or more infection events on bacterial host genomes. If one of these counterparts is deleted or silenced, the other will be activated to produce virions or even a toxic component. This feature is of crucial importance for the host-population because it can kill host organisms and therefore also has consequences for host population density as well as symbiotic partners. On the other hand, it is important to notice that host identity is extended in an evolutionary sense because hosts of the same species, which are not infected by lysogenic phages, do not possess this immune function which protects them from infection by competing parasites. The investigation of biocommunicative competences of soil bacteria therefore crucially depends on identifying the persistent invaders of soil bacteria which determine their phenotypic features.

1.7.3 *Addiction Modules Function as Counterbalanced Viral Competences*

Addiction modules can be defined as features that consist in general of a stable toxic component which is counterbalanced by an unstable component which inhibits and suppresses the toxic component (Lehnherr et al. 1993; Yarmolinsky 1995; Lehnherr and Yarmolinsky 1995; Engelberg-Kulka and Glaser 1999; Rawlings 1999). Both are necessary to transfer a feature to the host without harming the host. In the case of a restriction/modification module, this means that, for example, 52 restriction enzymes are counterbalanced by 52 modification enzymes. This indicates how complex addiction modules are constructed and how difficult it can be to understand the evolution of such phenotypes (Villarreal 2009).

Several kinds of addiction modules are known. First of all and most prominent in bacteria life is the aforementioned restriction/modification addiction module, which is a common feature in the immune function. One part consists of an antitoxic modification enzyme, which is an unstable beneficial (protective) agent (Hayes 2003). The counterpart consists of a toxic restriction enzyme component, which is a stable but harmful (destructive) agent. Another kind of addiction modules consists of two related features. There is an antitoxic antipore-toxin which represents the unstable protective agent and a toxic component with a toxic pore which represents the stable but destructive agent. A third kind of addiction module consists of the antitoxic viral immunity component and the toxic component of viral-mediated lysis. This third kind is the most obvious viral-derived immune function because it necessarily consists of a persistent genetic parasite and an external lytic phage. Interestingly, one of the most common episomal phages, P1, is much more complex than other phages in that it assembles up to 100 genes and several addiction modules. This is necessary because of plasmid stability in daughter host bacteria and for the coordination of both cellular and host DNA replication (Lehnherr et al. 1993). This involves the integration of three different immune regions. One of these is a very efficient restriction-modification addiction module. Daughter cells which have lost the P1 episome will undergo postsegregational killing as a result of the toxic restriction enzyme not being balanced any further by a modification enzyme, which is lacking (Hazan et al. 2001; Villarreal 2009).

As mentioned above, the P1 prophage is interesting because it must express genes that orchestrate both replication of the viral DNA and of the host DNA. This is also seen in the P1 competence of highly coordinated DNA replication and segregation control. P1 achieves plasmid stability by coordinating plasmid and host chromosome replication (Villarreal 2009). In parallel, P1 has to partition stable viral chromosomes and additionally host daughter cells. This ability indicates that P1 can differentiate between self (viral DNA) and nonself (host DNA). In the case of the host E. coli, it also prevents competing genetic parasites. This clearly contradicts the selfish DNA hypothesis in that these behavioral patterns are symbiotic as they benefit both the virus and its host.

An important feature of these bacterial colonizers is that they are often colonized themselves. Secondary and tertiary genetic parasites (parasites of parasites of parasites) are further important features because they represent more complex capabilities which are transferred to the host that would otherwise not be present (e.g., introns of T-even phages). Interestingly, this type of virus–virus interaction is an important contributor to bacterial features.

In the case of P1, the restriction/modification addiction module is interconnected with methylated imprinting. The imprinted DNA is protected against this immune system whereas nonimprinted (unmarked) copies are destroyed. The imprinting must be transferred actively after replication to daughter cells by an epigenetic identity tag. If this epigenetic imprinting is disturbed by other parasitic elements, the counterbalanced addiction module will become out of balance. The antitoxin will not function and the stable encoded toxic part of the module will kill the cell (Villarreal 2009). Although a great number of restriction/modification addiction modules depend on site-specific methylated DNA against site-specific endonucleases, this is a rather costly process in terms of energy.

1.7.4 The Persistent Viral Lifestyle of Plasmids and the Role of tRNAs

Plasmids have a similar lifestyle to that of episomal persistent phages (Blaisdell et al. 1996; Brüssow and Hendrix 2002; Ding and Hynes 2009). Both are derived from viruses and both share a nonlytic but persistent lifestyle (Oshima et al. 2001). Both lack genes coding for virion production and most of them require a helper virus (satellite or another hyperparasite) for their mobilization. Both provide an advantage to their host that identifies and wards off competing parasites. Additionally, most of them share the coding of a specific integrase that leads DNA integration at specific sequence sites (syntax-identification competence) which is associated – interestingly – with specific tRNA genes. Both persistent lifestyles also transfer and incorporate virulence factors to their host, which are in most cases transferred as a singular transfer event. This can include the transfer of gene blocks with dozens of complete genes, with the result of a great variety of phage-related toxins. These virus-derived toxin genes have no host counterparts which clearly indicates their viral origin. Such toxins and other virulence factors are identity markers because they reproductively isolate their hosts from host counterparts or host relatives by postsegregational killing (Villarreal 2009).

Bacteriocins are well-studied plasmid encoded toxins (Riley 1998; Bull and Regoes 2006), which are highly active against related bacterial strains that lack this plasmid. This makes evolutionary sense because the generation of new and perhaps better strains in most cases serves as an advantage in colonization or even for adaptational purposes. Additionally, plasmids can code for restriction/modification addiction modules. Interestingly, some plasmids serve as acquisition sites (traps) for other plasmids, transposable elements, addiction modules, and also immune modules. Such large plasmids are sometimes a kind of second chromosome. Transposable elements, which are clearly derived from viruses, are ancestors of plasmids (Cohen 1976; Villarreal 2009).

Let us return to the relationship between plasmids and phages to host tRNA. Interestingly, both the integrases of mobile plasmids as well as those from phages use the same tRNA integration sites. The pathogenicity islands, which are of great medical importance and represent a specific plasmid-mediated gene word order, affect immune identification and alter regulation of cell physiology (Hacker and Kaper 2000; Hayes 2003). More than 50% of them are associated with tRNA at a sequence junction at the site of integration. This seems to be an indicator that persistent phages are also involved, because phage integrases target tRNA DNA sequences (Villarreal 2009).

If we look at bacterial speciation, for example the relationship between E. coli and B. subtilis, then we can identify 230 regions which are quite dissimilar. Most of these regions are flanked by tRNA sequences which mark integration events. This means that the majority of speciation of these two species is the result of genetic infection events. Additionally, this is an indicator that tRNA not only plays a role in the transfer to translation, but that it also plays a role in higher order regulatory functions in general (Wegrzyn et al. 2001). Until recently the role of this kind of noncoding RNA, with its ancient and long lasting evolutionary history, was clearly underestimated (Maizels and Weiner 1993).

1.8 Swarming Group Behavior and Group Identity

Over decades bacteria were investigated as single organisms. Since the 1990s, this has changed significantly. The common behavioral patterns such as biofilm organization, mating, virulence, movement (fruiting body for sporulation), feeding, and colonization demonstrated that bacterial groups (colonies) are the rule rather than the exception (living in blooms, mats, biofilms). This indicates communication processes in which bacteria exchange signaling molecules (or tactile experiences) with the sensing of population density, nutrition availability, temperature, and light. Competition between related bacterial strains drives evolution and diversity. A group identification competence is necessary and will be the result of different modes of stable (persistent) infection by phages and plasmids – often mixed – via addiction modules. These phenotypes, such as restriction/modification, pore-toxins/antipore-toxins, endonuclease/antisense RNA, and holins/antiholins, serve as identity modules as well as immune modules and help to exclude bacterial strains which do not share these features (Villarreal 2009).

Viral infection changes molecular identity, immunity, and group identity of bacterial swarms. Quorum sensing as one of the sign-mediated coordination processes based on chemotactic competences (Nadell et al. 2008a, b; NG and Bassler 2009; Mehta et al. 2009) is associated with memory and learning and coordinated movement (Ben Jacob 2009). Besides the addiction modules, specific surface receptors are necessary to sense small pheromone molecules and coordinate programmed cell death as a common strategy of nonselfish group behavior (Villarreal 2009). Programmed cell death in prokaryotes is a strategic behavior in that it protects bacterial strains from group (self) members which are successfully attacked and lose their group identity (Yarmolinsky 1995; Engelberg-Kulka and Glaser 1999). This results in the addiction modules being out of balance whereby the unstable antitoxic part does not function and leads to toxic results.

According to the abundance of prokaryotes in all ecological habitats and the highly dynamic gene flux, which depends on the infection rates with phages and plasmids, the relationship between viruses and prokaryotes is one of the most intense living processes on earth (Hendrix 2002). In particular, the complex multigene solutions and the high rates of natural genome editing (generation, combination, recombination, repair) serve for a wide variety of issues of host identity. Additionally, the viral inhabitants of prokaryotes are in constant interaction with competing parasites and their satellites, such as multiple hyperparasites. The focus of investigations on prokaryotes in general should be upon these interrelationships and the highly dynamic interactions between prokaryotes and the network of genetic parasites and hyperparasites. The prokaryotic social competences on which their group identity, group signaling, group behavior, and immunity depend are part of these interactional networks with viruses. An important fact is that these features cannot be separated into single investigations, just as the investigation of the syntax of a single word cannot result in an appropriate description of the function of a language. This indicates that the most basic natural genome editing processes are not the result of simple solutions but in contrast include highly relevant genetic order arrangements and dynamic rearrangements according to the adaptational purposes of their host organisms (Villarreal 2009).

Interestingly, the ability to identify self (members of a population which share same addiction modules) and nonself (members sensing others than self identity) can be lost when their identity is temporarily modified, such as when they are attacked by competing parasites, resulting in a slight genetic rearrangement. This results in (self) members killing other (self) members. Even a slightly changed genetic syntax can initiate attack upon (self) members (Villarreal 2009).

Addiction modules, such as restriction/modification, are also used as toxins against lytic phages in that lytic DNA is fragmented and (self) DNA is protected against degradation. Interestingly, some sporulating bacteria and unicellular eukaryotes (dinoflagellates) use controlled DNA fragmentation as a normal step in the cellular differentiation process. This is an indicator for a further addiction module wherein the previously destructive role of special toxins is used as an advantageous process for differentiation (Villarreal 2009).

1.9 Genetic Content Operators and Viral Gene Factories

These interactional networks may have serious consequences on our understanding of early evolution of RNA replicators into the cellular era of last universal common ancestors (LUCAs) (Witzany 2010). Both the coding competences to encode information about the genetic content operators and their catalytic function for synthesis (the catalytic phenotype and its genotype) need a high density of swarms that compete and interact. The former picture of a single LUCA which serves as matrix for replicating agents is not useful because coding competences in every natural language are no *solus ipse* features but depend on an interactional network. This is the precondition for creating sequence data and sequence abundance (Witzany 2010). Sequence generating creativity is the result of competing genetic parasites in that the creation of de novo sequences is a crucial factor for host group identity and group immunity. "As soon as the first replicator evolves parasitic replicators could also have evolved" (Villarreal 2009). The emergence of competent nucleic acid editors is interconnected with slightly different replicators which parasitize each other. This sign-biocommunicative community is highly selective and in parallel dependent on one another. This seems to be in agreement with the fact that viruses not only create and acquire new genes in very high rates within various viral lineages but the viruses within their host can act as gene factories. It seems most likely that the adaptational competences in the prokaryotic world are the result of this creative force of viral natural genome editing.

On the other hand, the vast viral genetic content available in the ocean as well as in soil habitats demonstrates that viruses both assemble and mix (arrange and rearrange) new genetic contents (Osborn and Boltner 2002; Kimura et al. 2008). In this respect, viruses are the masters of genetic content innovation and they fulfill this important role by creating interactional networks with complementing genetic parasites, epiparasites, and hyperparasites (Weitz et al. 2008). Some viral inhabitants of prokaryotes need helper functions by other parasites, which then parasitize the virus in order to make a specific phenotype available. It is necessary to look at a large pool of phages which colonize all of the prokaryotes and represent one of the most abundant genes for proteins on this planet, such as receptors and pore-forming proteins. Some phages change the expression of encoded phenotypes and create a diversity of receptors. In the case of the Bordatella phage, it has been shown that two variable gene regions are used to create a wide variety of receptors by a phage-encoded reverse transcriptase, which is an indicator of retroviral infection of the phage itself. The Bordatella phage not only provides a benefit for the host in its capability to create variable gene products for both the host and phage self-identification, but is also clearly a forerunner of the vertebrate adaptive immune system (Villarreal 2009).

If these interconnected networks in real-life habitats are reduced to laboratory settings for the purpose of study, the conditions for interactions are changed dramatically. We then investigate simple prokaryotes without the interactional context of their persistent parasites, competing parasites, or complementary epi- as well as hyperparasites. We would then have no idea how these viral agents contribute to host fitness.

If we look at a special kind of such parasites of phages, we can identify RNA sequences which act as introns in the phage genetic content. They themselves are encoded in DNA. The three groups of introns are group I introns, group II, and group III introns. They date back to the early RNA world and play important roles as transposable elements that act as regulatory elements and ward off genetic parasites. They are now recognized as stable persistent virus-like agents (homing endonucleases) that serve as identity parts for host immunity. Additionally, these homing endonucleases serve as sequence specific toxins (holins) that can kill related bacterial strains which have not been colonized by these particular parasites (Young 2002; Ziedaite et al. 2005; Villarreal 2009).

1.10 Conclusion

Bacteria, which in former times were viewed as lower life-forms, have now been recognized as masters of monitoring, computing, interpretation, coordination, and organization. Bacterial communicative competences are sign-mediated interactions between the same or related species, but also between nonrelated species according to different situational contexts (pragmatic level of analyses) and the coherent combinatorial patterns of signals according to the molecular syntax (syntactic level of analyses).

The situational context determines the content of the messages (semantic level of analyses), the meaning of signaling molecules for a bacterial community which shares a common background memory and a competence for culture-dependent interpretation which is an advantage for adaptational purposes. Maybe this concept of investigations can be applied to other kingdoms in the future to reach a unifying perspective for transdisciplinary research beyond the borders of increasing specialization.

Acknowledgment Thanks to Luis P. Villarreal, Director of Center for Virus Research, University of California, Irvine, for helpful comments and suggestions.

References

Abedon ST, Duffy S, Turner PE (2009) Bacteriophage ecology. In: Schaechter M (ed) Encyclopedia of microbiology. Elsevier, Oxford, pp 42–57

Ackermann HW (2007) 5500 Phages examined in the electron microscope. Arch Virol 152:227–243

Ackermann HW (2009) Phage classification and characterization. Method Mol Biol 501:127–140

Bais HP, Park SW, Weir TL, Callaway RM, Vivanco JM (2003) How plants communicate using the underground information superhighway. Trends Plant Sci 9:26–32

Bassler BL (1999) How bacteria talk to each other: regulation of gene expression by quorum sensing. Curr Opin Microbiol 2:582–587
Bassler BL (2002) Small talk: cell-to-cell communication in bacteria. Cell 109:421–424
Bassler BL, Losick R (2006) Bacterially speaking. Cell 125:237–246
Batut J, Andersson GE, O'Callaghan D (2004) The evolution of chronic infection strategies in the α-proteobacteria. Nat Rev Microbiol 2:933–945
Bauer WD, Robinson JB (2002) Disruption of bacterial quorum sensing by other organisms. Curr Opin Biotechnol 13:234–237
Bell JL (2001) Viral eukaryogenesis: was the ancestor of the nucleus a complex DNA virus? J Mol Evol 53:251–256
Bell PJL (2006) Sex and the eukaryotic cell cycle is consistent with a viral ancestry for the eukaryotic nucleus. J Theor Biol 243:54–63
Ben Jacob E (2003) Bacterial self-organization: co-enhancement of complexification and adaptability in a dynamic environment. Phil Trans A Mathem Phys Engin Sci 361:1283–1312
Ben Jacob E (2009) Learning from Bacteria about Natural Information Processing. Annals of The New York Academy of Sciences 1178:78–90
Ben Jacob E, Becker I, Shapira Y, Levine H (2004) Bacterial linguistic communication and social intelligence. Trends Microbiol 12:366–372
Ben Jacob E, Levine H (2006) Self-engineering capabilities of bacteria. J Roy Soc Interf 3:197–214
Berg OG, Kurland CG (2002) Evolution of microbial genomes: sequence acquisition and loss. Mol Biol Evol 19:2265–2276
Bertani G, Deho G (2001) Bacteriophage P2: recombination in the superinfection preprophage state and under replication control of phage p4. Mol Genet Genomics 266:406–416
Blaisdell BE, Campbell AM, Karlin S (1996) Similarities and dissimilarities of phage genomes. Proc Natl Acad Sci USA 93:5854–5859
Blech J (2000) Leben auf dem Menschen. Die Geschichte unserer Besiedler. Hamburg, Rowohlt Taschenbuch Verlag
Bonfante P (2003) Plants, mycorrhizal fungi and endobacteria: a dialog among cells and genomes. Biol Bull 204:215–220
Bonfante P, Anca IA (2009) Plants, mycorrhizal fungi, and bacteria: a network of interactions. Ann Rev Microbiol 63:363–383
Bordenstein SR, Reznikoff WS (2005) Mobile DNA in obligate intracellular bacteria. Nat Rev Microbio l3:688–699
Brockhurst M, Buckling A, Racey D, Gardner A (2008) Resource and the evolution of public-goods cooperation in bacteria. BMC Biol 6:20
Brown JR (2003) Ancient horizontal gene transfer. Nat Rev Genet 4:122–132
Brüssow H, Hendrix RW (2002) Phage genomics: small is beautiful. Cell 108:13–16
Bull JJ, Regoes RR (2006) Pharmacodynamics of non-replicating viruses, bacteriocins and lysins. Proc Biol Sci 273:2703–2712
Burmølle M, Hansen LH, Sørensen SJ (2007) Establishment and early succession of a multispecies biofilm composed of soil bacteria. Microb Ecol 54:352–362
Camara M, Williams P, Hardman A (2002) Controlling infection by tuning in and tuning down the volume of bacterial small-talk. Lanc Infect Dis 2:667–676
Chen I, Christie PJ, Dubnau D (2005) The ins and outs of DNA transfer in bacteria. Science 310:1456–1460
Chiang CC, Lambowitz AM (1997) The Mauriceville retroplasmid reverse transcriptase initiates cDNA synthesis de novo at the 3′end of tRNAs. Mol Cell Biol 17:4526–4535
Chibani-Chennoufi S, Bruttin A, Dillmann M-L, Brüssow H (2004) Phage–host interaction: an ecological perspective. J Bacteriol 186:3677–3686
Cohen SN (1976) Transposable genetic elements and plasmid evolution. Nature 263:731–738
Crespi BJ (2001) The evolution of social behavior in microorganisms. Trends Ecol Evol 16:178–183

Danhorn T, Fuqua C (2007) Biofilm formation by plant-associated bacteria. Annu Rev Microbiol 61:401–422

Daniels R, Vanderleyden J, Michiels J (2004) Quorum sensing and swarming migration in bacteria. FEMS Microbiol Rev 28:261–289

Davis DG, Parsek MR, Pearson JP, Iglewski BH, Costerton JW, Greenberg EP (1998) The involvement of cell-to-cell signals in the development of a bacterial biofilm. Science 280:295–298

Defeu-Soufo JH, Graumann PL (2004) Dynamic movement of actin-like proteins within bacterial cells. EMBO Rep 5:789–794

Desplats C, Krisch HM (2003) The diversity and evolution of the T4-type bacteriophages. Res Microbiol 154:259–267

Ding H, Hynes MF (2009) Plasmid transfer systems in the rhizobia. Can J Microbiol 55:1–11

Diggle SP, Gardner A, West SA, Griffin AS (2007a) Evolutionary theory of bacterial quorum sensing: when is a signal not a signal? Philos Trans R Soc Lond B Biol Sci 362:1241–1249

Diggle SP, Griffin AS, Campbell GS, West SA (2007b) Cooperation and conflict in quorum-sensing bacterial populations. Nature 450:411–414

Dunn AK, Handelsman J (2002) Toward an understanding of microbial communities through analysis of communication networks. Antonie van Leeuwenhoeck 81:565–574

Engelberg-Kulka H, Glaser G (1999) Addiction modules and programmed cell death and anti death in bacterial cultures. Annu Rev Microbiol 53:43–70

Federle MJ, Bassler BL (2003) Interspecies communication in bacteria. J Clin Invest 112:1291–1299

Fierer N, Jackson RB (2006) The diversity and biogeography of soil bacterial communities. Proc Natl Acad Sci USA 103:626–631

Fierer N, Bradford MA, Jackson RB (2007) Toward an ecological classification of soil bacteria. Ecology 88:1354–1364

Forterre P (2002) The origin of DNA genomes and DNA replication proteins. Curr Opin Microbiol 5:525–532

Forterre P (2005) The two ages of the RNA world, and the transition to the DNA world: a story of viruses and cells. Biochimie 87:793–803

Forterre P (2006) The origin of viruses and their possible roles in major evolutionary transitions. Virus Res 117:5–16

Frost LS, Leplae R, Summers AO, Toussaint A (2005) Mobile genetic elements: the agents of open source evolution. Nat Rev Microbiol 3:722–732

Fuqua C, Greenberg EP (2002) Listening in on bacteria: acylhomoserine lactone signalling. Nat Rev Mol Cell Biol 3:685–695

Fuqua C, Winans SC, Greenberg EP (1996) Census and consensus in bacterial ecosystems: the LuxR-LuxI family of quorum-sensing transcriptional regulators. Ann Rev Microbiol 50:727–751

Gerdes K, Wagner EG (2007) RNA antitoxins. Curr Opin Microbiol 10:117–124

Gitai Z (2005) The new bacterial cell biolgy: moving parts and subcellular architecture. Cell 120:577–586

Gogarten JP, Townsend JP (2005) Horizontal gene transfer, genome innovation and evolution. Nat Rev Microbiol 3:679–687

Goyal SM, Gerba CP, Bitton G (1987) Phage ecology. Wiley, New York

Graumann PL (2004) Cytoskeletal elements in bacteria. Curr Opin Microbiol 7:565–571

Graumann PL, Defeu-Soufo HJ (2004) An intracellular actin motor in bacteria? Bioessays 26:1209–1216

Gray KM (1997) Intercellular communication and group behavior in bacteria. Trends Microbiol 5:184–188

Greenberg EP (2003) Bacterial communication: tiny teamwork. Nature 424:134

Guerrero R, Berlanga M (2007) The hidden side of the prokaryotic cell: rediscovering the microbial world. Intern Microbiol 10:157–168

Hacker J, Kaper JB (2000) Pathogenicity Islands and the evolution of microbes. Annu Rev Microbiol 54:641–679
Hambly E, Suttle CA (2005) The viriosphere, diversity, and genetic exchange within phage communities. Curr Opin Microbiol 8:444–450
Hayashi M (2005) Plastid proteins crucial for symbiotic fungal and bacterial entry into plant roots. Nature 433:527–531
Hayes F (2003) Toxins-antitoxins: plasmid maintenance, programmed cell death and cell cycle arrest. Science 301:1496–1499
Hazan R, Sat B, Reches M, Engelberg-Kulka H (2001) Postsegregational killing mediated by the P1 phage "addiction module" phd-doc requires the Escherichia coli programmed cell death system mazEF. J Bacteriol 183:2046–2050
Hendrix RW, Lawrence JG, Hatfull GF, Casjens S (2000) The origins and ongoing evolution of viruses. Trends Microbiol 8:504–508
Hendrix RW (2002) Bacteriophages: evolution of the majority. Theor Popul Biol 61:471–480
Hendrix RW (2008) Phage evolution. In: Abedon ST (ed) Bacteriophage ecology. Cambridge University Press, Cambridge, UK, pp 177–194
Henke JM, Bassler BL (2004) Bacterial social engagements. Trends Cell Biol 14:648–656
Holmes AJ, Gillings MR, Nield BS, Mabutt BC, Nevalainen KM, Stokes HW (2003) The gene cassette metagenome is a basic resource for bacterial genome evolution. Environ Microbiol 5:383–394
Hughes DT, Sperandio V (2008) Inter-kingdom signalling: communication between bacteria and their hosts. Nat Rev Microbiol 6:111–120
Imaizumi-Anraku H, Takeda N, Cherpentier M, Perry J, Miwa H, Umehara Y, Kouchi H, Murakami Y, Mulder L, Vickers K, Pike J, Downie A, Wang T, Sato S, Asamizu M, Tabata S, Yoshikawa M, Murooka Y, Wu GJ, Kawaguchi M, Kawasaki S, Parniske M, Hayashi M (2005) Plastid proteins crucial for symbiotic fungal and bacterial entry into plant roots. Nature 433:527–531
Jain R, Rivera MC, Lake JA (1999) Horizontal gene transfer among genomes: the complexity hypothesis. Evolution 96:3801–3806
Jain R, Rivera MC, Moore JE, Lake JA (2003) Horizontal gene transfer accelerates genome innovation and evolution. Mol Biol Evol 20:1598–1602
Kaiser D, Losick R (1993) How and why bacteria talk to each other. Cell 73:873–885
Kaiser D, Welch R (2004) Dynamics of fruiting body morphogenesis. J Bacteriol 186:919–927
Keller L, Surette MG (2006) Communication in bacteria: an ecological and evolutionary perspective. Nat Rev Microbiol 4:249–258
Kent AD, Triplett EW (2002) Microbial communities and their interactions in soil and rhizosphere ecosystems. Annu Rev Microbiol 56:211–236
Khayat R, Tang L, Larson ET, Lawrence CM, Young M, Johnson JE (2005) Structure of an archaeal virus capsid protein reveals a common ancestry to eukaryotic and bacterial viruses. Proc Natl Acad Sci USA 102:18944–18949
Kiers ET, Rousseau RA, West SA, Denison RF (2003) Host sanctions and the legume-Rhizobium mutualism. Nature 425:78–81
Kimura M, Jia Z, Nakayama N, Asakawa S (2008) Viral ecology in soil: past, present, and future perspectives. Soil Sci Plant Nutr 53:1–32
Kohlenbrander PE, Andersen RN, Blehert DS, Egland PG, Foster JS, Palmer RJ (2002) Communication among Oral Bacteria. Microbiol Mol Biol Rev 66:486–505
Kohlenbrander PE, Egland PG, Diaz PI, Palmer RJ (2005) Genome–genome interactions: bacterial communities in initial dental plaque. Trends Microbiol 13:11–15
Konaklieva MI, Plotkin BJ (2006) Chemical communication – do we have a quorum? Mini Rev Med Chem 6:817–825
Koonin EV, Senkevich TG, Dolja VV (2006) The ancient virus world and evolution of cells. Biology Direct 1:29
Koonin EV, Wolf YI (2008) Genomics of bacteria and archaea: the emerging dynamic view of the prokaryotic world. Nucleic Acids Res 36:6688–6719

Koonin EV (2009) On the origin of cells and viruses: primordial virus world scenario. Ann NY Acad Sci 1178:47–64

Kreft JU (2004) Biofilms promote altruism. Microbiol 150:2751–2760

Kroes I, Lepp PW, Relman DA (1999) Bacterial diversity within the human subgingival crevice. Proc Natl Acad Sci USA 96:14547–14552

Kulakauskas S, Lubys A, Ehrlich SD (1995) DNA restriction-modification systems mediate plasmid maintenance. J Bacteriol 177:3451–3454

Kunin V, Goldovsky L, Darzentas N, Ouzounis CA (2005) The net of life: reconstructing the microbial phylogenetic network. Genome Res 15:954–959

Lehnherr H, Maguin E, Jafri S, Yarmolinsky MB (1993) Plasmid addiction of bacteriophage P1: doc, which causes cell death on curing of prophage, and phd, which prevents host death when prophage is retained. J Mol Biol 233:414–428

Lehnherr H, Yarmolinsky MB (1995) Addiction protein Phd of plasmid prophage P1 is a substrate of the ClpXP serine protease of Escherichia coli. Proc Natl Acad Sci USA 92:3274–3277

Lenz DH, Mok KC, Lilley BN, Kulkarni RV, Wingreen NS, Bassler BL (2004) The small RNA Chaperone Hfq and multiple small RNAs control quorum sensing in vibrio harveyi and vibrio cholerae. Cell 118:69–82

Lerat E, Moran NA (2004) The evolutionary history of quorum-sensing systems in bacteria. Mol Biol Evol 21:903–913

Loh J, Pierson EA, Pierson LS 3rd, Stacey G, Chatteriee A (2002) Quorum sensing in plant associated bacteria. Curr Opin Plant Biol 5:285–290

Losick R, Kaiser D (1997) Why and how bacteria communicate. Sci Am 276:68–73

Maizels A, Weiner AM (1993) The genomic tag hypothesis: modern viruses as molecular fossils of ancient stretegies for genomic replication. In: Gesteland RF, Atkins JF (eds), The RNA World. Cold Spring Harbor Laboratory Press, Cold Spring Harbor NY, pp 577–602

Majdalani N, Vanderpool CK, Gottesman S (2005) Bacterial small RNA regulators. Crit Rev Biochem Mol Biol 40:93–113

Manefield M, Turner SL (2002) Quorum sensing in context: out of molecular biology and into microbial ecology. Microbiol 148:3762–3764

Margulis L (1996) Archaeal-eubacterial mergers in the origin of Eukarya: phylogenetic classification of life. Proc Natl Acad Sci USA 93:1071–1076

Margulis L (1999) Die andere evolution. Spektrum Akademischer Verlag, Heidelberg

Margulis L (2004) Serial endosymbiotic theory (SET) and composite individuality. Transition from bacterial to eukaryotic genomes. Microbiol Tod 31:173–174

Margulis L, Sagan D (2002) Acquiring genomes. A theory of the origin of species. Basic Books, New York

Masse E, Gottesman S (2002) A small RNA regulates the expression of genes involved in iron metabolism in *Escherichia coli*. Proc Natl Acad Sci USA 99:4620–4625

McNab R, Lamont RJ (2003) Microbial dinner-party conversations. The role of LuxS interspecies communication. J Med Microbiol 52:541–545

Mehta P, Goyal S, Long T, Bassler B, Wingreen NS (2009) Information processing and signal integration in bacterial quorum sensing. Mol Syst Biol 5:325

Moore WEC, Moore LVH (1994) The bacteria of periodontal diseases. Periodontol 2000 5:66–77

Morris C (1946) Signs, language, and behavior. Prentice Hall, New York

Nadell CD, Xavier JB, Levin SA, Foster KR (2008a) The evolution of quorum sensing in bacterial biofilms. PLoS Biol 6:e14

Nadell CD, Bassler B, Levin SA (2008b) Observing bacteria through the lens of social evolution. J Biol 7:27

Nandhagopal N, Simpson AA, Gurnon JR, Yan X, Baker TS, Graves MV, Van Etten JL, Rossmann MG (2002) The structure and evolution of the major capsid protein of a large, lipid-containing DNA virus. Proc Natl Acad Sci USA 99:14758–14763

Ng WL, Bassler BL (2009) Bacterial quorum-sensing network architectures. Ann Rev Genet 43:197–222

Olendzenski L, Gogarten JP (2009) Evolution of genes and organisms. The tree/web of life in light of horizontal gene transfer. Ann NY Acad Sci 1178:137–145

Osborn AM, Boltner D (2002) When phage, plasmids, and transposons collide: genomic islands, and conjugative- and mobilizable-transposons as a mosaic continuum. Plasmid 48:2002–2212

Oshima K, Kakizawa S, Nishigawa H, Kuboyama T, Miyata S, Ugaki M, Namba S (2001) A plasmid of phytoplasma encodes a unique replication protein having both plasmid- and virus-like domains: clue to viral ancestry or result of virus/plasmid recombination? Virology 285:270–277

Pal C, Papp B, Lercher MJ (2005) Adaptive evolution of bacterial metabolic networks by horizontal gene transfer. Nat Genet 37:1372–1375

Pandey DP, Gerdes K (2005) Toxin–antitoxin loci are highly abundant in free-living but lost from host-associated prokaryotes. Nucl Acids Res 33:966–976

Parsek MR, Greenberg EP (2005) Sociomicrobiology: the connections between quorum sensing and biofilms. Trends Microbiol 13:27–33

Paster BJ, Boches SK, Galvin JL, Ericson RE, Lau CN, Levanos VA, Sahasrabudhe A, Dewhirst FE (2001) Bacterial diversity in human subgingival plaque. J Bacteriol 183:3770–3783

Rawlings DE (1999) Proteic toxin-antitoxin, bacterial plasmid addiction systems and their evolution with special reference to the pas system of pTF-FC2. FEMS Microbiol Lett 176:269–277

Rice G, Tang L, Stedman K, Roberto F, Spuhler J, Gillitzer E, Johnson JE, Douglas T, Young M (2004) The structure of a thermophilic archaeal virus shows a double-stranded DNA viral capsid type that spans all domains of life. Proc Natl Acad Sci USA 101:7716–7720

Rickard AH, Palmer JR, Blehert DS, Campagna SR, Semmelhack MF, Egland PG, Bassler BL, Kolenbrander PE (2006) Autoinducer 2: a concentration-dependent signal for mutualistic bacterial biofilm growth. Mol Microbiol 60:1446–1456

Riley MA (1998) Molecular mechanisms of bacteriocin evolution. Annu Rev Genet 32:255–278

Rivera MC, Lake JA (2004) The ring of life provides evidence for a genome fusion origin of eukaryotes. Nature 431:152–155

Rohwer F, Edwards R (2002) The phage proteomic tree: a genome-based taxonomy for phage. J Bacteriol 184:4529–4535

Rohwer F (2003) Global phage diversity. Cell 113:141

Ryan F (2006) Genomic creativity and natural selection: a modern synthesis. Biol J Linn Soc 88:655–672

Sahasrabudhe A, Dewhirst FE (2001) Bacterial diversity in human subgingival plaque. Journal of Bacteriology 183:3770–3783

Samaj J, Baluska F, Voigt B, Schlicht M, Volkmann D, Menzel D (2004) Endcytosis, actin cytoskeleton, and signalling. Plant Physiol 135:1150–1161

Sandoz KM, Mitzimberg SM, Schuster M (2007) Social cheating in Pseudomonas aeruginosa quorum sensing. Proc Natl Acad Sci USA 104:15876–15881

Schauder S, Bassler BL (2001) The languages of bacteria. Genes Dev 15:1468–1480

Shapiro JA (1998) Thinking about bacterial populations as multicellular organisms. Ann Rev Microbiol 52:81–104

Shapiro JA (2007) Bacteria are small but not stupid: cognition, natural genetic engineering and sociobacteriology. Stud Hist Phil Biol Biomed Sci 38:807–819

Sharma A, Sahgal M, Johri BN (2003) Microbial communication in the rhizosphere: operation of quorum sensing. Curr Sci 85:1164–1172

Solomon JM, Grossman AD (1996) Who's competent and when: regulation of natural genetic competence in bacteria. Trends Genet 12:150–155

Sperandio V, Torres AG, Jarvis B, Nataro JP, Kaper JB (2003) Bacteria-host communication: the language of hormones. Proc Natl Acad Sci USA 100:8951–8956

Sonenshein AL (2006) Bacteriophages: how bacterial spores capture and protect phage DNA. Curr Biol 16:14–16

Spoering AL, Gilmore MS (2006) Quorum sensing and DNA release in bacterial biofilms. Curr Opin Microbiol 9:133–137

Stanley NR, Lazazzera BA (2004) Environmental signals and regulatory pathways that influence biofilm formation. Mol Microbiol 52:917–924

Sun J, Daniel R, Wagner-Dobler I, Zeng AP (2004) Is autoinducer-2 a universal signal for interspecies communication: a comperative genomic and phylogenetic analysis of the synthesis and signal transduction pathways. BMC Evol Biol 4:36

Sutherland IW, Hughes KA, Skillman LC, Tait K (2004) The interaction of phage and biofilms. FEMS Microbiol Lett 232:1–6

Swift S, Bainton NJ, Winson MK (1994) Gram-negative bacterial communication by N-acyl homoserine lactones: a universal language? Trends Microbiol 2:193–198

Taga ME, Bassler BL (2003) Chemical communication among bacteria. Proc Natl Acad Sci USA 100:14549–14554

Takemura M (2001) Poxviruses and the origin of the eukaryotic nucleus. J Mol Evol 52:419–425

Teplitski M, Robinson JB, Bauer WD (2000) Plants secrete substances that mimic bacterial N acylhomoserine lactone signal activities and affect population density dependent behaviors in associated bacteria. Mol Plant Microbe Interact 13:637–648

Tetart F, Desplats M, Kutateladze M, Monod C, Ackermann HW, Krisch HM (2001) Phylogeny of the major head and tail genes of the wide-ranging T4-type bacteriophages. J Bacteriol 183:358–366

Tettelin H, Masignani V, Cieslewicz MJ, Donati C, Medini D, Ward NL, Angiuoli SV, Crabtree J, Jones AL, Durkin AS, DeBoy RT, Davidsen TM, Mora M, Scarselli M, Ros IM, Peterson JD, Hauser CR, Sundaram JP, Nelson WC, Madupu R, Brinkac LM, Dodson RJ, Rosovitz MJ, Sullivan SA, Daugherty SC, Haft DH, Selengut J, Gwinn ML, Zhou L, Zafar N, Khouri H, Radune D, Dimitrov G, Watkins K, O'Connor KJB, Smith S, Utterback TR, White O, Rubens CE, Grandi G, Madoff LC, Kasper DL, Telford JL, Wessels MR, Rappuoli R, Fraser CM (2005) Genome analysis of multiple pathogenic isolates of Streptococcus agalactiae: implications for the microbial "pan-genome". Proc Natl Acad Sci USA 102:13950–13955

Thomas CM, Nielsen KM (2005) Mechanisms of, and barriers to, horizontal gene transfer between bacteria. Nat Rev Microbiol 3:711–721

Tu KC, Bassler BL (2007) Multiple small RNAs act additively to integrate sensory information and control quorum sensing in *Vibrio harveyi*. Gen Dev 21:221–233

Velicer GJ, Kroos L, Lenski RE (2000) Developmental cheating in the social bacterium Myxococcus Xanthus. Nature 404:598–601

Velicer GJ (2003) Social strife in the microbial world. Trends Microbiol 11:330–337

Velicer GJ, Yu YT (2003) Evolution and novel cooperative swarming in the bacterium Myxococcus Xanthus. Nature 425:75–78

Villarreal LP (1999) DNA virus contribution to host evolution. In: Domingo E, Webster RG, Holland JF (eds) Origin and evolution of viruses. Academic Press, London, pp 301–420

Villarreal LP, DeFilippis VR (2000) A hypothesis for DNA viruses as the origin of eukaryotic replication proteins. J Virol 74:7079–7084

Villarreal LP (2004) Can viruses make us humans? Proc Americ Phil Soc 148:296–323

Villarreal LP, DeFilippis VR, Gottlieb KA (2000) Acute and persistent viral life strategies and their relationship to emerging diseases. Virol 272:1–6

Villarreal LP (2005) Viruses and the evolution of life. American Society for Microbiology Press, Washington

Villarreal LP (2009) Origin of group identity. Springer, New York

Villarreal LP, Witzany G (2010) Viruses are essential agents at the stem and roots of the tree of life. J Theor Biol 262:698–710

Visick KL, Fuqua C (2005) Decoding microbial chatter: cell–cell communication in bacteria. J Bacteriol 187:5507–5519

Vogel J, Sharma CM (2005) How to find small non-coding RNAs in bacteria. J Biol Chem 386:1219–1238

Voloshin SA, Kaprelyants AS (2004) Cell–cell interactions in bacterial populations. Biochem (mosc) 69:1268–1275

Wadhams GH, Armitage JP (2004) Making sense of it all: bacterial chemotaxis. Nat Rev Mol Cell Biol 5:1024–1037

Walker TS, Bais HP, Grotewold E, Vivanco JM (2003) Root exudation and rhizosphere biology. Plant Physiol 132:44–51

Wang LH, He Y, Gao Y, Wu JE, Dong YH, He C, Wang SX, Weng LX, Xu JL, Tay L, Fang RX, Zhang LH (2004) A bacterial cell–cell communication signal with cross-kingdom structural analogues. Mol Microbiol 51:903–912

Wassarman KM (2002) Small RNAs in bacteria: diverse regulators of gene expression in response to environmental changes. Cell 109:141–144

Waters CM, Bassler BL (2005) Quorum sensing: cell to cell communication in bacteria. Ann Rev Cell Dev Biol 21:319–346

Waters CM, Bassler BL (2006) The Vibrio harveyi quorum-sensing system uses shared regulatory components to discriminate between multiple autoinducers. Genes Dev 20:2754–2767

Wegrzyn RD, Bapat K, Newnam GP, Zink AD, Chernoff YO (2001) Mechanism of Prion Loss after Hsp104 Inactivation in Yeast. Mol Cell Biol 21:4656–4669

Weitz JS, Mileyko Y, Joh RI, Voit EO (2008) Collective decision making in bacterial viruses. Biophys J 95:2673–2680

West SA, Griffin AS, Gardner A (2007) Social semantics: altruism, cooperation, mutualism, strong reciprocity and group selection. J Evol Biol 20:415–432

Witzany G (2005) Natural history of life: history of communication logics and dynamics. SEED J 5:27–55

Witzany G (2010) Biocommunication and natural genome editing. Springer, Dordrecht

Woese C, Kandler O, Wheelis M (1990) Towards a natural system of organisms: proposal for the domains Archaea, Bacteria, and Eucarya. Proc Natl Acad Sci USA 87:4576–4579

Wuertz S, Okabe S, Hausner M (2004) Microbial communities and their interactions in biofilm systems: an overview. Water Sci Technol 49:327–336

Xavier JB, Foster KR (2007) Cooperation and conflict in microbial biofilms. Proc Natl Acad Sci USA 104:876–881

Xie G, Bonner CA, Song J, No K, Jensen RA (2004) Inter-genomic displacement via lateral gene transfer of bacterial trp operons in an overall context of vertical genealogy. MBC Biol 2:15

Yarmolinsky MB (1995) Programmed cell death in bacterial populations. Science 267:836–837

Young R (2002) Bacteriophage holins: deadly diversity. J Mol Microbiol Biotechnol 4:21–36

Ziedaite G, Daugelavicius R, Bamford JK, Bamford DH (2005) The Holin protein of bacteriophage PRD1 forms a pore for small-molecule and endolysin translocation. J Bacteriol 187:5397–5405

Part I
Intracellular Biocommunication

Chapter 2
Communication Among Phages, Bacteria, and Soil Environments

Stephen T. Abedon

2.1 Introduction

Bacteriophages (phages) are viruses that infect members of domain *Bacteria*, contrasting the *viruses* of domains *Archaea* or *Eukarya*. Ecology is the interaction of organisms with environments, and those interactions, broadly speaking, may be depicted as forms of communication – the movement of information between or among entities, or at least the presentation of potentially detectable signals. Reception of these signals can result in organism modification, including in terms of their behavior. Alternatively, signals can impact the survival and reproduction of genes, giving rise to natural selection, a consequence of environment-to-organism communication.

Organism modification, in turn, can create detectable signals that may be received by the originally emanating entity, resulting, for instance, in environmental feedback loops. For example, the presence of lactose provides a signal that can result in alteration of gene expression by *Escherichia coli*, as in Jacob and Monod's classic operon model. This change can be viewed as an environment-induced reconfiguration of the bacterium's "behavior," one that results in lactose being used as a carbon and energy source. With this use the environment's lactose signal may become modified, and *E. coli* in response might again revise its behavior. "Communication," though, does not strictly require such direct, two-way aspects.

Within body tissues, communication occurs as a consequence of the expression of genes that are genetically linked and which therefore are less free to evolve toward antagonistic (fitness-reducing) interactions. Contrast microbiomes, communities of microorganisms found within specific environments such as soils. There, genes are much less likely to be genetically linked so have much greater potentials

S.T. Abedon
Department of Microbiology, The Ohio State University, 1680 University Dr., Mansfield OH 44906, USA
e-mail: abedon.1@osu.edu

G. Witzany (ed.), *Biocommunication in Soil Microorganisms*, Soil Biology 23,
DOI 10.1007/978-3-642-14512-4_2, © Springer-Verlag Berlin Heidelberg 2011

to evolve independently, such as toward reducing the fitness of other components. Thus, while our body's homeostasis can be viewed as predominantly a consequence of cooperative actions between clonally related entities (cells), the relative stasis seen within microbiomes is a consequence, to a much larger extent, of exploitative interactions. The latter includes both competition and direct, physical antagonism, such as phage lytic infection of bacterial cells.

Here I consider the many ways in which phages communicate with biotic as well as abiotic components of their environments, and *vice versa*. The results are environments that differ from what would exist given an absence of phage action. By and large I concentrate more on behavioral, ecological, or evolutionary ecological (i.e., adaptive) aspects of communication rather than genomic, metagenomic, or phylogenetic attributes (see contributions of Srividhya and S. Krishnaswamy, Riley in this volume). Where possible, I supply data or insights which pertain especially to soil environments.

Phage–soil reviews include those of Williams et al. (1987) and Day and Miller (2008). Reviews of phage–bacterial interactions within an artificial, spatially structured environment, i.e., phage plaques, are also available (Krone and Abedon 2008; Abedon and Yin 2008, 2009; Abedon 2010). Note the potential commonalities between phage population growth within soils, within biofilms, and as plaques: In each case, the predominant phage-exploitable units presumably are immobilized bacterial microcolonies (Fig. 2.1) rather than the planktonic, individual cells more typically envisioned as phage targets (O'Donnell et al. 2007; Abedon 2010; Abedon and Thomas-Abedon 2010).

2.2 General Concepts

In this section I review basic principles of environments and communication, especially as viewed from the perspective of microflora and particularly from the standpoint of phages.

2.2.1 Microbe-Containing Environments

One can categorize microorganism-containing environments or microenvironments in terms of levels of water saturation, degrees of spatial structure (i.e., tendencies to inhibit free movement of materials, especially as via mixing), porosity, permeability, and the extent to which the materials making up an environment are organic vs. inorganic. From this point of view we can envisage at least ten variations: (1) aqueous solutions; (2) sediments found within aqueous environments; (3) flocculations; (4) soils; (5) surfaces of nonparticulate solids such as rocks but also, e.g., of medical devices; (6) porous interior surfaces of otherwise nonparticulate solids, e.g., fluid-filled microfractures (Gomez and Laubach 2006) within rocks, or both micropores and macropores found within or between aggregates of materials

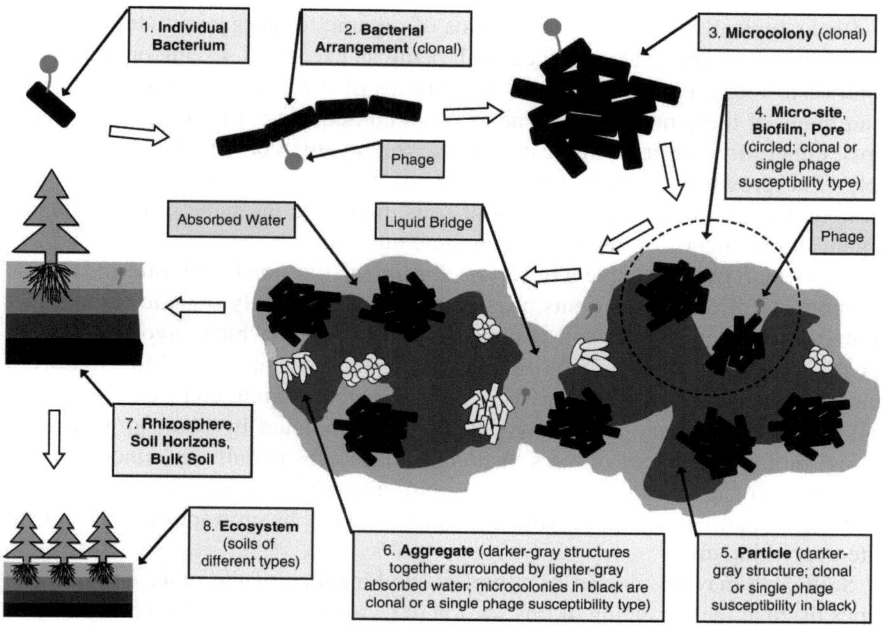

Fig. 2.1 Hierarchy of soil targets of phage attack. Spatially clumped, related groups of bacteria can serve as targets for phage attack in the sense that the infection of one bacterium within a group results in an increase in the likelihood of infection of other members. Groups, as shown, form a nested hierarchy that begins with individual cells and then includes arrangements of not-yet separated bacteria (not necessarily biofilm associated) through microcolonies and collections of microcolonies forming biofilms. Biofilm adherence is to particles which collectively form aggregates of such particles, collections of which represent soils, etc. Phage infection of a single bacterium found on, for example, an aggregate, thus increases the odds that other bacteria found on the same aggregate will also become phage infected, particularly to the degree that aggregates hold absorbed water through which phage movement via diffusion might occur. Phage movement likely is enhanced with increasing soil saturation such as occurs due to rain

(Ranjard and Richaume 2001); (7) living organisms; (8) decaying (no longer living) organisms; (9) organismal excreta; and (10) air.

Soils can be viewed as intermittently aqueous, colloidal suspensions that possess substantial spatial structure and which consist, in their dried form, of particulate mixtures of components that are either inorganic (i.e., sand, silt, or clay) or organic (living as well as decaying organisms). Aqueous sediments, by contrast, are continuously and fully saturated with water while the overlying planktonic world displays much less spatial structure than either sediments or soils. Many rocks are less particulate as well as less readily water saturated or porous within their interiors, though rocks such as sandstone are quite permeable, allowing large amounts of water transmission to occur. Living organisms tend to remain fully water saturated, like sediments and planktonic aqueous environments, but at the same time are only selectively permeable, plus display extensive spatially structured habitats (e.g., our skin and mucous membranes).

Rocks typically also have lower ratios of organic to inorganic materials, those consisting of organic materials, i.e., coal, being an exception. Living organisms, in contrast to rocks, consist of much higher ratios of organic to inorganic materials than many of these other environments. As a consequence, the addition of organisms, either alive or decaying, can increase the organic content of soils considerably. The presence of rocks and their breakdown products instead usually will directly *decrease* the organic content of soils, though at the same time can impact organism prevalence by supplying inorganic nutrients.

The overall nutrient content of soils is dependent, thus, on both biotic and abiotic components. Biotic components provide nutrients, especially organic carbon, that contribute particularly to soil secondary productivity (which involves animal, fungal, protist, and bacterial growth). Abiotic components, especially dissolved minerals such as phosphorous as well as inorganic nitrogen, contribute to soil or soil-associated primary productivity (which is dominated by plant growth). The living, biotic component of soils also contributes substantially to mineralization, i.e., the freeing up of especially inorganic nutrients, which then are available to both autotrophs (particularly photosynthetic organisms) and heterotrophs (which constitute everything else).

Soils also may be viewed as complex combinations of the other environment types listed at the beginning of this section (e.g., air, rocks, etc.). However, it is not possible for all of these environment types to exist simultaneously. Instead, soils can be temporarily fully saturated (and therefore aqueous like), can overlie permanently wet sediment-like environments, can contain flocculations of organic material, and can certainly contain or overlie large, solid, inorganic materials (porous or otherwise). Unless sterilized, soils will contain living organisms, and sterile or not will contain dead organisms and excretions. In addition, when dehydrated, either partially or fully, soils will become impregnated with air, plus, through the action of wind and other forces, can become suspended in air. Soils thus are organic and inorganic, biotic and abiotic, nutrient poor or more nutrient rich, solid as well as liquid and gaseous, etc. Soils also contain numerous interfaces between their various environmental and microenvironmental aspects.

Soils can be aerobic or anaerobic, with levels of molecular oxygen ranging from atmospheric concentrations to effectively nonexistant. Deeper, water-saturated soils, or water-filled microsites, for example, can tend toward the anaerobic end of the spectrum. Soils also can be differentiated in terms of texture (sand vs. silt vs. clay content), structure (aggregation properties along with porosity between aggregates, which together can greatly influence movement of water), and in terms of horizons. The latter term is a description of a soil's horizontal layering, which often varies predictably starting from the surface and going downward, e.g., organic matter, surface soil, subsoil, substratum, and then underlying bedrock. One can also differentiate soils, vis-à-vis plants, into the rhizosphere (soil immediately surrounding plant roots) and those soils which are not associated with plant roots (known as bulk soil).

Soils thus may be too complex to yield to general principles regarding phage–to-bacteria-to-environment, etc., communication. One can simplify things, however, by assuming a perspective of soils as alternatingly somewhat water soaked, during which biological processes such as movement, reproduction, and decay occur, and more air-like environments, during which desiccation survival is emphasized. Soil heterogeneity furthermore results in different microsites experiencing different degrees of desiccation within the same soil sample. From this vantage, the biology of soils perhaps can be viewed as analogous to that of sediments, though typically much more complex and heterogeneous (O'Donnell et al. 2007), and with the important property that often they are less than fully water saturated. Alternatively, from the perspective of individual phages, soils may be envisaged as heterogeneous mazes within which susceptible bacteria may be periodically encountered (Fig. 2.1).

2.2.2 *Communication and Microorganisms*

Communication is an important concept with many facets, depending upon both one's discipline of study as well as personal proclivity. A lay definition, anchoring one end of a spectrum, might entail two-way, abstract exchanges involving verbal or written language. That is, for example, where one individual speaks, a second (or more) listens, ponders, responds, and so on, with perhaps various verbal and nonverbal interactions occurring simultaneously. At its basis, though, communication involves simply signals emanating from one or more entity that are received by other entities. These other entities then can respond in some manner by modifying their physical, chemical, or behavioral state.

In addition to visual as well as vocal interactions between animals, communication can involve touch, plus extensive forms of communication exist that occur via chemical signals, whether airborne (e.g., pheromones, the sweet smell of a flower, etc.), fluid-based, or found in soils, such as secondary metabolites associated with soil bacteria (Karlovsky 2008). The latter also includes the chemically mediated cell-to-cell signaling that coordinates the growth and behavior of multicellular organisms. Indeed, extensive communication continuously takes place within our own bodies, maintaining, in a process collectively described as homeostasis, a perturbation away from an otherwise inevitable decay.

The coordination seen with homeostasis is the product of an evolution that is possible only because the genes involved are genetically linked, that is, they are found within the same genomes and the survival (and reproductive success) of one such gene is not independent of nor necessarily even possible without the success of others. Such coevolution of genes is much less likely without genetic linkage (e.g., see Hyman and Abedon 2008 for discussion), and the default state among most of the genes found within individual ecosystems, except those found within the genomes of individual organisms, is one of competition, and even antagonism, rather than coordination (or cooperation). An ecosystem thus may be viewed as consisting of partially isolated islands of active perturbation away from an otherwise inevitable decay toward chemical equilibrium, that is, islands consisting of individual organisms. These islands are connected by regions that either are at chemical equilibrium or are in the process of moving toward that equilibrium. Collectively, these ecosystems do not display homeostasis so much as attract or retain organisms that are able to exploit existing conditions, all toward their own, individual gain. Their interactions, absent environmental perturbation, give rise approximately to steady states.

Though interactions within ecosystems, or microbiomes if we are focusing predominantly on the microorganisms present, are not necessarily coordinated or cooperative, they nevertheless are communicative. Whether through soluble chemicals, intentionally or inadvertently released, or entities physically touching, signals of various sorts can emanate from one individual and be received by others. The result can be either a passive response by the signal's recipients, such as is the case at least initially when damaging agents are conveyed from one individual to another, or a more active response by the recipient, e.g., as seen with quorum sensing (Bonfante and Anca 2009; Dessaux and Faure in this volume). One can view the generation of signals also as passive vs. active, such as the passive release of chemicals upon decay versus the active exporting of signal molecules to the extracellular environment. Situations also can occur where both signals and responses are intentional and occur expressly for the sake of conveyance of information, phenomena that can be more easily appreciated as a form communication from a human-centered perspective. Irrespective of such details, natural selection favors those organisms whose survival and reproduction tends to *not* decline in response to environmentally common signals. Thus, we can envisage microbiomes as consisting not just of myriad organisms, but also of myriad communicative pathways, which to recipients can be beneficial, benign, or detrimental. Here I consider especially those pathways involving phages.

2.2.3 *Bacteriophages, Bacteria, and Environments*

Originally, bacteriophages were macroscopically observed as consumers of bacterial cultures, either the clearing of broth cultures or as plaques on solid media (Abedon 2008). Thus, while the word virus, once meaning poison, fittingly describes the macroscopic consequences of viral infection of plants and animals, the term phage – from Greek meaning to eat or devour – apparently seemed more appropriate for the viral infection of bacteria. Consistently, this potential for phages to clear pure bacterial cultures, that is, their "phage"-like lytic ability, provides us with clues as to the phage potential to impact bacterial communities (Abedon 2009c). Similarly, impediments on this lytic ability are indicative of the negative impact of bacteria, or environments in general, on phages. All of these negative impacts, as well as interactions between entities that are positive, i.e., that are fitness enhancing, result from communicative processes.

Phages are often described as the most numerous organisms on Earth, though all estimations of phage prevalence are just that: estimations. For instance, there are many who note that viruses tend to be present in ratios of about ten-to-one to cellular organisms. Since prokaryotic organisms, a combination of both *Bacteria* and *Archaea*, are prevalent with a total of about 10^{30} individual cells (Whitman et al. 1998), then the global phage total count logically would be about 10^{31}. This estimation is similar to the product of the volume of the world's oceans and estimations of 10^7 virus-like-particles per ml of sea water, which is not too excessive a possibility as an average density (Abedon 2008). Such virus total-count determinations typically involve electron or epifluorescence microscopy (Abedon et al. 2009), but generally are easier to obtain from water vs. within soil. This is because, with soils, the presence of debris and potential for virion absorption to particles (and debris) seems to obscure direct counts. Nevertheless, phage density estimations range up to and over 10^8 g^{-1} in soil, as extrapolated from transmission electron microscopy total counts (Ashelford et al. 2003) or as determined via epifluorescence microscopy (Williamson et al. 2003). Interestingly, though ratios of viruses to bacteria in forested soils can also be in the range of ten-to-one ($>10^9$ virus-like particles to $>10^8$ bacteria), agricultural soils containing a measured bacterial prevalence of $>10^5$ g^{-1} can still have densities of virus-like particles of $\sim 10^9$ g^{-1} (Williamson et al. 2005). Numerous additional studies have characterized temperate phages isolated from soil or lysogeny in soil-associated bacterial species (e.g., Fink and Zahler 2006; Williamson et al. 2007, 2008; Ghosh et al. 2008; see contributions of Kimura and Srividhya/S. Krishnaswamy in this volume).

Another bulk means of viewing phages within environments is in terms of their diversity, which in modern times has been analyzed especially via metagenomic techniques. Metagenomics takes a snapshot of a portion of the genetic diversity found within a given environment, e.g., such as that associated (seemingly) with the viral fraction (Casas and Rohwer 2007), or of the entire soil metagenome itself (Vogel et al. 2009). Alternatively, it is possible to be more directed in one's analysis by employing specific PCR primers, particularly in conjunction with sequencing (Jia et al. 2007). From such analyses, in light of the high number of sequences found that fail to match any currently present within databases, some have argued that phages are the most diverse organisms on Earth, which is in addition to being the most numerous (Pedulla et al. 2003; Weinbauer et al. 2007). A good take-home message from these musings is simply that phages can do many things within environments (i.e., they possess many different genes) and, based on analysis of measurements of phage turnover rates (in aquatic environments), must do those things fairly often (Hendrix 2008).

Bacteria in soil environments may be similarly considered. For instance, both bacterial diversity and density in soils can be high, with numbers of bacterial phylotypes – a species-like taxonomic designation applied to microorganisms – varying between locations especially as a function of pH (Fierer and Jackson 2006), with species number per 10 g of soil reportedly well in excess of 10^4 (Gans et al. 2005; Burmølle et al. 2007), and with prokaryote counts often in the range of 10^9 g^{-1} (Gans et al. 2005; Burmølle et al. 2007; O'Donnell et al. 2007). An additional consideration is that many bacteria within soils grow within biofilms (Burmølle et al. 2007; O'Donnell et al. 2007) and their arrangement within those biofilms likely impacts the phage potential to penetrate to and otherwise affect bacteria. Bacteria which phages can use as hosts thus will have certain densities, certain properties, and will be dispersed in certain ways, all of which will affect phages and, in turn, the phage impact, that is, phage communication with both bacteria and the larger environment.

2.3 Pathways of Communication in Soil

In this section I distinguish among phage-associated communication processes in terms of directions of communication, i.e., from specific categories of originators of signals (such as phages) to specific categories of recipients (such as bacteria). These various lines of communication are summarized in Fig. 2.2. An additional category, not presented in that figure, nor otherwise discussed, is that since bacteria can serve as prey for soil-dwelling animals, protists, predatory bacteria, and phages, then communication in the form of interorganismal competition will occur among these disparate organism types.

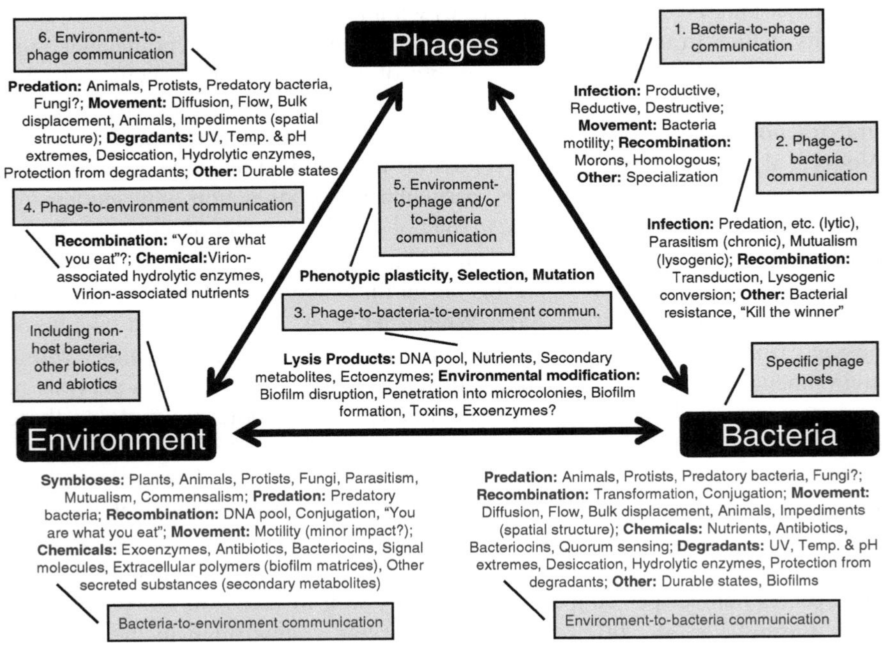

Fig. 2.2 Mechanisms of communication between phages, their hosts ("Bacteria"), and the non-host environment. Numbers refer to the order of discussion in the text. Text in *gray* (bottom) does not involve phages and is not otherwise discussed in the chapter

2.3.1 Bacteria-to-Phage Communication

Communication that can be viewed as coming from bacteria and being received by phages may be differentiated into a number of distinct facets. These include (1) transfer of DNA sequences (i.e., so-called moron acquisition by phages, as discussed in Sect. 2.3.1.2), (2) physical and chemical interactions between phage- and bacteria-sourced molecules (e.g., protein–protein or protein–DNA interactions), or (3) more general bacteria-mediated modifications of phage phenotype. The latter may be differentiated into modifications that impact the replication, survival, or movement of individual phages (their ecology), on the one hand, or, on the other hand, may be the product of bacteria-mediated natural selection on phage populations (their evolution). Except for phage interaction with bacteria-sourced, decay-mediating molecules such as extracellular proteases (Nasser et al. 2002), these mechanisms of bacteria-to-phage communication are, by and large, associated with phage infection of bacteria. I consider them in terms of bacterial impact on phage phenotype, genotype, and movement.

2.3.1.1 Bacterial Impact on Phage Phenotype

Phage infection can occur in a number of distinguishable modes: productive, reductive, and destructive (Abedon et al. 2009). Which mode is displayed during a given infection is determined in the course of communication between molecules produced by phages and bacteria. Productive infections are ones which produce and release phage virions. Reductive infections, such as lysogeny, are ones in which the phage survives but the infection does not immediately gear up to produce phage virions. Destructive means that the phage does not survive. It seems reasonable to suppose that all nondefective phages must be capable of productive infections and likely that all can also be subjected to phage-destructive infections.

Destructive Infection: Antagonism, Deception, and Primitive Immunity

There are a number of obvious situations in which bacteria noticeably impact phage phenotype. These include when bacteria effect phage restrictive or abortive infections, which are phage-destructive infections that may be distinguished in terms of whether bacteria survive or don't survive, respectively (Abedon 2008; Abedon et al. 2009; Hyman and Abedon 2010). These are clear examples of bacteria-to-phage communication, particularly ones with negative consequences for the phage.

In addition to destroying phages, restriction-modification systems can also modify phage genomes in a manner that alters their susceptibility to cutting by the same restriction-modification system. Notably, the latter is a rare occurrence in terms of its ability to forestall phage restriction, though important in that it imparts on the progeny of so-modified phages the ability to display bacteria-killing infections (Korona and Levin 1993). These inadvertent modifications confer to phages the power to communicate a *lie*; that is, to mislead bacteria into treating a phage genome as self.

There also exists a type of destructive infection that may be viewed as a form of phage-mediated bacterium-to-bacterium communication (Sumby and Smith 2002). In this instance phages are modified during productive infection such that subsequent infection of a clonally related bacterium, here the soil bacterium, *Streptomyces coelicolor*, results in phage restriction. The previously infected bacterium thus seems to be communicating, through the released phages, that those phage genomes represent foreign DNA that therefore should be eliminated. This "Phage growth limitation system" may be viewed as analogous to animal immune system functions in which the molecules of parasites are tagged, as foreign, and subsequently eradicated. Also analogous to animals, it is the passage of parasites among clumped, clonally related tissues, in this case bacterial tissue, which fosters the utility of this mechanism.

Reductive Infection: Sleeping with the Enemy

Phages may be packaged into bacterial endospores, increasing phage survival within soils (Pantastica-Caldas et al. 1992; Sonenshein 2006). Since bacterial genes are responsible for the sporulation phenotype (Errington 2003), phages may be viewed as modifying their infection outcome in response to bacterial signals, from obligately productive to at least temporarily reductive (Abedon 2009b). Similarly, phage survival may be extended via lysogenic infection (Stewart and Levin 1984). Continuing this theme, arguably any bacterial mutation that affects phage infection characteristics or any differences in phage phenotype when infecting different bacterial strains – such as display or lack of display of lysogenic infections – can be described as examples of bacterial gene impact on phage phenotype, and which therefore is illustrative of bacteria-to-phage communication. Furthermore, it seems logical that these interactions between phage- and bacteria-encoded molecules, interactions which are potentially disruptable by bacterial mutation, are products of phage evolution rather than of bacterial evolution, unless those interactions are antagonistic to phages or, alternatively, enhance the overall fitness of a lysogen.

2.3.1.2 Bacterial Impact on Phage Genotype (Evolution)

The ability of bacterial genes to modify phage phenotype underscores the obvious potential of bacteria to impact phage evolution. A large fraction of this impact can be described as resulting from bacterial mechanisms of resistance to phages, e.g., restriction-modification systems as well as evolved abortive infection mechanisms (Hyman and Abedon 2010). These and other phenomena collectively make up the "bacteriophage 'resistome'" (Hoskisson and Smith 2007), which has been explored by Coberly et al. (2009) in terms of its impact on phage evolution within spatially structured environments. However, to the extent that phages can form mutualisms with bacteria, then phage–bacterial coevolution presumably can occur other than as a consequence of antagonistic interactions (Sect. 2.3.2.2).

Another general mechanism of phage evolution involves their specialization on specific bacteria, which can be viewed as a form of bacteria-communicated phage adaptation. Biological adaptation, however, often involves tradeoffs where improvement in one character gives rise to reduced function in another, a process described as antagonistic pleiotropy (Duffy et al. 2006; Heineman et al. 2008). An extreme of such antagonistic pleiotropy is seen with host-range switching, where phage interaction with one bacterial moiety (e.g., the phage receptor molecule) is lost upon acquisition of phage affinity for a different bacterial molecule (the new phage receptor molecule). Alternatively, some phages have been found which display fairly wide host ranges, spanning multiple bacterial genera (Hyman and Abedon 2010). These broader host ranges presumably would facilitate access to greater host numbers but perhaps at the expense of more effective infection of any one host type.

Phage evolution also occurs as a consequence of acquisition of bacterial genetic material. The "Moron-accretion hypothesis" in fact posits that all phage genetic material had its ultimate origin from bacterial DNA (Hendrix et al. 2000). That DNA which increases phage fitness, or at least does not decrease phage fitness too greatly, is retained and subject to subsequent mutational modification, which again is subject to natural selection (Hendrix 2008). A more specific form of recombination between phages and bacterial chromosomes is the homologous recombination that can occur between phages and prophages found in the infected bacterium; see Abedon (2009a) for discussion and references. Here phage-evolved genes can be transferred intact to a second phage, though it's a matter of semantics whether this represents bacteria-to-phage communication vs. phage-to-phage communication.

2.3.1.3 Bacterial Impact on Phage Location

Bacteria can impact phage movement. If bacteria are immobile, such as are bacteria found in association with biofilms, then phage infection can have the consequence of temporarily halting phage movement. This temporary immobility presumably is equivalent to that which occurs upon bacterial infection during phage plaque formation (Abedon and Yin 2008, 2009; Abedon and Thomas-Abedon 2010). Bacteria also may obstruct phage movement by being tightly packed, perhaps even if those bacteria are not a phage's host (Yin and McCaskill 1992), plus can secrete substances, such as extracellular polymers, which bind together bacterial biofilms but which also may inhibit or at least slow the movement of phage particles (Abedon 2010; Abedon and Thomas-Abedon 2010). On the other hand, if bacteria are filamentous, then phage adsorption to one end presumably could result in virion release at the other. Another means by which phage movement can occur during infection is when bacteria themselves are motile, which may move infecting phages along faster than can diffusion alone, though only so long as the infected bacterium has not yet lysed.

In soils, we thus can expect a number of ways by which bacteria can impact phage movement, ranging from substantial inhibitors to highly effective enhancers. One can even speculate that environmental conditions, such as degree of water saturation, may modify the relative impact of these various mechanisms. I now turn from bacteria-to-phage communication to the seeming reverse, i.e., phage-to-bacteria communication.

2.3.2 Phage-to-Bacteria Communication

Phage-to-bacteria communication varies in terms of the costs, and benefits, experienced by recipient bacteria. These impacts range from depletion of bacterial fitness to zero, i.e., bacteria are killed by phage action, or worse (Sect. 2.3.2.1), to where bacterial fitness is enhanced in the course of lysogenic conversion (Sect. 2.3.2.2).

2.3.2.1 The Many Costs of Phage

The worst thing that could happen to a bacterium as effected by a phage is a noninduced lytic infection, that is, a productive infection terminating in lysis that does not follow lysogeny. This is because not only does the infected bacterium die but so too, upon subsequent infection, may the bacterium's fellow clone mates. With lysogeny, by contrast, clone mates generally do not also die, following induction and subsequent phage release, because of the expression of superinfection immunity by these fellow lysogens (Sect. 2.3.2.2). An abortive infection, which is bactericidal but nonetheless does not produce phages, should in these terms also be preferable to a productive, lytic infection.

Chronic infections, too, would be preferable to lytic ones. However, though not necessarily lethal, chronic infections still have the potential to slow bacterial growth and therefore reduce bacterial fitness, plus can increase a bacterium's susceptibility to environmental toxins such as antibiotics (Hagens et al. 2006). In each case, the negative effects are consequences of phage infection, though with abortive infections the actual effector of bacterial death can be the bacterium (Hyman and Abedon 2010). That bacteria would encode such a mechanism, however, makes sense so long as productive lytic infections are more costly than abortive ones, to clonal bacterial populations. With restrictive infections, by contrast, the phage dies but the bacterium survives, so, on that basis, phage restriction should be preferable to the bacterium than either abortive or chronic phage infections.

2.3.2.2 Phage Infection as Symbiosis

In reductive infections neither phage nor bacterium is killed, plus phages do not produce virions. The more familiar reductive infections are lysogenic, where infecting phages exist as intracellular prophages (Miller and Day 2008; Abedon 2009b). Lysogeny can be viewed as a symbiosis, as too can phage chronic infections, since in both cases the phage and bacterium are independent organisms that nonetheless are intimately associated over multiple generations. Myriad intraspecific communication can occur within symbiotic relationships, some protective while others are antagonistic to the relationship's existence. Prophage induction as well as nonlethal metabolic demands, for example, can be detrimental to host bacteria, and bacteria possibly possess antiprophage mechanisms (Lawrence et al. 2001).

Prophage–bacterial interactions can also be mutualistic, especially within the context of so-called lysogenic conversion, where bacterial phenotype is modified by phage gene expression (Miller and Day 2008; Hyman and Abedon 2008; Paul 2008). The most common such benefit is superinfection immunity (a.k.a. homoimmunity), which protects bacteria from exploitation by additional phages, ones of the same immunity type. In another example, though not necessarily mutualistic, prophages can down-regulate the metabolic activities of bacteria, which possibly provides lysogens with increased survival potential (Chen et al. 2005). As perhaps an extreme example of such tendencies, infection by various phages of soil bacteria has been shown to have positive impacts on host sporulation rates (Silver-Mysliwiec and Bramucci 1990).

Prophage induction itself can lead to the destruction of neighboring, potentially lysogen-competing bacteria, thereby providing a benefit to unlysed lysogen clones. A phage of *Bacillus aneurinolyticus*, ϕBA1, in fact, displays bacteriocin-like activity on some hosts (an apparently phage-DNA-independent bacterial death that occurs also without phage survival) but normal phage infection on other hosts (Ito et al. 1986). This process of temperate-phage induction and subsequent killing of nonlysogenic bacterial neighbors has been dubbed as a form of allelopathy by Stewart and Levin (1984) and more recently has been described as "kill the relatives" (Paul 2008). See Abedon and LeJeune (2005) and Brown et al. (2006) for additional discussion of "lysogen allelopathy" along with possible consequences. Temperate phages also have been implicated as contributors to bacterial biofilm formation (Rice et al. 2009; Abedon 2010), plus lysogenic conversion can provide metabolic functions that can be useful to bacteria only under certain circumstances, as considered in Abedon and LeJeune (2005).

Contrasting prophages, chronically infecting phages may be described as classically parasitic organisms since they display ongoing, detrimental, but typically sublethal infections. Lytic phages are viewed as parasites as well, as in "obligately intracellular parasites," but due to their propensity to kill their host bacteria they often are described as predators instead. In ecology, however, predators typically represent a higher trophic level, i.e., a feeding position one level above that of prey, while phages fail to achieve the molecular assimilation of bacteria that "feeding" would imply, since most of the consumed bacterium is discarded as waste (Thingstad et al. 2008). Phagotrophic protists, by contrast, clearly can be viewed as predators that feed on bacteria. It has been argued, alternatively, that lytic phages may be described as parasitoids (Forde et al. 2004), parasites that consume their hosts, often from the inside out. The analogy is that phages similarly consume their bacterial hosts from the inside out. However, the same lack of molecular assimilation that can be used to the criticize the labeling of phages as predators similarly could be applied to the labeling of phages as parasitoids. It is important, though, to not lose sight of the fact that lytic phage infections are equivalently detrimental to their bacterial hosts regardless of how we choose to label those interactions.

2.3.2.3 Phage-Mediated Horizontal Gene Transfer (Transduction)

The above considerations are ecological in terms of the consequences of phage-to-bacterium communication. Alternatively, one can view these communications from an evolutionary perspective, that is, in terms of mutation, sampling error (i.e., genetic drift), and migration, as well as, of course, natural selection (Duffy and Turner 2008; Abedon 2009a). In this section I emphasize the phage impact especially on the migration of genetic material between bacteria.

Migration can be viewed as the physical movement of organisms or, more pertinently with regard to evolutionary biology, the movement of individuals into or out of populations. While bacteria are capable of migrating into and out of populations, it is important to realize that bacterial genes, independent of bacteria themselves, also are capable of such movement. In microbiology, one typically describes this migration of genes between populations as horizontal gene transfer or lateral gene flow, though another term which seems equally applicable is introgression (Cohan et al. 1991; Campbell 1994; Lawrence and Ochman 1997; Brown et al. 2001; Colegrave 2002; Johnson et al. 2004; Abedon 2009a); that is, the low-level gene flow between otherwise minimally genetically interacting populations. Such gene flow is readily mediated by phages, including among soil bacteria, in a process termed transduction: the packaging of bacterial DNA within phage capsids and subsequent delivery of that DNA to a second bacterium. The recipient bacterium both survives and integrates the DNA into its genome via various forms of recombination (see contribution of Armon in this volume). A pertinent recent study is that of Ghosh et al. (2008) who showed that transducing particles could be induced, via treatment with mitomycin C, from soil-isolated lysogens.

Bacteria additionally may be recipients of what are better described as phage genes, i.e., genes that normally are found within a phage's genome rather than ones which are accidentally packaged into phage capsids. Lysogenic conversion is the most familiar context within which such genes are observed (Sect. 2.3.2.2). More generally, phages might serve as "the proving ground of choice for evolutionary innovation" for potential bacterial genes, "the critical motive force for the evolution of the entire biosphere" (Krisch 2003). Phages also can inactivate bacterial genes, most prominently via prophage insertion into bacterial chromosomes, e.g., as by phage Mu (Paolozzi and Ghelardini 2006).

2.3.2.4 Kill the Winner

In terms of natural selection, as mediated by phages upon bacteria, one can clearly differentiate between positive and negative impacts (above). In this section, I consider two interesting negative impacts. The first is the tendency for *tradeoffs*, where phage-resistant bacteria, in the absence of phages, tend to be less fit than otherwise isogenic, phage-sensitive bacteria (Gill and Abedon 2003; Kerr et al. 2008). The consequence of this effect is that in a world lacking in phages, the dominating bacteria will be inclined to be those that display the least tendency toward phage resistance.

The second impact is a frequency-dependent selection (Abedon 2009a), typically described in the phage literature as "Kill the winner" (Thingstad et al. 2008). That is, bacterial winners, i.e., phage-susceptibility types displaying the greatest densities, will be more susceptible to phage attack simply due to those higher bacterial densities. The result is a potential for greater fitness for lower-density phage-susceptibility types along with a resulting selection for greater overall bacterial diversity. So goes the concept of kill the winner as it was formulated to explain phage–bacterial community dynamics within aquatic environments (Thingstad 2000), but it is a valid question whether kill the winner is similarly applicable to soil environments (Day and Miller 2008).

While one should always expect higher bacterial densities to support greater phage densities and therefore greater bacteria killing potential, all other factors held constant, in fact a shortcoming of the idea of kill the winner, in terms of validation, is that dramatic bacterial killing will only occur if dramatic phage densities can be achieved, such as in the range of 10^7 or more phages of a single type per ml (Appunu and Dhar 2008; Abedon and Thomas-Abedon 2010). Alternatively, we can view kill the winner from the perspective of bacterial microcolonies as a phage target (Fig. 2.1). For two bacterial types, otherwise identical, the microcolony that is larger overall – in a sense, a winner among microcolonies – will be more likely to encounter a phage and therefore more likely to be reduced or eliminated by phage infection. Of course, the more microcolonies of a given type, too, the more prevalent will be those would-be microcolony-encountering phages.

The less water that is present in soil, the greater the effective phage density. Indeed, when calculating phage densities within soils, often conveniently determined on a per gram basis, it is highly relevant just what volume of free water for phage diffusion is available within that gram, and how that water is arranged with regard to the potential for phage diffusion between bacteria (Fig. 2.1) (see contribution of Armon in this volume). In particular, while water saturation of soils should allow greater potential for phage diffusion between bacteria, it also may have the effect of reducing phage densities such that phage–bacterial encounter and therefore kill the winner is less likely. In addition, regardless of the level of water saturation, the rate of impact of phages on bacteria in soils will be expected to be slower than that observed within well-mixed broth (Tan and Reanney 1976; Pantastica-Caldas et al. 1992), presumably also reducing the efficiency of kill the winner. A few experiments exist, however, which provide results that are at least consistent with the possibility that kill the winner could function at least in some soils, at some scales, against some bacteria.

Keel et al. (2002), in what represents essentially an augmented kill the winner scenario, added phages to target bacteria in soil at a ratio of 1.6:1.0 and observed a one-thousandfold drop in bacterial densities. The added phage-to-bacterium ratio, though, is insufficient to explain the resulting decline in bacterial densities, which at best would provide a reduction of 80%, i.e., $e^{-1.6} = 0.2$ (20%) = the expected number of surviving bacteria following phage attack (Carlson 2005). Instead, the phage-to-bacterium ratio must have reached at least ~7 fold (Abedon 2009d; Abedon and Thomas-Abedon 2010), which is about what was experimentally observed in terms of *in situ* increase in phage density. Such results are very much consistent with a kill the winner effect: the ability of high bacterial densities (there 10^8 g^{-1} soil) to support phage population growth to sufficient densities to substantially reduce bacterial numbers. It must be stressed, however, that phages were applied to soils at their initially high densities, $>10^8$ g^{-1}, rather than amplified to those levels while *in situ*. In addition, phages were added soon after target bacteria were mixed into soils, suggesting more or less uniform bacterial dispersion within soils along with a lack of biofilm formation prior to phage addition. Conditions thus were likely ideal, and artificially so, for both phage amplification and subsequent bacterial killing. Importantly, the phage densities involved approached or even exceeded total phage densities that have been reported for soils in general (Sect. 2.2.3).

Zeph and Casida (1986) provide additional, arguably less artificial evidence that kill the winner might potentially take place, at least upon soil manipulation. They added bacteria to soil as prey for indigenous bacterial predators. As a means of monitoring for enrichment of these predator bacteria, they assayed for increases in the prevalence of indigenous phages active against the predator bacteria. The soil, however, appears to have been well mixed prior to the start of experiments, implying some loss of spatial structure. In addition, typically no more than about 10^6 phages ml^{-1} of "percolate" were produced (except for one experiment in which peak phage densities instead were 10^7 ml^{-1}). It is uncertain, though, how these titers translate into within-soil bacteria killing potential, and therefore whether kill the winner could have been effected. Importantly, less artificial studies of phage–bacterial dynamics in soil have yielded even lower phage densities, i.e., at best little more than about 10^4 g^{-1} (Campbell et al. 1995; Ashelford et al. 1999, 2000). In other studies, however, counts well in excess of 10^6 phages g^{-1} have been observed in which soil had been amended with nutrients and/or host-bacteria (Reanney and Marsh 1973; Tan and Reanney 1976; Germida 1986; Pantastica-Caldas et al. 1992; Hussein et al. 1994; Burroughs et al. 2000).

In short, there is no evidence that kill the winner cannot, in principle, function in soil ecosystems, but there also is little evidence that it in fact does. My opinion is that it probably does operate on microscales, at least to the extent that soil phages are able to effectively reach, penetrate into, and otherwise fully lyse bacterial microcolonies. However, even on gram scales a modest proportion of local bacteria (one-third or two-fifths) appear to be sensitive to local phages while phages can be slightly more infective of more-local bacteria than of less-local bacteria (Vos et al. 2009). These observations suggest a failure of at least some local phages to drive at least some local bacteria to extinction. Two key questions therefore remain vis-à-vis our understanding of the degree to which phages can negatively impact bacterial soil communities: (1) Is phage exploitation of soil microcolonies efficiently accomplished, i.e., are phages highly virulent on a per-microcolony basis? (2) Do phages possess the means to effectively move from microcolony to microcolony, also within soils? Principally, if phages cannot destroy microcolonies, then they cannot eliminate microcolony-forming bacterial winners even on microscales. Furthermore, if phages cannot efficiently locate new microcolonies, then even if they are highly virulent against individual microcolonies, they will not succeed in effecting kill the winner on macroscales. See Williams et al. (1987) for additional complications on our potential to comprehend phage–bacterial population dynamics in soil.

2.3.3 *Phage-to-Bacteria-to-Environment Communication*

In addition to phage-to-bacteria communication, phages can impact aspects of environments other than hosts, but nonetheless as mediated through their impact on bacteria. This can be accomplished via phage-induced bacterial lysis but also can be effected through prophage gene expression. Among lysis-related effects is the likely phage impact on the quantity and quality of free DNA found within environments.

2.3.3.1 Lysis-Mediated Phage-Environment Communication

Other than in terms of their own existence as virions found in the extra-bacterial environment, phages can influence environments specifically through bacterial lysis and do so in at least four basic ways. First, by effecting lysis, phages convert bacteria-associated insoluble nutrients into soluble ones that are then available primarily to heterotrophic bacteria, i.e., as described within aquatic environments (Weinbauer 2004; Suttle 2007). Within soils, though, these nutrients may also be available to eukaryotic heterotrophic absorbers, such as fungi, whose soil presence can be vast (O'Donnell et al. 2007). Another interesting break from the experience of aquatic environments is that in soils the primary ecological process carried out by microorganisms is decay, rather than the photosynthesis of many aquatic bacteria. Since bacterial lysis too represents a form of decay, of bacteria, we can describe phage lytic action not just as interfering with ecosystem productivity but also as contributing to it. Indeed, phages, by lysing bacteria, could very well play a key role in the mineralization process that decay ultimately represents. In addition, phages could play a role in the release of secondary metabolites from bacteria, adding to those released into soils through bacterial secretion (Karlovsky 2008).

A second means by which phages affect ecosystems, in the course of lysing bacteria, is by releasing internal bacterial enzymes which can hydrolyze otherwise nutritiously unavailable substrate. While I find it compelling that these so-called ectoenzymes probably play relevant and even important roles within relatively simple ecosystems (Abedon and LeJeune 2005), I am less sanguine toward accepting that possibility within soils, especially more complex, less disturbed soils containing large diversities of decay-mediating microorganisms. Nonetheless, there remains at least a possibility that phage-released enzymes could play relevant roles in soil ecosystem ecology (see, e.g., Sect. 2.3.3.2).

The third mechanism of phage-to-environment communication, as mediated through bacterial lysis, involves the release of DNA. While this DNA can be viewed as another solubilized nutrient, released DNA if not too fragmented also can serve as a source of genetic material for bacterial transformation (Pietramellara et al. 2009). That is, certain bacteria are able to pick up environmental (naked) DNA into their cytoplasms and subsequently incorporate these "snippets" into their genomes (Day 2004). It is conceivable that the DNA pool available for this transformation is larger than it would be absent phage-mediated bacterial lysis (Abedon 2009a). To the extent that kill the winner operates, the DNA available upon phage-mediated lysis may either be more diverse owing to a greater assortment of bacteria which are available for lysis over time or less diverse owing to a bias toward lysis of particular (winner) populations. In addition, not only is bacterial DNA potentially released upon phage-mediated lysis, but so too is unencapsidated phage DNA, which also may be available to bacteria for transformation. In a study on DNA extraction from soils, however, *Hin*dIII-digested phage λ DNA was found to be difficult to recover, which at least in part was a consequence of DNA absorption to soil colloids such as clay (Frostegård et al. 1999).

Yet another potential consequence of phage-mediated bacterial lysis is the disruption of bacterial biofilms, microcolonies, or arrangements. Though such disruption is readily demonstrated in the laboratory (Abedon 2010; Abedon and Thomas-Abedon 2010), and there may even be augmented, i.e., as reviewed by Azeredo and Sutherland (2008), it is an open question how significant a role phages play in these processes in natural environments, such as soil. Furthermore, I am uncertain what might be the consequence of such disruption other than in terms of reduction in the competitive ability of these phage-susceptible bacteria vs. those which instead are phage resistant.

2.3.3.2 Prophage-Mediated Environmental Modification

Phages might also modify soil environments toward their own ends. This consideration is based on speculative analogies to phage encoding of bacterial virulence factors. These virulence factors, and encoding phages, are associated with many bacterial pathogens of animals (Abedon and LeJeune 2005; Hyman and Abedon 2008). Some phage-associated virulence factors, particularly exotoxins such as Shiga, diphtheria, or cholera toxins, can be viewed as environment-modifying enzymes. That is, the disease symptoms that these toxins effect are a consequence of their disruptive modifications of the body environment. In addition, the resulting environmental modifications might serve to enhance a phage's replication or dissemination (Abedon and LeJeune 2005).

It is conceivable that similar factors are encoded by soil phages and expressed perhaps chiefly during lysogenic infection. These factors might modify soil environments and do so in some manner that benefits the producing phage. For example, phages might encode enzymes whose release from bacteria can lead to the digestion of local substrate either for the sake of providing nutrients to host bacteria or to enhance the potential for phage diffusion, such as away from the lysed parental bacterium and toward not yet phage-infected bacteria. Indeed, phage-encoded depolymerases perhaps could be viewed as such enzymes (Barnet and Humphrey 1975; Abedon 2010). Phage DNA extracted from soil samples has been shown to carry exotoxins *sensu stricto*, such as Shiga toxin (Casas et al. 2006). Whether additional factors exist among soil phages, acting within soils, is however an open question.

2.3.4 Phage-to-Environment Communication

Bell (1992) discusses five general properties of environments, which roughly can be translated to that they are (1) highly variable and (2) complex, (3) that organisms respond to them inconsistently, (4) that they self regulate, and (5) that they "tend continually to deteriorate" (p. 34). The impact of phages on environments corresponds to at least the last three of these properties. That is, infections vary and have unexpected outcomes (e.g., transduction), bacterial density can be limited also by phage infection (i.e., environment self regulation), and, by lysing bacteria, phages contribute directly to environmental deterioration, in this case decay of what otherwise would be living entities (bacteria). Each of these mechanisms acts via phage infection of bacteria.

Can phage-to-environment communication also occur *without* a bacterium intermediary other than the bacteria needed to produce the phages themselves? I posit three mechanisms. First, phages can be consumed by certain eukaryotes, at least in lakes (González and Suttle 1993; Bettarel et al. 2005), and perhaps under some rare circumstances contribute to the genetic material of those organisms, i.e., the "you are what you eat" hypothesis (Doolittle 1998). Second, there exist certain virion-associated hydrolytic enzymes which, for example, can disrupt bacteria-secreted extracellular polysaccharides (Sutherland et al. 2004). Finally, phages themselves can serve as nutrients. This latter role can follow their decay and thereby solubilization, follow their consumption by eukaryotes, or, indeed, follow their adsorption to and restriction by otherwise phage-susceptible bacterial hosts (Fuhrman 1999).

2.3.5 Environment-to-Phage and/or to-Bacteria Communication

For phage biologists with an interest in phenotype, perhaps the most fascinating aspect of communication, as it occurs between phages, bacteria, and environments, is how modifying environmental parameters can result in modification of phage phenotype. This concept we can loosely describe as a phenotypic plasticity. Basically, it represents a physiological modification of a phage infection that may or may not be adaptative. Perhaps the most familiar phage response to environmental stimulus is the induction of prophages following lysogen exposure to DNA damaging agents, such as UV or mitomicin C (Campbell 2006). This response happens to be one that is both stimulated and mediated while a phage is associated with its bacterial host. Another and similar phenomenon is the resolution of the otherwise reductive pseudolysognic type state (Miller and Day 2008; Abedon 2009b) in which increasing host metabolic activity results in increased phage metabolic activity, producing a productive, lysogenic, or perhaps even phage-destructive infection.

With both induction of prophages and activation of pseudolysogens, the phage is potentially displaying an adaptive response to changes in environmental conditions, in both cases increasing infection activity. Alternatively, the response to a worsening of environmental conditions can be a reduction in the activity of a phage infection, or even an avoidance of adsorption altogether (Kutter et al. 1994). At an extreme, the phage could display the above-noted pseudolysogeny, a nonlysogenic, nonproductive, nondestructive, *and* nonreproductive delay in initiation of a more active infection (Abedon 2009b). Less extreme responses can involve decreases in phage burst sizes, lengthening in phage infection periods, or both. This has been observed in laboratory culture following nutrient limitation (Hadas et al. 1997) including with soil phages (Webb et al. 1982). These responses may be relevant especially to the extent that soils are oligotrophic or alternate between eutrophic and oligotrophic (feast-famine; Williams et al. 1987). Also with soil phages, Williams et al. (1987) review similar results associated with changes in temperature. Environments also can supply mutagens, some of which interact with phages while they are free, that is, not yet infecting, with mutations then observed only upon bacterial infection (Drake and Ripley 1994). Day and Miller (2008) provide a more general discussion of virion perturbability.

Another phage infection response to environmental stimulus involves detection of the presence of free phages, which make their presence known by adsorbing infected bacteria. The phage's response, when it occurs, can be an extension in the phage latent period and an associated *increase* in phage burst size (Abedon 1990). Perhaps more commonly, the response involves instead an increase in phage burst number, i.e., as can be subsequently mediated by multiple inductions of populations of clonally related phage lysogens (Abedon 2009a; Abedon et al. 2009). That is, higher phage multiplicities of infection can give rise to higher likelihoods of temperate phage reduction to lysogeny (Weitz et al. 2008) which, via binary fission, can produce a population of clonally related, potentially productive phage infections. In terms of the study of phages in soil, or any environment for that matter, the key take-home message is that phage characteristics can change in response to changes in environmental conditions, whether *in situ* or in comparing *in situ* conditions to those in the laboratory. In addition, even when held in constant, well-defined environments, phages and bacterial hosts can be expected to evolve, and often not independently of each other (Brockhurst et al. 2007).

2.3.6 Environment-to-Phage Communication

Soil components, in communicating their presence, can impact phage movement, survival, and reproduction. For this final discussion, however, I ignore two environmental aspects: First is the impact on phages of their bacterial hosts, which I specifically considered in Sect. 2.3.1. Second is phage–phage communication, e.g., such as via genetic recombination, which I mostly avoid here as genomic or phylogenetic concerns, or, as involving various sorts of antagonistic interactions which are reviewed elsewhere (Abedon 1994; Turner and Duffy 2008; Abedon et al. 2009). I instead concentrate on phenomena originating from nonhost and nonphage environmental aspects. That is, predation of phages along with mechanisms giving rise to phage virion movement, degradation, and durability.

2.3.6.1 Predation of Phages

Soil or soil-contaminated materials ingested by animals likely will contain phages, given phage ubiquity. The ingested phages may be digested along with other organic matter, except to the extent that they are resistant to digestive processes. One can envisage, for example, phage loss as a consequence of earthworm action, as well as from protist engulfment (Sect. 2.3.4). It seems unlikely, though, that phages would contribute greatly to either's nutrient needs. Predatory bacteria, such as *Myxococcus*, secrete hydrolytic enzymes as groups to obtain nutrients from soil-associated organisms, materials, and, potentially, even phages (Berleman et al. 2006; Evans et al. 2007); of course, as *Myxococcus* spp. are bacteria, there also exist phages which are capable of infecting them (e.g., Azuaga et al. 1990), and see also Zeph and Casida (1986). Other, more sedentary organisms, such as fungi, though not necessarily a major contributor of soil proteases (Watanabe and Hayano 1994), nonetheless could potentially affect phages.

2.3.6.2 Phage Movement

Many things can contribute to phage movement, though in soils mixing is limited or, at least, extremely slow. Diffusion and fluid flow, by contrast, should be major contributors to phage movement in soil. Nonetheless, as soil colloids, phage movement from location to location within soil, or in terms of penetration into partially closed off volumes, should be expected to be *limited* in comparison with dissolved materials or even soluble enzymes (McKay et al. 2002). These limitations are especially as compared with the potential for phage movement given less spatially structured environments, or if a phage should happen to infect a motile bacterium.

Given sufficient water, then fluid flow also can take place through soils, such as through "cracks" (Choi et al. 2004), "fractures," or "root holes" (Blanc and Nasser

1996), i.e., so-called bypass flow and finger flow (McLeod et al. 2001). The consequence can be phage transport from place to place without substantial interaction between phages and the soil matrix. At an extreme, groundwater, like surface water (Ferguson et al. 2007), may allow phage movement over meters, at least (Bales et al. 1995; McKay et al. 2002). Both diffusion and flow, however, may be inhibited by various obstructions (e.g., Davis et al. 2006; Van Cuyk and Siegrist 2007; Wong et al. 2008). These obstructions include especially colloidal substances to which phages can be absorbed, particularly clays (Duboise et al. 1979; Williams et al. 1987; Chattopadhyay and Puls 2000; Hassen et al. 2003; Day and Miller 2008; but also see Bixby and O'Brien 1979, and Armon chapter in this volume).

Animals also can play a role in the movement of phages, in part if phages can survive ingestion, but perhaps more likely with phages nonspecifically adhering to animals which are moving through soils, or through specific or nonspecific adherence between phage-infected bacteria and animals (Dennehy et al. 2006). Bulk movement of soils also occurs, such as a consequence of mechanical action, e.g., due to trees falling, the surface locomotion of relatively large animals, burrowing by animals in general, growth of plant roots, freeze-thaw cycles, gravity (mediating, for example, landslides), anthropogenic disturbances, etc. Soil-borne phages probably also spread from location to location through the air (Weinbauer 2004) and a number of papers consider such movement especially from the perspective of sewage aerosolization, e.g., Brooks et al. (2004) who point out (p. 8), "It is known that a bioaerosol is subject to intense physical pressures from the environment (specifically low humidity, ultraviolet and temperature extremes), which tend to inactivate microbes during transport of bioaerosols over long distances." See also (Clark 2005).

2.3.6.3 Phage Survival

Soils can be protective of phages such as from UV irradiation or otherwise as a function of phage absorption onto soil particles (Vettori et al. 2000 along with references cited). In addition, temperature and pH extremes can be buffered in soils, such as in comparison to water (pH; Tan and Reanney 1976) or air (temperature). That phages must display at least some virion durability in soil environments is speculated by Williams et al. (1987) who suggest (p. 162), "There is evidence that most soil bacteria have only spasmodic periods of activity in micro-sites dispersed within the soil mass... Therefore, it is likely that virulent phage in the absence of an active susceptible host must exist for considerable periods as free virions in the soil and be subjected to environmental factors and fluctuations." They then go on to provide a review of the evidence from which this position is derived. Phage durability also may be enhanced through lysogenic or psuedolysogenic infection (Stewart and Levin 1984; Williams et al. 1987; Miller and Day 2008; Abedon 2009b).

In many cases, by contrast, the impact of soil residence on phages can be negative (Williams et al. 1987; Day and Miller 2008). The latter can be a consequence of temperature or pH extremes and can vary with soil type, organic matter prevalence, electrolyte concentrations, and what organisms are present; see Williams and Lanning (1984), Song et al. (2005), and Davies et al. (2006) for data and references. Notwithstanding tendencies toward phage instability at higher temperatures, phages have been detected in surface sands of the Sahara Desert, though survival characteristics there were not determined (Prigent et al. 2005). Desiccation, of course, can also be a concern, especially for those phages which are less desiccation resistant, though there is evidence that for phage PRD1, a coliphage, there may exist an optimal soil-moisture content that confers extended survival (Song et al. 2005). Hydrolytic enzymes excreted by soil microorganisms, as noted (Sect. 2.3.6.1), also may degrade soil phages. McKay et al. (2002) suggests lodging of phage particles within tight pores or fractures as another possible phage loss mechanisms (so-called "straining") plus lists various other blocks on phage movement.

2.4 Conclusion

In this chapter, I strive to educe an expansive view of phages, bacteria, and soil, and how these entities communicate, especially as pathways of communication involve phages. I have attempted to identify broad principles rather than narrowing in on mechanisms that are specific to particular soil types, horizons within soils, biomes, phage–bacterial systems, etc. This has been done in part because those specifics, as derived from a very soil-centered point of view, as opposed to bulk phage or bacteria-centered perspectives, are somewhat lacking in the phage literature. However, I have also taken this broader perspective for the sake of presenting a guide to the range of questions that may be addressed for phages and communication within soils, or indeed any environment. That is, to achieve anything close to a comprehensive understanding of the phage role in soil environments there needs to be an understanding of phage biology, as it occurs within soils, at microscales, and in terms of the spatial and temporal complexity as well as heterogeneity which are the hallmarks of soil ecosystems.

Acknowledgment Thank you to Dawn Ferris who commented on the early, soils-specific portion of the chapter and to Kurt Williamson who provided a number of helpful comments on the penultimate version. This work was supported by an Ohio State intramural grant awarded to Jeff LeJeune, Brian McSpadden Gardener, and myself.

References

Abedon ST (1990) Selection for lysis inhibition in bacteriophage. J Theor Biol 146:501–511

Abedon ST (1994) Lysis and the interaction between free phages and infected cells. In: Karam JD (ed) The molecular biology of bacteriophage T4. ASM Press, Washington, DC, pp 397–405

Abedon ST (2008) Phages, ecology, evolution. In: Abedon ST (ed) Bacteriophage ecology. Cambridge University Press, Cambridge, UK, pp 1–28

Abedon ST (2009a) Bacteriophage evolution and ecology. Adv Appl Microbiol 67:1–45

Abedon ST (2009b) Disambiguating bacteriophage pseudolysogeny: an historical analysis of lysogeny, pseudolysogeny, and the phage carrier state. In: Adams HT (ed) Contemporary trends in bacteriophage research. Nova Science Publishers, New York

Abedon ST (2009c) Impact of phage properties on bacterial survival. In: Adams HT (ed) Contemporary trends in bacteriophage research. Nova Science Publishers, New York

Abedon ST (2009d) Kinetics of phage-mediated biocontrol of bacteria. Foodborne Pathog Dis 6 (7):807–815

Abedon ST (2010) Bacteriophages and Biofilms. In: Bailey WC (ed) Biofilms: Formation, Development and Properties. Nova Science Publishers, New York

Abedon ST, LeJeune JT (2005) Why bacteriophage encode exotoxins and other virulence factors. Evol Bioinf Online 1:97–110

Abedon ST, Thomas-Abedon C (2010) Phage therapy pharmacology. Curr Pharm Biotechnol 11:28–47

Abedon ST, Yin J (2008) Impact of spatial structure on phage population growth. In: Abedon ST (ed) Bacteriophage ecology. Cambridge University Press, Cambridge, UK, pp 94–113

Abedon ST, Yin J (2009) Bacteriophage plaques: theory and analysis. Methods Mol Biol 501:161–174

Abedon ST, Duffy S, Turner PE (2009) Bacteriophage ecology. In: Schaecter M (ed) Encyclopedia of microbiology. Elsevier, Oxford, pp 42–57

Appunu C, Dhar B (2008) Isolation and symbiotic characteristics of two Tn5-derived phage-resistant *Bradyrhizobium japonicum* strains that nodulate soybean. Curr Microbiol 57:212–217

Ashelford KE, Day MJ, Bailey MJ, Lilley AK, Fry JC (1999) In situ population dynamics of bacterial viruses in a terrestrial environment. Appl Environ Microbiol 65:169–174

Ashelford KE, Norris SJ, Fry JC, Bailey MJ, Day MJ (2000) Seasonal population dynamics and interactions of competing bacteriophages and their host in the rhizosphere. Appl Environ Microbiol 66:4193–4199

Ashelford KE, Day MJ, Fry JC (2003) Elevated abundance of bacteriophage infecting bacteria in soil. Appl Environ Microbiol 69:285–289

Azeredo J, Sutherland IW (2008) The use of phages for the removal of infectious biofilms. Curr Pharm Biotechnol 9:261–266

Azuaga MJ, Munoz J, Gonzalez F, Arias JM (1990) Isolation and characterization of bacteriophages from *Myxococcus virescens*. Microbios 61:83–88

Bales RC, Li S, Maguire KM, Yahya MT, Gerba CP, Harvey RW (1995) Virus and bacteria transport in a sandy aquifer, Cape Cod, MA. Ground Water 33:653–661

Barnet YM, Humphrey B (1975) Exopolysaccharide depolymerases induced by *Rhizobium* bacteriophages. Can J Microbiol 21:1647–1650

Bell G (1992) Five properties of environments. In: Grant PR, Horn HS (eds) Molds, molecules and metazoa: growing points in evolutionary biology. Princeton University Press, Princeton, pp 33–54

Berleman JE, Chumley T, Cheung P, Kirby JR (2006) Rippling is a predatory behavior in *Myxococcus xanthus*. J Bacteriol 188:5888–5895

Bettarel Y, Sime-Ngando T, Amblard C, Bouvy M (2005) Low consumption of virus-sized particles by heterotrophic nanoflagellates in two lakes of the French Massif central. Aquat Microb Ecol 39:205–209

Bixby RL, O'Brien DJ (1979) Influence of fulvic acid on bacteriophage adsorption and complexation in soil. Appl Environ Microbiol 38:840–845

Blanc R, Nasser A (1996) Effect of effluent quality and temperature on the persistence of viruses in soil. Water Sci Technol 33:237–242

Bonfante P, Anca IA (2009) Plants, mycorrhizal fungi, and bacteria: a network of interactions. Ann Rev Microbiol 63:363–383

Brockhurst MA, Morgan AD, Fenton A, Buckling A (2007) Experimental coevolution with bacteria and phage. The *Pseudomonas fluorescens* – Φ2 model system. Infect Genet Evol 7:547–552

Brooks JP, Tanner BD, Josephson KL, Gerba CP, Pepper IL (2004) Bioaerosols from the land application of biosolids in the desert southwest USA. Water Sci Technol 50:7–12

Brown EW, LeClerc JE, Kotewicz ML, Cebula TA (2001) Three R's of bacterial evolution: how replication, repair, and recombination frame the origin of species. Environ Mol Mutagen 38:248–260

Brown SP, Le Chat L, De Paepe M, Taddei F (2006) Ecology of microbial invasions: amplification allows virus carriers to invade more rapidly when rare. Curr Biol 16:2048–2052

Burmølle M, Hansen LH, Sørensen SJ (2007) Establishment and early succession of a multispecies biofilm composed of soil bacteria. Microb Ecol 54:352–362

Burroughs NJ, Marsh P, Wellington EMH (2000) Mathematical analysis of growth and interaction dynamics of streptomycetes and a bacteriophage in soil. Appl Environ Microbiol 66:3868–3877

Campbell A (1994) Comparative molecular biology of lambdoid phages. Ann Rev Microbiol 48:193–222

Campbell AM (2006) General aspects of lysogeny. In: Calendar R, Abedon ST (eds) The bacteriophages. Oxford University Press, Oxford, pp 66–73

Campbell JIA, Albrechtsen M, Sorensen J (1995) Large *Pseudomonas* phages isolated from barley rhizosphere. FEMS Microbiol Ecol 18:63–74

Carlson K (2005) Working with bacteriophages: common techniques and methodological approaches. In: Kutter E, Sulakvelidze A (eds) Bacteriophages: biology and application. CRC Press, Boca Raton, Florida, pp 437–494

Casas V, Miyake J, Balsley H, Roark J, Telles S, Leeds S, Zurita I, Breitbart M, Bartlett D, Azam F, Rohwer F (2006) Widespread occurrence of phage-encoded exotoxin genes in terrestrial and aquatic environments in Southern California. FEMS Microbiol Lett 261:141–149

Casas V, Rohwer F (2007) Phage metagenomics. Meth Enzymol 421:259–268

Chattopadhyay D, Puls RW (2000) Forces dictating colloidal interactions between viruses and soil. Chemosphere 41:1279–1286

Chen Y, Golding I, Sawai S, Guo L, Cox EC (2005) Population fitness and the regulation of *Escherichia coli* genes by bacterial viruses. PLoS Biol 3:e229

Choi C, Song I, Stine S, Pimentel J, Gerba C (2004) Role of irrigation and wastewater reuse: comparison of subsurface irrigation and furrow irrigation. Water Sci Technol 50:61–68

Clark JM (2005) Microbe-laden aerosols. Microbiol Today Nov:172–173

Coberly LC, Wei W, Sampson K, Millstein J, Wichman H, Krone SM (2009) Spatial structure and host evolution facilitate coexistence of competing bacteriophages: theory and experiment. Am Nat 173:E121–E138

Cohan FM, Roberts MS, King EC (1991) The potential for genetic exchange by transformation within a natural-population of *Bacillus subtilis*. Evolution 45:1393–1421

Colegrave N (2002) Sex releases the speed limit on evolution. Nature 420:664–666

Davies CM, Logan MR, Rothwell VJ, Krogh M, Ferguson CM, Charles K, Deere DA, Ashbolt NJ (2006) Soil inactivation of DNA viruses in septic seepage. J Appl Microbiol 100:365–374

Davis JA, Farrah SR, Wilkie AC (2006) Adsorption of viruses to soil: impact of anaerobic treatment. Water Sci Technol 54:161–167

Day MJ, Miller RV (2008) Phage ecology of terrestrial environments. In: Abedon ST (ed) Bacteriophage ecology. Cambridge University Press, Cambridge, UK, pp 281–301

Day MJ (2004) Transformation. In: Miller RV, Day MJ (eds) Microbial evolution: gene establishment, survival, and exchange. ASM Press, Washington DC, pp 158–172

Dennehy JJ, Friedenberg NA, Yang YW, Turner PE (2006) Bacteriophage migration via nematode vectors: host-parasite-consumer interactions in laboratory microcosms. Appl Environ Microbiol 72:1974–1979

Doolittle WF (1998) You are what you eat: a gene transfer ratchet could account for bacterial genes in eukaryotic nuclear genomes. Trends Genet 14:307–311
Drake JW, Ripley LS (1994) Induced mutagenesis and isolation of T4 mutants. In: Karam JD (ed) Molecular biology of bacteriophage T4. ASM Press, Washington, pp 447–451
Duboise SM, Moore BE, Sorber CA, Sagik BP (1979) Viruses in soil systems. In: Isenberg HD (ed) CRC critical reviews in microbiology. CRC Press, Boca Raton, FL, pp 245–285
Duffy S, Turner PE (2008) Introduction to phage evolutionary biology. In: Abedon ST (ed) Bacteriophage ecology. Cambridge University Press, Cambridge, UK, pp 147–176
Duffy S, Turner PE, Burch CL (2006) Pleiotropic costs of niche expansion in the RNA bacteriophage Phi6. Genetics 172:751–757
Errington J (2003) Regulation of endospore formation in *Bacillus subtilis*. Nat Rev Microbiol 1:117–126
Evans KJ, Hobley L, Lambert C, Sockett RE (2007) *Bdellovibrio*: lone hunter "cousin" of the "pack hunting" myxobacteria. In: Whitworth DE (ed) Myxobacteria: multicellularity and differentiation. ASM Press, Washington, DC, pp 351–362
Ferguson CM, Davies CM, Kaucner C, Krogh M, Rodehutskors J, Deere DA, Ashbolt NJ (2007) Field scale quantification of microbial transport from bovine faeces under simulated rainfall events. J Water Health 5:83–95
Fierer N, Jackson RB (2006) The diversity and biogeography of soil bacterial communities. Proc Natl Acad Sci USA 103:626–631
Fink PS, Zahler SA (2006) Temperate bacteriophages of *Bacillus subtilis*. In: Calendar R, Abedon ST (eds) The bacteriophages. Oxford University Press, Oxford, pp 557–571
Forde SE, Thompson JN, Bohannan BJM (2004) Adaptation varies through space and time in a coevolving host–parasitoid interaction. Nature 431:841–844
Frostegård A, Courtois S, Ramisse V, Clerc S, Bernillon D, Le Gall F, Jeannin P, Nesme X, Simonet P (1999) Quantification of bias related to the extraction of DNA directly from soils. Appl Environ Microbiol 65:5409–5420
Fuhrman JA (1999) Marine viruses and their biogeochemical and ecological effects. Nature 399:541–548
Gans J, Wolinsky M, Dunbar J (2005) Computational improvements reveal great bacterial diversity and high metal toxicity in soils. Science 309:1387–1390
Germida JJ (1986) Population dynamics of *Azospirillum brasilense* and its bacteriophage in soil. Plant Soil 90:117–128
Ghosh D, Roy K, Williamson KE, White DC, Wommack KE, Sublette KL, Radosevich M (2008) Prevalence of lysogeny among soil bacteria and presence of 16S rRNA and trzN genes in viral-community DNA. Appl Environ Microbiol 74:495–502
Gill JJ, Abedon ST (2003) Bacteriophage ecology and plants. APSnet Feature. http://www.apsnet.org/online/feature/phages/
Gomez LA, Laubach SE (2006) Rapid digital quantification of microfracture populations. J Struct Geol 28:408–420
González JM, Suttle CA (1993) Grazing by marine nanoflagellates on viruses and virus-sized particles: ingestion and digestion. Mar Ecol Prog Ser 94:1–10
Hadas H, Einav M, Fishov I, Zaritsky A (1997) Bacteriophage T4 development depends on the physiology of its host *Escherichia coli*. Microbiology 143:179–185
Hagens S, Habel A, Blasi U (2006) Augmentation of the antimicrobial efficacy of antibiotics by filamentous phage. Microb Drug Resist 12:164–168
Hassen A, Jamoussi F, Saidi N, Mabrouki Z, Fakhfakh E (2003) Microbial and copper adsorption by smectitic clay – an experimental study. Environ Technol 24:1117–1127
Heineman RH, Springman R, Bull JJ (2008) Optimal foraging by bacteriophages through host avoidance. Am Nat 171:E150–E157
Hendrix RW (2008) Phage evolution. In: Abedon ST (ed) Bacteriophage ecology. Cambridge University Press, Cambridge, UK, pp 177–194

Hendrix RW, Lawrence JG, Hatfull GF, Casjens S (2000) The origins and ongoing evolution of viruses. Trends Microbiol 8:504–508

Hoskisson PA, Smith MCM (2007) Hypervariation and phase variation in the bacteriophage 'resistome'. Curr Opin Microbiol 10:396–400

Hussein ME, El-Hawa MEA, El Dydamony G (1994) Population and persistence of Zag-1 phage and cowpea *Rhizobium* in two sterile soils. Egypt J Microbiol 29:271–283

Hyman P, Abedon ST (2008) Phage ecology of bacterial pathogenesis. In: Abedon ST (ed) Bacteriophage ecology. Cambridge University Press, Cambridge, UK, pp 353–385

Hyman P, Abedon ST (2010) Bacteriophage host range and bacterial resistance. Adv Appl Microbiol 70:217–248

Ito S-I, Nishimune T, Abe M, Kimoto M, Hayashi R (1986) Bacteriocinlike killing action of a temperate bacteriophage ϕBA1 of *Bacillus aneurinolyticus*. J Virol 59:103–111

Jia Z, Ishihara R, Nakajima Y, Asakawa S, Kimura M (2007) Molecular characterization of T4-type bacteriophages in a rice field. Environ Microbiol 9:1091–1096

Johnson LJ, Koufopanou V, Goddard MR, Hetherington R, Schafer SM, Burt A (2004) Population genetics of the wild yeast *Saccharomyces paradoxus*. Genetics 166:43–52

Karlovsky P (2008) Secondary metabolites in soil ecology. In: Karlovsky P (ed) Secondary metabolites in soil ecology. Springer, Berlin, pp 1–19

Keel C, Ucurum Z, Michaux P, Adrian M, Haas D (2002) Deleterious impact of a virulent bacteriophage on survival and biocontrol activity of *Pseudomonas fluorescens* strain CHAO in natural soil. Mol Plant Microbe Interact 15:567–576

Kerr B, West J, Bohannan BJM (2008) Bacteriophage: models for exploring basic principles of ecology. In: Abedon ST (ed) Bacteriophage ecology. Cambridge University Press, Cambridge, UK, pp 31–63

Korona R, Levin BR (1993) Phage-mediated selection and the evolution and maintenance of restriction-modification. Evolution 47:556–575

Krisch HM (2003) The view from Les Treilles on the origins, evolution and diversity of viruses. Res Microbiol 154:227–229

Krone SM, Abedon ST (2008) Modeling phage plaque growth. In: Abedon ST (ed) Bacteriophage ecology. Cambridge University Press, Cambridge, UK, pp 415–438

Kutter E, Kellenberger E, Carlson K, Eddy S, Neitzel J, Messinger L, North J, Guttman B (1994) Effects of bacterial growth conditions and physiology on T4 infection. In: Karam JD (ed) The molecular biology of bacteriophage T4. ASM Press, Washington, DC, pp 406–418

Lawrence JG, Hendrix RW, Casjens S (2001) Where are the pseudogenes in bacterial genomes? Trends Microbiol 9:535–540

Lawrence JG, Ochman H (1997) Amelioration of bacterial genomes: rates of change and exchange. J Mol Evol 44:383–397

McKay LD, Harton AD, Wilson GV (2002) Influence of flow rate on transport of bacteriophage in shale saprolite. J Environ Qual 31:1095–1105

McLeod M, Aislabie J, Smith J, Fraser R, Roberts A, Taylor M (2001) Viral and chemical tracer movement through contrasting soils. J Environ Qual 30:2134–2140

Miller RV, Day M (2008) Contribution of lysogeny, pseudolysogeny, and starvation to phage ecology. In: Abedon ST (ed) Bacteriophage ecology. Cambridge University Press, Cambridge, UK, pp 114–143

Nasser AM, Glozman R, Nitzan Y (2002) Contribution of microbial activity to virus reduction in saturated soil. Water Res 36:2589–2595

O'Donnell AG, Young IM, Rushton SP, Shirley MD, Crawford JW (2007) Visualization, modelling and prediction in soil microbiology. Nat Rev Microbiol 5:689–699

Pantastica-Caldas M, Duncan KE, Istock CA, Bell JA (1992) Population dynamics of bacteriophage and *Bacillus subtilis* in soil. Ecology 73:1888–1902

Paolozzi L, Ghelardini P (2006) The bacteriophage Mu. In: Calendar R, Abedon ST (eds) The bacteriophages. Oxford University Press, Oxford, pp 469–496

Paul JH (2008) Prophages in marine bacteria: dangerous molecular time bombs or the key to survival in the seas? ISME J 2(6):579–589

Pedulla ML, Ford ME, Houtz JM, Karthikeyan T, Wadsworth C, Lewis JA, Jacobs-Sera D, Falbo J, Gross J, Pannunzio NR, Brucker W, Kumar V, Kandasamy J, Keenan L, Bardarov S Jr, Kriakov J, Lawrence JG, Jacobs WR Jr, Hendrix RW, Hatfull GF (2003) Origins of highly mosaic mycobacteriophage genomes. Cell 113:171–182

Pietramellara G, Ascher J, Borgogni F, Ceccherini MT, Guerri G, Nannipieri P (2009) Extracellular DNA in soil and sediment: fate and ecological relevance. Biol Fertil Soils 45:219–235

Prigent M, Leroy M, Confalonieri F, Dutertre M, DuBow MS (2005) A diversity of bacteriophage forms and genomes can be isolated from the surface sands of the Sahara Desert. Extremophiles 9:289–296

Ranjard L, Richaume AS (2001) Quantitative and qualitative microscale distribution of bacteria in soil. Res Microbiol 152:707–716

Reanney DC, Marsh SCN (1973) The ecology of viruses attacking *Bacillus stearothermophilus* in soil. Soil Biol Biochem 5:399–408

Rice SA, Tan CH, Mikkelsen PJ, Kung V, Woo J, Tay M, Hauser A, McDougald D, Webb JS, Kjelleberg S (2009) The biofilm life cycle and virulence of *Pseudomonas aeruginosa* are dependent on a filamentous prophage. ISME J 3:271–282

Silver-Mysliwiec T, Bramucci MG (1990) Bacteriophage-enhanced sporulation: comparison of the spore converting bacteriophages PMB12 and SP10. J Bacteriol 172:1948–1953

Sonenshein AL (2006) Bacteriophages: how bacterial spores capture and protect phage DNA. Curr Biol 16:R14–R16

Song I, Choi CY, O'Shaughnessy S, Gerba CP (2005) Effects of temperature and moisture on coliphage PRD-1 survival in soil. J Food Prot 68:2118–2122

Stewart FM, Levin BR (1984) The population biology of bacterial viruses: why be temperate. Theor Pop Biol 26:93–117

Sumby P, Smith MCM (2002) Genetics of the phage growth limitation (Pgl) system of *Streptomyces coelicolor* A3(2). Mol Microbiol 44:489–500

Sutherland IW, Hughes KA, Skillman LC, Tait K (2004) The interaction of phage and biofilms. FEMS Microbiol Lett 232:1–6

Suttle CA (2007) Marine viruses – major players in the global ecosystem. Nat Rev Microbiol 5:801–812

Tan JSH, Reanney DC (1976) Interactions between bacteriophages and bacteria in soil. Soil Biol Biochem 8:145–150

Thingstad TF (2000) Elements of a theory for the mechanisms controlling abundance, diversity, and biogeochemical role of lytic bacterial viruses in aquatic systems. Limnol Oceanogr 45:1320–1328

Thingstad TF, Bratbak G, Heldal M (2008) Aquatic phage ecology. In: Abedon ST (ed) Bacteriophage ecology. Cambridge University Press, Cambridge, UK, pp 251–280

Turner PE, Duffy S (2008) Evolutionary ecology of multi-phage infections. In: Abedon ST (ed) Bacteriophage ecology. Cambridge University Press, Cambridge, UK, pp 195–216

Van Cuyk S, Siegrist RL (2007) Virus removal within a soil infiltration zone as affected by effluent composition, application rate, and soil type. Water Res 41:699–709

Vettori C, Gallori E, Stotzy G (2000) Clay minerals protect bacteriophage PBS1 of *Bacillus subtilis* against inactivation and loss of transducing ability by UV radiation. Can J Microbiol 46:770–773

Vogel TM, Simonet P, Jansson JK, Hirsch PR, Tiedje JM, van Elsas JD, Bailey MJ, Nalin R, Philippot L (2009) TerraGenome: a consortium for the sequencing of a soil metagenome. Nat Rev Microbiol 7:252

Vos M, Birkett PJ, Birch E, Griffiths R, Robert I, Buckling A (2009) Local adaptation of bacteriophages to their bacterial hosts in soil. Science 325:833

Watanabe K, Hayano K (1994) Estimate of the source of proteases in upland fields. Biol Fertil Soils 18:341–346

Webb V, Leduc E, Spiegelman GB (1982) Burst size of bacteriophage SP82 as a function of growth rate of its host *Bacillus subtilis*. Can J Microbiol 28:1277–1280

Weinbauer MG (2004) Ecology of prokaryotic viruses. FEMS Microbiol Rev 28:127–181

Weinbauer MG, Agis M, Bonilla-Findji O, Malits A, Winter C (2007) Bacteriophage in the environment. In: McGrath S, van Sinderen D (eds) Bacteriophage: genetics and molecular biology. Caister Academic Press, Norfolk, UK, pp 61–92

Weitz JS, Mileyko Y, Joh RI, Voit EO (2008) Collective decision making in bacterial viruses. Biophys J 95:2673–2680

Whitman WB, Coleman DC, Wiebe WJ (1998) Prokaryotes: the unseen majority. Proc Natl Acad Sci USA 95:6578–6583

Williams ST, Lanning S (1984) Studies of the ecology of streptomycete phage in soil. In: Ortiz-Ortiz L, Bojalil LF, Yakoleff V (eds) Biological, biochemical and biomedical aspects of actinomycetes. Academic Press, London, pp 473–483

Williams ST, Mortimer AM, Manchester L (1987) Ecology of soil bacteriophages. In: Goyal SM, Gerba CP, Bitton G (eds) Phage ecology. John Wiley & Sons, New York, pp 157–179

Williamson KE, Wommack KE, Radosevich M (2003) Sampling natural viral communities from soil for culture-independent analyses. Appl Environ Microbiol 69:6628–6633

Williamson KE, Radosevich M, Wommack KE (2005) Abundance and diversity of viruses in six Delaware soils. Appl Environ Microbiol 71:3119–3125

Williamson KE, Radosevich M, Smith DW, Wommack KE (2007) Incidence of lysogeny within temperate and extreme soil environments. Environ Microbiol 9:2563–2574

Williamson KE, Schnitker JB, Radosevich M, Smith DW, Wommack KE (2008) Cultivation-based assessment of lysogeny among soil bacteria. Microb Ecol 56:437–447

Wong TP, Byappanahalli M, Yoneyama B, Ray C (2008) An evaluation of the mobility of pathogen indicators, *Escherichia coli* and bacteriophage MS-2, in a highly weathered tropical soil under unsaturated conditions. J Water Health 6:131–140

Yin J, McCaskill JS (1992) Replication of viruses in a growing plaque: a reaction-diffusion model. Biophys J 61:1540–1549

Zeph LR, Casida LJ (1986) Gram-negative versus gram-positive (actinomycete) nonobligate bacterial predators of bacteria in soil. Appl Environ Microbiol 52:819–823

Chapter 3
Soil Bacteria and Bacteriophages

Robert Armon

3.1 Soil Bacteria Types, Characteristics, Prevalence, Genetic Diversity, and Source

3.1.1 Soil (General)

The soil environment is basically made of three major types of parent materials: igneous rock (cooled molten magma), sedimentary rock (deposition and cementation of loose mud /sand materials), and metamorphic rock (igneous and sedimentary material exposed to intensive heat, pressure, or chemical action) (Sumner 2000). Surface soil is the final product of weathered parent materials by climate and biological processes. Beside organic matter, soil mineral composition is sand, silt, and clay at various ratios. The percentage of each component defines the soil texture and its performance including water and nutrients retention (Anonymous 1993). Water content is also used to define different soil layers according to saturation ratio (from top to depth): (1) surface soil (unsaturated); (2) vadose zone (no spontaneous water movement); (3) capillary fringe (nearly saturated) and (4) saturated zone containing water table (spontaneous water movement).

3.1.1.1 Soil Bacteria Characteristics, Prevalence, Genetic Diversity, and Source

In this environment, a large variety of bacteria live, multiply, and act as the biological factor on soil texture and chemistry.

R. Armon
Faculty of Civil & Environmental Engineering, Division of Environmental, Water & Agricultural Engineering, Technion-Israel Institute of Technology, Haifa 32000, Israel
e-mail: cvrrobi@tx.technion.ac.il

G. Witzany (ed.), *Biocommunication in Soil Microorganisms*, Soil Biology 23,
DOI 10.1007/978-3-642-14512-4_3, © Springer-Verlag Berlin Heidelberg 2011

The new phylogenetic tree of living organisms at kingdom level (now a new taxon called domain) is composed of three domains: Bacteria, Archaea, and Eucarya, each containing two or more kingdoms (Woese et al. 1990). The Bacteria domain contains the following phyla (based on 16S rRNA genes) (Lebedinsky et al. 2007; Ludwig W and Klenk 2001): *Proteobacteria* [*α-Proteobacteria*, *β-Proteobacteria*, *γ-Proteobacteria*, *δ-Proteobacteria*, *ε-Proteobacteria*], *Bacteroidetes*, *Chlorobi*, *Planctomycetes*, *Verrucomicrobia*, *Chlamydiae*, *Defferibacteres*, *Spirochaetes*, *Fibrobacteres*, *Acidobacteria*, *Nitrospirae*, *Cyanobacteria*, *Fusobacteria*, *Actinobacteria*, *Firmicutes*, *Cloroflexi*, *Deinococcus-Thermus*, *Dictyoglomi*, *Thermotogae*, and *Aquificae*. A supplementary detailed information on actual members of each *Proteobacteria* group is presented in Table 3.1.

The diversity of soil bacteria is clearly much higher compared with aquatic bacteria. Due to low cultivability of soil bacteria (99.5–99.9% cannot be grown on laboratory media) (Torsvik et al. 1990), only molecular methods can detect and specify the extended diversity of these microorganisms (Kent and Triplett 2002). In his introductory book on soil microbiology, Alexander (1977) had suggested nine genera members to be significant in soils: *Agrobacterium*, *Alcaligenes*, *Arthrobacter*, *Bacillus*, *Flavobacterium*, *Micromonospora*, *Nocardia*, *Pseudomonas*, and *Streptomyces*.

Briefly, soil bacteria can be divided in two groups (at genus level): Eubacteriales and Myxobacteriales. The Eubacteriales group includes the following genera: *Arthobacter*, *Azotobacter*, *Bacillus*, *Clostridium*, *Micrococcus*, *Nitrosomonas*, *Pseudomonas*, and *Rhizobium*, while the Myxobacteriales group comprises genera such as *Chondromyces* and *Cytophaga*. Among these bacteria, *Bacillus* and *Clostridium* sp. are spore formers and *Azotobacter* and *Chondromyces* cysts formers. As such, both spore and cyst formers reveal a higher survival potential under adverse environmental conditions. Another important bacterial group frequently isolated from soil environments belongs to the *Actinomycetes* genera. At genus level, we can find in this genera: *Micromonospora*, *Nocardia*, *Streptomyces*, *Streptosporangium*, and *Thermoactinomycetes*.

The main characteristics of soil bacteria but not exclusively are: growth potential at low water activity (a_w), survival under extreme conditions (temperature and pH), interactions with rhyzosphere and mineral milieu, exposure to unexpected pollution, sunlight irradiation, capability to utilize soil organic matter, and finally to compete or collaborate (symbiotic or antagonistic relationships) with other microorganisms and plants.

Concisely, the most recent common methods used for soil microbial diversity/prevalence analysis are based on molecular fingerprinting techniques of the DNA, RNA, phospholipids, and hopanoids (Pearson et al. 2007; Ogram and Sharma 2002), such as mole % G + C composition, DNA reassociation rate, phospholipid fatty acid (PLFA) analysis (Ogram and Sharma 2002), polymerase chain reaction (PCR), denaturating/temperature gradient gel electrophoresis (DGGE/TGGE) (Nakatsu 2007), amplified rDNA restriction analysis (ARDRA) (Vaneechoutte M et al. 1995), terminal restriction fragment length polymorphism (T-RFLP) (Liu et al. 1997) and ribosomal intergenic spacer analysis (RISA/ARISA) (Cardinale

Table 3.1 The phylum Proteobacteria (Bacteria domain) and its proposed classes and orders (Lebedinsky et al. 2007; Ludwig and Klenk 2001; Gupta 2005; Dworkin et al. 2007; Gupta 2000; Gupta 2006)

α-Proteobacteria	β-Proteobacteria	γ-Proteobacteria	δ–Proteobacteria	ε-Proteobacteria
Novosphingomonadales	Spaerotilacea	Pseudomonadales	Bdellovibrionales	Campylobacterales
Rhodospirillales	Alcaligenaceae	Acinetobacteriacea	Desulfobacterales	Helicobacterales
Rhizobiaceae	Burkholderiaceae	Acidithiobacillaceae	Desulfovibrionales	Nautiliales
Brucellaceae	Comamonadaceae	Aeromonadaceae	Desulfurellales	
Phyllobacteriaceae	Gallionellaceae	Alcanivoraceae	Desulfarcales	
Bradyrhizobiaceae	Hydrogenophilaceae	Alteromonadaceae	Desulfuromonadales	
Methylobacteriaceae	Methylophilaceae	Cardiobacteriaceae	Myxococcales	
Rhodobacterales	Neisseriaceae	Chromatiaceae	Syntrophobacterales	
Caulobacterales	Nitrosomonadaceae	Coxiellaceae		
Rickettsiaceae	Oxalobacteraceae	Ectothiorhodospiraceae		
Anaplasmataceae	Rhodocyclaceae	Enterobacteriaceae		
	Spirillaceae	Francisellaceae		
		Halomonadaceae		
		Legionellaceae		
		Methylococcaceae		
		Moraxellaceae		
		Oceanospirillaceae		
		Pasteurellaceae		
		Piscirickettsiaceae		
		Pseudomonadaceae		
		Succinivibrionaceae		
		Thiotrichaceae		
		Vibrionaceae		
		Xanthomonadaceae		

et al. 2004), RNA dot/slot blot hybridization, and fluorescence in situ hybridization (FISH)(Griffiths et al. 2002; Lynch et al. 2004). Using the above molecular methods, it is achievable today to understand in depth the diversity and prevalence of the domain "Bacteria" in a large variety of soil types (Wang et al. 2008, Oren 2004).

In addition, primers have been identified for specific groups of Bacteria domain, such as: the β-subdivision NH_3 oxidizer (Kowalchuk et al. 1997), *Actinomycetes* (Heuer et al. 1997), sulfate-reducing bacteria (Teske et al. 1996), methanotrophs (Jensen et al. 1998; Seghers et al. 2003; Wise et al. 1999), cyanobacteria (Boutte et al. 2006), the genera *Acinetobacter* (Vanbroekhoven et al. 2004), *Sphingomonas* (Leys et al. 2004), *Pseudomonas* (Bergsma-Vlami et al. 2005), *Burkholderia* (Salles et al. 2004), and *Bacillus* species (Garbeva et al. 2003) (Table 3.2).

For example, Tarlera et al. (2008) in a study on subtropical aeolian dunes from Altamaha field and Ohoopee River Valley (of cc. 77000 years of aeolian deposition) in southeast Georgia, USA found that (quote) "bacterial community composition changed with soil age, and the overall diversity, richness and evenness of the communities increased as the soil habitat matured." The authors applying a multivariate Bray–Curtis ordination technique analysis of ribotype distribution showed an orderly pattern of bacterial community development that was clearly associated with soil and ecosystem development. Across all soils, *Acidobacteria* sequences were most abundant (54.5%), followed by *Proteobacteria* (19.2%), *Planctomycetes* (13.1%), and *Chloroflexi* (8.2%). Among the *Proteobacteria*, the α-proteobacteria were the most abundant (10%) followed by β-proteobacteria (3.8%), γ-proteobacteria (3.8%), and the δ-proteobacteria (1.7%). *Acidobacteria* kingdom abundance in soil environment was also shown by Barns et al. (1999). Additional studies around the world on various soils and their bacterial content are presented in Table 3.3. In general, *Proteobacteria* were present in almost all soils tested with high variability (Wang et al. 2008; Lehman 2007).

Generally speaking there are six soil types according to their formation trail: alluvial, marine, lacustrine, aeolian, glacial, and colluvial. Another way to look upon soil bacterial variety is to categorize their diversity linked to soil formation types (Table 3.4). Water is one of the main factors that augment soil with new bacteria as well their survival in hostile environments.

The diversity of soil bacterial communities is influenced by several factors listed bellow:

1. Soil type and mineral composition
2. Water content
3. Ionic strength, pH, and redox potential
4. Clay and nutrients content
5. Physical factors as: temperature, sunlight, wind, rain, and agricultural activity
6. Plant type and its rhisosphere
7. Livestock grazing
8. Presence of bacteriophages

Table 3.2 Primers identified for detection of specific Bacteria domain found in soil

Phyla	Primers	Ref.	Genera	Primers	Ref.
β-proteobacteria (NH_3 oxidizers)	CTO189fA-GC, CTO189fBGC, and CTO189fC-GC (f) CTO654r (r)	12	*Acinetobacter*	Ac436(f) Ac676(r)	Vanbroekhoven et al. (2004)
Actinomycetes	F243, F984GC, F27 (f), R513GC, R1378 (r)	13	*Sphingomonas*	Sphingo108(f), Sphingo420 (r), 40GC clamp	Leys et al. (2004)
Sulfate-reducing bacteria (SRB)	GM5F-GC-clamp, Probe 804, Probe 687, Probe 385 (f) DS907 (r)	14	*Pseudomonas*	DGGE292(f), DGGE618(r) and 6DGGE618 (r), 40-bp GC clamp	Bergsma-Vlami et al. (2005)
Methanotrophs	*E. coli* positions 338-358-GC clamp, *E. coli* position 517-534	15	*Burkholderia*	Burk3, universal eubacterial primer R1378	Salles et al. (2004)
	P197(f) and P1378 (r), P142(f) and P1378(r), P338 (f) and P518(r)	16			
	MethT1b(r), MethT1c(r), MethT1d(f), MethT2(r)	17			
Cyanobacteria	CYA359(f), CYA359(f)GC, CYA784(r), CYA781(r), CYA781(r) RGC, CYA781 (r), CYA781 (rGC, 23S30(r)	18	*Bacillus* sp.	Bac(f), (r)1378 universal bacterial reverse primer	Garbeva et al. (2003)

r reverse transcription; *f* forward transcription

To examine genetic polymorphisms of functional genes within soil communities, nitrogenase reductase (Diallo et al. 2004; Rosado et al. 1998), denitrifying genes (Throback et al. 2004), ammonia monooxygenase (Avrahami et al. 2002; Avrahami and Conrad 2003; Hornek et al. 2006), methane monooxygenase (Henckel et al. 1999, 2000; Hoffmann et al. 2002), and hydrogenase genes (Wawer and Muyzer 1995) had been targeted (Table 3.5). Before selecting primers for any study, their specificities and limitations should be understood (Baker and Cowan, 2004; Baker et al., 2003).

Table 3.3 Examples of soil microbial diversity from different selected studies

Soil type	Main soil bacteria	Detection method	Ref.
Seminatural grassland (Andosol) in Shiriyazaki, northern Honshu, Japan	Not specified	DNA extraction and DGGE[a] analysis	Sugiyama et al. (2008)
Queensland soil (Australia)	*Proteobacteria*	16S rDNA[b] clone libraries using template DNA	Stackebrandt et al. (1993)
Arlington sandy loam (from the top 15 cm of a fallow field) University of California Riverside Agricultural Experiment Station (USA)	*Pseudomonas reactans*, *Pseudomonas putida*, *Holiothrix oregonensis*, *Acidobacterium capsulatum*, *Bacillus subtilis*, *Sphingomonas* sp. strain CD, *Legionella sainthelensi*, α-subclass *Proteobacterium*	(Biolog)[c], PLFA[d], DGGE of 16S rDNA fragments amplified directly from soil DNA	Ibekwe et al. (2001)
Wisconsin pasture soil (USA)	*Proteobacteria*, *Fibrobacter* and low G + C gram-positive bacteria	rDNA library construction, PCR[e], pGEM-T vector	Borneman et al. (1996)
Siberian tundra soil (Russia)	Proteobacteria and Fibrobacter	Analysis of 16S rDNA clones	Zhou et al. (1997)
Long-term sewage sludge application on soil with heavy metals contamination (Germany)	α-*Proteobacteria*, *Proteobacteria*: the *Cytophaga–Flavobacterium* division, Gram-positive high G + C and low G + C	Dot blot hybridization	Sandaa et al. (2001)
Chihuahuan Desert region of West Texas (USA)	*Actinomycetes* and Bacteria (undefined)	PCR, DGGE and FAME[f]	Clark et al. (2009)
Soil drill from lower reach of Heihe river basin an arid environment (northwest) (China)	a-*Proteobacteria*, b-*Proteobacteria*, g-*Proteobacteria*, Gram + high G + C, Gram − low G + C, and *Cytophaga–Flexibacter–Bacteroides* group	16S rDNA amplification, sequencing and molecular identification	Su et al. (2004)
Alpine and arctic soils, Niwot Ridge LTER site (located in the Front Range of the Colorado Rocky Mountains), Canadian and Siberian tundra soil (USA, Canada, Russia)	*Acidobacteria*, *Verrucomicrobia*, γ-*Proteobacteria*, δ-*Proteobacteria*, *Chloroflexi*, *Bacteroidetes*	[SSU[j]-rRNA) amplified 16S rRNA gene clone library sequencing, fluorescence in situ hybridization (FISH)[k]	Nemergut et al. (2005)
Aeolian deposition, subtropical Altamaha and Ohoopee River Valley dunes of southeast Georgia (USA)	*Planctomycetes*, *Firmicutes*, α-*Proteobacteria*, *b-Proteobacteria*, γ-*Proteobacteria*, *Acidobacteria*,	16S rRNA gene clone libraries and FAME	Tarlera et al. (2008)

	Chloroflexi, *Thermomicrobia*, *Gemmatimonadetes*,*Verrumicrobia*, *Nitrospira*, *Deltaproteobacteria*, *Bacteroidetes*, *Actinobacteria*		
Soils along a climatic transect in the Judean Desert (Israel)	G+and G− bacteria (unidentified)	PLFA and FAME	Steinberger et al. (1999)
Top soil, 10 cm of the surface (Brazil, Canada and USA-Florida and Illinois)	*β-Proteobacteria*, *γ-Proteobacteria*, *Bacteroidetes*, *Acidobacteria*, *Actinobacteria*, *Firmicutes*, *Nitrospira*, *gemmatimonadetes*, *Verrucomicrobia*	DNA extraction, PCR and pyrosequencing[g]	Roesch et al. (2007)
Alaska and Minnesota soil (USA)	*Proteobacteria*, *Acidobacteria*, *Bacteroidetes*, *Actinobacteria*, *Gemmimomonas*, *Planctomyces*, *Verrucomicrobia*, *Nitrospirae*, *Firmicutes*, WCHB1, OP10, ACE, Bd[i], *Thermomicrobia*, *Coprothermobacter*, *Chloroflexi*, *Fibrobacter*, *Chlorobi*	16S rRNA genes amplified in a single reaction by PCR	Schloss and Handelsman (2006)
Topsoil samples (to a depth of 0 to 20 cm) from the Ultuna long-term field experiment with different fertilizers treatment (Sweden)	*a-Proteobacteria*, *Holophaga*/*Acidobacterium*	T-RFLP[h] analysis, rDNA clone libraries and PCR	Sessitsch et al. (2001)
Pristine and petroleum hydrocarbon-contaminated Alpine soils from Tyrol (Austria)	*Actinobacteria*, *α-Proteobacteria*, *β-Proteobacteria*, *γ-Proteobacteria*	PCR-amplified 16S rRNA gene fragments, DGGE	Labbe et al. (2007)

[a]Denaturing Gradient Gel Electrophoresis
[b]Ribosomal RNA
[c]Biolog™
[d]Phospholipid fatty acid analysis
[e]Polymerase chain reaction
[f]Fatty acid methyl ester analysis
[g]Method of DNA sequencing based on the sequencing by synthesis principle
[h]Terminal Restriction Fragment Length Polymorphism
[i]Candidate phyla
[j]Small subunits
[k]Fluorescence in situ hybridization

Table 3.4 Various soil types according to their geophysical formation and intrinsic bacterial diversity

Soil	Process	Bacteria	Ref.
Alluvial	Streams with decreasing velocity toward the lowland and formation of delta	*Spirilla, Myxobacteria, Cytophages-Flavobacteria, Azotobacters, Azospirilla, Bacilli, Streptomycetes, Arthrobacters,* Yellow coryneform bacteria, *Rhodococci*	Golovchenko et al. (2001)
Marine	Materials carried into the seas by streams and beach tidal corrosion	*Desulfotalea* sp., *Desulfuromonas palmitatis, Myxobacteria spp.* and *Bdellovibrio* spp., α-, β-, γ-, δ-*Proteobacteria, Cytophaga-Flavobacterium* group, *Marinobacter, Methylophaga, Coxiella, Desulfobacula, Desulfosarcina, Holophaga-Acidobacterium, Bacteroidetes, Chloroflexi, Nitrospira, Deferribacteres, Firmicutes, Acidobacteria, Actinobacteria*	Ravenschlag et al. (2006); Lanbais et al. (2008)
Lacustrine	Fresh water lake deposits originating from streams	α-, β-, δ-, and γ-*Proteobacteriaceae* and *Firmicutes*	Nelson et al. (2007)
Aeolian	Sand grains deposits carried by winds to form dunes and fine grains to form Loess	*Planctomycetes, Firmicutes,* α-*Proteobacteria,* β-*Proteobacteria,* γ-*Proteobacteria, Acidobacteria, Chloroflexi, Thermomicrobia, Gemmatimonadetes, Verrumicrobia, Nitrospira, Deltaproteobacteria, Bacteroidetes, Actinobacteria*	Tarlera et al. (2008)
Glacial	Drift formed from glaciers melting process	*Flavobacterium, Cyanobacteria, Actinobacteria/Myxococcales, Actinomycetales, Acidiphilium, Janthinobacterium, Alcaligenes, Holophaga, Caulobacteriales/ Bradyrhizobium, Janthinobacterium/Aquaspirillum, Nitrosococcus/Methylocystaceae,* β-*Proteobacteria, Planctomycetes Acidobacteria/Holophaga, Flexibacteraceae, Oxalobacteraceae Aquicella, Bradyrhizobium, Acidiphilium, Pedobacter Polyangium, Mesorhizobium/Rhodanobacter, Holophaga, Caulobacteriales, Caulobacter, Flavobacterium, Nitrosomonas/Comamonadaceae, Methylocystis*	Noll and Wellinger (2008)
Colluvial	Deposits of soils originating from gravitational accumulation at the steep slopes	*Actinobacterium,* α-*Proteobacteria, Bacillus*(bacterium genus), *Rhizobiaceae, Rhodococcus, Streptomyces*	Reith and Rogers (2008)

Table 3.5 Genetic polymorphism in functional genes within soil communities

Enzyme	Primer	Ref.
Nitrogenase reductase	FGPH19 (f), PolR (r), AQER (r), PolF, GC-clamp(f), NHA1 and NHA2, degenerate primers (f, r), GC-clamp	Diallo et al. (2004) Rosado et al. (1998)
Denitrifying genes [nitrate reductases, nitrite reductases, nitrous oxide reductase]	nirS1F, F1acd, F1bcd, Heme 832F, Fldcd, cd3aF, cd3bF, nirS3F, nirS3Fa, F3nirS, nirS4F, nirS4R, R3cd, R4cd, R4bcd, nirS6R, Cunir3, nirK1F, F1aCu, nirK3R, R3Cu, nirK5R, Nos661F, nosZ661b, nosLb, Nos1527F, nosZ-F, Nos1527R, Nos1773R, nosZ1773b, nosRb, nosZ1622R, nosZ-R	Throback et al. (2004)
Ammonia monooxygenase	amoA-1F, amoA-2R-GG, GC clamp; *amoA*, GC clamp amoA-1F-GC clamp, amoAf-i (inosine), amoA-2R, amoAr New2.0, amoAr-i(inosine)	Avrahami et al. (2002) Avrahami and Conrad (2003), Hornek et al. (2006)
Hydrogenase	Hyd2F, Hyd1F, Hyd4F,Hyd5R, Hyd7R, GC clamp	Wawer and Muyzer (1995)
Methane monooxygenase	Universal (537f and 907r), MB10γ (197f and 533r), MB9α (142f and 533r), pmoA (a189 and A682), mxaF (f1003 and r1561), mmoB (f77 and 369r) SSU rRNAbased primer set, pmoA *pmoA* primers (A189 with GC clamp and A682)	Henckel et al. (1999) Henckel et al. (2000) Hoffman et al. (2002)

3.1.2 *Soil Bacteria (Pathogens, Phytopathogens) and Nonpathogenic*

Human bacterial pathogens reach soil through water, wastewater, manure, fertilizers, and refuse application. In general, soil-transmitted diseases include those caused by dormant forms (spores) of infectious agents owing their long survival capability, though freshly contaminated soil by sewage or manure can harbor also nonspore forming bacteria known as human pathogens. The main infection route is by direct contact with broken skin but also gastrointestinal track may be involved through direct intake of soil particles (i.e., food and water). At the moment, some of the bacterial pathogens are definitely known to be transmitted through contaminated soil, but others are only suspected as such and further scientific research is needed.

3.1.2.1 Human Pathogens

Human bacterial pathogens are defined as bacteria that cause infection and clinical manifestation in human beings. The major bacterial pathogens are listed below;

however, some other bacteria not listed can cause human diseases under certain condition therefore called opportunistic pathogens (Szponar and Mordarska 1997).

Salmonella: *Salmonella* spp. can be divided into two fairly distinct groups: the typhoidal species/serovars (*Salmonella typhi* and *S. paratyphi*) and the remaining nontyphoidal species/serovars. The nontyphoidal were grouped into more than 2,000 species (serotypes) according to their somatic (O) and flagellar (H) antigens (Kauffmann-White classification). **Shigella**: [*S. dysenteriae*, *S. flexneri*, *S. boydii* and *S. sonnei*]; **Campylobacter**: [*C. jejuni*, *C. coli and C. laridis*]; **Yersinia**: *Y. enterocolitica*; **Legionella**: *Legionella* spp.; **Burkholderia:** *B. pseudomallei*; **Acinetobacter**: [*A. calcoaceticus baumannii* complex, *A. israelii*, *A. naeslundii*, *Arachnia propionica*, *A. eriksonii*, *A. viscosus*, *Rhodococcus*, *Nocardia*, *Streptomyces*]; **Aeromonas** spp., **Bacillus**: [*B. anthracis*, *B. cereus*, *B. thuringiensis*, *B. subtilis*, *B. licheniformis*, *B. pumilus*, *B. megaterium*, *B. sphaericus*, *B. clausii*, *B. halodurans*]; **E. coli**: [enteropathogenic *E. coli* identified on the basis of different virulence factors: enterohaemorrhagic *E. coli* (EHEC), enterotoxigenic *E. coli* (ETEC), enteropathogenic *E. coli* (EPEC), enteroinvasive *E. coli* (EIEC), enteroaggregative *E. coli* (EAEC) and diffusely adherent *E. coli* (DAEC)]; ***Helicobacter pylori***; ***Klebsiella***: [*K. pneumoniae* and *K. oxytoca*]; ***Mycobacterium***: *M. avium* complex has been used to describe a group of pathogenic species including *M. avium* and *M. intracellulare and other atypical Mycobacterium;* ***Pseudomonas***: *Pseudomonas aeruginosa*, ***Staphylococcus aureus;*** ***Vibrio***: [*V. cholerae*, *V. parahaemolyticus* and *V. vulnificus*]; ***Nocardiaceae***: [*Tsukamurella genus*]; ***Spirochaetes***: [*Leptospira*].

Some human pathogens are called emergent or re-emergent. According to WHO recent publications, the emergency or re-emergency can be attributed to the following intercrossing factors (Epstein 1995; Bartlett 1996; Murray et al. 1995): (1) new environments (climate shifts/deforestation), water resources development projects (dams and irrigation), water-cooled air conditioning plants, industrial and agricultural practices shifts, e.g. intensive livestock rearing, piped water systems with inadequate design and increased number of humanitarian emergencies, (2) new technologies and scientific advances (water resources development projects; excessive use of antibiotics/antiparasitic drugs/public health insecticides, new industrial and agricultural practices, detection and analysis improved methodology at molecular scale, and inappropriate usage of new generation of insecticides/herbicides); and (3) changes in human behavior and vulnerability (worldwide human travel and rapid transportation, demographic changes, rise in high risk populations, and accidental/deliberate release of pathogens to water). Agricultural and wastewater management practices related to human behavior have a continuous impact on soil presence of these pathogens (Quintana et al. 2008).

3.1.2.2 Phytopathogens

Among a large variety of microorganisms that are plant pathogens (fungi, bacteria, viruses, oomycetes, viroids, and protozoa), bacterial phytopathogens encompass ~100 species that is relatively a small number. Globally, these phytopathogens are prevalent mainly in the subtropical and tropical regions. Bacterial phytopathogenicity is expressed by cell wall degrading enzymes, host- and nonhost-specific toxins, phytohormones level alteration and exopolysaccharides (EPS) that block xylem vessels. All these pathogenicity factors are under control of phytopathogens quorum sensing system.

The main bacterial phytopathogens known to us are: ***Agrobacterium*** (*tumefaciens, vitis, radiobacter*), ***Burkholderia*** *cepacia*, ***Ralstonia*** (*solanacearum* pvs. and *metalidurans*), ***Erwinia*** (*amylovora, carotovora atroseptica, carotovora, chrysanthemi*), ***Pectobacterium*** *athrosepticum*, ***Pseudomonas*** *syringae* (pv. *syringae* pv. *tomato*, pv. *phaseolicola*), ***Xanthomonas*** *axonopodis* (*citri, aurantfolli*), *Xanthomonas campestris* (*campestris* pvs.), *Xanthomonas oryzae* (*oryzae, oryzicole*), ***Xylella*** *fastidiosa* (pvs.), ***Clavibacter*** *michiganensis* (*michiganensis, sepedonicus*), ***Leifsoni****a xyli* (*xyli*), ***Spiroplasma*** *kunkelii* and ***Phytoplasma*** groups (previously known as "*Mycoplasma*-like organisms" or MLO-s), and **Coryneform** (*Clavibacter, Curtobacterium, Arthrobacter, Rhodococcus,* and *Rathayibacter*).

3.1.2.3 Nonpathogenic Bacteria

Nonpathogens are divided among a broad numbers of ecosystems: animals, plants, water, and soil. The definition of nonpathogens is somehow tricky, as opportunism is one of the main characteristics of bacteria. As such, the so-called nonpathogenic bacteria can turn to be pathogenic upon host and environmental changes. In the strict sense, animals and humans harbor a large variety of bacteria that do not cause infection and diseases and are commonly called commensals. Some are external, colonizing the skin, while others are internal colonizing mainly the gut. Among these bacteria it can be found: the coliforms group (nonpathogenic *E. coli* sp., *Klebsiella* sp., *Enterobacter* sp., *Citrobacter* sp.), typical representatives of the genera *Bacteroides, Bifidobacterium, Lactobacillus, Streptococci, Clostridium, Fusobacterium, Eubacterium, Peptococcus, Peptostreptococcus,* and *Veillonella*. Other nonpathogens are more related to different biogeochemical cycles such as nitrogen [fixation (genera *Azotobacter, Beijerinckia, Azospirillum, Clostridium*), ammonium assimilation-ammonification, nitrification (*Nitrosomonas, Nitrosococcus, Nitrospira, Nitrosolobus, Nitrosovibrio, Nitrobacter, Nitrospina, Nitrococcus*) and denitrification (*Alcaligenes, Agrobacterium, Aquaspirillum, Azospirillum, Bacillus, Blastobacter, Flavobacter, Halobacterium, Rhizobium, Alcaligenes, Bradyrhizobium*)], sulfur [sulfate reduction, sulfur oxidation/reduction (*Thiobacillus, Desulfuromonas, Desulfovibrio, Thiomicrospira, Achromatium, Beggiatoa, Thermotrix, Chlorobium, Chromatium, Ectothiorhodospira, Thiophedia, Rhodopseudomonas, Desulfotomaculum, Thermodesulfovibrio*], and phosphorous [*Pseudomonas* sp., *Rhodopseudomonas, Trichodesmium*] Some of the above bacteria can cross biogeochemical cycles by being involved in one or more cycles pending environmental conditions (Maier *et al.* 2009; Lambais *et al.* 2008). In plants, we find an entire group of rhizophytes (*Pseudomonas*, *Rhizobium leguminosarum* bv. *viciae* and *Sinorhizobium meliloti*, *Azospirillum*, *Bacillus* sp. *Paenibacillus*), epiphytes (*Pseudomonas*, *Stenotrophomonas*, *Bacillus*, *Xanthomonas*, and *Arthrobacter*), and endophytes (*Gluconacetobacter*, *Azospirillum* sp., *Herbaspirillum*, and *Azoarcus*) commensals that are known to benefit plants.

3.1.3 *Bacteriophages (Abbr. Phages) (Systematics, Life Cycle, and Genetics)*

3.1.3.1 Phages Systematic

In the light of the tremendous advancement in the last decades in molecular genomics, bacteriophages' systematic is in turmoil. The fundamental taxonomy has been based on general morphology as detected by electron microscopy and additional parameters such as: DNA/RNA configuration (single or double stranded, linear, superhelical, circular, or segmented). In the last decade, comparative genomics introduced a new way to look on bacteriophages' systematic (Brüssow et al. 2004; Brüssow and Hendrix 2003; Choi et al. 2009). Comparative genomic studies of bacteriophages reveal some of the comprehensive mechanisms by which these prokaryotic viruses evolved and shaped their bacterial and archaeal hosts evolution. The widespread horizontal exchange of sequences among genomes, mediated by both homologous and nonhomologous recombination, is one of the main pathways. Phages from similar ecological niches revealing high frequency exchange of genome sequences brought to a high degree of mosaic diversity in local populations. Horizontal exchange at lower frequency was observed across the entire span of phage sequence domain (Hendrix 2003).

Here is the place to mention, that in spite of our expanding knowledge, we are only at the beginning of advanced comparative genomics able to change the whole picture. At present, we encompass only ~500 complete phage genomes. This fraction is extremely small considering that in many environments phages outnumber bacteria by a factor of 10 reaching numerically 10^{31} tailed phage particles that is the largest share of biological material on earth (Wommack and Colwell 2000). For example, 10^7 bacteriophage particles/ml were found in ocean water and sediment and taking in consideration the vast number of bacterial hosts present in these niches, it has been estimated that 10^{25} infections/second occur simultaneously around the globe (Brüssow et al. 2004; Wommack and Colwell 2000; Breitbart et al. 2002) through a time span of 3 billion years. For that reason, the actual bacteriophages classification shown in Table 3.6, based on the "International Committee on Taxonomy of Viruses" (ICTV) nomenclature, is still accepted among virologists (Brüssow et al. 2004; van Regenmortel MHV et al. 2000; Ackermann 2007, 2009; Hendrix 2003).

Table 3.6 Bacteriophages classification according to ICTV (van Regenmortel et al. 2000) and adapted from (Ackermann 2003, 2007, 2009)

Family	Nucleic acid composition	Morphology (capsid and tail)	Shape	Schematic representation
Myoviridae	Linear dsDNA[a]	Nonenveloped, contractile tail	Tailed	
Siphoviridae	Linear dsDNA	Nonenveloped, long noncontractile tail	Tailed	
Podoviridae	Linear dsDNA	Nonenveloped, short noncontractile tail	Tailed	
Tectiviridae	Linear dsDNA	Nonenveloped, isometric	Polyhedral	
Corticoviridae	Circular superhelical dsDNA	Nonenveloped, isometric	Polyhedral	

(*continued*)

Table 3.6 (continued)

Family	Nucleic acid composition	Morphology (capsid and tail)	Shape	Schematic representation
Lipothrixviridae	Linear dsDNA	Enveloped, rod-shaped	Filamentous	
Plasmaviridae	Circular dsDNA	Enveloped, pleomorphic	Pleomorphic	
Rudiviridae	Linear dsDNA	Nonenveloped, rod-shaped	Filamentous	
Fuselloviridae	Circular dsDNA	Nonenveloped, lemon-shaped	Pleomorphic	
Inoviridae	Circular ssDNA	Nonenveloped, filamentous	Filamentous	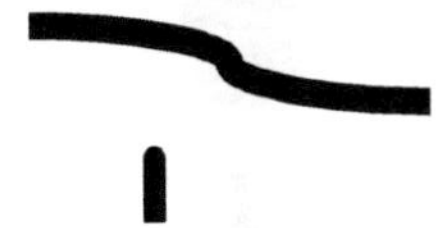
Microviridae	Circular ssDNA[b]	Nonenveloped, isometric	Polyhedral	
Leviviridae	Linear ssRNA[c]	Nonenveloped, isometric	Polyhedral	
Cystoviridae	Segmented dsRNA[d]	Enveloped, spherical	Polyhedral	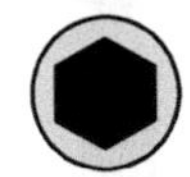
SH1 group[e]	Linear dsDNA	Nonenveloped, isometric, inner lipid vesicle	Polyhedral	
STV1 group[e]	Circular dsDNA	Nonenveloped, isometric, turret-shaped protrusions	Polyhedral	

Salterprovirus	Linear superhelical dsDNA	Nonenveloped, lemon-shaped	Pleomorphic	
Guttaviridae	Circular superhelical dsDNA	Nonenveloped, droplet-shaped	Pleomorphic	
Ampullaviridae[e]	Linear dsDNA	Nonenveloped, bottle-shaped	Pleomorphic	
Bicaudaviridae[e]	Circular dsDNA	Nonenveloped, two-tailed	Pleomorphic	
Globuloviridae[e]	Linear dsDNA	Spherical, lipid-containing envelope and a helical	Pleomorphic	

[a]dsDNA: double stranded DNA
[b]ssDNA: single stranded DNA
[c]ssRNA: single stranded RNA
[d]dsRNA: double stranded RNA
[e]Not yet classified

3.1.3.2 Phages Life Cycles

The life cycle of bacteriophages is related to their parasitic characteristic and prokaryotic host, specifically bacteria and archaea domains. Bacteriophages can be categorized into two types accordingly: virulent and temperate. The virulent bacteriophages are also called lytic due to their potential to lyse the host cell following infection and release of new progeny phages that in turn infect new uninfected adjacent host cells. The temperate bacteriophages can undergo a lytic cycle under some circumstances; however, more frequently they integrate within the bacterial genome (or as extrachromosomal plasmid), and the inserted form (called prophage) is simultaneously replicated along with bacterial chromosome, a process called lysogeny (Bertani 2004). Under these circumstances, the host cell containing a prophage is called lysogenic. Lysogenic hosts can multiply and transfer the prophage through many generations including immunity to superinfection. Spontaneously or by exposure to UV irradiation (Liu et al. 2005), mutagenic compounds (i.e., mitomycin C) (Raya and Hebert 2009), temperatures >19°C (Cochran and Paul 1998), and response to higher trophic conditions (Bongiorni et al. 2005; Williamson et al. 2005), prophages are released from their former state as progeny phages able to infect and cause lysis in any nonlysogenic host cells present in the same culture. Schematic lytic and lysogenic cycles are presented in Fig. 3.1.

The vast majority of studies on lytic and lysogenic bacteriophages were performed in aquatic ecosystems (fresh, marine, brackish water, etc.) (Wommack and Colwell 2000; Breitbart et al. 2002); Choi et al. 2009; Bongiorni et al. 2005; Cochran and Paul 1998) and only few in terrestrial habitats (Dhritiman et al. 2008; Danovaro et al. 2001; Danovaro et al. 2002; Williamson et al. 2008). Initial studies indicate that soil viral communities are more abundant and diverse than their aquatic counterparts (Williamson et al. 2005). Soil moisture and organic matter richness are significantly correlated with viral abundance but not to the viral to bacteria ratio (VBR) (Williamson et al. 2005; Williamson et al. 2008).

In a recent study, performed on Delaware six soil types, forest soil revealed higher abundance of phages [the authors used the expression of virus like particles-VLP based on epifluorescence microscopy (EFM) and transmission electron microscopy (TEM) analysis] compared with agricultural soil (Williamson et al. 2005). In this particular ecosystem, the measurement of the overall free bacteriophages revealed that the moist and organic matter rich forested soils had high viral abundance and a VBR value of 10, while the dry and organic matter poor agricultural soils showed lower viral abundance and staggeringly high VBR values of ~3,000. Past studies showed that marine waters have a typical VBR value of 10 (Wommack and Colwell 2000), while its value in soil and marine sediments is 1 (Danovaro et al. 2001, 2002). Viral abundance vs. VBR is somehow confusing because it is dependent on successful extraction of both phages and bacteria from soil particles (Williamson et al. 2005). The most important findings of this study were: (a) soil moisture and organic content increase soil bacteriophages abundance and (b) tillage decrease bacterial abundance therefore impacting bacteriophages abundance and diversity.

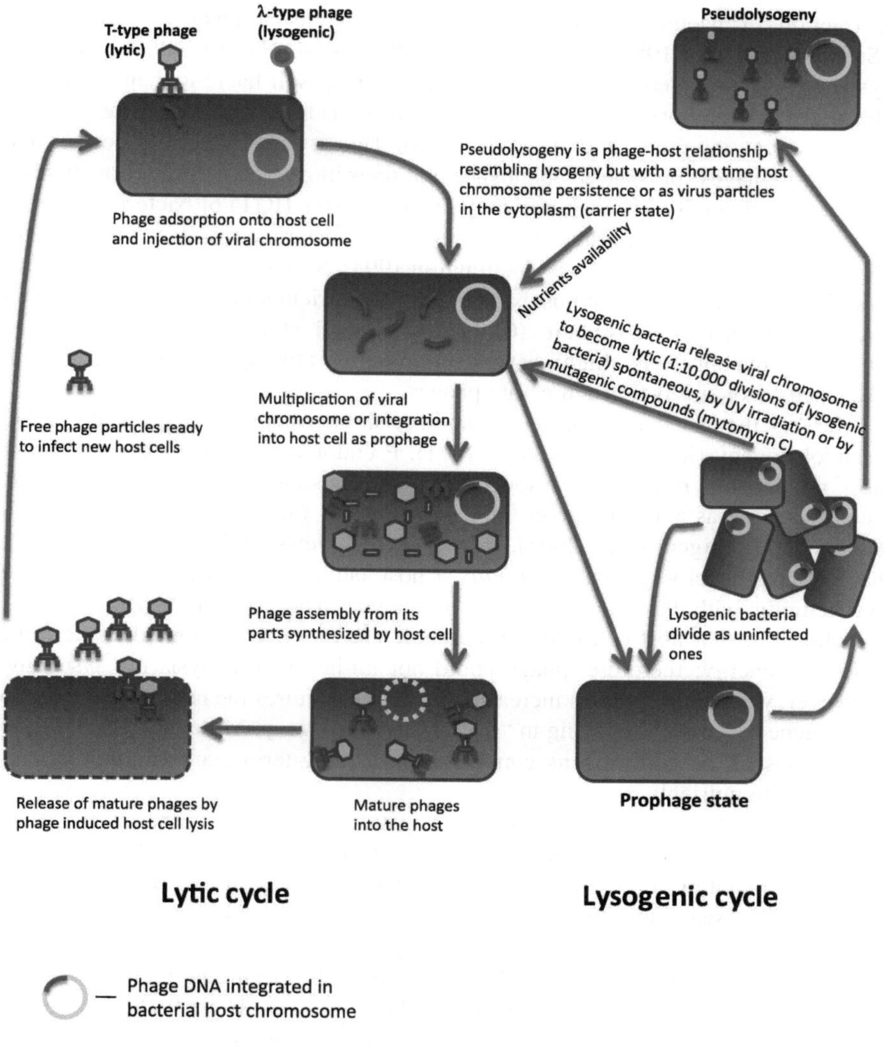

Fig. 3.1 Bacteriophage life cycle

Another recent study suggests that lysogeny is a prevalent reproductive strategy among soil bacteriophages and that the potential for horizontal gene transfer via transduction is most significant among soil microbial communities (Dhritiman et al. 2008). In an elegant experiment, employing atrazine-impregnated Bio-Sep beads as cell immobilization matrix, the authors sampled active microbiota from soils with prior pesticide exposure history. Bead grown communities were further induced with mitomycin C to evaluate the incidence of inducible prophage of soil bacteria. The inducible calculated fraction within bead communities was high (ca. 85%) relative to other studies in aquatic and sedimentary environments that showed a fraction of bacterioplankoton-containing inducible prophages of 0.07–42% (Stewart and Levin 1984; Weinbauer and Suttle 1996). These results suggest that lysogeny is a prevalent reproductive strategy among soil bacteriophages and that the potential for horizontal gene transfer via transduction is significant in soil microbial communities (Nielsen et al. 1998). This process is very important due to bacteriophages' biospheric abundance and their impact on activity (host fitness) and genetic diversity (through horizontal gene transfer-HGT) of bacterial communities. Older studies suggested that soil environment prone to select for lysogeny (Jiang and Paul 1996; Marsh and Wellington 1994). Soil as a heterogeneous matrix with irregular distribution of host bacteria, due to nutrient and water constraints and presence/absence of rizosphere (Chibani-Chennoufi et al. 2004), aims to host abundant diversity with highly even communities although at variable growth rates that favor survival of temperate phages.

Here is the place to mention an additional intermediary phage state called pseudolysogeny (Miller 2001; Baess 1971). Pseudolysogeny describes a phage–-host relationship resembling lysogeny but with a short time host chromosome persistence or as virus particles in the cytoplasm (also called "carrier state") (Baess 1971; Prigent et al. 2005). Romig and Brodetsky (1961), while isolating bacteriophages active against *B. subtilis* as host had not been able to obtain a stable lysogenicity with the so-called "only temperate phage isolate." Miller (2001) hypothesized that host cell under remarkable starvation conditions does not owe sufficient energy; therefore, phage could not initiate neither lysis nor lysogeny. However, when nutrient level increases, the phage acquires the necessary energy to allow gene expression resulting in "real" lysis or lysogeny. Miller and Ripp (1998) observed such pseudolysogens conversion with long-term starved cultures augmented with nutrients.

The pseudolysogenity is an interesting feature of phages but the exact role in ecology and evolution is still unclear. Starvation, desiccation, elevated temperature, and UV irradiation are extreme conditions found mainly in desert soils. Prigent et al. (2005) searching for extracellar (lytic) and intracellular (lysogenic and pseudolysogenic) bacteriophages possibly present in Saharan desert sand soil particles, found a large variety of phage-like particles resembling morphologically groups such as *Myoviridae*, *Siphoviridae*, and *Podoviridae*. These findings give us a clue about the role of pseudolysogenity in defending bacteriophages against harsh environmental conditions.

3.1.3.3 Phages Genetic

Finally, perhaps the most important genetic role of bacteriophages beside their own replication is the process called "transduction." Transduction is a process of bacterial DNA transfer from one bacterium to another through bacteriophage infection (lytic and lysogenic) due to low fidelity in the process of phage packaging (see lysogenic integration in Fig. 3.2). Viruses with RNA genomes are not able to wrap up DNA therefore transduction does not occur. There are generally two kinds

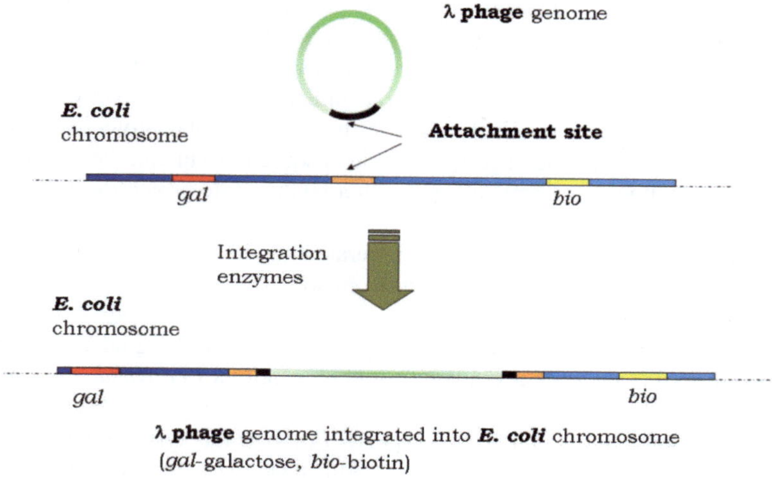

Fig. 3.2 Phage λ integration in *E. coli* chromosome as an example of lysogenic cycle

of transduction: general and specialized. The general transduction is a process of recombination and headful packaging. Lytic phages entering a bacterium will take control over host's machinery to replicate its own DNA. Inadvertently, parts of host DNA can be packed together with phage DNA due to head spare capacity, leading to generalized transduction. The specialized transduction is more related to lysogenic phages that during the process of exit as a lytic phages (i.e., after UV irradiation), the prophage will displace itself from the host DNA by seizing small fragments of bacterial chromosome.

Beside natural recombination, transformation, and conjugation, transduction is one of the most important processes in genetic material exchange among bacteria in all ecosystems. Horizontal gene transfer as a natural genetic tool is further dealt with, in other contexts.

3.1.4 Interaction Between Phages and Soil-Bacteria

As already mentioned earlier the soil is an excellent matrix for phage–bacteria interaction. Soil water content is a crucial factor that brings free phages in contact with resident bacteria and minerals (Chu et al. 2003). From the metabolic point of view, soil bacteria need water for growth and survival. Soil water is measured by two components: moisture content and moisture holding capacity. Both components depend on soil type and are greatly influenced by climatic conditions (Williamson et al. 2005).

Beside these factors, it should be noted that rizosphere contributes also to water content and soil consolidation. Under these circumstances, bacteria and phages interact. Considering phages as obligatory parasites that have to multiply and transfer their genome to future generations, survival under such circumstances is critical. As pointed earlier in this chapter, phages can be transferred through lytic and lysogenic pathways following bacterial infection. The lytic pathway is more promiscuous as beside a relatively short term of intracellular protection, the new progeny phages are excluded again to the harsh soil environment. On the contrary, the lysogenic pathway warrants phages with a relative safe protection as long they reside as prophages.

Ashelford et al. (1999) isolated six temperate and virulent phages active against *Serratia liquefaciens strain* CP6 as host, from soil surrounding sugar beets phytosphere. Among these six isolated phages, the most abundant were of two types: ΦCP6-1 a temperate, morphologically a *Siphoviridae* (with long latent period and big burst size) and ΦCP6-4 a virulent, morphologically a *Podoviridae* (with short latent period and small burst size). After 6 months, the authors observed a shift from the predominant *Siphoviridae* type to *Podoviridae* type as the dominant phage population. It seems that one or another phage population assertiveness is a result of rhizosphere different niches bound to availability and physiological status of host bacteria and plants as a function of year's season as further demonstrated in a subsequent study, using assigned soil plots with seeded sugar beet plants. Ashelford et al. (1999, 2000) hypothesis that short latent period + small burst size is in bacteriophage advantage when a physiological suitable host is abundant, while long latent period + large burst size imparts advantage when the host is scarce. This in situ scenario was consistent with the optimal foraging theory applied to phage ecology as already described by Wang (2006) and Abedon (1989). Both authors, using theoretical models, concluded that when the number of physiologically suitable host bacteria is high, their specific phages with short latent period and small burst size (like the above *Podoviridae*) would outcompete phages with long latent period and big burst size (like the above *Siphoviridae*). Perhaps, the most important finding by Ashelford et al. (1999, 2000) is the repeated temporal dynamics of specific phage populations in soil over successive years.

Burroughs et al. (2000) studied the growth and interaction dynamics of *Streptomycetes* and their bacteriophages. Their experimental set-up was a sterile soil microcosm, in which the infection cycle of a temperate actinophage KC301 in relation to the growth of its host *Streptomyces lividans* TK24 was studied. The main conclusion of this study was the significant role of soil environment in host–phage interactions. Among the more specific conclusions were: (a) newly germinated spores are more susceptible to phage lysis in comparison with hyphae of the developed mycelia; (b) protection of young hyphae from infection by ~98% adsorption of total present phages by mycelia of mature colonies; (c) phage burst size in soil was significantly larger compared with liquid culture and (d) no measurable impact on host reduced growth by the phage was observed. This host–phage interaction was attributed to spatial effects observed within soil in comparison with liquid environments, obliterating these effects as a result of high diffusion and mixing. Here we can report (unpublished data) that a similar phenomenon was observed in our laboratory during production of high titer stock of

F-male-specific coliphages grown on the host *E. coli* F_{amp}^{+}. Liquid culture resulted in two orders of magnitude less progeny phages compared with overlay agar collected from agar plates (Armon et al. 1997, Gino et al. 2007).

Generally speaking, there are several soil characteristics that impact host–phage interaction. Among these factors are: pH, temperature, organic matter (OM) water and clay content, ionic strength, hydrophobicity/hydrophilicity balance, rhizosphere characteristics, etc., but perhaps the major one is soil organic matter content in its dual form: mineral-associated and dissolved. Zhuang and Jin (2003) studied the retention and transport of two coliphages: MS2 (group *Leviviridae*) and ϕX174 (group *Microviridae*) in two porous media (sand and soil) as a function of organic matter content. These authors found a significant difference between the two phages in their adsorption and movement in porous media. Mineral-associated humics strongly interfered with MS2 phage promoting its faster movement with increased organic matter concentration with no observed difference with ϕX174. These interactions pointed toward two possible mechanisms: (1) for dissolved humics adsorbed onto phage surface resulting in alteration of its chemical properties and (2) for mineral-associated humics adsorbed onto mineral surface resulting in increased electrostatic repulsion. Their final conclusion was that "as a general trend, the effect of organic matter was dominated by electrostatic rather than hydrophobic interactions."

Among the high variability of natural organics present in soil environment such as acetate, lactate, formate, oxalate, citrate, malate, polygalacturonate, humic acids (HA), fulvic acids (FA), benzoate, phthalate, phytate, syringate, vanillate, ferulate, tartrate, and protocatechuate (Guppy et al. 2005), there are also a large variety of xenobiotic organics introduced by human activity or indirect pollution such as persistent organic pollutants (POP) that include dichlorodiphenyltrichloroethane (DDT), toxaphene, as well as the polychlorinated biphenyls (PCBs), dioxins, furans, pesticides (atrazine, aldicarb, chlorpyrifos, simazine, terbutryn, prometryn, diuron, etc., while the first three types are of special interest as they are applied in large quantities over a broad area and have a range of toxicological effects), polyaromatic hydrocarbons (PAH), methyl tert-butyl ether (MTBE), benzene-toluene-ethylbenzene-xylene (BTEX), and many others including combination like crude oil.

Our knowledge about the influence of these xenobiotics on host–phage interaction is still limited; however, several publications showed adverse effects. Roslycky (1982) studied the influence of selected herbicides on phages of some soil bacteria. Various concentrations of paraquat, atrazine, simazine, linuron, diuron, paraquat, and combinations did not inhibit lytic activity of four bacteriophages of *Agrobacterium radiobacter*, three bacteriophages of *Rhizobium meliloti*, three bacteriophages of *Rhizobium trifolii*, and two bacteriophages of *Streptomyces chrysomallus*. Paraquat (20–400 μg/mL) gradually reduced the adsorption and the average burst size of two phages; however, at same concentrations, it showed no effect on the phages attachment or the length of the latent period.

Toure and Stenz (1977) studied the effect of nine herbicides on the multiplication of *Escherichia coli*, strains W1665F^{+} and C600 and bacteriophages M12 (RNA phage) and λ (DNA phage) by one-step growth experiments. *E. coli* was inhibited by seven of the herbicides at ~0.1–1 mM, but promoted by some compounds at lower concentrations. The multiplication of the M12 phage was inhibited in a wide concentration range of 4 herbicides but stimulated by 2 herbicides (NES and amitrole). 2,4-D strongly inhibited the plaque-forming ability of M12 phages prior to their contact with the host cells. The λ phage multiplication was also inhibited by 4 herbicides while phenylacetic acid interfered with the lysogenization of the host, increasing its lytic activity.

Kaszubiak (1968) using lethal doses of two herbicides Afalon and Aretit revealed mutagenic activity on a *Rhizobium* strain. Afalon increased the frequency of mutants resistant to erythromycin and bacteriophages, but decreased the frequency of mutants resistant to streptomycin. Aretit had a similar action increasing the frequency of mutants resistant to erythromycin and streptomycin but decreased the frequency of bacteriophages resistant mutants. The author concluded that these mutation rate alterations induced by herbicides were caused by mutagenesis and not by selection. The mutagenic action of both herbicides was dependent on the medium's pH and no correlation was observed between resistance to herbicides and resistance to UV radiation.

Friedrich et al. (1983) used *E. coli* JMP397 harboring the plasmid pJP4 (encoding the ability to degrade the herbicide 2,4-dichlorophenoxyacetic acid) to conjugate and transfer this phenotypic characteristic to *Alcaligenes eutrophus* and *Pseudomonas oxalaticus*. The herbicide-degrading function of the plasmid was phenotypically expressed in all of the recipients. The majority of transconjugants also exhibited additional plasmid-encoded properties such as 3-chlorobenzoate degradation, resistance to mercuric ions, and sensitivity to the male-specific bacteriophage PR11.

These and other studies do not conclusively reveal a certain impact of herbicides on bacteriophage–host interaction but only contradictory results due to the variability of the systems studied. It should be pointed out that the majority of the studies on bacteriophages and their host in soil environment were performed with well defined phages (mainly coliphages such as: M12, λ, MS2, f2, ΦX-174, T series, etc.) that limits our broader knowledge on the variability of these processes among different bacterial genera (Zhuang and Jin 2003; Toure and Stenz 1977; Chu et al. 2003).

The soil host–phage interaction is a highly complex system that has to be further studied in order to understand this milieu in depth. Studies performed with soil columns or lysimeters are common among soil scientists and supply only partial information, therefore more detailed studies in particular at the surface chemistry level should be performed in the future.

3.1.5 Mutual Effect of Microbial Activity in Soil and Effect on Bacteriophages

By and large, vadose zone is the most important part of the subsurface soil extending from the terrestrial surface to groundwater table harboring a large variety and abundance of prokaryotic organisms such as bacteria, archea, and bacteriophages (Holden and Fierer 2005). The vadose zone is also the major matrix in which plants rizosphere develops and bacterial colonization is the most prevalent (Van Elsas et al. 2007). Vadose zone is also called the unsaturated soil part as related to water content, where fluctuating capillary forces supply water necessary for microbial activity. The main factors affecting microbial physiology in the vadose zone are similar to those affecting microbes everywhere: water availability, carbon and nitrogen sources for energy and metabolic processes, terminal electron acceptors, rhizosphere, pH, and temperature. Nutrients (more specifically carbon sources) mobility in the subsurface is governed mainly by their phase partition: dissolved (DOC) and particulate (POC). When in dissolved phase (DOC), the mobility is affected by water content and flow, consequently less available to soil particles colonized by bacterial communities, while in particulate phase the mobility is affected by soil porosity and adsorption characteristics. Sobczak et al. (1998) observed along a soil–stream transect (riparian zone) during a 2-year study that POC has a potential significance controlling bacterial productivity and biomass in near-stream sediments. Their study can give us a clue on many still unanswered questions on bacterial colonization, activity and heterogeneity in the subsurface. Microbial enzymatic activity expressed in rates per soil mass such as dehydrogenase activity, β-galactosidase induction, CO_2 evolution from substrate-induced respiration, N mineralization, incorporation of $^{32}PO_4$ and FDA hydrolysis beside other enzymatic activities involved in C, N, P, and S transformations decreases with depth in vadose zone (Holden and Fierer 2005; Taylor et al. 2002). Other enzymatic activities such as: cellulase, β-fructofuranosidase (invertase), β-glucosidase, protease, urease, alkaline and acid phosphatases, arylsulfatase, and catalase were also used to measure soil bacterial activity (Bastida et al. 2008).

The rhizosphere can be divided into three components related to each other: the rhizosphere (soil), rhizoplane, and roots (Barea et al. 2005). The rhizosphere is the soil around the roots that contain microorganisms and substances released by plants, the rhizoplane is the roots surface that harbors microorganisms and soil particles and the roots that harbor certain endophytes able to colonize root tissue (Danhorn and Fuqua 2007). In this environment, bacteria decompose organic detritus, promote plant growth (by plant growth promoting rhizobacteria, PGPR), and defend against root pathogens (Kent and Triplett 2002). Among PGPR, the most abundant are *Pseudomonas* and *Bacillus* genera. Another important activity of bacteria in soil is symbiotic N_2-fixation by the key enzyme nitrogenase, which specifically reduces atmospheric N_2 to ammonia in the symbiotic root nodules. PGPR-*Rhizobium* cooperation improves N_2-fixation by PGPR enhancing nodules formation and reducing acetylene production (Chebotar et al. 2001).

Syntropic activity of bacteria in soil has been showed to play a far larger role in carbon cycling than was previously thought (McInerney et al. 2008). The most abundant syntropic bacteria belong to δ-*Proteobacteria* and the low G+C gram-positive bacteria (see Table 3.1). A recent discovery of three genes in the sulfate reducing bacteria *Desulfovibrio vulgaris* that showed altered expression during a shift from syntrophic metabolism to sulfate reduction might provide an imminent clue on horizontal gene transfer from archaea bacteria (like methanogenic archaea *Methanosarcina barkeri* able to grow on lactate) (Scholten et al. 2007). Here, it can be very much speculated that the horizontal gene transfer was mediated by bacteriophages due to their abundance and presence in extreme environments (Santos et al. 2007). If we include bacteriophages as an additional biotic parameter to the above "ecological equation" the situation becomes much more complex.

In an extensive review, Weinbauer (2004) suggested that bacteriophages have a substantial role on bacterial abundance and diversity in a variety of environments (including soil) based on predation through lysis resulting in release of cell lysis products contributing to the microbial food web processes and biogeochemical cycles and on gene transfer mediation between prokaryotes. In this scenario, host diversity is also affected by "killing the winner" therefore retaining the capability of other competitively dominant species or populations. To emphasize this important aspect, Vos et al. (2009) in an elegant experimental set-up of two 625 cm^2 soil grids (from grazed flood plain) isolated 24 isomorphic bacterial colonies (75% of them identified as *Stenotrophomonas*) and 25 bacteriophage suspensions. Lytic susceptibility was tested with 600 bacterial clones originating from the selected grids and bacteriophage suspensions. Bacteriophages cross-lytic activity on the different bacterial clones revealed an average of 9% more infectivity to local than to foreign bacterial clones. These results were surprising as the soil was under flood and grazing continuous activity that was expected to daze these differences. The main result of this study emphasizes the importance of biotic interactions in shaping natural microbial communities in addition of physical environment variations previously mentioned.

Nasser et al. (2002) tried to determine the antiviral microbial activity of soil saturated with secondary effluents. One of the viruses used in this study was the RNA containing coliphage MS2. MS2 revealed durability to introduced protease (pronase) and *Pseudomonas aeruginosa* extracellular enzymes (as proteolytic factors) at 15°C. However, saturation of soil with secondary effluents and incubation at elevated temperature (30°C) reduced MS2's T_{99} from 21 to 2 days, thus emphasizing that high organic load and elevated temperatures negatively impact this specific bacteriophage survival in soil.

Normand et al. (2007) studied the genome characteristics of *Actinobacteria* from the genus *Frankia* (a facultatively symbiont of plants involved in formation of N_2-fixing root nodules on diverse and globally distributed angiosperms in the "actinorhizal" symbioses). Sequenced genome of three clades of *Frankia* sp., which colonize three subsets from among eight angiosperm families, revealed high divergence from a 5.43 Mbp (for a narrow host range strain) through 7.50 Mbp (for a medium host range strain) to 9.04 Mbp (for a broad host range strain). Their main conclusion was that host plant isolation favored genome contraction, whereas host plant diversification favored genome expansion. Following analysis of IS elements and prophage of the three representative *Frankia* sp. clades, the authors suggested prophage-induced plasticity of the genome expressed in its variable size to fit the host global diversification through millions of years.

Schuch and Fischetti (2009) studied the ecological and genetic factors that govern anthrax reservoir occurrence and persistence. According to their findings, bacteriophages provide *B. anthracis* with alternatives to sporulation that involve the activation of soil-survival (by blocking or promoting sporulation, exopolysaccharide expression, biofilm formation and long-term colonization of artificial soil environment, and intestinal tract of invertebrates) and endosymbiotic capabilities. Interestingly, all bacteriophages tested existed in a pseudolysogenic-like state in both environments, revealing a capability of continuous infection conferring survival phenotypes.

Bogopolśkii and Korneeva (1940) investigated the effect of specific bacteriophages infecting *Bacillus denitrificans* II in soil. They found that the denitrification process have been significantly diminished by these phages, suggesting on a possible approach to conserve soil nitrates.

Althauser et al. (1972) isolated bacteriophages infective against saprophytic strains of the *Bdellovibrio bacteriovorus*. The interest in this case is that *B. bacteriovorus* is by itself a periplasmic parasite of gram-negative bacteria present in soil environment, such as: *Erwinia amylovora*, *Proteus mirabilis*, *Serratia marcescens*, *Spirillum serpenis*, *Salmonella typhimurium*, and *E. coli*. Phages lysis susceptibility was observed only in the presence of *B. bacteriovorus* and its bacterial host. From the environmental aspect, the parasite–parasite–host interaction is remarkable pointing on the extremely complex biological interactions as related to horizontal gene transfer (Rendulic et al. 2004).

Keitarou and Yoshifumi (2004) reported on two physiological role of poly-γ-glutamate capsule of *Bacillus subtilis*: (1) to serve as a nutrient reserve during stationary phase starvation and (2) to protect vegetative cells from bacteriophages attack by obstructing their access to host cell surface receptors.

Biocontrol of plant pathogens was reported on *Pseudomonas fluorescens* CHA0 and other *Pseudomonas* strains (Keel and Défago 1997; Keel et al. 2002). These strains produce the antifungal polyketides 2,4-diacetylphloroglucinol and pyoluteorin that inhibit fungal growth. However, Keel et al. (2002) reported on ΦGP100 lytic bacteriophage that decimates *Pseudomonas fluorescens* CHA0 population (by more than 100-fold) thereby impairing its beneficial effect. Using a variant of *Pseudomonas fluorescens* CHA0 resistant to phage ΦGP100, root colonization and disease suppression were not affected.

An important feature of all bacteria is formation of biofilms on solid substrates and soil as a porous media is no exception (Danhorn and Fuqua 2007). Since the early 1980s, biofilms were recognized as bacteria aggregates colonizing solid surfaces and being in close contact with each other where adherent cells are usually embedded in a self-produced extracellular polymeric substance (EPS) commonly referred as slime, glycocalix, etc. (Lewandowski and Beyenal 2007). The collective life of bacterial cells in biofilm offers a major advantage over planktonic life form by conferring better defense against continuous environmental changes, close extra and intercellular molecular "conversation," better utilization of soluble or particulate organic/inorganic matter, horizontal gene transfer due to bacterial heterogeneity, and increased pathogenicity. Rudrappa et al. (2008) in a recent review assessed the biological importance of biofilm association with plants. These authors emphasized the rhizosphere, with its complex surface chemistry, as capable of directing various interactions with the colonizing biofilm through its formation, beneficial interactions, and pathogens suppression by gene modulation, all involving complex regulatory mechanisms (Narula et al. 2009; Lambers et al. 2009). According to these reviews, our present knowledge on root-biofilm interactions is still very limited. Here it should be mentioned that biofilm cells are able to communicate with each other through the quorum sensing mechanism based on autoinducers, firstly discovered in luminous bacteria (*Photobacterium fischeri*) (Nealson et al. 1970; Eberhard et al. 1981). These luminous bacteria produce an autoinducer (by gene *lux* I) now recognized universally as acyl-homoserine lactone (AHL) that further links to cell receptors (produced by gene *lux* R). The autoinducer triggers a series of *lux* genes (lux ICDABE) that in the final result produce bioluminescence. Quorum mechanism of bacterial cells communication in biofilms was further detected in a variety of microorganisms with different kinds of autoinducers (AHL, palmitic acid methyl-PAME, protein com X, g-butryolactone-A factor, 2-heptyl-3-hydroxy-4quinolone-PQS, etc.) (Van Elsas et al. 2007).

Our knowledge on the effect or interconnection between quorum sensing products and their interaction with bacteriophages is very limited. The only publication that links bacteriophages with quorum sensing was published by Ripp et al. (2006, 2009) that used λ bacteriophage as biosensing reporter of pathogenic agents. Briefly, the authors introduced *lux* I gene to a λ bacteriophage able to transduce this gene to *E. coli* pathogens (in their case *E. coli* XL1-Blue as model pathogen) to produce the AHL autoinducer in case of their presence in the tested sample. As a bioluminescent bioreporter, they used *E. coli* OHHLux that under certain AHL concentration can emit detectable bioluminescence. The authors suggested that this molecular manipulation based on binary phage-based reporter assay can be applied to other pathogenic bacteria detection. It is clear that if such a short laboratory manipulation was successful on the biosphere long time scale of millions of years such a process certainly can also occur in environment (water and soil) impacting rhizosphere quorum sensing communities. The effect of such interactions is not well understood and further research is needed.

Another newly reported interaction between plants and bacteria is the penetration and colonization of host internal tissues through root system. The human pathogen *Escherichia coli* O157:H7 was shown to be transmitted from manure-contaminated soil and irrigation water to lettuce plants. Experiments demonstrated that *E. coli* O157:H7 can enter the lettuce plant through the root system and migrate internally throughout the edible portion of the plant (Solomon et al. 2002). This group also tested another human pathogen, *Salmonella enterica* sv. *typhimurium*, for its ability to enter parsley plant roots or phyllosphere and colonize its internal tissues. Two phenotypical characteristics of the experimental bacterial pathogen were revealed to play a major role in this process: cellulose synthesis and curli (described as fibronectin-binding flexible fibrillar structure, fimbriae like). Curli mutants were more affected in their colonization potential of internal plant parts compared with cellulose mutants (Lapidot and Yaron 2009).

Providing lysogeny of these specific and other soil bacteria, it will not be a pure speculation that in penetration, transmission and colonization of internal plant tissues, bacteriophages play an evolutionary role through HGT.

3.1.6 Genetic Transfer Involving Bacteriophages and Bacteria in Soil Environment

At present, there is no doubt on the major role of bacteriophages in horizontal genetic transfer in and from different environments and across a large variety of bacterial species. From the environmental aspect, soil is no except as it harbors both phages and bacteria in close proximity under various biological interactions including plant rhizosphere. Horizontal gene transfer occurs through three fundamental mechanisms: conjugation, transformation, and transduction (by phages). Phages may transfer genetic information in several ways. First is capturing a genomic fragment of one bacteria and transferring it to another (various types of transduction) provided that adsorption and injection occur. If adsorption and injection do not occur, the phage genome may be efficiently transferred into a bacterial cell by a hybrid plasmid. Other pathway is conversion, specifically, expression of phage's own genes (the simplest cases are cell resistance to superinfection with a homoimmune phage and mutual exclusion of related or unrelated phages) (Krylov 2003). Here we will discuss only the third mechanism, namely transduction as being related to bacteriophages.

There is a considerable mass of evidences that bacterial virulence genes are mediated by phages or plasmids carrying bacteriophage genes which insert into tRNA genes. Virulent bacteria of clinical importance such as *Dichelobacter nodosus*, *Neisseria gonorrhoeae*, *Escherichia coli*, *Actinobacillus actinomycetemcomitans*, and *Treponema denticola* have been shown to harbor integrase genes originating from bacteriophages and plasmids. A number of these genes (involved in the virulence of bacterial pathogens) were found on integrated bacteriophages and possibly associated with tRNA genes and/or integrase genes derived from bacteriophages. The use of tRNA genes as integration sites for many bacteriophages and plasmids may favor intergeneric transmission, as tRNA genes are highly conserved (Brian et al. 1995).

There are several factors that can impact transduction in soil such as nutrients, host cell density, temperature, pH, ionic strength, and moisture content (Dröge et al. 1999). Nutritional conditions are expected to increase the transduction due to enhanced growth and division of host cells. Zeph et al. (1988) using *E. coli* as host and a genetically marked derivative of phage P1 in sterile soil supplemented with nutrient-rich medium found significantly higher numbers of phages, recipient cells, and transductants compared with saline supplemented soil (control); however, transduction frequency was similar. In nonsterile soils, the transductants numbers was only slightly higher and transduction frequency was unaffected.

However, Burroughs et al. (2000) in their study on *Streptomycetes* and their phages in a soil microcosm revealed little correlation between gross nutritional status vs. burst size and frequency of lysogeny establishment. These authors suggested that in soil microcosm nutrient availability should be related at the micro-scale due to its particulate nature.

Host cell density significance was tested by Marsh and Wellington (1992) on *Streptomyces lividans* KT24 and its temperate phage ϕC31 in soil at 40% moisture content. They found that host cells density bellow 2×10^3 CFU/g neither phage propagation nor lysogenisation were detected.

Abiotic parameters such as temperature, pH, and moisture content can have a significant effect mainly on the host cells. Seeley and Primrose (1980) reported on temperature influence on the adsorption of phages onto host cells as well host capability to grow and divide a prerequisite for phage multiplication. Another parameter involved in phage adsorption to host is the pH that beside this effect, it plays a significant role on phage adsorption to clay particles (Lipson and Stotzky 1987) and for some phages inactivation following adsorption (Armon and Cabelli 1988). Soil clay content has also an effect on phage and host survival through adsorption on, however, above a certain concentration it may become an inhibitory factor by limiting the contact between these microorganisms (Stotzky 1989). Zeph et al. (1988) showed that clay type is also important as phage survival in soil amended with montmorillonite was significantly longer compared with kaolinite amended soil. Ionic strength expressed as concentration of bi-valent ions (Ca^{+2}, Mg^{+2} and $Fe^{+2/+3}$) is important in phages–host interaction through bridging the negative surface charge of both components in nature, therefore alternatively increasing transduction. Finally, Kidambi et al. (1994) showed the importance of surfaces in transduction process by coinoculation of two *P. aeruginosa* strains on phylloplane: one a lysogenic recipient (with phage F116) strain and a second donor strain containing the nonconjugative antibiotic resistance plasmid Rms149. Transductants harboring plasmid Rms149 was observed on 90% of the leaves studied.

An additional interesting factor involved in soil microcosms are genetically modified plant (GMP), firstly introduced in 1987 (Vaeck et al. 1987). Since then, >15,000 field trials were performed in different locations that raised the important question on risk assessment of HGT from plants to soil bacteria. Transduction is an important mechanism in HGT and was shown in soil microcosm (Zeph et al. 1988; Germida and Khachatourians 1988; Herron and Wellington 1994), but is still not well understood and difficult to prove (Herron 2004; Nielsen and Townsend 2004). Only recently it was shown that HGT can occur between plants and soil bacteria. Monier et al. (2007) used *Acinetobacter baylyi* strain BD143 and transplastomic tobacco plants harboring the *aadA* gene (streptomycin and spectinomycin resistance) found that sequences identical to the flanking regions containing as few as 55 nucleotides were sufficient for recombination to occur. Therefore, they screened a collection of bacteria able to colonize tobacco plant tissue infected by *Ralstonia solanacearum* strain K60 for DNA sequence similarity with the chloroplastic genes accd and rbcL that flank the transgene. The uptake of extracellular DNA was tested by natural or electro-transformation. 8% of the isolates were found to have DNA sequence similarity with one or both chloroplastic regions flanking the transgene, among which *Acinetobacter* sp. was able to integrate exogenous plasmid DNA by natural transformation. The obtained data by these authors suggest that transplastomic plant DNA recipients might be present in soil bacterial communities.

Recently, Pontiroli et al. (2009) showed that plant DNA released during degradation of plant tissue together with other nutrients may create a copitrophic environment favoring opportunistic bacterial growth. These authors used a naturally transformable *Acinetobacter baylyi* strain BD413 (carrying a promotorless *aadA*::*gfp* fusion) and a transplastomic tobacco plant harboring an *aadA* gene (conferring resistance to spectinomycin and streptomycin) with the objective to determine whether specific niches could be shown to foster bacterial growth on intact or decaying plant tissues, to develop a competence state, and to possibly acquire exogenous plant DNA by natural transformation. Their results unequivocally revealed both experimentally and visually that both antibiotic resistance and green fluorescence phenotypes were restored in recombinant bacterial cells after homologous recombination with transgenic plant DNA. This case is very important in spite of being based on transformation process but further transduction by bacteriophages cannot be excluded in such environments where plants and bacteria act together.

As already pointed out in Sect. 3.1.7, Schuch and Fischetti (2009) on their ecological and genetic factors study that govern anthrax reservoir occurrence and persistence demonstrated that during its environmental phase, bacteriophages provide *B. anthracis* with alternatives to sporulation that involve the activation of soil-survival and endosymbiotic capabilities.

Perhaps another aspect of HGT that cannot be overlooked is the experience gained from clinical microbiology. Hiramatsu et al. (2001) reviewed the methicilline resistance acquired by *Staphylococcus aureus* transforming it in the notorious hospital pathogen *S. aureus* (MRSA). Integration of a staphylococcal cassette chromosome *mec* (SCCmec) element (that encodes resistance to all β-lactam antibiotics) together with *ccr* gene complex (that encodes recombinases responsible for its mobility) conveys resistance to a variety of other non β-lactam antibiotics. The authors strongly suggest that with the identification of new types of SCC*mec*, the number of new clonotypes will increase in the future, making the use of antibiotics almost impossible. The uncontrolled use of antibiotics in hospital is one of the main mechanisms in genetic selection of these "unbeatable pathogens." If we take this example to soil environment and integrate it with the variety of xenobiotics applied to soil in juxtaposition with the present bacteriophages, a continuous evolutionary HGT is expected to crop up.

3.1.7 Bacteriophage Transport in the Subsurface and Soil Bacteria

Bacteriophages transport in subsurface is mainly impacted by water content and flow in this soil zone (Schijven and Hassanizadeh 2000). Under saturated conditions, other abiotic parameters are involved such as pH, temperature, organic load, and soil type. Poletika et al. (1995) used phage MS2 beside Br and simazine (a herbicide from the triazine group) to study their transport during unsaturated flow in a field lysimeter of undisturbed loamy sand. MS2 was found to reside into the lysimeter in contrast to Br due to adsorption. Yield coefficients of MS2 implied strong adsorption and rapid inactivation in contrast to laboratory scale that indicated high mobility in soil. The discrepancy between the laboratory and field studies was explained by these authors due to air–water interfacial forces present in unsaturated field experiments. Schulze-Makuch et al. (2003) studied the effects of pH on coliphage MS2 in a one meter model aquifer as a simulation of a shallow aquifer. According to their results, MS2 transport was affected by pH of groundwater and the isoelectric points of mineral matrix. Higher pH increased the transport of these coliphages through adsorption reduction, a reality already known for many years in environmental virology (Hurst et al. 1980).

Guan et al. (2003) using a model aquifer of 1 m deep investigated the effect of various pH-s (4.6 to 8.3) on the transport of MS2 and ϕX174 bacteriophages in groundwater (soil saturated zone). The authors reported on a critical pH at which bacteriophage behavior altered sharply (0.5 pH unit bellow the highest isoelectric point of the phages), where the phage had an opposite charge to at least one component of the porous medium and as a result removed from water phase.

Sinton et al. (1997) investigated the movement of bacteria (*E. coli*) and bacteriophage (MS2) compared with rhodamine WT dye tracer into and through an alluvial gravel aquifer at large scale (approximately 60 and 445 m downstream of the center of the inoculated strips). According to their results, there was evidence of rapid microbial transport through the vadoze zone upstream, with coliphages exhibiting greater attenuation and longer travel times compared with bacteria. In the aquifer due to its higher homogenity, only a small difference was found between the bacteria and phages attenuation.

Soil matrix mineral composition is also important in the transport of bacteriophages. McLeod et al. (2004) used three soil types in lysimeters: granular (gravel, sand, or silt soil with little or no clay content), ultic (quartz-rich sediments which have weathered to clay or sandy clays found in New Zealand), and recent (a soil that has been undisturbed long enough to develop a clear topsoil horizon, but not long enough for any differentiation of the subsoil to occur) (Hewitt 2009) to follow soil transport of chemical tracer along with microbial ones. A *Salmonella* bacteriophage used in this study revealed a rapid movement with an early peak and then a tail off indicating bypass flow in all soils.

Transport of bacteriophages in the subsurface may be influenced also by xenobiotics of different origins. Chattopadhyay et al. (2000) examined the transport and survival of three bacteriophages (MS2, T2 and ϕX174) in the presence of cationic (hexadecyl trimethylammmonium bromide-CTAB), anionic (sodium dodecylbenzene sulfonate-SDBS), and nonionic (Triton X100) surfactants. In terms of inactivation, the highest rate was observed in the presence of cationic surfactant followed by anionic and nonionic as last. Inactivation was found to be also related to surfactant concentrations and bacteriophages type. Thompson and Yates (1999) used a dynamic batch system (tubes loaded with glass and Teflon beads simulating soil hydrophobic/hydrophilic particles) to examine the transport and survival of three bacteriopages (MS2, R17 and ϕX174) under different ionic strengths and a nonionic surfactant (Tween 80) at various concentrations. Their conclusions related to this specific system were: that viral inactivation is dependent upon (1) the presence of a dynamic air–water–solid interface (where the solid is a hydrophobic surface), (2) the ionic strength of the solution, (3) the concentration of surface active compounds in the solution, and (4) the type of virus used.

Gupta et al. (2009) used pilot scale natural soil columns fed with a variety of microorganisms among them MS2 and PRD1 bacteriophages to assess their removal during Ohio river water percolation. Their results revealed much greater retention for the smallest microorganisms namely the two bacteriophages compared with bacteria and protozoan parasite oocysts being in qualitative agreement with colloid filtration theory (CFT), which predicts the least removal for micrometer size colloids, suggesting that the respective sizes of the organisms were a dominant control on their transport despite expected differences in their surface characteristics.

Using similar bacteriophages (MS2 and PRD1), Corapcioglu et al. (2006) conducted field studies to trace large-scale (34 m) and small-scale (10 m) virus transport under natural and forced-gradients through a sand and gravel aquifer at a ground water research site at Texas A&M University. Their results indicated fast transport of MS2 over PRD1, where MS2 was capable of traveling at least 34 m in a heterogeneous sandy aquifer. The authors hypothesized that the faster transport velocity of MS2 may be a result of the bacteriophage negative electrical charges and the clay minerals present in the heterogeneous aquifer, which cause the viruses to remain in the center of the preferential flow paths where the largest microscopic velocity exists, a phenomenon known as hydrodynamic chromatography. Their explanation of PRD1 retarded transport was its better adsorption capability onto aquifer particles.

Bacteriophages MS2 and PRD1 inactivation in solution phase vs. attached to mineral surfaces was tested by Ryan et al. (2002) under field and laboratory conditions. In field studies, PRD1 radiolabeled with ^{32}P and injected into a ferric oxyhydroxide-coated sand aquifer was detected as a small fractions of infective and ^{32}P-labeled PRD1 brake through with the bromide tracer, followed by the slow release of 84% of the ^{32}P activity and only 0.011% of the infective PRD, in a zone where the aquifer was contaminated by secondary sewage infiltration. Under laboratory conditions, the inactivation of phages PRD1 (protein capsid radiolabeled with ^{35}S) and MS2 (protein capsid radiolabeled with ^{35}S dual and RNA with ^{32}P) was monitored in the presence of groundwater and sediment from the field contaminated area. Infective bacteriophages release declined at faster rate compared with radiolabeled biomaterial, indicating inactivation of attached bacteriophages onto surfaces. ^{32}P and ^{35}S release discrepancy suggested on disintegration of experimental bacteriophages. Comparison of the experimental bacteriophages inactivation rates in solution and on surfaces revealed a three times faster inactivation of phages in solution; however, these authors stated that due to slow release of inactivated bacteriophages, the surface inactivation might be underestimated.

Cheng et al. (2007) investigated the influence of DOC on the transport of phage MS2 in sandy soil following effluent application. There was no difference between presence and absence of DOC on phage recovery. In addition, using a mathematical model, the authors did not find any differences between the two optimized attachment rate coefficients concluding that MS2 transport in the subsurface is not affected by DOC. However, using mass-balance analysis, they found that MS2 attachment occurs in the early stages of the experiment with late inactivation.

Chetochine et al. (2006) studied the leaching and transport of MS2 phages from biosolids Class B (biosolids that are treated to reduce pathogens but still contain detectable levels of them and can be land applied with formal site restrictions and strict record keeping)(USEPA 2002) mixed with soil (93.1% sand, 5.6% silt, 1.3% clay, and 0.01% TOC content) and saturated with water (1:1). The authors reported an 8% of the indigenous coliphages leaching out of the biosolid-soil matrix with a fraction subsequently transported with minimal retention through the sandy porous matrix.

McLeod et al. (2003) experimented the transport of phage 28B (host *Salmonella typhimurium* type 5) together with various coliforms and a nonreactive tracer (Br-) in soil core lysimeters. The two soils tested were typical soils from southland (New Zealand) with two main characteristics: poorly and well drained. The movement of phage 28B was rapid in both soils with an early peak followed by tail off indicating bypass flow.

Van der Wielen et al. (2008) determined the removal of two coliphages, MS2 and ϕX174 in an anoxic sandy aquifer field study, with relatively high pH. Interestingly, both phages were removed at lower rate compared with laboratory studies obviously due to anoxic redox condition in this experimental aquifer. The authors explained their results based on the absence of oxidized metal ions like ferric oxihydroxides that increase phages adsorption and consequently their removal. Twenty years ago, the presence/absence of oxygen in soil represented by aerobic/anaerobic bacteria has been investigated by Hurst (1988) while studying the survival of a human virus (Poliovirus type 1) in a sandy loam soil. In this case, the presence of aerobic bacteria was deleterious to Poliovirus survival. As already mentioned above, the interaction between viruses and oxidized ferric oxihydroxides could play in this case a similar role due to irreversible adsorption resulting in very low recovery interpreted as inactivation. More accurate and elegant experiments will be needed to uncover this phenomenon in soil subsurface.

Carlander et al. (2000) studied transport of phage 28B (host *Salmonella typhimurium*) in two types of willow-cropped field lysimeters containing clay or sand soil through irrigation. Phage 28B was rapidly transported in clay soil indicating macropore flow while in sandy soil it varied but was in general slower with higher retention. Interestingly, the willow plants did not facilitate phages transport and the opposite was found, namely the rhizosphere is playing a role in retention of these phages.

Summarizing the above selected data several conclusions can be drawn on the following parameters (Schijven and Hassanizadeh 2000):

Adsorption: (a) Phages adsorption parameters obtained in batch experiments or small scale soil columns seem to be of little use in predicting large scale/field studies results; (b) pH seems to be the main abiotic factor affecting adsorption, at pH-s > 7 the net surface charge of most bacteriophages is negative therefore repulsion forces unfavor attachment; (c) water with increased ionic strength (divalent cations preferentially) can cancel the repulsion forces and enhance adsorption to soil particles; (d) solid surfaces containing iron, aluminum, and manganese oxides have positive charges favoring adsorption; (e) studies support the concept of "patch adsorption" in soil based on heterogeneous nature of this matrix; (f) in general dissolved organic matter (DOC) compete with bacteriophages for solid attachment sites or in case of surfactants group disrupting hydrophobic bonds increasing desorption; (g) soil adsorbed DOC can provide hydrophobic sites for bacteriophages attachment, and (h) the large variability of DOC chemistry makes bacteriophages transport, attachment and inactivation very unpredictable.

Inactivation: (a) Enhanced/reduced inactivation is bacteriophage-specific and almost independent of attachment/detachment coefficients; (b) solid attachment of bacteriophages may significantly reduce inactivation by preventing contact with air–water or air–water–solid interfaces; (c) reversible attachments of bacteriophages to soil particles can increase inactivation; (d) temperature plays a cardinal role in bacteriophages inactivation by direct action or by increasing microbial activity under aerobic conditions, however some bacteriophages can be insensitive to temperature changes.

Bacteriophage type: (a) MS2 was shown to be the worst-case scenario among the many phages used as models of transport in soil, due to its stability, lower hydrophobicity, and less "gluey"; (b) PRD1 can be considered as the second best as being more resistant to higher temperature; (c) ϕX174 could be used as a conservative model in high organic content soils, due to its low hydrophobicity and stability, however easily impacted by pH.

Soil structure and chemistry: (a) Saturation/unsaturation impacts bacteriophages transport and close contact with surrounding bacterial communities; (b) bacterial diversity depends on soil chemistry therefore affecting bacteriophages specificity too; (c) soil porosity affects bacteriophages transport with preferential flow; (d) metal ions oxides enhance bacteriophages irreversible adsorption decreasing their release and in some cases increasing their inactivation; (e) dissolved organic content of soil would increase bacterial growth providing a larger host population exposed to bacteriophages infection.

Rhizosphere: (a) rhizosphere provides a chemosphere of natural organic components that enhance bacteriophage infection and multiplication pending on bacterial host welfare; (b) rhizosphere is a biosurface that facilitate bacteriophage adsorption therefore increasing soil retention; (c) plant roots can create preferential flow through channeling; (d) compared with surrounding soil, rhizosphere is the main bacterial colonization site (biofilm) where phages can be entrapped and reduce their mobility.

3.1.8 Discussion, Remarks and Thoughts

In 1992, Moll and Vestal (1992) published a more "philosophical" rather "applicable" aspect on bacteriophages survival entitled: "Survival of microorganisms in smectite clays: implications for Martian exobiology." In their manuscript, *Bacillus subtilis*, *Azotobacter chroococcum*, and the enteric bacteria lytic phage MS2 were tested for survival in two soil types. The authors simulated terrestrial and Martian soils using Wyoming type montmorillonite and Fe^{3+}-montmorillonite, respectively. These microorganisms were exposed to terrestrial and Martian environmental conditions for 112 days, excluding UV irradiation and redox conditions. Martian atmosphere has low pressure (5–10 mbar) containing: ~95% CO_2, 2.7% N_2, 1.6 % Ar, 0.13% O_2, 0.13% CO as major components and 210 ppm water, with temperatures from −140 to 20°C (average −63°C), which in respect to our biosphere is very inhospitable, but do not exclude possible life forms. Under these conditions, both bacteria survived well in Wyoming montmorillonite soils but less in Fe^{+3}- montmorillonite under Martian conditions, attributed to lower pH of the last. MS2 phage survived better in Fe^{3+}-montmorillonite compared to Wyoming montmorillonite under Martian conditions, explained as stabilization of the phage under cold and dry conditions. The authors suggested that phages/viruses may be able to survive in Martian type soils environment, having implication on planetary protection for future Mars missions. At present, after 17 years of additional acquired knowledge, it is not surprising that phages play a pivotal role in biogenesis.

While reviewing the most recent literature on soil bacteria and their bacteriophages, it is obvious that phages due to their abundance and diversity impact soil bacteria through a variety of interactions. Viewing soil environment as a "bouillabaisse" of a variety of microorganisms (bacteria, phages, fungi, protozoa, nematodes, plants, algae, etc.), it seems logically to assume that a variety of biological interactions can take place at molecular levels. Some of these reactions do not happen at high rate, requiring special environmental conditions and our detection methods are still limited for detection. However, we are able to detect these interactions when they are prominent or generated at laboratory scale. Molecular biology massively expanded our potential to understand the life genomics although with present confusion due to presence of many gaps in our knowledge (only 554 phages genomes were sequenced according to NCBI phage database) (Hatfull 2008). Here is the place to mention some pivotal works related to bacteriophages and soil bacteria, such as d'Herelle (1917) and Twort (1915) on the discovery of phages, Zinder and Lederberg (1952) on phage transduction, soil bacteria as recipients of transgenic plant DNA (Monier et al. 2007), adaptation of bacteriophages to their hosts in soil (Vos et al. 2009), transmission of *E. coli* O157:H7 into plant tissue (Solomon et al. 2002), microbial diversity in soil (Torsvik and Øvreås 2002; Øvreås and Torsvik 1998), etc. that opened new areas of research and raising endless new questions.

- One of the major questions is: where the human pathogens came from and how extensive is bacteriophages role in their evolution? If we look at the biosphere from the evolutionary point of view, archaea/bacteria and/or viruses/bacteriophages were in existence before human appearance and evolution. Today we know that many bacterial virulence factors are transmitted by temperate and lytic bacteriophages in a variety of bacteria (*E.coli* O157:H7, *Salmonella typhimurium*, *Streptococcus pyogenes*, *Clostridium boutulinum*, *Vibrio cholerae*, etc.) that gained alternative niches under continuous evolutionary pressure (Cheetham and Katz 1995; Brüssow and Hendrix 2002).

- In a recent publication Koonin et al. (2006) based on viruses genomics, hypothesized a scenario on the origin of viruses including bacteriophages. According to these authors, selfish genetic elements ancestral to viruses evolved prior to typical cells that once archaea and bacteria appeared on the scene turned into intracellular parasites. The authors emphasized that paradoxically, selection favored evolution of temperate viruses (lysogenic phages in our case) and other primitive defense mechanisms. Today, after the evolution of eukaryotic cells, this scenario is more complicated also as a result of increasing human activity. As we deal with soil environment, the chemistry involved in the various processes is one of the cardinal factors impacting soil bacteria and bacteriophages' interactions. Therefore, the variability of soil bacteria is large and depends on climatic, agricultural, and mineralogical factors. To add new components to this already complicated equation, we should pay attention to large quantities and various pollutants (some xenobiotics) produced by industrial activity that impact soil bacteria in favor of a continuous adaptation to new stresses in soil. Bacteriophages in this context, being also considered genetic mobile elements, can transfer (through transduction) different phenotypic characteristics to soil bacteria (nonpathogenic as well pathogenic) that in turn could evolve in different directions. Bacterial pathogenicity was already proved to be related to phages (Brüssow et al. 2004; Casas et al. 2006) in both terrestrial and aquatic environments, and soil rhizosphere is an excellent fertile matrix for such interactions. The vast literature on bacteriophages is based mainly on aquatic environments and much less on terrestrial one. Only in the last 20 years, terrestrial phages were looked in depth and proved as an important factor in this environment. In the evolution time scale span, it is only a tiny fraction, therefore we can only deduct from present observations the possible interactions between phages and soil bacteria. Another problem, brought up during the preparation of this chapter, is the limited information of phages transport in soil subsurface due to narrow phage spectrum used by scientists. This is understandable, as everybody in the scientific community likes to experiment with a known well defined organism (see phages like MS2, PRD1 and ϕX174 in soil transport) instead of a more complicated system with a less well identified phage.
- As bacteriophages do not have proper metabolic pathways compared with bacteria, it is expected that their interactions and evolution are mainly based on genetic transfer, which will express itself later in a bacteria and further in eukaryotic cells (humans, plants, algae, protozoa, etc.). The area of horizontal gene transfer is now evolving much faster since molecular methods became the global standard of most laboratories. The "biological role" of bacteriophages in ecology is still unknown and will remain as such, as long we are not able to define the origin of life; however, their effects are observed in our everyday life. Beside prions, bacteriophages are the most austere microorganisms, but still

complicated in their interaction with their hosts and the surrounding environment. It took over 3 billion of years to reach this point in time and is almost impossible to reproduce it in the laboratory at a short time span of years, however we start to see genetical patterns that cross borders among the three domains (based on rRNA): Archaea, Bacteria, and Eukaryota that are linked to bacteriophages genomic and function.

- In an interesting experimental study, Lu and Collins (2007) engineered coliphage T7 with DspB gene that codes an enzyme, produced by the bacterium *Actinobacillus actinomycetemcomitans*, able to hydrolyze 1,6-*N*-acetyl-D-glucosamine, an adhesin glycoprotein involved in biofilm formation and integrity of a variety of bacteria. Expression of this gene, carried by the lytic T7 phage and applied to an *E. coli* biofilm, revealed 4.5 orders of magnitude reduction in biofilm cells. This experimental manipulation cannot be excluded to happen evolutionary and therefore existent somewhere in the environment, waiting for discovery. The above example just emphasizes the fact that bacteria-bacteriophage interaction is a continuous process that if we are able to reproduce it under laboratory conditions, nature by itself will certainly do it along billions of years time span.
- Soil through its complexity and heterogenity sets hurdles in bacteriophage research, but also opens new horizons in the primordial and present interactions between organic matter and minerals. Different theories and models were published on the minerals and their role in life formation that according to thermodynamic principles are plausible (Wächtershäuser 1988; Hanczyc et al. 2007; Pietramellara et al. 2007; Gallori et al. 2004; Franchi and Gallori 2004). The relevance of these models is still highly important to understand already existing microorganisms and their interrelations in soil habitats.
- Finally, one of the recently raised questions was on bacteriophages small genome and their morphological variability and functions. The genomic diversity of phages ranges from 4 kb to up to 600 kb according to already published literature (Brüssow and Hendrix 2002; Gordon et al. 1999). The genome size does not reflect the exact genomic functions; therefore, the analyses of natural viral assemblages based on genome size severely underestimate genetic diversity and viral community dynamics (Holmfeldt et al. 2007).

Acknowledgment The author would like to thank Prof. Em. H-W Ackermann for his inspiration, knowledge, and life time dedication to phages consequently "infecting" other scientists with his élan. The author would like also to thank Ms. Miri Offer for her support in bibliographic searches during the preparation of this chapter.

References

Abedon ST (1989) Selection for bacteriophage latent period length by bacterial density: a theoretical examination. Microb Ecol 18:79–88

Ackermann H-W (2003) Bacteriophage observations and evolution. Res Microbiol 154:245–251

Ackermann H-W (2007) 5500 Phages examined in the electron microscope. Arch Virol 152: 277–243

Ackermann H-W (2009) Phage classification and characterization. Method Mol Biol 501:127–140

Alexander M (1977) Introduction to soil microbiology, 2nd edn. John Wiley and Sons, New York, NY

Althauser M, Samsonoff WA, Anderson C, Conti SF (1972) Isolation and preliminary characterization of bacteriophages for Bdellovibrio bacteriovorus. J Virol 10:516–523

Anonymous (Soil Survey Division Staff) (1993) Soil survey manual. Soil Conservation Service. U.S. Department of Agriculture Handbook 18.

Armon R, Araujo R, Kott Y, Lucena F, Jofre J (1997) Bacteriophages of enteric bacteria in drinking water, comparison of their distribution in two countries. J Appl Microbiol 83:627–633

Armon R, Cabelli VJ (1988) Phage f2 desorption from clay (bentonite and kaolinite) using non-ionic detergents, beef extract and chaotropic agent. Can J Microbiol 34:1022–1024

Ashelford KE, Fry JC, Bailey MJ, Jeffries AR, Day MJ (1999) Characterization of six bacteriophages of Serratia liquefaciens CP6 isolated from the sugar beet phytosphere. Appl Environ Microbiol 65:1959–1965

Ashelford KE, Norris SJ, Fry JC, Bailey MJ, Day MJ (2000) Seasonal population dynamics and interactions of competing bacteriophages and their host in the rhizosphere. Appl Environ Microbiol 66:4193–4199

Avrahami S, Conrad R (2003) Patterns of community change among ammonia oxidizers in meadow soils upon long-term incubation at different temperatures. Appl Environ Microbiol 69:6152–6164

Avrahami S, Conrad R, Braker G (2002) Effect of soil ammonium concentration on N2O release and on the community structure of ammonia oxidizers and denitrifiers. Appl Environ Microbiol 68:5685–5692

Baess I (1971) Report on a pseudolysogenic mycobacterium and a review of the literature concerning pseudolysogeny. Acta Pathol Microbiol Scand 79:428–434

Baker GC, Cowan DA (2004) 16S rDNA primers and the unbiased assessment of thermophile diversity. Biochem Soc T 32:218–221

Baker GC, Smith JJ, Cowan DA (2003) Review and re-analysis of domain-specific 16S primers. J Microbiol Meth 55:541–555

Barea JM, Pozo MJ, Azcón R, Azcón-Aguilar C (2005) Microbial co-operation in the rhizosphere. J Exper Botany 56:1761–1778

Barns SM, Takala SL, Ruske R (1999) Wide distribution and diversity of members of the bacterial kingdom Acidobacterium in the environment. Appl Environ Microbiol 65:1731–1737

Bartlett CLR (1996) An overview of emerging foodborne and waterborne diseases. East Medit Health J 2:51–60

Bastida F, Zsolnay A, Hernández T, García C (2008) Past, present and future of soil quality indices: a biological perspective. Geoderma 147:159–171

Bergsma-Vlami M, Prins ME, Staats M, Raaijmakers JM (2005) Assessment of genotypic diversity of antibiotic-producing Pseudomonas species in the rhizosphere by denaturing gradient gel electrophoresis. Appl Environ Microbiol 71:993–1003

Bertani G (2004) Guest commentary: lysogeny at mid-twentieth century: P1, P2, and other experimental systems. J Bacteriol 186:595–600

Bogopolśkii MD, Korneeva NP (1940) An investigation of the effect of bacteriophages against Bacillus denitrificans II. Mikrobiologichnii Zhurnal (1934–1977) 7:75–83

Bongiorni L, Magagnini M, Armeni M, Noble R, Danovaro R (2005) Viral production, decay rates, and life strategies along a trophic gradient in the north Adriatic sea. Appl Environ Microb 71:6644–6650

Borneman J, Skroch PW, Osullivan KM, Palus JA, Rumjanek NG et al (1996) Molecular microbial diversity of an agricultural soil in Wisconsin. Appl Environ Microbiol 62:1935–1943

Boutte C, Grubisic S, Balthasart P, Wilmotte A (2006) Testing of primers for the study of cyanobacterial molecular diversity by DGGE. J Microbiol Methods 65:542–550

Breitbart M, Salamon P, Andresen B, Mahaffy JM, Segall AM, Mead D, Azam F, Rohwer F (2002) Genomic analysis of uncultured marine viral communities. Proc Natl Acad Sci USA 99:14250–14255
Brian F, Cheetham BF, Katz ME (1995) A role for bacteriophages in the evolution and transfer of bacterial virulence determinants. Mol Microbiol 18:201–208
Brüssow H, Hendrix RW (2002) Phage genomics: small is beautiful. Cell 108:13–16
Brüssow H, Canchaya C, Hardt W-D (2004) Phages and the evolution of bacterial pathogens: from genomic rearrangements to lysogenic conversion. Microbiol Molec Biol Rev 68: 560–602
Burroughs NJ, Marsh P, Wellington EM (2000) Mathematical analysis of growth and interaction dynamics of streptomycetes and a bacteriophage in soil. Appl Environ Microbiol 66:3868–3877
Cardinale ML, Brusetti P, Quatrini S, Borin AM, Puglia A et al (2004) Comparison of different primer sets for use in automated ribosomal intergenic spacer analysis of complex bacterial communities. Appl Environ Microbiol 70:6147–6156
Carlander A, Aronsson P, Allestam G, Stenström TA, Perttu K (2000) Transport and retention of bacteriophages in two types of willow-cropped lysimeters. J Environ Sci Health A Tox Hazard Subst Environ Eng 35:1477–1492
Casas V, Miyake J, Balsley H, Roark J, Telles S, Leeds S et al (2006) Widespread occurrence of phage-encoded exotoxin genes in terrestrial and aquatic environments in SouthernCalifornia. FEMS Microbiol Lett 261:141–149
Chattopadhyay S, Breidenbach GP, Lyon WG, Wilson JT (2000) Effect of environmental factors in transport of viruses in subsurface system. Preprints of Extended Abstracts presented at the ACS National Meeting, American Chemical Society, Division of Environmental Chemistry, vol 40, pp 375–377
Chebotar VK, Asis CA, Akao S (2001) Production of growth promoting substances and high colonization ability of rhizobacteria enhance the nitrogen fixation of soybean when inoculated with Bradyrhizobium japonicum. Biol Fert Soils 34:427–432
Cheetham BF, Katz ME (1995) A role for bacteriophages in the evolution and transfer of bacterial virulence determinants. Mol Microbiol 18:201–208
Cheng L, Chetochine AS, Pepper IL, Brusseau ML (2007) Influence of DOC on MS-2 bacteriophage transport in a sandy soil. Water Air Soil Pollut 178:315–322
Chetochine AS, Brusseau ML, Gerba CP, Pepper IL (2006) Leaching of phage from class B biosolids and potential transport through soil. Appl Environ Microbiol 72:665–671
Chibani-Chennoufi S, Bruttin A, Dillmann M-L, Brüssow H (2004) Phage–host interaction: an ecological perspective. J Bacteriol 186:3677–3686
Choi S, Jeliazkov I, Jiang SC (2009) Lysogens and free viruses in fresh, brackish, and marine waters: a Bayesian analysis. FEMS Microbiol Ecol 69:243–254
Chu Y, Jin Y, Baumann T, Yates MV (2003) Effect of soil properties on saturated and unsaturated virus transport through columns. J Environ Qual 32:2017–2025
Clark JS, Campbell JH, Grizzle H, Acosta-Martìnez V, Zak JC (2009) Soil microbial community response to drought and precipitation variability in the Chihuahuan Desert. Microb Ecol 57:248–260
Cochran PK, Paul JH (1998) Seasonal abundance of lysogenic bacteria in a subtropical estuary. Appl Environ Microb 64:2308–2312
Corapcioglu MY, Vogel JR, Munster CL, Pillai SD, Dowd S, Wang S (2006) Virus transport experiments in a sandy aquifer. Water Air Soil Pollut 169:47–65
Danhorn T, Fuqua C (2007) Biofilm formation by plant-associated bacteria. Annu Rev Microbiol 61:401–422
Danovaro R, Dell'anno A, Trucco A, Serresi M, Vanucci S (2001) Determination of virus abundance in marine sediments. Appl Environ Microbiol 67:1384–1387
Danovaro R, Manini E, Dell'Anno A (2002) Higher abundance of bacteria than of viruses in deep Mediterranean sediments. Appl Environ Microbiol 68:1468–1472

D'Herelle F (1917) Sur un microbe invisible antagonistic des bacilles dysenteriques. C R Acad Sci Paris 165:373–375

Dhritiman G, Krishnakali R, Williamson KE, White DC, Sublette WKE, KL RM (2008) Prevalence of lysogeny among soil bacteria and presence of 16S rRNA and trzN genes in viral-community DNA. Appl Environ Microbiol 74:495–502

Diallo MD, Willems A, Vloemans N, Cousin S, Vandekerckhove TT, de Lajudie P, Neyra M, Vyverman W, Gillis M, Van der Gucht K (2004) Polymerase chain reaction denaturing gradient gel electrophoresis analysis of the N2-fixing bacterial diversity in soil under Acacia tortilis ssp. raddiana and Balanites aegyptiaca in the dryland part of Senegal. Environ Microbiol 6:400–415

Dröge M, Pühler A, Selbitschka W (1999) Horizontal gene transfer among bacteria in terrestrial and aquatic habitats as assessed by microcosm and field studies. Biol Fertil Soils 29:221–245

Dworkin M, Falkow S, Rosenberg E, Schleifer K-H, Stackebrandt E (eds) (2007) The prokaryotes: a handbook on the biology of bacteria, vol 1–7, 3rd edn. Springer-Verlag, New York

Eberhard AA, Burlingame L, Eberhard C, Kenyon GL, Nealson KH, Oppenheimer NJ (1981) Structural identification of autoinducer of Photobacterium fischeri luciferase. Biochem 20:2444–2449

Epstein PR (1995) Emerging diseases and ecosystem instability: new threats to public health. Am J Public Health 85:168–172

Franchi M, Gallori E (2004) Origin, persistence and biological activity of genetic material in prebiotic habitats. Orig Life Evolut Biosph 34:133–141

Friedrich B, Meyer M, Schlegel HG (1983) Transfer and expression of the herbicide-degrading plasmid pJP4 in aerobic autotrophic bacteria. Arch Microbiol 134:92–97

Gallori E, Biondi E, Franchi M (2004) Mineral surfaces as a cradle of primordial genetic material. Cell Orig Life Extr Habit Astrobiol 7:145–148

Garbeva P, van Veen JA, van Elsas JD (2003) Predominant Bacillus spp. in agricultural soil under different management regimes detected via PCR-DGGE. Microb Ecol 45:302–316

Germida JJ, Khachatourians GG (1988) Transduction of Escherichia coli in soil. Can J Microbiol 43:190–193

Gino E, Starosvertsky J, Armon R (2007) Bacteriophages ecology in a small settlement sewer system in northern part of Israel as related to their indicative role in sewage pollution of drinking water. Environ Microbiol 9:2407–2416

Golovchenko AV, Dobrovol'skaya TG, Fedoritenko MS, Dobrovol'skaya NG, Zvyagintsev DG (2001) The structure of bacterial complexes in the Protva River floodplain. Microbiol (Mosc, Russian Federation) 70:600–605

Gordon SV, Heym B, Parkhill J, Barrell B, Cole ST (1999) New insertion sequences and a novel repeated sequence in the genome of Mycobacterium tuberculosis H37Rv. Microbiol (Reading Engl) 145:881–892

Griffiths AJF, Gelbart WM, Lewontin RC, Miller JH, Freeman WH (eds) (2002) Modern genetic analysis: integrating genes and genomes, 2nd edn. W.H. Freeman & Company (Pub), New York

Guan H, Schulze-Makuch D, Schaffer S, Pillai SD (2003) The effect of critical pH on virus fate and transport in saturated porous medium. Ground Water 41:701–708

Guppy CN, Menzies NW, Moody PW, Blamey FPC (2005) Competitive sorption reactions between phosphorus and organic matter in soil: a review. Austr J Soil Res 43:189–202

Gupta RS (2000) The phylogeny of Proteobacteria: relationships to other eubacterial phyla and eukaryotes. FEMS Microbiol Rev 24:367–402

Gupta RS (2005) Protein signatures distinctive of alpha Proteobacteria and its subgroups and a model for α-proteobacterial evolution. Crit Rev Microbiol 31:101–135

Gupta RS (2006) Molecular signatures (unique proteins and conserved indels) that are specific for the epsilon proteobacteria (Campylobacterales). BMC Genomics 7:167. doi:10.1186/1471-2164-7-167

Gupta V, Hohnson WP, Shafieian P, Ryu H, Alum A, Abbaszdegan M, Hubbs SA, Rauch-Williams T (2009) Riverbank filtration: comparison of pilot scale transport with theory. Environ Sci Technol 43:669–676
Hanczyc MH, Mansy SS, Szostak JW (2007) Mineral surface directed membrane assembly. Orig Life Evol Biosph 37:67–82
Hatfull GF (2008) Bacteriophage genomics. Curr Opin Microbiol 11:447–453
Henckel T, Friedrich M, Conrad R (1999) Molecular analyses of the methane-oxidizing microbial community in rice field soil by targeting the genes of the 16S rRNA, particulate methane monooxygenase, and methanol dehydrogenase. Appl Environ Microbiol 65: 1980–1990
Henckel T, Jackel U, Schnell S, Conrad R (2000) Molecular analyses of novel methanotrophic communities in forest soil that oxidize atmospheric methane. Appl Environ Microbiol 66:1801–1808
Hendrix RW (2003) Bacteriophage genomics. Curr Opin Microbiol 6:506–511
Herron PR, Wellington EMH (1994) Population dynamics of phage-host interactions and phage-conversion of streptomyces in soil. FEMS Microbiol Ecol 14:25–32
Herron PR (2004) Phage ecology and genetic exchange in soil. In: Kowalchuk GA, de Bruijn FJ, Head IM, Akkermans AD, van Elsas JD (eds) Molecular microbial ecology manual (2nd ed). Kluwer Academic, Dordrecht, Section 5.07:1173–1186
Heuer H, Krsek M, Baker P, Smalla K, Wellington EMH (1997) Analysis of actinomycete communities by specific amplification of genes encoding 16S rRNA and gel electrophoretic separation in denaturing gradients. Appl Environ Microbiol 63:3233–3241
Hewitt A (2009) 'Soils', Te Ara – the Encyclopedia of New Zealand, updated 1 Mar 2009. URL: http://www.TeAra.govt.nz/en/soils/6/2
Hiramatsu K, Cui L, Kuroda M, Ito T (2001) The emergence and evolution of methicillin-resistant Staphylococcus aureus. Trends Microbiol 9:486–493
Hoffmann T, Horz HP, Kemnitz D, Conrad R (2002) Diversity of the particulate methane monooxygenase gene in methanotrophic samples from different rice field soils in China and the Philippines. Syst Appl Microbiol 25:267–274
Holden PA, Fierer N (2005) Microbial processes in the vadose zone. Vadose Zone J 4:1–21
Holmfeldt K, Middelboe M, Nybroe O, Lasse Riemann L (2007) Large variabilities in host strain susceptibility and phage host range govern interactions between lytic marine phages and their Flavobacterium hosts. Appl Environ Microbiol 73:6730–6739
Hornek R, Pommerening-Roser A, Koops H-P, Farnleitner AH, Kreuzinger N, Kirschner A, Mach RL (2006) Primers containing universal bases reduce multiple amoA gene specific DGGE band patterns when analyzing the diversity of beta-ammonia oxidizers in the environment. J Microbiol Methods 66:147–155
Hurst CJ (1988) Influence of aerobic microorganisms upon virus survival in soil. Can J Microbiol 34:696–699
Hurst CJ, Gerba CP, Cech I (1980) Effects of environmental variables and soil characteristics on virus survival in soil. Appl Environ Microbiol 40:1067–1079
Ibekwe AM, Papiernik SK, Gan J, Yates SR, Yang CH, Crowley DE (2001) Impact of fumigants on soil microbial communities. Appl Environ Microbiol 67:3245–3257
Jensen S, Ovreas L, Daae FL, Torsvik V (1998) Diversity in methane enrichments from agricultural soil revealed by DGGE separation of PCR amplified 16S rDNA fragments. FEMS Microbiol Ecol 26:17–26
Jiang SC, Paul JH (1996) Occurrence of lysogenic bacteria in marine microbial communities as determined by prophage induction. Mar Ecol Prog Ser 142:27–38
Kaszubiak H (1968) Effect of herbicides on Rhizobium. III. Influence of herbicides on mutation. Acta Microbiol Pol 17:51–57
Keel C, Défago G (1997) Interactions between beneficial soil bacteria and root pathogens: mechanisms and ecological impact. In: Gange AC, Brown VK (eds) Multitrophic interactions in terrestrial systems. Blackwell Science, London, pp 27–46

Keel C, Ucurum Z, Michaux P, Adrian M, Haas D (2002) Deleterious impact of a virulent bacteriophage on survival and biocontrol activity of Pseudomonas fluorescens strain CHA0 in natural soil. Mol Plant Microbe Interact 15:567–576

Keitarou K, Yoshifumi I (2004) Physiological roles of capsule poly-γ-glutamate in Bacillus subtilis. Curr Trends Microbiol 1:59–67

Kent AD, Triplett EW (2002) Microbial communities and their interactions in soil and rhizosphere ecosystems. Annu Rev Microbiol 56:211–236

Kidambi SP, Ripp S, Miller RV (1994) Evidence for phage-mediated gene transfer among Pseudomonas aeruginosa strains on the phylloplane. Appl Environ Microbiol 60:496–500

Koonin EV, Senkevich TG, Dolja VV (2006) The ancient virus world and evolution of cells. Biol Direct 2006, 1: no pages given. 29doi:10.1186/1745-6150-1-29

Kowalchuk GA, Stephen JR, Deboer W, Prosser JI, Embley TM, Woldendorp JW (1997) Analysis of ammonia-oxidizing bacteria of the beta subdivision of the class Proteobacteria in coastal sand dunes by denaturing gradient gel electrophoresis and sequencing of PCR-amplified 16S ribosomal DNA fragments. Appl Environ Microbiol 63:1489–1497

Krylov VN (2003) The role of horizontal gene transfer by bacteriophages in the origin of pathogenic bacteria. Russ J Genetics 39:483–504

Labbe D, Margesin R, Schinner F, Whyte LG, Greer CW (2007) Comparative phylogenetic analysis of microbial communities in pristine and hydrocarbon-contaminated Alpine soils. FEMS Microbiol Ecol 59:466–475

Lambais MR, Otero XL, Cury JC (2008) Bacterial communities and biogeochemical transformations of iron and sulfur in a high saltmarsh soil profile. Soil Biol Biochem 40:2854–2864

Lambers H, Mougel C, Jaillard B, Hinsinger P (2009) Plant–microbe–soil interactions in the rhizosphere: an evolutionary perspective. Plant Soil 321:83–115

Lapidot A, Yaron S (2009) Transfer of Salmonella enterica serovar typhimurium from contaminated irrigation water to parsley is dependent on curli and cellulose, the biofilm matrix components. J Food Protect 72:618–623

Lebedinsky AV, Chernyh NA, Bonch-Osmolovskaya EA (2007) Phylogenetic systematics of microorganisms inhabiting thermal environments. Biochemistry (Mosc) 72:1299–1312

Lehman R (2007) Microbial distributions and their potential controlling factors in terrestrial subsurface environments. In: Franklin R, Mills A (eds) The spatial distribution of microbes in the environment. Springer, Dordrecht, The Netherlands, pp 135–178

Lewandowski Z, Beyenal H (2007) Fundamentals of biofilm research. CRC Press, Taylor & Francis Group, Boca Raton

Leys N, Ryngaert A, Bastiaens L, Verstraete W, Top EM, Springael D (2004) Occurrence and phylogenetic diversity of Sphingomonas strains in soils contaminated with polycyclic aromatic hydrocarbons. Appl Environ Microbiol 70:1944–1955

Lipson SM, Stotzky G (1987) Interactions between viruses and clay minerals. In: Rao VC, Melnick JL (eds) Human viruses in sediments, sludges and soils. CRC, Boca Raton, FL, pp 198–229

Liu Y, Zhang Q, Fang C, Zhu S, Tang Y, Huang S (2005) Effect of glutathione on UV induction of prophage lambda. Arch Microbiol 183:444–449

Liu WT, Marsh TL, Cheng H, Forney LJ (1997) Characterization of microbial diversity by determining terminal restriction fragment length polymorphisms of genes encoding 16S rRNA. Appl Environ Microbiol 63:4516–4522

Ludwig W, Klenk HP (2001) In: Garrity GM et al. (eds) Overview: a phylogenetic backbone and taxonomic framework for prokaryotic systematics. Bergey's manual of systematic bacteriology, 2nd edn, vol 1. Springer Verlag, New York, Berlin, Heidelberg, pp 49–65

Lu TK, Collins JJ (2007) Dispersing biofilms with engineered enzymatic bacteriophage. Proc Natl Acad Sci USA 104:11197–11202

Lynch JM, Benedetti A, Insam H, Nuti MP, Smalla K, Torsvik V, Nannipieri P (2004) Microbial diversity in soil: ecological theories, the contribution of molecular techniques and the impact of transgenic plants and transgenic microorganisms. Biol Fertil Soils 40:363–385

Maier RM, Pepper IL, Gerba CP (eds) (2009) Environmental microbiology, 2nd edn. Academic Press, San Diego
Marsh P, Wellington EMH (1992) Interactions between acinophage and their streptomycete hosts in soil and the fate of phage-borne genes. In: Gauthier MJ (ed) Gene transfers and environment. Springer, Berlin Heidelberg New York, pp 135–142
Marsh P, Wellington EMH (1994) Phage–host interactions in soil. FEMS Microbiol Ecol 15:99–107
McInerney MJ, Struchtemeyer CG, Sieber J, Mouttaki H et al (2008) Physiology, ecology, phylogeny, and genomics of microorganisms capable of syntrophic metabolism. Ann NY Acad Sci 1125:58–72
McLeod M, Aislabie J, Ryburn J, McGill A (2004) Microbial and chemical tracer movement through granular, ultic, and recent soils. New Zeal J Agr Res 47:557–563
McLeod M, Aislabie J, Ryburn J, McGill A, Taylor M (2003) Microbial and chemical tracer movement through two Southland soils, New Zealand. Aust J Soil Res 41:1163–1169
Miller RV (2001) Environmental bacteriophage–host interactions: factors contribution to natural transduction. Anton Leeuw Int J G 79:141–147
Miller RV, Ripp S (1998) The importance of pseudolysogeny to in situ bacteriophage-host interactions. In: Syvanen M, Kado CI (eds) Horizontal gene transfer. Chapman and Hall, London, pp 78–191
Moll DM, Vestal JR (1992) Survival of microorganisms in smectite clays: implications for Martian exobiology. Icarus 98:233–239
Monier J-M, Bernillon D, Kay E, Faugier A, Rybalka O, Dessaux Y, Simonet P, Vogel TM (2007) Detection of potential transgenic plant DNA recipients among soil bacteria. Environ Biosafety Res 6:71–83
Murray PR, Baron EJ, Pfaller MA, Tenover FC, Yolken RH (eds) (1995) Manual of clinical microbiology, 6th edn. ASM Press, Washington, DC
Nakatsu CH (2007) Soil microbial community analysis using denaturing gradient gel electrophoresis. Soil Sci Soc Am J 71:562–571
Narula N, Kothe E, Behl RK (2009) Role of root exudates in plant–microbe interactions. J Applied Botany Food Qual 82:122–130
Nasser AM, Glozman R, Nitzan Y (2002) Contribution of microbial activity to virus reduction in saturated soil. Water Res 36:2589–2595
Nealson KH, Platt T, Hastings JW (1970) Cellular control of synthesis and activity of the bacterial luminescent system. J Bacteriol 104:313–322
Nelson DM, Ohene-Adjei S, Hu FS, Cann IKO, Mackie RI (2007) Bacterial diversity and distribution in the holocene sediments of a northern temperate lake. Microb Ecol 54:252–263
Nemergut DR, Costello EK, Meyer AF, Pescador MY, Weintraub MN, Schmidt SK (2005) Structure and function of alpine and arctic soil microbial communities. Res Microbiol 156:775–784
Nielsen KM, Bones AM, Smalla K, van Elsas JD (1998) Horizontal gene transfer from transgenic plants to terrestrial bacteria – a rare event? FEMS Microbiol Rev 22:79–103
Nielsen KM, Townsend JP (2004) Monitoring and modeling horizontal gene transfer. Nature Biotechnol 22:1110–1114
Noll M, Wellinger M (2008) Changes of the soil ecosystem along a receding glacier: testing the correlation between environmental factors and bacterial community structure. Soil Biol Biochem 40:2611–2619
Normand P, Lapierre P, Tisa LS et al (2007) Genome characteristics of facultatively symbiotic Frankia sp. strains reflect host range and host plant biogeography. Genome Res 17:7–15
Ogram A, Sharma K (2002) Methods of soil microbial community analysis. In: Hurst CJ (ed) Manual of environmental microbiology, 2nd edn. ASM Press, Washington, DC, pp 554–563
Oren A (2004) Prokaryote diversity and taxonomy: current status and future challenges. Phil Trans R Soc Lond B 359:623–638

Øvreås L, Torsvik V (1998) Microbial diversity and community structure in two different agricultural soil communities. Microb Ecol 36:303–315

Pearson A, Page SRF, Jorgenson TL, Fischer WW, Higgins MB (2007) Novel hopanoid cyclases from the environment. Environ Microbiol 9:2175–2188

Pietramellara G, Ascher J, Ceccherini MT, Nannipieri P, Wenderoth D (2007) Adsorption of pure and dirty bacterial DNA on clay minerals and their transformation frequency. Biol Fertil Soils 43:731–739

Poletika NN, Jury WA, Yates MV (1995) Transport of bromide, simazine and MS-2 coliphage in a lysimeter containing undisturbed, unsaturated soil. Water Resour Res 31:801–810

Pontiroli A, Rizzi A, Simonet P, Daffonchio D, Vogel TM, Monier J-M (2009) Visual evidence of horizontal gene transfer between plants and bacteria in the phytosphere of transplastomic tobacco. Appl Environ Microbiol 75:3314–3322

Prigent M, Leroy M, Confalonieri F, Dutertre M, DuBow MS (2005) A diversity of bacteriophage forms and genomes can be isolated from the surface sands of the Sahara Desert. Extremophiles 9:289–296

Quintana ET, Wierzbicka K, Mackiewicz P, Osman A, Fahal AH, Hamid ME, Zakrzewska-Czerwinska J, Maldonado LA, Goodfellow M (2008) Streptomyces sudanensis sp. nov., a new pathogen isolated from patients with actinomycetoma. Anton Leeuw Int J G 93:305–313

Ravenschlag K, Sahm K, Pernthaler J, Amann R (2006) High bacterial diversity in permanently cold marine sediments. Appl Environ Microbiol 72:4596–4603

Raya RR, Hébert EM (2009) Isolation of Phage via Induction of Lysogens. In: Clokie MRJ, Kropinski AM (eds) Bacteriophages, (Methods and Protocols, Volume 1: Isolation, Characterization, and Interactions), Series: Methods in Molecular Biology, Vol. 501, Sec. 1.3, 23–32

Reith F, Rogers SL (2008) Assessment of bacterial communities in auriferous and non-auriferous soils using genetic and functional fingerprinting. Geomicrobiol J 25:203–215

Rendulic S, Jagtap P, Rosinus A, Eppinger M, Baar C et al (2004) A predator unmasked: life cycle of Bdellovibrio bacteriovorus from a genomic perspective. Science 303(5658):689–692

Ripp S, Jegier P, Birmele M, Johnson CM, Daumer KA, Garland JL, Sayler GS (2006) Linking bacteriophage infection to quorum sensing signaling and bioluminescent bioreporter monitoring for direct detection of bacterial agents. J App Microbiol 100:488–499

Ripp S, Jegier P, Lopes N, Sayler G (2009) Phage reporter biosensing of pathogenic agents. In: Laudon M, Romanowich B (eds) Nanotech Conference & Expo 2009: An Interdisciplinary Integrative Forum on Nanotechnology, Biotechnology and Microtechnology, Vol. 2. CRC Press Boca Raton, Fla, Houston, TX, United States, 3–7 May 2009, pp 119–122

Roesch LFW, Fulthorpe RR, Riva A, Casella G, Hadwin AKM, Kent AD, Daroub SH, Camargo FAO, Farmerie WG, Triplett EW (2007) Pyrosequencing enumerates and contrasts soil microbial diversity. ISME J 1:283–290

Romig WR, Brodetsky AM (1961) Isolation and preliminary characterization of bacteriophages for Bacillus subtilis. J Bacteriol 82:135–141

Rosado AS, Duarte GF, Seldin L, van Elsaa JD (1998) Genetic diversity of nifh gene sequences in Paenibacillus azotofixans strains and soil samples analyzed by denaturing gradient gel electrophoresis of PCR amplified gene fragments. Appl Environ Microbiol 64:2770–2779

Roslycky EB (1982) Influence of selected herbicides in phages of some soil bacteria. Can J Soil Sci 62:217–220

Rudrappa T, Biedrzycki ML, Bais HP (2008) Causes and consequences of plant-associated biofilms. FEMS Microbiol Ecol 64:153–166

Ryan JN, Harvey RW, Metge D, Elimelech M, Navigato T, Pieper AP (2002) Field and laboratory investigations of inactivation of viruses (PRD1 and MS2) attached to iron oxide-coated quartz sand. Environ Sci Technol 36:2403–2413

Salles JF, van Veen JA, van Elsas JD (2004) Multivariate analyses of Burkholderia species in soil: effect of crop and land use history. Appl Environ Microbiol 70:4012–4020

Sandaa R-A, Torsvik V, Enger Ø (2001) Influence of long-term heavy-metal contamination on microbial communities in soil. Soil Biol Biochem 33:287–295

Santos F, Meyerdierks A, Peña A, Rosselló-Mora R, Amann R, Antón J (2007) Metagenomic approach to the study of halophages: the environmental halophage1. Environ Microbiol 9:1711–1723
Schijven JF, Hassanizadeh SM (2000) Removal of viruses by soil passage: overview of modeling, processes, and parameters. Crit Rev Env Sci Tec 30:49–127
Schloss PD, Handelsman J (2006) Toward a census of bacteria in soil. PLoS Comput Biol 2:786–793
Scholten JC, Culley DE, Brockman FJ, Wu G, Weiwen Z (2007) Evolution of the syntrophic interaction between Desulfovibrio vulgaris and Methanosarcina barkeri: involvement of an ancient horizontal gene transfer. Biochem Biophys Res Commun 352:48–54
Schuch R, Fischetti VA (2009) The secret life of the Anthrax agent Bacillus anthracis: bacteriophage-mediated ecological adaptations. PLoS ONE 4:e6532. doi:10.1371/journal.pone.0006532
Schulze-Makuch D, Guan H, Pillai SD (2003) Effects of pH and geological medium on bacteriophage MS2 transport in a model aquifer. Geomicrobiol J 20:73–84
Seeley ND, Primrose SB (1980) The effect of temperature on the ecology of aquatic bacteriophages. J Gen Virol 46:87–95
Seghers D, Bulcke R, Reheul D, Siciliano SD, Top EM, Verstraete W (2003) Pollution induced community tolerance (PICT) and analysis of 16S rRNA genes to evaluate the long-term effects of herbicides on methanotrophic communities in soil. Eur J Soil Sci 54:679–684
Sessitsch A, Weilharter A, Gerzabek MH, Kirchmann H, Kandeler E (2001) Microbial population structures in soil particle size fractions of a long-term fertilizer field experiment. Appl Environ Microbiol 67:4215–4224
Sinton LW, Finlay RK, Pang L, Scott DM (1997) Transport of bacteria and bacteriophages in irrigated effluent into and through an alluvial gravel aquifer. Water Air Soil Pollut 98:17–42
Sobczak WV, Lars O, Hedin LO, Klug MJ (1998) Relationships between bacterial productivity and organic carbon at a soil–stream interface. Hydrobiol 386:45–53
Solomon EB, Yaron S, Matthews KR (2002) Transmission of Escherichia coli O157:H7 from contaminated manure and irrigation water to lettuce plant tissue and its subsequent internalization. Appl Environ Microbiol 68:397–400
Stackebrandt E, Liesack W, Goebel BM (1993) Bacterial diversity in a soil sample from a subtropical Australian environment as determined by 16S rDNA analysis. FASEB J 7:232–236
Steinberger Y, Zelles L, Bai QY, von Lützow M, Munch JC (1999) Phospholipid fatty acid profiles as indicators for the microbial community structure in soils along a climatic transect in the Judean Desert. Biol Fertil Soils 28:292–300
Stewart FM, Levin BR (1984) The population biology of bacterial viruses – why be temperate. Theor Popul Biol 26:93–117
Stotzky G (1989) Gene transfer among bacteria in soil. In: Levy SB, Miller RV (eds) Gene transfer in the environment. McGraw-Hill, New York, pp 165–222
Su J, Wu Y, Ma X, Zhang G, Feng H, Zhang Y (2004) Soil microbial counts and identification of culturable bacteria in an extreme by arid zone. Folia Microbiol 49:423–429
Sugiyama S, Zabed HM, Okubo A (2008) Relationships between soil microbial diversity and plant community structure in seminatural grasslands. Grassland Sci 54:117–124
Sumner ME (ed) (2000) Handbook of soil science. CRC Press, Boca Raton
Szponar B, Mordarska H (1997) Rhodococcus equi – human and animal opportunistic pathogen. Postepy Mikrobiologii 36:353–368
Tarlera S, Jangid K, Ivester HA, Whitman WB, Williams MA (2008) Microbial community succession and bacterial diversity in soils during 77,000 years of ecosystem development. FEMS Microbiol Ecol 64:129–140
Taylor JP, Wilson B, Mills MS, Burns RG (2002) Comparison of microbial numbers and enzymatic activities in surface soils and subsoils using various techniques. Soil Biol Biochem 34:387–401
Teske A, Wawer C, Muyzer G, Ramsing NB (1996) Distribution of sulfate-reducing bacteria in a stratified fjord (Mariager Fjord, Denmark) as evaluated by most-probable-number counts and

denaturing gradient gel electrophoresis of PCR-amplified ribosomal DNA fragments. Appl Environ Microbiol 62:1405–1415

Thompson SS, Yates MV (1999) Bacteriophages inactivation at the air–water–solid interface in dynamic batch systems. Appl Environ Microbiol 65:1186–1190

Throback IN, Enwall K, Jarvis A, Hallin S (2004) Reassessing PCR primers targeting nirS, nirK and nosZ genes for community surveys of denitrifying bacteria with DGGE. FEMS Microbiol Ecol 49:401–417

Torsvik V, Goksoyr J, Daae FL (1990) High diversity in DNA of soil bacteria. Appl Environ Microbiol 56:782–787

Torsvik V, Ovreas L (2002) Microbial diversity and function in soil: from genes to ecosystems. Curr Opin Microbiol 5:240–245

Toure IM, Stenz E (1977) The effect of selected herbicides on bacteriophages and Escherichia coli. Zentr Bakter Paras Infekt Hygiene 2:163–177

Twort FW (1915) An investigation on the nature of ultra-microscopic viruses. Lancet 186:1241–1243

USEPA, Untied States Environmental Protection Agency (2002) Biosolids applied to land: advancing standards and practices. http://www.epa.gov/waterscience/biosolids/nas/complete.pdf

Vaeck M, Reynaerts A, Hofte H, Jansens S, De Beuckeleer M, Dean C, Zabeau M, Van Montagu M, Leemans J (1987) Transgenic plants protected from insect attack. Nature 328:33–37

van der Wielen PWJJ, Senden WJMK, Medema G (2008) Removal of bacteriophages MS2 and fX174 during transport in a sandy anoxic aquifer. Environ Sci Technol 42:4589–4594

Van Elsas JD, Tam L, Finlay RD, Kilham K, Trevors JT (2007) Microbial interactions in soil. In: Van Elsas JD, Jansson JK, Trevors JT (eds) Modern soil ecology. CRC Press, Taylor & Francis Group, Boca Raton, FL, pp 1177–210

van Regenmortel MHV, Fauquet CM, Bishop DHL, Carstens EB, Estes MK, Lemon SM, Maniloff J, McGeoch DJ, Pringle CR, Wickner RB (eds) (2000) Virus Taxonomy, Seventh Report of the International Committee on Taxonomy of Viruses, Academic Press, San Diego, pp 63–136, 267–284, 389–393, 645–650

Vanbroekhoven K, Ryngaert A, Wattiau P, De Mot R, Springael D (2004) Acinetobacter diversity in environmental samples assessed by 16S rRNA gene PCR-DGGE fingerprinting. FEMS Microbiol Ecol 50:37–50

Vaneechoutte M, Dijkshoorn L, Tjernberg I, Elaichouni A, De vos P et al (1995) Identification of acinetobacter genomic species by amplified ribosomal DNA restriction analysis. J Clin Microbiol 33:11–15

Vos M, Birkett PJ, Birch E, Griffiths RI, Buckling A (2009) Local adaptation of bacteriophages to their bacterial hosts in soil. Science 325:833

Wächtershäuser G (1988) Before enzymes and templates: theory of surface metabolism. Microbiol Rev 52:452–484

Wang I-N (2006) Lysis timing and bacteriophage fitness. Genetics 172:17–26

Wang J, Wu Y, Jiang H, Li C, Dong H, Wu Q, Soininen J, Shen J (2008) High beta diversity of bacteria in the shallow terrestrial subsurface. Environ Microbiol 10:2537–2549

Wawer C, Muyzer G (1995) Genetic diversity of Desulfovibrio spp. in environmental samples analyzed by denaturing gradient gel electrophoresis of [NiFe] hydrogenase gene fragments. Appl Environ Microbiol 61:2203–2210

Weinbauer MG (2004) Ecology of prokaryotic viruses. FEMS Microbiol Rev 28:127–181

Weinbauer MG, Suttle CA (1996) Potential significance of lysogeny to bacteriophage production and bacterial mortality in coastal waters of the Gulf of Mexico. Appl Environ Microbiol 62:4374–4380

Williamson KE, Radosevich M, Wommack KE (2005) Abundance and diversity of viruses in six Delaware soils. Appl Environ Microbiol 71:3119–3125

Williamson KE, Schnitker JB, Radosevich M, Smith DW, Wommack KE (2008) Cultivation-based assessment of lysogeny among soil bacteria. Microb Ecol 56:437–447

Wise MG, McArthur JV, Shimkets LJ (1999) Methanotroph diversity in landfill soil: isolation of novel type I and type II methanotrophs whose presence was suggested by culture-independent 16S ribosomal DNA analysis. Appl Environ Microbiol 65:4887–4897

Woese CR, Kandler O, Wheelis ML (1990) Towards a natural system of organisms: proposal for the domains Archaea, Bacteria, and Eucarya. Proc Natl Acad Sci USA 87:4576–4579

Wommack KE, Colwell RR (2000) Virioplankton: viruses in aquatic ecosystems. Microbiol Mol Biol Rev 64:69–114

Zeph LR, Onaga MA, Stotzky G (1988) Transduction of Escherichia coli by bacteriophage P1 in soil. Appl Environ Microbiol 54:1731–1737

Zhou J, Davey ME, Figueras JB, Rivkina E, Gilichinsky D, Tiedje JM (1997) Phylogenetic diversity of a bacterial community determined from Siberian tundra soil DNA. Microbiol 143:3913–3919

Zhuang J, Jin Y (2003) Virus retention and transport as influenced by different forms of soil organic matter. J Environ Qual 32:816–823

Zinder ND, Lederberg J (1952) Genetic exchange in Salmonella. J Bacteriol 64:679–699

Chapter 4
Soil Phage Ecology: Abundance, Distribution, and Interactions with Bacterial Hosts

Kurt E. Williamson

4.1 Introduction

The discovery of viruses really begins with soil. The sequential efforts of Mayer (1886), Iwanowski (1892), and Beijerinck (1898) led to the gradual realization that the agent responsible for a prominent plant disease was something quite different from a bacterium (Zaitlin 1998). This submicron agent, dubbed a virus (from the Latin *virus* or poison), has completely changed the face of microbiology. The first known virus, tobacco mosaic virus (TMV), as we know it today, has enjoyed a glorious history of firsts in the viral world: the first virus to be positively identified, the first virus to be chemically purified, the first virus to be viewed under electron microscopy, and the first virus to have its protein capsid sequenced [(Creager et al. 1999) and references therein]. As a direct consequence of these seminal achievements, research on TMV has pushed forward new frontiers in the fields of genetics, molecular biology, and structural biochemistry. But while TMV has gone on to enjoy the scientific spotlight, its humble origins have been all but forgotten. For tobacco mosaic virus is not transmitted to new plants by nematode or insect vectors, but rather depends on soil as a repository and vehicle for transmission (Creager et al. 1999). The great irony of the situation is that this autochthonous soil virus led to such expansion and growth in the biological sciences, and yet so little is known regarding the ecology of its brethren: viruses native to soil.

While the discovery of viruses fostered an initial exploration phase of soil viral ecology [e.g., (Demolon and Dunez 1935)], such studies apparently languished for decades, perhaps in the wake of more exciting phage-based discoveries in medicine and molecular biology (Mann 2005). The fairly recent discovery of high viral abundance in the marine water column (Borsheim et al. 1990; Heldal and Bratbak 1991; Proctor and Fuhrman 1990) and sediments (Bergh et al. 1989; Paul et al. 1993)

K.E. Williamson
Department of Biology, College of William and Mary, Williamsburg, VA 23187, USA
e-mail: kewilliamson@wm.edu

G. Witzany (ed.), *Biocommunication in Soil Microorganisms*, Soil Biology 23,
DOI 10.1007/978-3-642-14512-4_4, © Springer-Verlag Berlin Heidelberg 2011

reignited curiosity about viruses in the environment. Since the early 1990s, the literature has reflected an explosion of interest in the prevalence of viruses in various environments and the effects of viral activity on microbial evolution and ecological processes [for reviews, see (Fuhrman 1999; Weinbauer 2004; Wommack and Colwell 2000)]. Indeed, viruses have been observed everywhere cellular life can be found: from oceans to soils (Ashelford et al. 2003; Williamson et al. 2005), from geothermal hot springs (Rice et al. 2001) to deep sea hydrothermal vents (Wommack et al. 2004), and from ice-covered Antarctic lakes (Madan et al. 2005) to parched Saharan sands (Prigent et al. 2005). But beyond sheer enumeration, we have only recently begun to place viruses within an ecological context.

John Andrews described the field of microbial ecology as a set of experiments in search of a theoretical framework (Andrews 1991). Sufficient experimental evidence has emerged from studies in marine microbial ecology to assemble a tentative framework that includes viruses (Bratbak and Heldal 2000; Bratbak et al. 1994; Fuhrman 1999; Weinbauer 2004; Wommack and Colwell 2000). It is now widely known that viruses significantly impact biogeochemical cycles in marine environments, particularly the carbon cycle (Azam et al. 1983; Bratbak et al. 1994; Fuhrman 1992; Middelboe et al. 1996). Indeed, features such as the viral shunt of the microbial loop (Li and Dickie 2001) and the “kill the winner” hypothesis (Thingstad 2000) are the reification of recurring motifs found within accumulated experimental data from marine ecosystems. By contrast, the extremely scanty experimental evidence available regarding autochthonous soil virus assemblages has precluded the construction of any sturdy, paradigmatic framework for soils. However, this situation is beginning to change. The development of appropriate methods specific to the study of viruses in soils has enabled us to take the first, tentative steps toward establishing the roles of viruses in soil ecosystems. This chapter will explore some of the emerging trends from data amassed thus far, including viral abundance, viral diversity, and viral impacts in soil ecosystems.

4.2 Measuring Viral Abundance in Soils

4.2.1 Targeted Assays

Historically, the study of viruses in soil has focused on economically significant pathogens of plants, animals, and insects. A large fraction of research on viruses in soil has been concerned with the fate and transport of land-applied human viral pathogens through wastewater or biosolids (Duboise et al. 1979; Santamaria and Toranzos 2003). Consequently, most efforts to enumerate viruses within soil samples have emphasized the quantification of virus particles infecting specific hosts of interest. Early efforts directed at the detection and enumeration of viruses in soils used plaque assays to quantify infectious virus particles (Grijns 1927). Such plaque assays not only provided infectious particle counts, but were also critical for isolation of new phage strains for laboratory use. However, a major shortcoming of these assays is that detection and enumeration are limited only to specific viruses capable of propagating within the selected target organism, under the specific conditions tested (temperature, humidity, etc.). Thus, plaque assays provide no information about total viral abundance in soils.

A similar scenario exists today with the targeted and highly sensitive detection and enumeration strategies available for specific viral agents. For example, the use of integrated cell culture PCR (Reynolds et al. 1996) or real-time PCR (Ratti et al. 2004) enables the accurate detection of extremely low numbers of virus particles in soil samples. But these strategies can only be employed in the enumeration of viruses for which the complete genome sequence or specific marker genes are known, and, like the plaque assay, cannot provide any indication of total viral abundance.

4.2.2 Total Direct Counts of Viruses in Soil

Total virus abundance in soils may be assessed through transmission electron microscopy (TEM) or epifluorescence microscopy (EFM), each of which enable the direct counting of virus particles. Both of these direct counting procedures require efficient dispersion of soil aggregates and extraction of virus particles from the soil matrix into an aqueous carrier phase. Physical dispersion has been accomplished through bead-beating (Ashelford et al. 2003), blending (Swanson et al. 2009), or sonication (Williamson et al. 2003) of soils suspended in buffer, such as 1% potassium citrate. For EFM enumeration, virus suspensions are passed through 0.02 μm filters, and the captured virus particles are stained with a fluorochrome that binds to nucleic acids (Williamson et al. 2003). Filters are viewed under an appropriate excitation wavelength for the fluorochrome stain, and the emitted fluorescence signals from the stained virus genomes are captured digitally using a charge-cooled device camera. For TEM enumeration, additional purification such as density gradient centrifugation may be necessary prior to loading samples onto EM grids, so as to clearly visualize virus particles. Grids are then stained with an electron-dense heavy metal salt, such as uranyl acetate or phosphotungstic acid, and virus particles are visualized in a transmission electron microscope.

While TEM allows for visualization of viral morphology, electron microscopy entails higher operating costs, increased sample processing time, and poorer capture efficiency generally leading to lower abundance estimates. Epifluorescence microscopy is cheaper and allows high sample throughput with much higher capture efficiency, but does not allow for visualization of fine structure. Furthermore, EFM may detect fluorescence signals from nonviral sources, thus care must be taken in processing images to obtain viral direct count data.

As of January, 2010, it is critical to note that the use of EFM to enumerate virus particles in environmental samples may be facing abrupt decline, due to several supply chain problems with the Whatman Anodiscs that are exclusively used in these procedures. While other filter types of the requisite pore size (0.02 μm) are commercially available, a viable substitute has not been identified with the porosity (throughput), rigidity (evenness of focal plane), or low background fluorescence associated with the Anodisc line. Many recently published studies of viral abundance in aquatic ecosystems have avoided this catastrophe by capitalizing on improvements in flow-assisted cell sorting technologies to enumerate viral particles (Personnic et al. 2009; Tijdens et al. 2008). However, flow-cytometric approaches have not been evaluated for enumeration of viruses in soil extracts. Without a reliable supply of filters suitable for EFM, continued determinations of viral abundance in soils may depend on the development of flow-cytometric approaches or a return to TEM.

4.3 Trends in Soil Viral Abundance and Distribution

Soils provide an incredible range of niches housing a diverse array of organisms, leading to a similarly diverse representation of viruses that parasitize these organisms. Soils have been shown to contain plant viruses (Delogu et al. 2003; Fillhart et al. 1998), insect viruses (Christian et al. 2006; Fuxa 2004), fungal viruses (Melzer and Bidochka 1998), animal viruses (Duboise et al. 1979; Pourcher et al. 2007; Santamaria and Toranzos 2003), and bacterial viruses (also known as bacteriophages, or simply "phages") (Ashelford et al. 1999a; Ashelford et al. 2003; Keel et al. 2002; Williamson et al. 2007; Williamson et al. 2003; Yin et al. 1997). However, more recent data, particularly from TEM-based direct count studies, indicate that the majority of naturally occurring soil viruses are tailed phages, most likely with dsDNA genomes (Swanson et al. 2009; Williamson et al. 2005). This is consistent with the overwhelming abundance of bacteria relative to other potential host organisms in soil ecosystems.

Trends in total viral abundance across soil ecotypes (Fig. 4.1a) are quite striking. First of all, it is important to note methodological differences in how viral direct counts were obtained for these soils. In particular, the bead-beating extraction approach used for the Cardiff and Oxford soils (black bars) was likely not as effective as sonication (gray bars) or blender extractions (white bars), leading to an apparent reduction in viral abundance for these soils. Beyond this methods-based difference, viral abundance appears to vary within approximately one order of magnitude (ca. 10^8–10^9 gdw^{-1}) across a wide variety of soil ecotypes, including samples from Antarctic, agricultural (bulk soil and rhizosphere), and wetland soils. A notable exception is the hot desert soils, in which viral abundance is several orders of magnitude lower (ca. 10^5 gdw^{-1}). In support of this reduced viral abundance in hot deserts, a study of viral ecology of Saharan sands found that viral abundance could not be determined for these samples, as particle counts were below the limit of detection (Prigent et al. 2005). In seeking to explain the observed trends in viral abundance across soils, one must consider several factors, including the rates of viral production and decay in these systems.

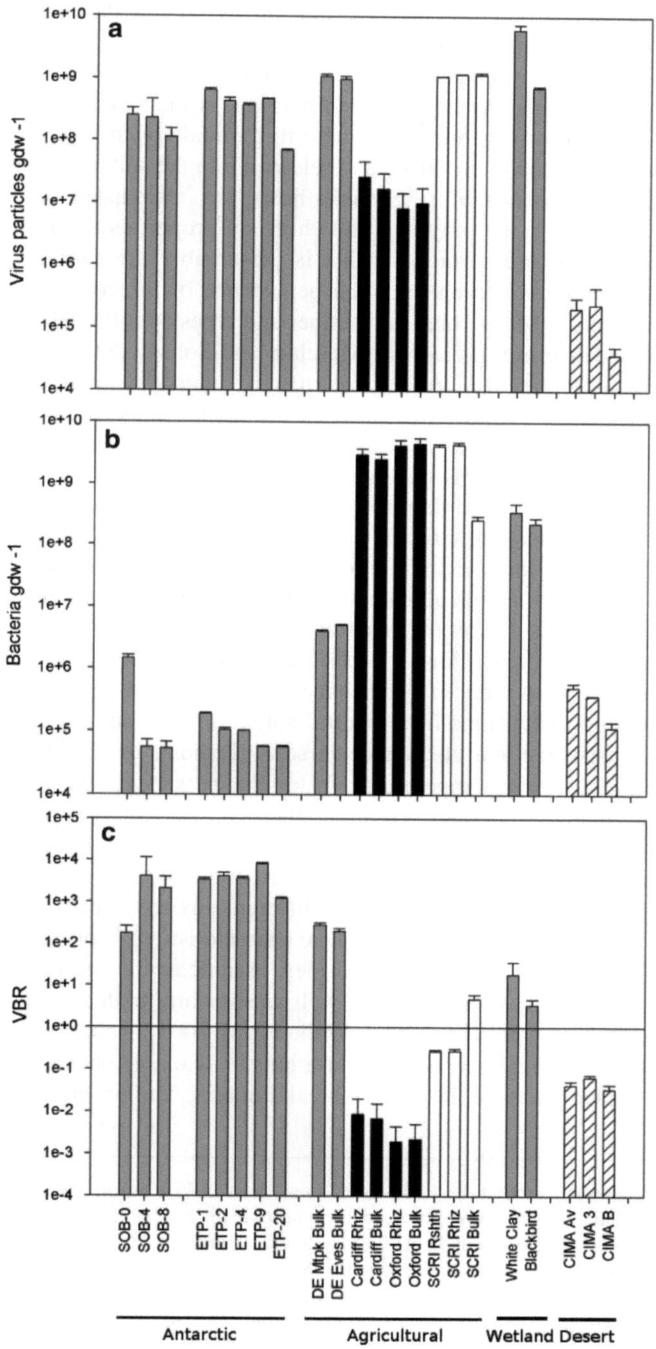

Fig. 4.1 (**a**) Viral abundance, (**b**) bacterial abundance, and (**c**) virus-to-bacterium ratio (VBR) across soil ecotypes. *Gray bars* denote abundance data that were gathered using sonication-based

In the temperate agricultural and wetland soils, viral abundance is significantly correlated to bacterial abundance (Fig. 4.1b) (Williamson et al. 2007). This supports the previous position that most viruses in these soils are phages and indicates that the presence and abundance of susceptible hosts is a key factor controlling viral abundance. When host organisms are present, particularly in high numbers, new viruses may be continually produced and released into the soil matrix through lytic infections. In the absence of susceptible hosts, the abundance of extracellular viruses is controlled by the physical and chemical properties of the soil environment. Adsorption to soil surfaces, which is presumably the natural state of soil viruses, has been shown to enhance the persistence of infectious virus particles (Lipson and Stotzky 1986). Thus, the number and types of colloidal surfaces affect total viral abundance in a given soil (Meschke and Sobsey 2003; Zhuang and Jin 2003). Soils with higher clay and organic matter contents generally provide larger surface area for adsorption of viral particles and harbor higher viral abundances.

Soil moisture content is a key factor controlling viral abundance and persistence in soils, as thicker water films have smaller interfacial areas (air–water interface, solid–water interface, and triple-phase boundary), and surface tension created at interfaces tends to inactivate viruses (Gerba 1984; Jin and Flury 2002 ; Zhuang and Jin 2003). The pH of the soil solution (Loveland et al. 1996) is another important abiotic factor controlling viral abundance, with higher abundances generally occurring in neutral pH soils. Lower pH generally favors stronger adsorption of viruses to soil surfaces (Gerba 1984); thus, it is unclear whether higher viral abundances in neutral soils are due to reduced persistence of intact virions in acid soils or more efficient extraction of virions from neutral soils. Temperature significantly influences viral persistence and abundance in soils (Duboise et al. 1979; Wen et al. 2004). Generally, lower temperatures foster the persistence or preservation of viruses while higher temperatures promote thermal decay of viral particles.

Currently, two studies have compared distribution of viral abundance between rhizosphere and bulk soil (Fig. 4.1a) (Ashelford et al. 2003; Swanson et al. 2009). The rhizosphere represents a completely different environment from bulk soil in terms of its chemistry, biology, and physical characteristics. Roots exude a range of organic molecules including polysaccharides, organic acids, and proteins (Bowen and Rovira 1999; Jones et al. 2004) as well as signaling molecules that attract or suppress specific microbes (Fray 2002; Walker et al. 2003). Root exudates may also significantly impact local pH and concentrations of available phosphorus (Pavinato et al. 2008). Due to these important local impacts, known collectively as the

Fig. 4.1 (Continued) extraction and EFM-based direct counts (Williamson et al. 2007). *Black bars* pertain to abundance data that were gathered using bead-beating extraction and TEM-based direct counts (Ashelford et al. 2003); and the *white bars* denote abundance data obtained through Waring blender extractions and TEM-based direct counts (Swanson et al. 2009). The *hatched bars* indicate unpublished data from the author, using sonication-based extraction and EFM-based direct counts. Sample names correspond to those provided in the original manuscripts: numbers for the Antarctic transects indicate meters from the shores of ephemeral ponds; Bulk, bulk soil; Rhiz, rhizosphere; Rshth, rhizosheath. *Error bars* indicate standard deviation

rhizosphere effect (Hartmann et al. 2007), bacterial abundance is generally higher in the rhizosphere as compared with the bulk soil (Jones et al. 2004; Nie et al. 2009; Rangarajan et al. 2002; Swanson et al. 2009). Yet, surprisingly, no significant differences in viral abundance were observed across these very distinct soil compartments (Ashelford et al. 2003; Swanson et al. 2009). The keys to interpreting these puzzling results may lie at the intersection between viral production, viral decay, and the persistence of virus particles in the soil environment (discussed below, in Sect. 4.4).

4.4 The Virus-to-Bacterium Ratio

The ratio of viral abundance to bacterial abundance (virus-to-bacterium ratio, or VBR) has been used as means to infer relationships between viral production and decay, and bacterial production and infection in aquatic ecosystems (Wommack and Colwell 2000). In aquatic ecosystems, two significant trends have emerged regarding VBR, regardless of large ranges in temperature, salinity, and trophic status across the systems of study: (1) a significant positive relationship exists between viral abundance and bacterial abundance, and (2) the range of VBRs is surprisingly narrow, typically varying from 1 to 50 with a mean value of about 10 (Weinbauer 2004; Wommack and Colwell 2000). Significant correlations between viral and bacterial abundance suggest a close coupling between viral and bacterial production in aquatic environments, while the narrow range of VBRs may indicate balanced rates of viral production and decay.

In contrast to the relatively narrow range of VBRs in aquatic systems, including marine and freshwater sediments, VBRs of soils encompass a range of several orders of magnitude (Fig. 4.1c). The low VBRs of wetland soils suggest that in perennially wet, temperate soils, viral and bacterial production are tightly linked and viral production and decay are closely balanced. The VBRs of agricultural soils vary greatly, depending upon extraction methods used for both virus particles and bacterial cells. Specifically, when bead-beating extractions were used, viral abundances were almost 100-fold lower than viral abundances obtained using sonication or blender extractions (Fig. 4.1a). Currently, it is unclear as to whether the observed differences in viral abundance across the surveyed agricultural soils are due to extraction efficiency or to actual differences in viral abundance between these soils. Similar disparities are apparent in bacterial direct counts across studies (Fig. 4.1b), and again, one cannot distinguish whether these differences arise from genuine biological properties of the soils in question or from methodological artifact. In the remaining studies (i.e., those which did not use bead-beating extractions), viral abundances are surprisingly consistent across soil ecotypes, and it appears that changes in VBR are due largely to differences in bacterial abundance across the soils. If we ignore, temporarily, concerns regarding extraction efficiency and assume that all data presented are reasonable estimates of actual abundance, the highly variable VBRs of agricultural soils would indicate a disconnect between viral production and decay. Either viruses are produced faster than they are destroyed or removed from these systems (Swanson et al. 2009; Williamson et al. 2007), or viruses are produced at very low rates compared with host proliferation (Ashelford et al. 2003).

The remarkably high VBRs recorded for Antarctic soils are due not to the fact that these soils have extraordinarily high viral loads compared with other soils, but rather, their relatively low bacterial densities. Indeed, while bacterial abundances are several orders of magnitude lower than those of agricultural soils, viral abundance in the Antarctic soils is comparable to those in temperate agricultural soils. This is most likely due to long-term storage of viruses in the Antarctic. A multiplicity of growth-limiting conditions, including low temperature, lack of free water, and scarcity of organic substrates for growth, explains the low bacterial abundances in these soils. It is unlikely that the sparse bacterial population is capable of supporting such a burgeoning population of viruses. Since viruses are known to persist longer when adsorbed to surfaces (Lipson and Stotzky 1986; Vettori et al. 2000) and kept at low temperatures (Duboise et al. 1979; Wen et al. 2004), it is likely that viral abundances obtained for Antarctic soils are comprised largely of stored viruses.

By contrast, the very low VBRs for hot desert samples indicate that extracellular viruses are relatively rare in these environments. The abundance of extracellular virus particles in Saharan sands was below the limit of detection by TEM, whereas bacterial abundances in the same samples were about 10^4 g^{-1} (Prigent et al. 2005). Epifluorescence-based direct enumeration of bacterial cells and virus particles in samples from the Mojave desert support these trends, with bacteria outnumbering viruses by almost 100-fold. Low bacterial abundance in hot deserts is conflated with a low growth rate, neither of which could be expected to support a large viral population. While the low bacterial abundance and production rates are also characteristic of Antarctic soils, the environmental conditions leading to this situation in hot deserts are completely different and likely explain the lack of extracellular viruses. Sandy, hot desert soils typically experience large diurnal fluxes of temperature, severe doses of UV radiation, near-perpetual arid conditions, and aeolian abrasion, all of which likely contribute to the destruction of extracellular virus particles shortly after they are released.

The higher VBR observed for bulk soil relative to rhizosphere soil (Swanson et al. 2009) is also likely due to differences between the two soil compartments in rates of host growth, viral production, and viral decay. While viral production would be higher in the rhizosphere due to larger populations of more active host cells, it is reasonable to expect that viral decay would also be elevated due to the reduced pH and higher concentrations of extracellular enzymes in this compartment. By comparison, one would predict that viral production in the bulk soil would be much lower since host densities and labile carbon pools for host growth are typically lower here than in the rhizosphere. However, once produced, extracellular viruses may persist longer due to more favorable physico-chemical conditions than those found in the rhizosphere.

While VBR may indicate linkages between host production, viral production, and viral decay in aquatic ecosystems, the value of VBR for soil ecosystems is highly contentious, as critical questions surround what VBR is measuring. What fraction of extracted soil viruses are infectious or active at the time of sampling? What fraction of extracted bacteria is active? Do the extracted viruses have anything to do with the extracted bacteria, whatsoever? And how accurate are the estimates of both bacterial and viral abundance used to compute this ratio? This speaks to the critical issue of extraction efficiency, and the adoption of standard extraction methods to enable direct comparisons across studies. To answer these questions and advance our understanding of soil viral ecology, additional experiments are needed, which explore (1) survival and/or decay rates for autochthonous soil viruses within various soil ecosystems; (2) site-specific factors leading to the destruction or removal of viruses from soil; and (3) improvements in the extraction of bacteria from soil and in estimations of bacterial extraction efficiency. VBR continues to be reported in studies of soil viral ecology, but these important questions should be carefully considered in any interpretation of VBR data.

4.5 Viral Diversity in Soil Ecosystems

4.5.1 Marker Genes

In conjunction with abundance, viral diversity can provide important clues to the activity and importance of viruses within microbial ecosystems. Most surveys of viral diversity within soils have focused on diversity within specific groups of agronomically or medically important viruses [e.g., yellow dwarf virus of wheat (Bisnieks et al. 2004), or viruses of the soil-borne opportunistic pathogen *Burkholderia cepacia* (Govan and Vandamme 1998; Langley et al. 2003)], but few studies have sought to quantify global viral diversity in a given soil. The lack of a universally conserved phylogenetic marker analogous to the prokaryotic 16S rRNA gene presents a strong impediment to constructing a comprehensive view of viral diversity and phylogeny. However, marker genes thought to be highly conserved among specific viral families may be used to assess phylogenetic relationships within these families. For example, PCR assays have shown that gene homologues of T7-like DNA polymerase can be commonly found in a broad range of environments, including soils (Breitbart et al. 2004). More recently, the T4-like major capsid protein, g23, has been successfully used as a marker to explore phylogenetic diversity of myoviruses within rice paddy soils (Fujii et al. 2008). However, many commonly used phage marker genes (e.g., g23, g20, and T7-like DNA polymerase) may not be routinely found in all soils (Srinivasiah et al. 2008). This limits the applicability of the present array of marker gene PCR-assays for examining viral diversity across soils.

4.5.2 Pulsed-Field Gel Electrophoresis

Pulsed field gel electrophoresis (PFGE) has been a useful method for fingerprinting viral communities and monitoring community-level changes in rumen (Klieve and Swain 1993) and in marine environments (Steward and Azam 1998; Steward et al. 2000; Wommack et al. 1999). In this approach, viral genomes appear as distinct bands within a gel, and comparative analyses of changes in viral richness within or across samples may be performed based on the distribution of viral genome sizes, and potentially changes in band intensity. However, PFGE has yet to be successfully applied to soil viral assemblages. Reasons for its problematic implementation remain unclear. It may be that the limit of detection for PFGE bands is not sufficient to resolve complex viral assemblages one might expect to find in soil environments. For example, PFGE typically requires a minimum of 10^5 virus particles with a nominal genome size of 50 kb in order for visible bands to resolve in the gel (Wommack et al. 1999). While it is possible that soil viral communities are so diverse that few genome size fractions contain enough DNA to resolve into a discrete band, this hypothesis includes a significant burden of proof. An alternative hypothesis is that, in spite of stringent purification of virus particles from soil samples, degradative enzymes or chelating compounds remain in solution and compromise the integrity of genomic DNA when viral capsids are lysed. Some support for this hypothesis may be found in attempts to purify bacterial DNA from soils (Liles et al. 2008). Regardless of the underlying cause, the optimization of PFGE for assessment of soil viral diversity remains an eminent challenge.

4.5.3 Randomly Amplified Polymorphic DNA-PCR

Perhaps the best hope for a cheap, rapid, molecular assay for assessing viral diversity in soil lies in randomly amplified polymorphic DNA-PCR (RAPD-PCR). This technique relies upon a single primer (typically 10 nucleotides long) to amplify DNA fragments from a heterogeneous mixture of templates. Subsequent separation of RAPD-PCR amplicons by gel electrophoresis generates banding patterns indicative of the underlying complexity of the original DNA template mixture. In the case of viral assemblages, the template mixture is viral genomic DNA, and the gel banding pattern serves as a proxy indication of viral genotype richness within a given sample. The approach is termed "random" because specific primer sequences are often selected at random and need not be based on actual sequence information from the target DNA. RAPD-PCR has been used to assess the richness of marine virioplankton assemblages (Winget and Wommack 2008) and recent experiments indicate the high potential utility of this approach in soils (Srinivasiah et al. 2008), as well as sediments (Helton and Wommack 2009). However, further methodological refinements are needed before RAPD-PCR can be routinely applied to surveying the composition and diversity of soil viral assemblages.

4.5.4 Transmission Electron Microscopy

In the absence of robust molecular fingerprinting methods for synecological studies, initial determinations of total viral diversity in soils have relied upon transmission electron microscopy (TEM) to examine virus particle morphology (Swanson et al. 2009; Williamson et al. 2005). Despite the relatively low resolution of TEM-based estimates, the distribution of virus morphotypes observed in six different Delaware soils (including agricultural, forest, and wetland soils) indicated that the viral communities of each soil were distinct (Williamson et al. 2005). While one of the agricultural soils was dominated by icosahedral particles lacking tails (56%), most soil virus assemblages were dominated by tailed viruses (ca. 80%) suggesting that bacteriophages are the most abundant virus type in most soils. Interestingly, filamentous viruses were observed in two soils, and phages with elongated capsids were observed in five of the six soils. Phage particles with elongated capsids have also been observed in samples of Saharan sands (Prigent et al. 2005). Apart from these few observations, this morphotype has been rarely reported among environmental samples and comprises fewer than 2% of all known phages (Ackermann 2001). Thus, assemblages of soil viruses present unique and novel features even at this level of resolution. The predominance of tailed viral morphotypes indicated that bacteriophages were the most abundant viral type within all six soil environments. The mean capsid diameter for soil viruses was about 50 nm, with a range of 20–160 nm (Williamson et al. 2005).

By contrast, a more recent TEM-based assessment of viral diversity in a Scottish agricultural soil indicated that viral assemblages were dominated by small (20 nm) spherical particles (ca. 60%), with filamentous viruses (ca. 18%) and larger (40–60 nm) spherical viruses (ca. 10%), the second and third most abundant morphotypes, respectively (Swanson et al. 2009). Tailed viruses only comprised about 5% of the observed morphotypes. Furthermore, comparisons across rhizosheath, rhizosphere, and bulk soil revealed that the distribution of viral morphotypes was identical for each soil compartment. As noted with trends in abundance, the uniformity of viral community composition across these soil compartments is strikingly unexpected. Due to the often dramatic differences in pH, carbon availability, and other physical and chemical features between the rhizosphere and bulk soil, the microbial community structures of these two soil compartments tend to be significantly different (Bowen and Rovira 1999). The magnitude and direction of these community shifts can vary with plant cultivar (Lemanceau et al. 1995; Offre et al. 2007) and the stage of plant development for a given cultivar (Mougel et al. 2006), depending on which microbial taxa are favored by the specific momentary conditions in the rhizosphere. Several studies have indicated that rhizosphere bacterial communities show reduced species richness relative to microbial communities in the surrounding bulk soil (Baudoin et al. 2002; Baudoin et al. 2003; Garbeva et al. 2008; Gomes et al. 2001). This apparent lack of shift in viral community structure across these soil compartments represents an interesting puzzle. While it is possible that TEM-based assessments of morphological diversity do not provide fine enough resolution to detect changes in viral community composition, when identical methods were used to compare viral assemblages across soil ecotypes, the observed differences were quite striking (Williamson et al. 2005). Although methodological artifact cannot be completely ruled out, for the time being, one must consider that viruses in soil ecosystems do not appear to strictly reflect the trends in abundance or diversity of their host populations.

4.5.5 Metagenomic Analysis of Soil Viral Assemblages

Ultimately, this question of resolving fine-scale differences in viral community structure in soil ecosystems over both time and space will be answered through the application of metagenomics: the sequencing of the collective genomic DNA from a microbial assemblage within a specific environmental sample. Metagenomic analysis is a particularly powerful – and increasingly popular – tool for studying the genetic diversity of viral assemblages within environmental samples because sequence data can be obtained for (theoretically) the entire viral assemblage without dependence upon cultivation techniques, marker genes, or prior sequence knowledge. In the first (and currently, only) published metagenomic analysis of soil viral assemblages, the bacterial, fungal, and viral communities of three soils were compared, representing tallgrass prairie, desert, and tropical rainforest ecotypes (Fierer et al. 2007). According to contig spectra analysis, assemblages of soil viruses were found to be extremely rich both on a local and global scale, with up to 10^8 estimated viral genotypes in one soil and only one overlapping sequence in a total of 4,577 sequences. These findings suggest that each soil viral assemblage is a distinct microcosm of unique sequences, sharing little in common with viral assemblages from other soils. This contrasts with a recent metagenomic characterization of marine virus assemblages, which found between 85 and 95% of the most abundant virus genomes were shared among four discrete sampling locations (Sargasso Sea, the Gulf of Mexico, coastal waters of British Columbia, and the Arctic Ocean) (Angly et al. 2006). While marine viral assemblages may be locally diverse but share much in common on a global scale, the available data regarding soil viral assemblages indicate that soil viruses are much more diverse than their marine counterparts.

This is consistent with initial results from RAPD-PCR-based assessments of viral genotype richness in soils (Srinivasiah et al. 2008). In spite of its incredible advantages, metagenomics has yet to be adopted as a common and cost-effective approach to document changes in a given viral community, most likely due to its substantial resource demands, particularly in terms of sequencing infrastructure and computing power. In addition to these practical constraints on metagenomics in general, soils remain highly underrepresented among viral metagenome projects: of the >1 Tbp of viral metagenomic sequence available in public databases, only about 2.5 Mbp are from soils, and all of these originate from a single study (Fierer et al. 2007). Thus, in terms of applying contemporary molecular methods to synecological studies of soil viral assemblages, we have scarcely scratched the surface: the future is wide open.

4.6 The Importance of Lysogeny in Soil Environments

The characterization of virus replication strategies is of critical importance to understanding of the ecological roles of viruses within soil microbial communities. In particular, phage replication strategy may be a significant controlling factor in the population dynamics (Mei and Danovaro 2004) and coevolution of virus and host communities (Weitz et al. 2005) in soils. Phages may replicate through four major life cycles: the lytic cycle, the lysogenic cycle, the carrier state, and chronic infection/shedding (Abedon 2009; Campbell 2006). In the carrier state, frequently referred to as pseudolysogeny, phages coexist in stable equilibrium with their hosts. The majority of the host population is resistant to phage attack, but a small subpopulation of sensitive cells enables the virus to replicate and maintain itself within the host population at large (Abedon 2009). The macroscopic effect is the consistent production of phage coincident with high host abundance. While phages that replicate through the carrier state or chronic infections may also play important roles in the viral ecology of soils, these mechanisms are poorly characterized in other systems and are impossible to differentiate within synecological studies. For simplicity, a basic assumption is made that most viruses replicate through one of two major life cycles: lytic or lysogenic.

Lytic phage infections are characterized by rapid production of intracellular virus particles followed by host cell lysis and release of progeny virions. Lytic infections tend to quickly eliminate local populations of susceptible hosts, and extracellular virus particles must survive long enough to propagate infection if the virus strain is to remain successful. In this regard, addressing the question of viral persistence in the soil environment is absolutely critical to our understanding the ecology of lytic soil phages. By contrast, temperate phages may switch between the lytic replication pathway described above and lysogenic replication. Lysogenic replication involves the silencing of lytic gene expression by a phage-encoded repressor protein, and frequently, the integration of the phage genome into the bacterial chromosome mediated by a phage-encoded integrase. The specific replication pathway of temperate phages, lytic or lysogenic, is determined primarily by host metabolic status (Campbell 2006; Ptashne 1991). The state in which the host cell has been colonized by a nonreplicating phage genome, termed lysogeny, can be maintained indefinitely. During cell division, the phage genome (prophage) is regularly assorted to daughter cells either as an integrated part of the host chromosome (Ptashne 1991) or as an extrachromosomal element (Bertani 2004). In this fashion, the phage is maintained within the host population.

While lysogeny is highly stable, this relationship is not necessarily terminal for the infecting phage. Environmental signals, particularly damage to host DNA, can initiate a cascade of cell responses that inactivate the phage repressor protein. Following this event (induction), the virus shifts into lytic replication and progeny virus particles are released into the environment through cell lysis. The spatial heterogeneity and temporal variability of soils can be extremely high and may select for the predominance of specific viral replication strategies. In particular, lysogenic replication appears to hold a distinct advantage for viruses in soils, as lysogeny supports the long-term maintenance of phages within host populations when host distribution is patchy (Stewart and Levin 1984), when local host densities are low (Marsh and Wellington 1994; Pantastico-Caldas et al. 1992), or when host cells are nutrient deplete (Marsh and Wellington 1994; Williamson et al. 2002). Recent investigations into the frequency of inducible lysogeny among soil bacteria isolated on laboratory media (Williamson et al. 2008) and within natural assemblages of soil bacteria (Williamson et al. 2007) indicate that inducible lysogens (i.e., bacteria carrying a prophage element capable of entering a productive infection cycle) account for approximately 30%, and in some cases over 60%, of the total bacterial population in a given soil. This is a significant fraction and suggests that many soil phages are temperate. This predominance of lysogeny may result in significant impacts of soil phages on the genetic diversity and evolution of hosts, because lysogenic relationships increase the probability of genetic exchange between phage and host, as well as horizontal gene transfer between hosts (Replicon et al. 1995; Saye and Miller 1989; Saye et al. 1987; Wilcox and Fuhrman 1994) (see also Sect. 4.7.2). However, soils also contain appreciable numbers of obligately lytic viruses, whose impacts will be discussed below (Sect. 4.7.1). Thus, the importance of lytic soil viruses to the ecology of soils should not be underestimated.

In terms of refining induction assays for soil bacteria, or indeed, environmental bacteria in general, the serious question remains as to what causes induction of prophages in the environment. Sublethal doses of antibiotics have been shown to induce prophages in laboratory strains of lysogenic bacteria (Maiques et al. 2006; Majtan and Majtanova 1997). This has particular importance to soil microbial ecosystems since many autochthonous soil bacteria are known to produce antibiotics. However, no known studies have assessed the efficacy of this induction approach on soil bacteria. Recent studies involving the induction of soil bacterial enrichments indicated that quorum-sensing compounds such as homoserine lactones generate stronger induction responses than the standard inducing agent, mitomycin C (Roy et al. 2006). Like antibiotics, homoserine lactones are naturally occurring in soils (as well as many other environments) and appear to serve as endogenous prophage inducing agents.

4.7 Viral Impacts in Soil Ecosystems

4.7.1 Bacterial Mortality, Clonal Diversity, and Community Succession

As with viruses in marine ecosystems, viruses in soils appear to play important roles in mediating bacterial mortality and controlling the clonal diversity of bacterial assemblages. Until recently, these impacts have only been demonstrated using specific phage–host systems. For example, rhizobia-specific phages (rhizobiophages) can exert selective pressures on the population of soil rhizobia, a broad class of nitrogen fixing bacteria that engage in symbiotic relationships with leguminous plants. Infection and lysis of sensitive rhizobia strains results in a prevalence of phage-resistant mutant rhizobia that tend to be poor at colonizing plant roots (Kleczkowska 1971). Apparently, the loss of phage receptor which results in resistance to infection also negatively impacts the microbe–plant interactions that are critical to productive symbioses. In addition, the presence or absence of specific rhizobiophages within a soil ecosystem can dictate which of many competing rhizobia strains are available for successful nodulation (Hashem and Angle 1990). Beyond the impacts on bacterial mortality and selection, the lytic activity of rhizobiophages influences the efficacy of the legume-rhizobia symbiosis and may contribute to rates of nitrogen fixation and overall soil fertility.

Field experiments focusing on the population dynamics of specific, cultivable bacteriophages in the sugar beet rhizosphere revealed that phage and host dynamics exhibit recurring annual cycles, with temporal dynamics of phage and host populations appearing almost identical over 3 consecutive years (Ashelford et al. 1999a, b). Thus, soil viruses appear to play important roles in the succession of microbial communities within soils. Alternatively, soil viral assemblages may be highly responsive to and indicative of changes in soil bacterial communities. Due to the nature of phage–host dynamics, the directionality of this relationship is difficult to determine.

Microcosm experiments utilizing naturally occurring microbial assemblages suggest that soil bacterial communities and associated viral assemblages are highly dynamic and responsive to perturbations. The addition of various carbon amendments to agricultural soil within microcosm incubations resulted in significant fluctuations in bacterial abundance. Furthermore, a distinct coupling effect was observed, wherein an increase or decrease in bacterial abundance co-occurred with a respective decrease or increase in viral abundance, albeit with a time lag between virus and host fluxes (Srinivasiah et al. 2008). These patterns of staggered fluctuations between soil viruses and soil bacteria closely follow conventional models of predator–prey dynamics.

Recent in-field microcosm experiments have provided additional evidence that viral-mediated mortality may be an important factor in the biogeochemistry of soils. Measurements taken at a field site in Barrow, Alaska, indicate that these arctic soils are rich in labile carbon, and microbial growth is not substrate-limited. Addition of carbon or nutrient amendments produced no significant impacts on soil microbial biomass or respiration. However, plots that were treated with a concoction of tea extract and $FeSO_4$ as an antiphage solution showed a marked decrease in ambient viral abundance, as well as concomitant increases in microbial biomass and microbial respiration (Allen et al. 2010). Thus, phage predation in these soils apparently held bacterial abundance and activity well below maxima that the ecosystem was capable of supporting. The collective results from both of these microcosm experiments would suggest that viral-mediated mortality is a significant factor controlling bacterial abundance, and by association, biomass turnover and rates of nutrient cycling in soil ecosystems. However, soils are notorious for their extremely high spatial heterogeneity and temporal variability, and such top-down control of soil bacterial assemblages by phage predation may not apply to all ecosystems. For example, the low abundance of extracellular phages in hot desert soils would not be expected to factor heavily into control of bacterial abundance or activity in these biomes. Additional studies are needed to determine whether the results of these microcosm experiments apply at global or even landscape scales.

4.7.2 Phage Conversion, Host Fitness, and Horizontal Gene Transfer

Temperate phages are known to affect host fitness through a process known as lysogenic conversion. Prophage elements may contain genes that are beneficial to host survival and proliferation under specific conditions, thus providing a selective advantage over nonlysogenized hosts (Barksdale and Arden 1974; Barondess and Beckwith 1995; Waldor and Mekalanos 1996). Prophages also frequently constitute the sole differences in specific gene complement among closely related strains of bacteria, and consequently, the range of relationships between the lysogenic host bacterium and other organisms (Desiere et al. 2002; Ohnishi et al. 2001; van der Mee-Marquet et al. 2006). Most of the data supporting these observations come from clinical strains of bacteria, and very little data are available regarding these same phage impacts on environmentally relevant strains. However, recent investigations indicate that lysogeny is highly prevalent among autochthonous communities of soil bacteria, with inducible lysogens (i.e., bacteria carrying a prophage element) accounting for 4–66% of the total population in a given soil (Williamson et al. 2007). Thus, if the impacts of temperate phages observed in clinical settings hold true for environmental bacteria, the high frequency of lysogeny among soil bacteria implies that temperate soil viruses significantly influence host fitness and evolution through the genetic process of lysogenic conversion. However, as is the case with most of soil phage ecology, more data are needed to confirm whether this hypothesis is true on the macro-scale.

Beyond the impacts of lysogenic conversion, which are confined solely to temperate phages, viruses in soils may have significant impacts on host population genetics by acting as vectors for the horizontal transfer of genes between dissimilar bacterial strains. Generally speaking, bacteria may horizontally acquire novel gene sets via three mechanisms: (1) transformation, the uptake of naked DNA from the environment by competent cells; (2) conjugation, the transfer of genetic material from one cell to another through a pilus; and (3) transduction, phage-mediated transfer of DNA. Transduction begins as an error during the phage replication process in which host genomic DNA is mistakenly packaged into virus capsids. Two types of transduction are possible, generalized or specialized, depending on the phenotype of the phage mediating the transfer.

Generalized transducing phages, which may be virulent or temperate, obtain essentially random fragments of the host genome through packaging errors.

By contrast, specialized transduction is mediated only by temperate phages that integrate into the host chromosome. Only specific bacterial genes flanking the prophage insertion site are transferred due to improper excision of the phage genome from the host chromosome (Campbell 2006; Levy and Miller 1989). In either case, these defective phages, now termed transducing particles, act as shuttles capable of transferring DNA fragments between bacterial hosts. While transduction is a rare event leading to a low frequency of gene transfer, the short generation times and fast growth rates of bacteria means that even rare recombinations may quickly alter the genotypic diversity of a bacterial community (Beumer and Robinson 2005).

In aquatic studies, the presence of particulate matter increased transduction frequencies by providing surface area on which phages and host cells could aggregate and interact (Ripp and Miller 1995). Soil, especially the clay and organic matter fractions, has an extremely high specific surface area [80–800 m^2 g^{-1} (Hillel 1998)], providing many potential sites for phage and host adsorption and interaction. However, natural transduction has never been observed in soil environments. Furthermore, only a handful of experiments using cultivable phage–host systems have been performed, all of which highlight the potential for transduction in soils (Deb et al. 2003; Vettori et al. 2000; Vettori et al. 1999). A more recent study of the importance of phage-mediated gene transfer in soils examined viral assemblages that were obtained from the induction of soil bacteria in atrazine-treated soils. Subsequent PCR screening detected the presence of the atrazine catabolic gene, *trzN* within the induced phage community, indicating that this important phenotypic trait could be transferred among bacterial populations through phage vectors (Ghosh et al. 2008). While much uncertainty remains regarding the frequency of transduction in soil environments, the possible impacts on bacterial genetic diversity and evolution are immense.

4.8 Conclusion

The global significance of soil ecosystems can hardly be overstated. With prokaryotic diversity estimated to be three orders of magnitude greater than in all other ecosystems combined (Curtis et al. 2002; Kemp and Aller 2004), soils represent the greatest reservoir of biodiversity on the planet. The activities of the microbial denizens of soil are critical to global biogeochemical cycles, including carbon, nitrogen, and phosphorus, which support all other forms of terrestrial diversity. With specific regard to human civilization, soils provide ecosystem services on a global scale estimated at greater than $20 trillion (Boumans et al. 2002). But the significance of viruses to maintaining these important features of the terrestrial biosphere remains a huge topic of future research.

Many questions remain surrounding even the basic studies of soil phage ecology described in this chapter. With specific regard to viral abundance, it appears that the total number of virus particles within a given soil results from the combination of viral production and viral storage or decay. Thus, the quantification of rates for these separate processes represents immediate targets for future studies in soil viral ecology. Similarly, the question of viral diversity in soils is ripe for the taking. It remains a puzzle as to why, more than 8 years after the first viral metagenome was constructed from marine viral assemblages (Breitbart et al. 2002), metagenomic approaches have yet to be earnestly applied to capturing the genetic diversity represented by soil viruses. This represents an immediate opportunity for the development of soil phage ecology. In addition, the implementation of rapid molecular assays to fingerprint viral assemblages will allow for relative changes in viral community composition to be compared over space and time and linked with specific biological, physical, or chemical properties of a given soil. The establishment of such linkages is vital to the expansion of our understanding of soil phage ecology.

The apparently high prevalence of lysogeny among soil bacteria suggests that prophages play important roles in host fitness and evolution in soil environments. Furthermore, available evidence strongly suggests that phage-mediated gene transfer is an important process for bacterial acquisition of novel gene sets among soil bacteria. The discovery of marine cyanophages that carry functional homologues of host genes for photosynthesis (Lindell et al. 2004) and phosphate transport (Sullivan et al. 2005) boldly underscore the importance of marine phages for genetic transfer – particularly for transfer of gene sets that may be crucial for host survival. What, then, is the significance of phages in soils: environment in which conjugation may be limited by spatial constraints, and transducing phage may be a primary mechanism of horizontal gene transfers?

Finally, recent evidence from both the field and laboratory scales indicates that soil viruses are important predators of bacteria, capable of altering microbial respiration and substrate utilization rates. As global carbon budgets are further refined, the impact of viral-mediated mortality and biomass turnover in soils will likely become an important factor in balancing the scales and in weighing the merits of proposed carbon reduction activities. Related to this, broader-scale assessments of the roles of soil viruses in other major nutrient transformations, including nitrogen, phosphorus, and sulfur, will aid in determining viral impacts on soil fertility and ecosystem health. The age of soil viral ecology is still experiencing its lingering dawn, and the future promises exciting discoveries yet to come.

References

Abedon ST (2009) Disambiguating bacteriophage pseudolysogeny: an historical analysis of lysogeny, pseudolysogeny, and the phage carrier state. In: Adams HT (ed) Contemporary trends in bacteriophage research. Nova Publishers, Hauppage, New York, pp 285–307

Ackermann HW (2001) Frequency of morphological phage descriptions in the year 2000. Arch Virol 146:843–857

Allen B, Willner D, Oechel WC, Lipson D (2010) Top-down control of microbial activity and biomass in an Arctic soil ecosystem. Environ Microbiol 12:642–648

Andrews JH (1991) Comparative ecology of microorganisms and macroorganisms. Springer-Verlag, New York

Angly FE, Felts B, Breitbart M, Salamon P, Edwards RA, Carlson C, Chan AM, Haynes M, Kelley S, Liu H, Mahaffy JM, Mueller JE, Nulton J, Olson R, Parsons R, Rayhawk S, Suttle CA, Rohwer F (2006) The marine viromes of four oceanic regions. PLoS Biol 4:2121–2131

Ashelford KE, Day MJ, Bailey MJ, Lilley AK, Fry JC (1999a) In situ population dynamics of bacterial viruses in a terrestrial environment. Appl Environ Microbiol 65:169–174

Ashelford KE, Day MJ, Fry JC (2003) Elevated abundance of bacteriophage infecting bacteria in soil. Appl Environ Microbiol 69:285–289

Ashelford KE, Fry JC, Bailey MJ, Jeffries AR, Day MJ (1999b) Characterization of six bacteriophages of serratia liquefaciens CP6 isolated from the sugar beet phytosphere. Appl Environ Microbiol 65:1959–1965

Azam F, Fenchel T, Field JG, Gray JS, Meyer Reil LA, Thingstad F (1983) The ecological role of water-column microbes in the sea. Mar Ecol Prog Ser 10:257–263

Barksdale L, Arden SB (1974) Persisting bacteriophage infections, lysogeny, and phage conversions. Annu Rev Microbiol 28:265–299

Barondess JJ, Beckwith J (1995) Bor gene of phage-lambda, involved in serum resistance, encodes a widely conserved outer-membrane lipoprotein. J Bacteriol 177:1247–1253

Baudoin E, Benizri E, Guckert A (2002) Impact of growth stage on the bacterial community structure along maize roots, as determined by metabolic and genetic fingerprinting. Appl Soil Ecol 19:135–145

Baudoin E, Benizri E, Guckert A (2003) Impact of artificial root exudates on the bacterial community structure in bulk soil and maize rhizosphere. Soil Biol Biochem 35:1183–1192

Beijerinck MJ (1898) Concerning a contagium vivum fluidum as cause of the spot disease of tobacco leaves. Verhandelingen der Koninkyke akademie Wettenschappen te Amsterdam 65:3–21

Bergh O, Borsheim KY, Bratbak G, Heldal M (1989) High abundance of viruses found in aquatic environments. Nature (London) 340:467–468

Bertani G (2004) Lysogeny at mid-twentieth century: P1, P2, and other experimental, systems. J Bacteriol 186:595–600

Beumer A, Robinson JB (2005) A broad-host-range, generalized transducing phage (SN-T) acquires 16S rRNA genes from different genera of bacteria. Appl Environ Microbiol 71:8301–8304

Bisnieks M, Kvarnheden A, Sigvald R, Valkonen JPT (2004) Molecular diversity of the coat protein-encoding region of Barley yellow dwarf virus-PAV and Barley yellow dwarf virus-MAV from Latvia and Sweden. Arch Virol 149:843–853

Borsheim KY, Bratbak G, Heldal M (1990) Enumeration and biomass estimation of planktonic bacteria and viruses by transmission electron microscopy. Appl Environ Microbiol 56: 352–356

Boumans R, Costanza R, Farley J, Wilson MA, Portela R, Rotmans J, Villa F, Grasso M (2002) Modeling the dynamics of the integrated earth system and the value of global ecosystem services using the GUMBO model. Ecol Econ 41:529–560

Bowen GD, Rovira AD (1999) The rhizosphere and its management to improve plant growth. Adv Agron 66:1–102

Bratbak G, Heldal M (2000) Viruses rule the waves: the smallest and most abundant members of marine ecosystems. Microbiol Today 27:171–173

Bratbak G, Thingstad F, Heldal M (1994) Viruses and the microbial loop. Microb Ecol 28:209–221

Breitbart M, Miyake JH, Rohwer F (2004) Global distribution of nearly identical phage-encoded DNA sequences. FEMS Microbiol. Lett. 236(2): 249–256

Breitbart M, Salamon P, Andresen B, Mahaffy JM, Segall AM, Mead D, Azam F, Rohwer F (2002) Genomic analysis of uncultured marine viral communities. Proc Nat Acad Sci USA 99:14250–14255

Campbell A (2006) General aspects of lysogeny. In: Calendar R (ed) The bacteriophages. Oxford University Press, New York, pp 66–73

Christian PD, Richards AR, Williams T (2006) Differential adsorption of occluded and nonoccluded insect-pathogenic viruses to soil-forming minerals. Appl Environ Microbiol 72:4648–4652

Creager ANH, Scholthof KG, Citovsky V, Scholthof HB (1999) Tobacco mosaic virus: pioneering research for a century. Plant Cell 11:301–308

Curtis TP, Sloan WT, Scannell JW (2002) Estimating prokaryotic diversity and its limits. Proc Natl Acad Sci USA 99:10494–10499

Deb C, Chakraborty R, Ghosh AN, Mandal NC, Mukherjee T, Roy P (2003) A generalized transducing thiophage (TPC-1) of a facultative sulfur chemolithotrophic bacterium, Bosea thiooxidans CT5, of alpha-Proteobacteria, isolated from Indian soil. FEMS Microbiol Lett 227:87–92

Delogu G, Faccini N, Alberici R, Gianinetti A, Stanca AM (2003) Soil-borne viruses of barley seriously affect plant growth and grain yield in a monocropping system. Cereal Res Commun 31:137–144

Demolon A, Dunez A (1935) Recherchhes sur la role du bacteriophage dans la fatigue des luzerniers. Ann Agron 5:89–111

Desiere F, Lucchini S, Canchaya C, Ventura M, Brussow H (2002) Comparative genomics of phages and prophages in lactic acid bacteria. Antonie Van Leeuwenhoek Int J Gen Mol Microbiol 82:73–91

Duboise S, Moore BE, Sorber CA, Sagik BP (1979) Viruses in soil systems. CRC Crit Rev Microbiol 7:245–285

Fierer N, Breitbart M, Nulton J, Salamon P, Lozupone C, Jones R, Robeson M, Edwards RA, Felts B, Rayhawk S, Knight R, Rohwer F, Jackson RB (2007) Metagenomic and small-subunit rRNA analyses reveal the genetic diversity of bacteria, archaea, fungi, and viruses in soil. Appl Environ Microbiol 73:7059–7066

Fillhart RC, Bachand GD, Castello JD (1998) Detection of infectious tobamoviruses in forest soils. Appl Environ Microbiol 64:1430–1435

Fray RG (2002) Altering plant-microbe interaction through artificially manipulating bacterial quorum sensing. Ann Bot 89:245–253

Fuhrman J (1992) Bacterioplankton roles in cycling of organic matter: the microbial food web. In: Falkowski PG, Woodhead AD (eds) Primary productivity and biogeochemical cycles in the sea. Plenum Press, New York, pp 361–383

Fuhrman JA (1999) Marine viruses and their biogeochemical and ecological effects. Nature 399:541–548

Fujii T, Nakayama N, Nishida M, Sekiya H, Kato N, Asakawa S, Kimura M (2008) Novel capsid genes (g23) of T4-type bacteriophages in a Japanese paddy field. Soil Biol Biochem 40:1049–1058

Fuxa JR (2004) Ecology of insect nucleopolyhedroviruses. Agric Ecosyst Environ 103:27–43

Garbeva P, van Elsas JD, van Veen JA (2008) Rhizosphere microbial community and its response to plant species and soil history. Plant Soil 302:19–32

Gerba CP (1984) Applied and theoretical aspects of virus adsorption to surfaces. Adv Appl Microbiol 30:133–168

Ghosh D, Roy K, Williamson KE, White DC, Wommack KE, Sublette KL, Radosevich M (2008) Prevalence of lysogeny among soil bacteria and presence of 16S rRNA and trzN genes in viral-community DNA. Appl Environ Microbiol 74:495–502

Gomes NCM, Heuer H, Schonfeld J, Costa R, Mendonca-Hagler L, Smalla K (2001) Bacterial diversity of the rhizosphere of maize (Zea mays) grown in tropical soil studied by temperature gradient gel electrophoresis. Plant Soil 232:167–180

Govan JRW, Vandamme P (1998) Agricultural and medical microbiology: a time for bridging gaps. Microbiology 144:2373–2375

Grijns A (1927) Clover-plants in sterile cultivation do not produce a bacteriophage od B. radicicola. Zentr Bakt Parasitenk Infek 2:248–251

Hartmann A, Rothballer M, Schmid M (2007) Lorenz Hiltner, a pioneer in rhizosphere microbial ecology and soil bacteriology research. 2nd International conference on rhizosphere, Springer, Montpellier, France, 2007, pp 7–14

Hashem FM, Angle JS (1990) Rhizobiophage effects on nodulation, nitrogen-fixation, and yield of field-grown soybeans (Glycine-Max L Merr). Biol Fertil Soils 9:330–334

Heldal M, Bratbak G (1991) Production and decay of viruses in aquatic environments. Mar Ecol Prog Ser 72:205–212

Helton RR, Wommack KE (2009) Seasonal dynamics and metagenomic characterization of estuarine viriobenthos assemblages by randomly amplified polymorphic DNA PCR. Appl Environ Microbiol 75:2259–2265

Hillel D (1998) Environmental soil physics. Harcourt Brace and Company, San Diego

Iwanowski D (1892) Concerning the mosaic disease of the tobacco plant. St Petersburg Acad Imperial Sci Bull 35:67–70

Jin Y, Flury M (2002) Fate and transport of viruses in porous media. Adv Agron 77:39–51

Jones DL, Hodge A, Kuzyakov Y (2004) Plant and mycorrhizal regulation of rhizodeposition. New Phytol 163:459–480

Keel C, Ucurum Z, Michaux P, Adrian M, Haas D (2002) Deleterious impact of a virulent bacteriophage on survival and biocontrol activity of Pseudomonas fluorescens strain CHA0 in natural soil. Mol Plant Microbe Interact 15:567–576

Kemp PF, Aller JY (2004) Bacterial diversity in aquatic and other environments: what 16S rDNA libraries can tell us. FEMS Microbiol Ecol 47:161–177

Kleczkowska J (1971) Genetic changes in rhizobium bacteria and in their bacteriophages during coexistence. Plant Soil 35(1):47:–56

Klieve AV, Swain RA (1993) Estimation of ruminal bacteriophage numbers by pulsed-field gel electrophoresis and laser densitometry. Appl Environ Microbiol 59:2299–2303

Langley R, Kenna DT, Vandamme P, Ure R, Govan JRW (2003) Lysogeny and bacteriophage host range within the Burkholderia cepacia complex. J Med Microbiol 52:483–490

Lemanceau P, Corberand T, Gardan L, Latour X, Laguerre G, Boeufgras JM, Alabouvette C (1995) Effect of two plant species, flax (Linum usitatissinum L.) and tomato (Lycopersicon esculentum Mill.), on the diversity of soilborne populations of fluorescent pseudomonads. Appl Environ Microbiol 61:1004–1012

Levy SB, Miller RV (1989) Gene transfer in aquatic environments. McGraw-Hill, New York

Li WKW, Dickie PM (2001) Monitoring phytoplankton, bacterioplankton, and virioplankton in a coastal inlet (Bedford Basin) by flow cytometry. Cytometry 44:236–246

Liles MR, Williamson LL, Rodbumrer J, Torsvik V, Goodman RM, Handelsman J (2008) Recovery, purification, and cloning of high-molecular-weight DNA from soil microorganisms. Appl Environ Microbiol 74:3302–3305

Lindell D, Sullivan MB, Johnson ZI, Tolonen AC, Rohwer F, Chisholm SW (2004) Transfer of photosynthesis genes to and from Prochlorococcus viruses. Proc Natl Acad Sci USA 101:11013–11018

Lipson SM, Stotzky G (1986) Effect of kaolinite on the specific infectivity of reovirus. FEMS Microbiol Lett 37:83–88

Loveland JP, Ryan JN, Amy GL, Harvey RW (1996) The reversibility of virus atachment to mineral surfaces. Colloid Surf A 107:205–221

Madan NJ, Marshall WA, Laybourn-Parry J (2005) Virus and microbial loop dynamics over an annual cycle in three contrasting Antarctic lakes. Freshw Biol 50:1291–1300

Maiques E, Ubeda C, Campoy S, Salvador N, Lasa I, Novick RP, Barbe J, Penades JR (2006) Beta-lactam antibiotics induce the SOS response and horizontal gene transfer of virulence factors in Staphylococcus aureus. J Bacteriol 188:2726–2729

Majtan V, Majtanova L (1997) Postantibiotic effects and postantibiotic sub-MIC effects of ciprofloxacin, perfloxacin and amikacin on the biological properties of Salmonella strains. Folia Microbiol 42:327–332

Mann NH (2005) The third age of phage. PLoS Biol 3:e182

Marsh P, Wellington EMH (1994) Phage–host interactions in soil. FEMS Microbiol Ecol 15:99–107

Mayer A (1886) Concerning the mosaic disease of tobacco. Die Landwirtshaftliche Versuchsstationen 32:451–467

Mei ML, Danovaro R (2004) Virus production and life strategies in aquatic sediments. Limnol Oceanogr 49:459–470

Melzer MJ, Bidochka MJ (1998) Diversity of double-stranded RNA viruses within populations of entomopathogenic fungi and potential implications for fungal growth and virulence. Mycologia 90:586–594

Meschke JS, Sobsey MD (2003) Comparative reduction of Norwalk virus, poliovirus type 1, F+ RNA coliphage MS2 and Escherichia coli in miniature soil columns. Water Sci Technol 47:85–90

Middelboe M, Jorgensen NOG, Kroer N (1996) Effects of viruses on nutrient turnover and growth efficiency of noninfected marine bacterioplankton. Appl Environ Microbiol 62:1991–1997

Mougel C, Offre P, Ranjard L, Corberand T, Gamalero E, Robin C, Lemanceau P (2006) Dynamic of the genetic structure of bacterial and fungal communities at different developmental stages of Medicago truncatula Gaertn. cv. Jemalong line J5. New Phytologist 170:165–175

Nie M, Zhang XD, Wang JQ, Jiang LF, Yang J, Quan ZX, Cui XH, Fang CM, Li B (2009) Rhizosphere effects on soil bacterial abundance and diversity in the Yellow River Deltaic ecosystem as influenced by petroleum contamination and soil salinization. Soil Biol Biochem 41:2535–2542

Offre P, Pivato B, Siblot S, Gamalero E, Corberand T, Lemanceau P, Mougel C (2007) Identification of bacterial groups preferentially associated with mycorrhizal roots of Medicago truncatula. Appl Environ Microbiol 73:913–921

Ohnishi M, Kurokawa K, Hayashi T (2001) Diversification of Escherichia coli genomes: are bacteriophages the major contributors? Trends Microbiol 9:481–485

Pantastico-Caldas M, Duncan KE, Istock CA, Bell JA (1992) Population dynamics of bacteriophage and *Bacillus subtilis* in soil. Ecology 73:1888–1902

Paul JH, Rose JB, Jiang SC, Kellogg CA, Dickson L (1993) Distribution of viral abundance in the reef environment of Key Largo. Florida Appl Environ Microbiol 59:718–724

Pavinato PS, Merlin A, Rosolem CA (2008) Organic compounds from plant extracts and their effect on soil phosphorus availability. Pesquisa Agropecuaria Brasileira 43:1379–1388

Personnic S, Domaizon I, Dorigo U, Berdjeb L, Jacquet S (2009) Seasonal and spatial variability of virio-, bacterio-, and picophytoplanktonic abundances in three peri-alpine lakes. Hydrobiologia 627:99–116

Pourcher AM, Francoise PB, Virginie F, Agnieszka G, Vasilica S, Gerard M (2007) Survival of faecal indicators and enteroviruses in soil after land-spreading of municipal sewage sludge. Appl Soil Ecol 35:473–479

Prigent M, Leroy M, Confalonieri F, Dutertre M, DuBow MS (2005) A diversity of bacteriophage forms and genomes can be isolated from the surface sands of the Sahara Desert. Extremophiles 9:289–296

Proctor LM, Fuhrman JA (1990) Viral mortality of marine bacteria and cyanobacteria. Nature (London) 343:60–62

Ptashne M (1991) A genetic switch. Blackwell Scientific Publications and Cell Press, Cambridge, MA

Rangarajan S, Saleena LM, Nair S (2002) Diversity of Pseudomonas spp. isolated from rice rhizosphere populations grown along a salinity gradient. Microb Ecol 43:280–289

Ratti C, Budge G, Ward L, Clover G, Rubies-Autonell C, Henry C (2004) Detection and relative quantitation of Soil-borne cereal mosaic virus (SBCMV) and Polymyxa graminis in winter wheat using real-time PCR (TaqMan®). J Virol Methods 122:95–103

Replicon J, Frankfater A, Miller RV (1995) A continuous culture model to examine factors that affect transduction among Pseudomonas aeruginosa strains in freshwater environments. Appl Environ Microbiol 61:3359–3366

Reynolds KA, Gerba CP, Pepper IL (1996) Detection of infectious enteroviruses by an integrated cell culture PCR procedure. Appl Environ Microbiol 62:1424–1427

Rice G, Stedman K, Snyder J, Wiedenheft B, Willits D, Brumfield S, McDermott T, Young MJ (2001) Viruses from extreme thermal environments. Proc Natl Acad Sci USA 98:13341–13345

Ripp S, Miller RV (1995) Effects of suspended particulates on the frequency of transduction among Pseudomonas aeruginosa in freshwater environments. Appl Environ Microbiol 61:1214–1219

Roy K, Ghosh D, Martin EC, White DC, Wommack KE, Williamson KE, Radosevich M (2006) Soil bacterial communities sampled with BioSep beads exhibit a high frequency of lysogeny. 11th international symposium on microbial ecology, Vienna, Austria, 2006

Santamaria J, Toranzos GA (2003) Enteric pathogens and soil: a short review. Int Microbiol 6:5–9

Saye DJ, Miller RV (1989) The aquatic environment: consideration of horizontal gene transmission in a diversified habitat. In: Levy SB, Miller RV (eds) Gene transfer in the environment. McGraw-Hill, New York, pp 223–259

Saye DJ, Ogunseitan O, Sayler GS, Miller RV (1987) Potential for transduction of plasmids in a natural freshwater environment: effect of plasmid donor concentration and a natural microbial community on transduction in *Pseudomonas aeruginosa*. Appl Environ Microbiol 53:987–995

Srinivasiah S, Bhavsar J, Thapar K, Liles M, Schoenfeld T, Wommack KE (2008) Phages across the biosphere: contrasts of viruses in soil and aquatic environments. Res Microbiol 159:349–357

Steward GF, Azam F (1998) Analysis of marine viral communities: the killer's fingerprints. Eighth international symposium on Microbial Ecology, Halifax, Canada, 1998, p 313

Steward GF, Montiel JL, Azam F (2000) Genome size distributions indicate variability and similarities among marine viral assemblages from diverse environments. Limnol Oceanogr 45:1697–1706

Stewart FM, Levin BR (1984) The population biology of bacterial viruses: why be temperate? Theor Pop Biol 26:93–117

Sullivan MB, Coleman ML, Weigele P, Rohwer F, Chisholm SW (2005) Three Prochlorococcus cyanophage genomes: signature features and ecological interpretations. PLoS Biol 3:790–806

Swanson MM, Fraser G, Daniell TJ, Torrance L, Gregory PJ, Taliansky M (2009) Viruses in soils: morphological diversity and abundance in the rhizosphere. Ann Appl Biol 155:51–60

Thingstad TF (2000) Elements of a theory for the mechanisms controlling abundance, diversity, and biogeochemical role of lytic bacterial viruses in aquatic systems. Limnol Oceanogr 45:1320–1328

Tijdens M, Hoogveld HL, Kamst-van Agterveld MP, Simis SGH, Baudoux AC, Laanbroek HJ, Gons HJ (2008) Population dynamics and diversity of viruses, bacteria and phytoplankton in a shallow eutrophic lake. Microb Ecol 56:29–42

van der Mee-Marquet N, Domelier AS, Mereghetti L, Lanotte P, Rosenau A, van Leeuwen W, Quentin R (2006) Prophagic DNA fragments in Streptococcus agalactiae strains and association with neonatal meningitis. J Clin Microbiol 44:1049–1058

Vettori C, Gallori E, Stotzky G (2000) Clay minerals protect bacteriophage PBS1 of Bacillus subtilis against inactivation and loss of transducing ability by UV radiation. Can J Microbiol 46:770–773

Vettori C, Stotzky G, Yoder M, Gallori E (1999) Interaction between bacteriophage PBS1 and clay minerals and transduction of Bacillus subtilis by clay-phage complexes. Environ Microbiol 1:347–355

Waldor MK, Mekalanos JJ (1996) Lysogenic conversion by a filamentous phage encoding cholera toxin. Science 272:1910–1914

Walker TS, Bais HP, Grotewold E, Vivanco JM (2003) Root exudation and rhizosphere biology. Plant Physiol 132:44–51

Weinbauer MG (2004) Ecology of prokaryotic viruses. FEMS Microbiol Rev 28:127–181

Weitz JS, Hartman H, Levin SA (2005) Coevolutionary arms race between bacteria and bacteriophage. Proc Natl Acad Sci USA 102:9535–9540

Wen K, Ortmann AC, Suttle CA (2004) Accurate estimation of viral abundance by epifluorescence microscopy. Appl Environ Microbiol 70:3862–3867

Wilcox RM, Fuhrman JA (1994) Bacterial viruses in coastal seawater: lytic rather than lysogenic production. Mar Ecol Prog Ser 114:35–45

Williamson KE, Radosevich M, Smith DW, Wommack KE (2007) Incidence of lysogeny within temperate and extreme soil environments. Environ Microbiol 9:2563–2574

Williamson KE, Radosevich M, Wommack KE (2005) Abundance and diversity of viruses in six Delaware soils. Appl Environ Microbiol 71:3119–3125

Williamson KE, Schnitker JB, Radosevich M, Smith DW, Wommack KE (2008) Cultivation-based assessment of lysogeny among soil bacteria. Microb Ecol 56:437–447

Williamson KE, Wommack KE, Radosevich M (2003) Sampling natural viral communities from soil for culture-independent analyses. Appl Environ Microbiol 69:6628–6633

Williamson SJ, Houchin LA, McDaniel L, Paul JH (2002) Seasonal variation in lysogeny as depicted by prophage induction in Tampa Bay, Florida. Appl Environ Microbiol 68:4307–4314

Winget DM, Wommack KE (2008) Randomly amplified polymorphic DNA PCR as a tool for assessment of marine viral richness. Appl Environ Microbiol 74:2612–2618

Wommack KE, Colwell RR (2000) Virioplankton: viruses in aquatic ecosystems. Microbiol Mol Biol Rev 64:69–114

Wommack KE, Ravel J, Hill RT, Colwell RR (1999) Population dynamics of Chesapeake Bay virioplankton: Total community analysis using pulsed field gel electrophoresis. Appl Environ Microbiol 65:231–240

Wommack KE, Williamson SJ, Sundbergh A, Helton RR, Glazer BT, Portune K, Cary SC (2004) An instrument for collecting discrete large-volume water samples suitable for ecological studies of microorganisms. Deep Sea Res Part I Oceanogr Res Pap 51:1781–1792

Yin X, Zeph LR, Stotzky G (1997) A simple method for enumerating bacteriophages in soil. Can J Microbiol 43:461–466

Zaitlin M (1998) The discovery of the causal agent of the tobacco mosaic disease. In: Kung SD, Yang SF (eds) Discoveries in plant biology. World Publishing Co, Hong Kong, pp 105–110

Zhuang J, Jin Y (2003) Virus retention and transport through Al-oxide coated sand columns: effects of ionic strength and composition. J Contam Hydrol 60:193–209

Chapter 5
Identification and Analysis of Prophages and Phage Remnants in Soil Bacteria

K.V. Srividhya and S. Krishnaswamy

5.1 Introduction

Large-scale comparative genome analysis has aided unraveling the impact of genomic features of bacteria leading to their adaptations in varying ecological niches and pathogenicity (Ravel and Fraser 2005). Bacteriophages are very abundant in the biosphere, and viral infection is believed to affect the activity and genetic diversity of bacterial communities in aquatic environments. Soil phage, like their aquatic counterparts, is likely to be important in controlling bacterial populations and mediating gene transfer in soil (Ashelford et al. 2003). Lysogenic conversion, for example, can improve host fitness and lead to phage-mediated horizontal gene transfer (HGT). However, little is known about prophages, lysogeny, and transduction in the soil environment. Prophages in soil microbes are still largely unexplored. This chapter deals with prophages associated with soil microbes and the possible phenotype contributions due to them.

5.1.1 Overview of Soil Microbes

Microbes are abundantly distributed in soil ranging from bacterial actinomycetes, fungi, algae, cyanobacteria, and soil protozoa. Hence normal soil contains enormous amount of microbial biomass. These microbes contribute many beneficial processes to the soil by taking part in cycling nutrients like carbon, sulfur, phosphorus, and nitrogen. One of the important processes carried out by soil microbes is

K.V. Srividhya and S. Krishnaswamy (✉)
Centre of Excellence in Bioinformatics School of Biotechnology, Madurai Kamaraj University, Madurai 625021, Tamil Nadu, India
e-mail: krishna@mrna.tn.nic.in

G. Witzany (ed.), *Biocommunication in Soil Microorganisms*, Soil Biology 23,
DOI 10.1007/978-3-642-14512-4_5,

nitrogen fixation. Symbiotic nitrogen fixation is mediated by *Rhizobium* and *Bradyrhizobium* (http://soils.usda.gov/sqi/concepts/soil_biology/bacteria.html).

In order to understand the soil biodiversity, methods to arrive at census of soil bacteria have been reported. The proposed statistical model takes into account the problem of estimating richness statistically from the characteristics of samples taken from simulated community distribution (Schloss and Handelsman 2006). Ecological classification of bacteria has been attempted (Fierer et al. 2007). Almost 71 unique soil samples from ecosystems near North America were collected and analyzed by authors. They classify soil bacteria into two major ecological communities as copiotroph (surviving in environments with greater nutritional opportunities) and oligotroph (which survive in environments with low levels of nutrients).

Viruses have been enumerated in soil samples, but their abundance and distribution in soils remain an open question. Initial studies indicate that soil viral communities are more abundant and diverse than their aquatic counterparts (Williamson et al. 2005). The existence of lysogeny in prokaryotic communities is estimated by exposing them to various inducing agents such as mitomycin C or UV radiation. In previous studies, the frequency of lysogeny among cultivable soil bacteria was estimated and approximately 30% contained inducible prophage (Williamson et al. 2008), while cultivation-independent assessments ranged from 22 to 68% in soils from Delaware and 4 to 20% in Antarctic soil, suggesting that lysogeny may be more prevalent in soil microbial communities than in their marine counterparts (Williamson et al. 2007).

It is suggested that lysogeny is a prevalent reproductive strategy among soil bacteriophages and that the potential for HGT via transduction is significant in soil microbial communities (Ghosh et al. 2008).The authors have also shown that viral preparations contain 16S rRNA and *trzN*, a gene that encodes atrazine chlorohydrolase. On examination, some of the sequences revealed hybrid 16S rRNA structures which are indicative of interfamily horizontal transfer and recombination of 16S rRNA genes (Ghosh et al. 2008).

The fluidic nature of microbial genomes, substantially contributed by lateral gene transfer, enables the adaptation of microbes to varying ecological niches. The genome of *Chromobacterium violaceum* revealed the existence of four prophages and twelve insertion elements. The possibility of gene transfer between *Salmonella*, *Ralstonia*, and *Xanthomonas* has been suggested. The lateral gene transfers have contributed to acquisition of new traits to the microbe (de Almeida et al. 2004).

5.1.2 Horizontal Gene Transfer in Bacteria

HGT is regarded as a fast mode of evolution wherein new genes are acquired by transduction, transposition, and transformation. Microbes therefore exploit ecological opportunities by such kind of transfer events. Moreover, lateral transfer mediated by phages seems to be an ideal vehicle for shuttling pathogenicity islands between species (Brussow et al. 2004). It was proposed that a high genomic deletion rate is instrumental in removing dangerous genetic parasites from the bacterial genome. Deletion process could explain why the bacterial genomes did not increase in size despite a constant bombardment with parasitic DNA over evolutionary time periods. The streamlined bacterial chromosome containing few pseudogenes might be the consequence of this deletion process of parasitic DNA (Brussow 2007).

Bacterial genomes evolve through a variety of process including HGT to survive under selective pressures exerted by the environment (Arber 1991). Internal modifications of genome by intergenomic homologous recombination and HGT (intragenic recombination) have been prime reasons for bacterial genome diversity (Chitra and Archana 2002). Mobile elements are responsible for the transfer of new functions to a bacterial cell and are recognized as important agents in bacterial evolution (Tinsley et al. 2006).

In trying to understand the importance of HGTs and mobile genetic elements (MGEs), databases were generated by groups. A database for the classification of MGEs has been developed (Leplae et al. 2004) emphasizing the central role played by the members of the prokaryotic MGEs in mobilizing and reorganizing genes. Meanwhile, classification of phages had been initiated by Edward's group (Rohwer and Edwards 2002). A database for horizontally transferred genes (HGT-DB) (Garcia-Vallve et al. 2003) includes genes, which tend to deviate in statistical parameters such as G+C content, codon, and amino-acid usage from their prokaryotic genomes. With this method, a significant number of prokaryotic genes have been proposed as having been acquired by HGT.

5.1.3 Significance of Prophages

Bacteriophages represent the biggest reservoir of sequence information in the biosphere. They are extremely common in the environment. They take up either of the two cycles, Lytic or Lyosgeny, after infecting the bacterial host. In lytic cycle, phages infect the host and program the synthesis of phage virions, which are then released after lysing the host. In lysogeny, the phage genome integrates with the bacterial genome and establishes a stable relationship with the host bacteria. The impact of prophages on the evolution of microbes is clearly becoming evident (Canchaya et al. 2004).

Temperate phages encode functions that increase the fitness of the lysogen and are repeatedly reported to encode virulence factor responsible for the pathogenicity of the corresponding microbe (Brussow et al. 2004). Prophages seem to be transient residents of bacterial genomes on the evolutionary time scale. Many prophages are no longer inducible. In silico analysis of the bacterial genomes suggested that the majority of prophages are defective and apparently in a process of gradual decay (Casjens 2003). Such cryptic prophages are predominantly observed on most of sequenced bacterial genomes. Prophage and prophage-related entities are classified into four additional types namely defective prophages, which are prophages in the state of mutational decay, satellite phages which do not carry their own virion structural protein genes, bacteriocins that resemble phage tails and are devices that kill bacteria, and gene transfer agents (GTAs), which are tailed phage-like particles that encapsidate random fragments of bacterial genome (Casjens 2003).

Prophages have been implicated in serotype conversion, pathogenesis (encode virulence genes) (Boyd et al. 2001), phage immunity, and interstrain genetic variability in several bacterial species (Baba et al. 2002; Banks et al. 2002; Boyd and Brussow 2002). Prophages, both intact and defective, have a special role in this context as they are resident elements and play a significant role in the physiology of host bacteria (Canchaya et al. 2003). As reviewed by Canchaya et al. (Canchaya et al. 2004), the impact of prophage DNA on the evolution and genome structure diversification of bacteria is evident.

5.1.4 Prophages and Associated Fitness Islands

Genomic islands are 10–100 kb in length. They frequently harbor phage- and/or plasmid-derived sequences, including transfer genes or integrases and insertion sequence (IS) elements. These blocks of DNA are most often inserted into tRNA genes and may be unstable. In many ways, they increase the fitness of the bacterium and are thereby called as "fitness islands." Such fitness islands are classified into several subtypes based on the life style of the microbes that harbor them, such as "ecological islands" in which they contribute to the survival of the microbe in an environment. "Saprophytic islands" in microorganisms can persist as a saprophyte in a host. In many cases, the fitness factor temporarily or permanently resides in the host either providing some benefits ("Symbiosis islands") or cause damage (pathogenicity islands – PAIs) by interacting with living hosts (Hacker and Carniel 2001).

Well-known insertion hotspots for genomic islands include tRNA and tmRNA regions. A tool for systematically investigating the contents and contexts of bacterial tRNA and tmRNA genes was devised (Ou et al. 2006). They have analyzed the genomes of *E coli* and *Shigella*. Prokaryotes often contain pathogenicity islands by the advent of site-specific integrases. Integration is mostly observed within the tRNA and tmRNA gene resulting in the disruption of gene. An algorithm developed based on this is called Islander (Mantri and Williams 2004). With a total of 106 genomes screened, 143 candidate integrative islands were identified and are available through a web interface. Genomic islands in prokaryotes have also been detected by another network service which incorporates multiple DNA signals and genome annotation data in aiding the detection of genome islands (Hsiao et al. 2003). A search for pathogenicity islands in 148 prokaryotic sequences resulted in the identification of 77 candidates PAIs by applying homology-based method combined with abnormalities detected in genomic composition (Yoon et al. 2005). Prophages were found to contribute to fitness islands (Brussow et al. 2004). Prophage regions also contain PAIs (Srividhya et al. 2007).

5.2 Soil Prophage Genomics

5.2.1 Prophage Existence in Soil Bacteria

The existence of prophages in soil bacteria was explored. Soil bacteria in existing genomes in prophage database (Srividhya et al. 2006) were taken. Prophages detected by PSA (Srividhya et al. 2006) and DRAD (Srividhya et al. 2007) in soil genomes were analyzed and compared with prophage reports from other methods namely Phage finder (Fouts 2006) and Prophinder (Lima-Mendez et al. 2008).

A total of 200 bacterial genomes with no prophage reports were taken up for the study. Employing proteome comparison namely PSA (Protein similarity approach) yielded 30 prophage-like elements. Using the genome comparison method, DRAD (Dinucleotide relative abundance approach), 52 prophage elements were identified. Comparative analysis of other available methods against the above approaches developed here will be discussed. Detailed analysis of locus will help understand the contribution of prophages to the microbial communities in soil. Tables 5.1 and 5.2 detail the prophage loci in soil bacteria identified by PSA and DRAD.

5.2.1.1 Prophages Detected by Protein Similarity Approach

A comprehensive bioinformatics analysis was earlier carried out for the e14 cryptic prophage sequence. This showed that the e14 is modular and shares a large part of its sequence with *Shigella flexneri* phage SfV. Based on this similarity, the regulatory region including the repressor and Cro proteins and their promoter binding sites were identified (Mehta et al. 2004). A protein-based comparative approach using the COG database as a starting point was carried out to detect new lambdoid prophage like elements in a set of completely sequenced genomes. This protein similarity approach (PSA) was extended by the use of BLAST similarity searches rather than limiting to the COG database (Rao et al. 2005). Detection of prophages based on proteomic signature comparison using e14 prophage proteome as reference set was carried out. The PSA method was tested with bacterial genomes having known reports of prophages and then extended to newly sequenced bacteria. This approach using BLAST helped identify lambdoid-like prophage elements in a representative set of completely sequenced bacterial genomes.

In a total of 14 soil bacteria, 30 prophage-like elements were identified. Table 5.1 details on the prophages and the respective genomes. Among the 30 prophage loci identified by PSA, *Nostoc* prophage encodes nitrogenase genes. Although other loci encode phage proteins and virulence factors, they do not have critical genes that could contribute for the host survival in soil environment. With PSA, 30 prophages detected have a total of 673 proteins, among which 332 are hypothetical. Figure 5.1 represents the distribution of functional proteins from prophages identified by PSA in soil bacteria. After hypothetical proteins, the next highest occurrence is for viral morphogenesis proteins.

Table 5.1 Prophages detected by PSA in soil bacteria

Host	Number of prophages	Comment on the nature of bacteria	Comment on the prophage encoded proteins
Acinetobacter sp. *ADP1*	2	Nutritionally versatile soil bacterium	MORON and Transposase, putative tellurium resistant protein (KlaA/kilA), encodes lytic transglycosylase
Bacillus anthracis str. Ames	3	Anthrax, facultative anaerobic, soil organism	Most functional prophage associated proteins. Annotated as LambdaBa04 prophage proteins
Bacillus cereus ATCC 10987	5	Aerobic food poisoning, Soil Saprophyte	Encode teminase packaging protein, recombinase
Bacillus thuringiensis serovar konkukian str. 97-27	1	Facultative, insect pathogen, soil organism	Encodes Integrase and sporulation proteins
Burkholderia pseudomallei K96243	4	Soil bacteria, melioidosis	Very crytic integrase and DNA binding proteins
Chromobacterium violaceum ATCC 12472	1	Facultative anaerobic, sepsis and liver abscesses, microbe soil and water	Has bacteriophage associated proteins
Desulfovibrio vulgaris subsp. *vulgaris str. Hildenborough*	4	Anaerobic, breaks down metals and can corrode machinery, soil bacterium	Encode teminase packaging protein, tail packaging
Methanococcus maripaludis S2	1	Methanogenic	Very cryptic with only recombinase
Methylococcus capsulatus str. Bath	1	Aerobic, thermophilic, methanotroph	Prophage MuMc02 proteins
Mycobacterium bovis	4	Aerobic-TB in cattle	Transposase, hypothetical protein, transcriptional regulatory protein, Probable phiRv1 phage protein
Nocardia farcinica IFM 10152	1	Nocardiosis, soil bacteria	Mostly hypothetical proteins with Integrase and terminase
Nostoc sp. *PCC 7120*	1	Nitrogen-fixing cyanobacteria	Nitrogenase, fdxN element site-specific recombinase, anerredoxin, encodes Integrase and excisionase
Streptomyces avermitilis MA-4680	1	Soil bacteria, antiparasitic agent avermectin and other secondary metabolites	Encodes only integrase
Yersinia pestis biovar Medievalis str. 91001	1	Pneumonic, septicemic, and the notorious bubonic plagues, soil	Encodes tail fiber and repressor protein
Yersinia pseudotuberculosis IP 32953	4	Tuberculosis-like symptoms soil bacterium	Integrase, Moron, tail fiber proteins and repressor protein

Table 5.2 Prophages detected by DRAD in soil bacteria

Bacterial genome	Number of prophages	Comment on the nature of bacteria	Comment on the prophage encoded proteins
Bacillus anthracis str. Ames	2	Anthrax, facultative anaerobic, soil organism	MORONS – glucosyl transferase
Bacillus cereus ATCC 10987	1	Aerobic food poisoning, Soil Saprophyte	MORONS – glucosyl transferase
Bacillus clausii KSM-K16	2	Alkaliphilic, soil and marine	Phage proteins, MORONS – beta glucosidase
Bacillus thuringiensis serovar konkukian str. 97-27	3	Facultative, insect pathogen, soil organism	Flagellar proteins (virulence), sporulation proteins
Bradyrhizobium japonicum USDA 110	3	Nitrogen fixing bacterium	Transposase, integrase
Burkholderia pseudomallei 1710b	4	Soil bacteria, melioidosis	Phage proteins, restriction protein
Burkholderia thailandensis E264	3	Nonfermenting motile soil bacteria	Phage protein, Transposase, and restriction protein
Chromobacterium violaceum ATCC 12472	1	Soil bacteria, bactericidal purple pigment, violacein	Glucosyl transferase and lysis protein
Corynebacterium diphtheriae NCTC 13129	1	Soil, diphtheria	Phage and HNH proteins
Corynebacterium efficiens YS-314	1	Thermostable, soil bacterium	Capsule proteins
Corynebacterium glutamicum ATCC 13032 (Bielefield)	3	Natural producer of glutamic acid, soil bacterium	Transposase and phage protein
Coxiella burnetii RSA 493	1	Q fever, soil, and water	Pilus proteins
Dehalococcoides ethenogenes 195	3	Dechlorinates tetrachloroethene to vinyl chloride, soil bacterium	Integrase, transposase, recombinase Virulence, and HNH proteins
Desulfovibrio desulfuricans G20	5	Sulfate-reducing soil bacterium	Phage proteins, restriction protein, and transposase
Desulfovibrio vulgaris subsp. *vulgaris str. Hildenborough*	5	Breaks down metals and can corrode machinery, soil bacteria	Phage proteins, transposase, and restriction systems
Frankia sp. *CcI3*	1	Nitrogen fixing filamentous soil bacteria	Excisionase
Geobacter sulfurreducens PCA	1	Metal-reducing soil bacteria	Transposase and glucosyl transferase
Leptospira interrogans serovar copenhageni str. Fiocruz L1-130	2	Soil highly invasive spirochete	Transposase, nuclease-like protein and RNA polymerase sigma factor, Purine/Pyrimidine kinase
Leptospira interrogans serovar Lai	2	Leptospirosis, soil bacterium	Putative glycosyl transferase, Fimbrial proteins (Fimh-like and Fmh-like protein)

(*continued*)

Table 5.2 (continued)

Bacterial genome	Number of prophages	Comment on the nature of bacteria	Comment on the prophage encoded proteins
Nitrosococcus oceani ATCC 19707	3	Ammonia oxidizing soil bacteria	Phage proteins and transposase
Nitrosospira multiformis ATCC 25196	2	Ammonia oxidizing soil bacteria	Transposase, recombinase
Pseudomonas fluorescens Pf-5	1	Soil bacteria	Phage proteins
Rhodopseudomonas palustris HaA2	1	Purple nonsulfur phototrophic bacterium, soil, and marine	Phage protein
Rhodospirillum rubrum ATCC 11170	1	Purple nonsulfur bacteria soil and marine	Resolvase, intergrase, capsid

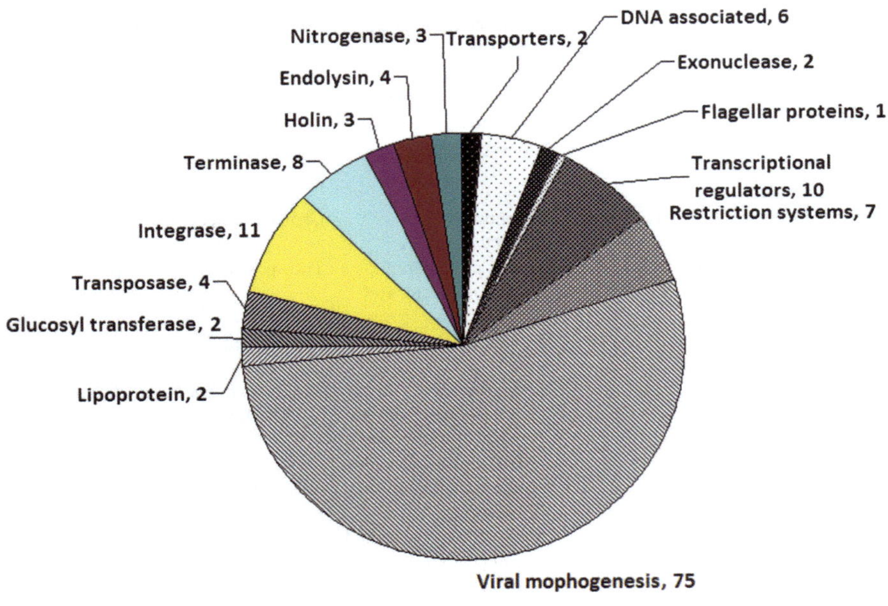

Fig. 5.1 Protein distribution in prophages detected by PSA in soil bacteria

Prophages bcpup4 (*Bacillus cereus ATCC 10987*), Dvpup3 (*Desulfovibrio vulgaris* subsp. *vulgaris str. Hildenborough*), Nopup1 (*Nostoc* sp. *PCC 7120*), bapup3 (*Bacillus anthracis str. Ames*), Cvpup1 (*Chromobacterium violaceum ATCC 12472*), Mbpup4 (*Mycobacterium bovis*), and Stapup1 (*Streptomyces avermitilis MA-4680*) are highly cryptic prophages with less than ten Open Reading frames (ORFs). Among the prophages detected, bcpup4 prophage from *Bacillus cereus* ATCC 10987 is smallest (3121 bp) and Ypsudopup4 from *Yersinia pseudotuberculosis* IP 32953 (43403 bp) is the largest prophage.

Acinetobacter sp. strain ADP1 is a nutritionally versatile soil bacterium and is highly competent for natural transformation. Genome analysis of the *Acinetobacter* ADP1 revealed genes for metabolic pathways involved in utilization with five major "islands of catabolic diversity," contributing to one-quarter of the complete genome. This genome is devoid of pyruvate kinase (Barbe et al. 2004). Acpup1 and Acpup2 prophages detected by PSA harbor MORON and transposase, putative tellurium resistant protein (KlaA/kilA), and lytic transglycosylase. DRAD method did not detect any new prophage in this genome. Phagefinder and Prophinder methods detect one new locus each.

Bacillus anthracis str. Ames encodes three prophages bapup1, bapup2, and bapup3 as identified by PSA. They encode mostly functional prophage-associated proteins and are as LambdaBa04 prophage proteins. However, bapup3 is very cryptic with hypothetical proteins. Three prophages among five detected by Prophinder are the same locus as reported by PSA. Among the four reported by Phagefinder three are the same as PSA.

Prophage detected by PSA in *Bacillus cereus* ATCC 10987 encodes terminase packaging protein, restriction-modification system, glycosyl transferase, recombinase, and exonuclease. Prophage bcpup1 is also detected by Phagefinder along with three new loci. Four new loci are reported by Prophinder. Btpup1 prophage detected by PSA in *Bacillus thuringiensis* serovar konkukian str. encodes sporulation, integrase, and repressor proteins. No prophages detected by other methods. In the genome of *Chromobacterium violaceum*, one prophage Cvpup1 detected was nothing but CvP1 reported in literature (de Almeida et al. 2004).

In *Desulfovibrio vulgaris* subsp. *vulgaris str.* Hildenborough, four prophages are identified by PSA. Among them Dvpup1 and Dvpup4 are also detected by Phagefinder. Dvpup2 and Dvpup3 are detected by Prophinder.

Methylococcus capsulatus can fix atmospheric nitrogen by the advent of its genes nifE, nifN, and nifX nitrogenase structural genes (Ward et al. 2004). *Methylococcus capsulatus str.* Bath encodes one prophage Mcpup1 prophage, which encodes MuMc02 proteins as detected by PSA. This prophage locus is reported by other methods phage finder and Prophinder along with another new locus by both the methods. *Mycobacterium bovis* has four prophages, which encode transposase, hypothetical protein, transcriptional regulatory protein, and phage proteins. Mbpup2 is also detected by Phagefinder program. *Methanococcus maripaludis* S2 encoded Mmpup1 prophage detected by PSA is very cryptic with only recombinase. No prophages are reported by other methods.

The genome of *Nocardia farcinica* revealed that the virulence factor Mce operon exists as six copies (Ishikawa et al. 2004). *Nocardia farcinica* IFM 10152 nfpup1 encodes mostly hypothetical proteins with integrase and terminase, which is also detected by phage finder along with two new prophage loci. Prophinder program reports three new loci. *Nostoc* sp. prophage nopup1 detected by PSA encodes nitrogenase subunits along with fdxN element site-specific recombinase, anerredoxin, Integrase, and excisionase. A new locus is reported by prophinder.

Streptomyces avermitilis MA-4680 prophage Stapup1 detected by PSA encodes only integrase and this locus is reported by both phage finder and prophinder.

Yersinia pestis biovar Medievalis str. 91001 prophage Ypespup1 detected by PSA encodes tail fiber and repressor protein and is identified by prophinder along with two more loci.

E coli O18:K1:H7 has filamentous phage CUS1 in its genome and CUS-2 another prophage which is homologous to CUS-1 is present in *Yersinia pestis* Biovar orientalis. This locus in *Yersinia* is responsible for virulence (Gonzalez et al. 2002). Type IV pilus biogenesis is regulated by ypm superantigen *pil* gene cluster. This locus has been reported to be acquired by lateral gene transfer by *Yersinia pseudotuberculosis* and is associated with 270 unrelated strains (Collyn et al. 2005). The clustered regularly interspaced short palindromic repeats called CRISPRs are repetitive elements present in both archaea and bacteria and are associated with DNA recombination and repair genes. They consist of repeats interspaced with nonrepetitive elements or spacers. Three distinct loci in *Yersinia pestis* genome have been reported (Pourcel et al. 2005). Majority of these spacers were found to be associated with inactive prophage loci. In *Yersinia pseudotuberculosis*, Ypsudopup1, Ypsudopup3, and Ypsudopup4 prophages detected by PSA are also detected by Prophinder. Additionally, phage finder also predicts Ypsudopup3 as prophage loci.

5.2.1.2 Prophages Detected by DRAD

Dinucleotide Relative abundance difference (DRAD) approach takes into account the local heterogeneity within the given bacterial genomes. DRA values are reported to remain relatively uniform within a genome and its closely related organisms. On this basis, the collection of 16 DRA values has been referred to as a genomic signature. Thus local heterogeneity in DRA values has been used to detect alien regions in bacterial genomes (Karlin 1998). This method has also been applied to phage genomes to understand similarities and dissimilarities associated with them (Blaisdell et al. 1996). This approach is modified to detect prophages in bacterial genomes. Since not all the dinucleotides show variation, an appropriate selection helps to further increase the discrimination of the prophage regions. Putative prophage regions could be identified by finding local regions of bacterial genomes that show significant deviation in dinucleotide abundance relative to the background. Bacterial genomes loci showing significant deviation in dinucleotide abundance relative to the background genome and at the same time have similar dinucleotide abundance relative to that of a reference set of nonredundant prophage sequences relevant for those bacteria were considered as hit and further based on annotation from the protein table file, the hit was annotated as putative prophage.

Out of 1302 proteins from 52 prophages identified by DRAD, 680 are hypothetical. As seen in Fig. 5.2, higher incidence of transposase is observed.

Among the 52 prophage loci, prophages burthapup3 (*Burkholderia thailandensis E264*), frankiapup1 (*Frankia* sp. *CcI3*), burthapup2 (*Burkholderia thailandensis E264*), rhodpalpup1 (*Rhodopseudomonas palustris HaA2*), leplai1 (*Leptospira interrogans serovar Lai*), lepcop2 (*Leptospira interrogans serovar copenhageni*

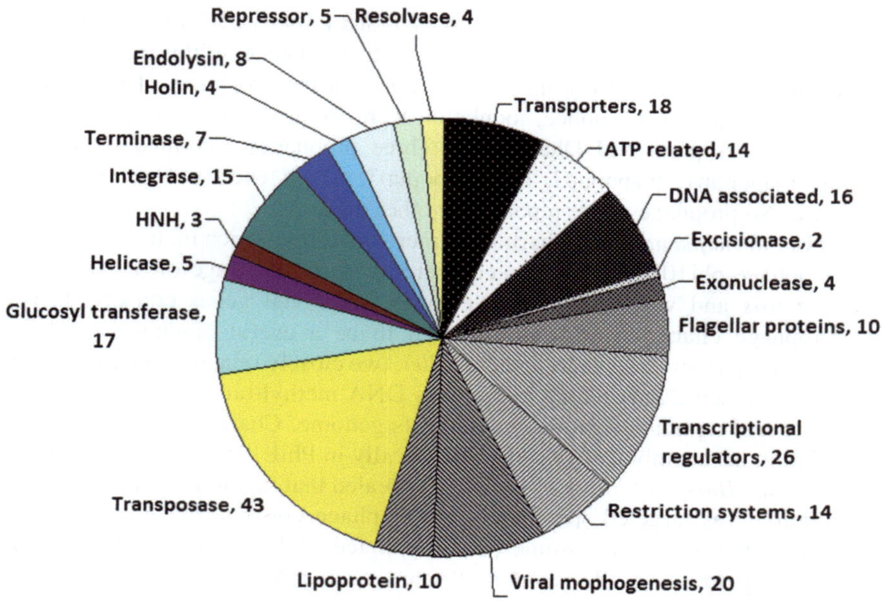

Fig. 5.2 Protein distribution in prophages detected by DRAD in soil bacteria

str. Fiocruz L1-130), bcerpup1 (*Bacillus cereus ATCC 10987*), burpsepup2 (*Burkholderia pseudomallei 1710b*), and leplai2 (*Leptospira interrogans serovar Lai*) are very cryptic and typical phage remnants with less than four ORFs. The largest prophage is bclaupup2 with 106 ORFs (109213 bp) is from *Bacillus clausii* KSM-K16 and it encodes phage proteins and morons. *Frankia*, nitrogen fixing filamentous soil bacteria, has one prophage frankiapup1 which encodes 2 ORFs with genome size of 1627 is the smallest of the lot.

In *Bacillus anthracis*, DRAD detects two new loci banthpup1 and banthpup2 which harbor MORONS, glucosyl transferase. DRAD scan on *Bacillus cereus* ATCC 10987 was found to encode bcerpup1 encodes morons, glucosyl transferase. bclaupup1 and bclaupup2 from *Bacillus clausii* KSM-K16 encodes morons beta-glucosidase, phage proteins, and morons. bclaupup2 is also detected by Phage finder and Prophinder.

Bacillus thuringiensis serovar israelensis is reported to encode extrachromosomal prophage on a plasmid (Verheust et al. 2003). The prophage is reported to harbor muramidases and DNA packaging protein. *Bacillus thuringiensis serovar konkukian* str. 97-27 genomes have three prophages bthuripup1, bthuripup2, and bthuripup3 detected by DRAD which encode flagellar and sporulation proteins hence encode virulence factors.

Phage of *Bradyrhizobium japonicum* Rhizophage V was found to be lytic for this host and its lytic activity varied between strains of the host (Abebe et al. 1992).

Two other temperate phages include L4-4 (V5) and L4-4 (V12). However, none of the phages were incorporated into any of the nodulation genes of this microbe. Two transconjugants of *Bradyrhizobium japonicum* with the properties of both phage resistance and ability to induce nodulation were isolated using transposons Tn5 mutagenesis (Appunu and Dhar 2008). Three prophages identified by DRAD include bjapopup1, bjapopup2, and bjapopup3, which encode transposase and integrase. No prophages were reported by other methods.

Burkholderia pseudomallei harbors numerous MGEs, which include temperate bacteriophage phi1026b. This was compared with E125 phage of *Burkholderia thailandensis* and was found to have many functional genes conserved. The bacteriophage characterized was proposed to be a useful diagnostic tool for differentiating *B. pseudomallei* and *B. mallei*, two closely related biological threat agents (DeShazer 2004). PhiE125 encodes DNA methyltransferases in both the lysogenic and replication modules within its genome. Characterization of DNA methylation in recombinant systems, specifically in PhiE125 lysogenic strains of *B. mallei* and *Burkholderia thailandensis*, revealed that, upon induction, cytosine methylation was targeted specifically to the phage episome but not the phage provirus or the host chromosome (Smith and Jeddeloh 2005). Genomic differences among strains of *B. pseudomallei* are predicted to be one of the major causes of the diverse clinical manifestations observed among patients with melioidosis. The existence of genomic islands (GIs) contributes to genomic diversity in this species. A total of 71 distinct GIs are present in this genome especially near tRNA gene loci (Tuanyok et al. 2008). Interestingly, *Burkholderia pseudomallei* 1710b prophages burpsepup1, burpsepup2, burpsepup3, and burpsekpup1 detected by DRAD encode phage and importantly restriction proteins. burpsepup3 is also reported by Phagefinder. burpsepup1, burpsepup2, and burpsepup3 are reported by Prophinder.

Burkholderia thailandensis E125 harbors temperate phage, which is specific to *Burkholderia mallei*. The phage is similar to lambda-like phages and prophages and belongs to Siphoviridae (Woods et al. 2002). *Burkholderia thailandensis* E264 prophage encodes phage and restriction proteins. burthapup2 and burthapup3 although very cryptic are also detected by Prophinder. Phagefinder detected three new prophage loci. Genomic sequences of four virulent myophages, Bcep1, Bcep43, BcepB1A, and Bcep781, whose hosts are soil isolates of the *Burkholderia cepacia* complex has been reported (Summer et al. 2006).

The prophage CvP1-4 is located on the genome of *Chromobacterium violaceum*. CvP1 (CV0337-CV0356) is Mu-like. CvP2 (CV0406-CV0432) is related to P2 and Mu. CvP3 from CV0645-CV0652 has higher incidence of hypothetical proteins and is related to P2 and P4 phages. CvP4 from CV 2114 to CV2150 is close to P2 and CT18 phages (de Almeida et al. 2004). DRAD approach detected only one prophage with locus CV4014-CV4034, chrviopup1 encoding glucosyl transferase and lysis protein and other reported prophages are not detected by this method. CvP3 and CvP4 are detected by Phagefinder. Prophinder found CvP2 and CvP4.

The existence of prophages was reported very early in corynebacteria. They include phages Φ304L, Φ304S and prophage Φ15/Φ16 from *Corynebacterium glutamicum* strains. All these phages belonged to Siphoviridae family (Moreau et al. 1995). corglupup1, corglupup2, and corglupup3 detected by DRAD harbor transposase and integrase. Other methods do not report prophages. *Corynebacterium glutamicum* genome has DNA regions with unusal composition indicating the presence of horizontally transferred genes. This region includes cg0415-cg0443, cg3280-cg3295, and many IS elements. This soil bacterium is nonpathogenic making it ideal for understanding the cell wall and mycolic acid synthesis. Also many genes are dedicated to the synthesis of aspartate-derived aminoacids and vitamins (Kalinowski et al. 2003). The genome of *Corynebacterium glutamicum* has three prophages CGP1, CGP2, and CGP3. The CGP1 and CGP2 are relatively smaller prophages. Prophage CGP3 contributes to the population heterogeneity in *Corynebacterium glutamicum* ATCC 13032. The CGP3 constitutes cg1893-cg2071 (Frunzke et al. 2008). However, this locus is not identified by prophage detection methods. *Corynebacterium diphtheriae* NCTC 13129 prophage identified by DRAD encodes phage and HNH proteins. Cordippup1 prophage is also detected by Phagefinder. In *Coxiella burnetii* RSA 493, DRAD identified prophage encodes pilus proteins. Again other methods do not report prophages.

Dehalococcoides ethenogenes strain 195 is the only bacterium known that is capable of reductively dechlorinating tetrachloroethene (PCE) and trichloroethene (TCE) to the nontoxic form ethane (Maymo-Gatell 2005). Reductive dehalogenase (RD) gene transcript levels in *Dehalococcoides ethenogenes* strain 195 were investigated using reverse transcriptase quantitative PCR during growth and reductive dechlorination of tetrachloroethene (PCE), trichloroethene (TCE), or 2,3-dichlorophenol (2,3-DCP) (Fung et al. 2007). Upregulation of reductive halogenases genes has been revealed by microarray analysis of *Dehalococcoides ethenogenes 195* (Johnson et al. 2008). The genome contains 18 copies of putative reductive dehalogenase genes wherein hierarchical classification of the atypical regions containing the reductive dehalogenase genes indicated that these regions were probably acquired by several gene transfer events (Regeard et al. 2005). A total of four prophages are identified in the genome of *Dehalococcoides ethenogenes* by DRAD approach. One of reductive halogenase gene namely DET0162 is encoded by prophage locus dehethpup2 and other DET0079 in dehethpup1. The dehethpup3 encodes a potential phage with most the phage genes being conserved. This region also reported to be a 22-kb integrated region (Seshadri et al. 2005) encodes site-specific recombinase along with reductive dehalogenase (RD) genes. This suggests their recent acquisition (or perhaps consignment for dissemination). The gene that encodes TceA-RD, which essentially defines this organisms ability to dechlorinate chloroethenes past dichloroethane (DCE), is located in an IE. Among the nine integrated element locus reported (Seshadri et al. 2005), three loci are identified by DRAD approach, which includes the IE-I, IE-II, and IE-VII. Also among the histidine kinases and regulators DET1064 and DET1063 is located in prophage loci namely, dehethpup3.

In a total of five prophages identified by DRAD in *Desulfovibrio vulgaris* subsp. *vulgaris str.* Hildenborough genome, desvulpup1 and desvulpup2 are identified as single prophage locus by Prophinder, which also detects desvulpup4 and desvulpup5. Prophage desvulpup5 is also identified by Phagefinder. Again all these prophage loci encode phage transposase and restriction systems. Prophages detected by DRAD also encode terminase and tail packaging proteins.

DRAD method identified five prophages in the genome of *Desulfovibrio desulfuricans G20*. Among them desdespup1, 2, and 3 are also counter identified by Prophinder. Interestingly, desdespup3 is equivalent of two prophage loci identified by Prophinder and is also identified by Phagefinder along with desdespup1. These prophages encode phage proteins, restriction systems, and transposase. *Frankia* sp. CcI3 encodes a cryptic prophage with excisionase protein. One prophage identified by DRAD in *Geobacter sulfurreducens* PCA encodes transposase and glucosyl transferase.

Leptospira is a soil highly invasive spirochete. The biflexa species of *Leptospira* harbors LE1, LE3, and LE4 prophage. LE1 prophage harbors head–tail structural proteins and immunity repressor proteins. It replicates autonomously with the genome of biflexa (Bourhy et al. 2005). *Leptospira interrogans serovar Lai* type strain 56601 encodes a 22-kb genomic island covering a cluster of 34 genes (i.e., genes LA0186 to LA0219). This genome island is a putative prophage-like remnant with at least 8 of 34 sequences encoding prophage-like proteins, of which the LA0195 protein is probably a putative prophage CI-like regulator (Qin et al. 2008). *Leptospira interrogans serovar copenhageni* str. Fiocruz L1-130 prophages detected by DRAD encode putative glycosyl transferase and fimbrial proteins (Fimh-like and Fmh-like protein). DRAD detects lepcop1 and lepcop2 from *Leptospira interrogans serovar Lai* and they encode transposase, nuclease-like protein, RNA polymerase sigma factor, and purine/pyrimidine kinase. Other methods are unable to detect any prophages.

Nitrosospira multiformis is an ammonia-oxidizing soil bacterium. The genome of this microbe has unique gene clusters, which include *amo* and *hao* gene clusters. Also the gene clusters encoding urease and hydrogenase, ribulose phosphate carboxylase/oxygenase encoding operon is distinct and is absent in other species of *Nitrosospira*. Lateral gene transfer segments are located in plasmids as well. Three plasmids are seen to encode phage proteins and post segregational killing system. The genome loci comprising of Num1_A0922-Num1_A0934 are reported to be laterally acquired (Norton et al. 2008). *Nitrosospira multiformis* ATCC 25196 prophages detected by DRAD encode transposase, recombinase, and resolvase. Phagefinder and Prophinder report one locus each. In *Nitrosococcus oceani* ATCC 19707, DRAD identified prophage encodes phage proteins and transposase. Among them nitocepup3 is also identified by Phagefinder.

Pseudomonas fluorescens Pf-5 is a rhizobacterium. By sequence analysis, genome of Pf-5 was found to be devoid of transposons and IS elements. But have MGEs in the form of prophages and genomic islands spanning about 260 kb. The prophages include an F-pyocin-like prophage 01, a chimeric prophage 03, a lambdoid prophage 06, and decaying prophages 02, 04, and 05 with reduced size and/or complexity. The genomic islands are represented by a 115-kb integrative conjugative element (ICE) PFGI-1, which shares plasmid replication, recombination, and conjugative transfer genes with those from ICEs found in other *Pseudomonas* spp., and PFGI-2, which resembles a portion of pathogenicity islands in the genomes of the plant pathogens *Pseudomonas syringae* and *P. viridiflava*. Almost all of the MGEs in the Pf-5 genome are associated with phage-like integrase genes and are integrated into tRNA genes (Mavrodi et al. 2009). Prophages identified by DRAD in the genomes of *Pseudomonas fluorescens Pf-5* and *Rhodopseudomonas palustris* HaA2 encode phage proteins. Among these pseflupup1 is also identified by Phage-finder and Prophinder programs. Bacteriophage D3112 is a transposable phage found in the clinically relevant of *Pseudomonas aeruginosa* and is very similar to that of enterobacteriophage Mu. The highly mosaic nature and the role played by extensive horizontal exchange of genetic material are probably important in the evolution of D3112 (Wang et al. 2004).

Rhodopseudomonas spheroides harbors temperate phage which is inducible by mitomycin treatment (Mural and Friedman 1974). This phage RΦ-1 does not infect related species of *Rhodopseudomonas* namely palustris and capsulata. DRAD method identified one prophage with locus RPB_3315-RPB_3318 encoding phage proteins. However, three new prophage loci are identified by Phagefinder and one by Prophinder respectively. *Rhodospirillum rubrum* ATCC 11170 encodes two prophages as identified by DRAD. These prophages encode phage protein resolvase, integrase, and capsid proteins.

5.2.1.3 Prophages Reported by Other Methods Compared with PSA and DRAD

The prophages identified by PSA and DRAD were compared against other prophage detection methods. Figure 5.3 gives the comparative plot of prophages identified in selected soil bacteria by PSA against other prophage identification approaches. A total of 26 and 35 prophage elements are reported by Phagefinder and Prophinder, respectively. No prophages are reported by both methods in genomes of *Bacillus thuringiensis serovar konkukian str. 97-27* and *Methanococcus maripaludis S2*. Phagefinder has no prophage reports on genomes of *Nostoc* sp. *PCC 7120* and *Yersinia pestis biovar Medievalis str.* 91001. In *Mycobacterium bovis*, no prophage locus is reported by Prophinder. In the soil bacterium *Desulfovibrio vulgaris Hildenborough*, Prophinder program reports a higher number of prophages.

In 24 genomes scanned by DRAD, 40 and 47 prophages are reported by methods Phagefinder and Prophinder, respectively. Figure 5.4 gives a plot of prophages identified in by DRAD and other prophage identification methods. No prophage hits are reported for genomes *Bacillus thuringiensis serovar*, *Bradyrhizobium japonicum Corynebacterium efficiens*, and *Corynebacterium glutamicum* where DRAD reports 3, 3, 1, and 3 prophages, respectively, encoding phage integrase, transposase, HNH, flagellar, sporulation, and capsule proteins. Also in the genomes of *Coxiella burnetii RSA 493*, *Leptospira copenhageni*, *Leptospira Lai*, and *Rhodospirillum rubrum*, no prophage loci are reported by Phagefinder and Prophinder programs where DRAD detects 1, 2, 2, and 1 prophages, respectively, encoding

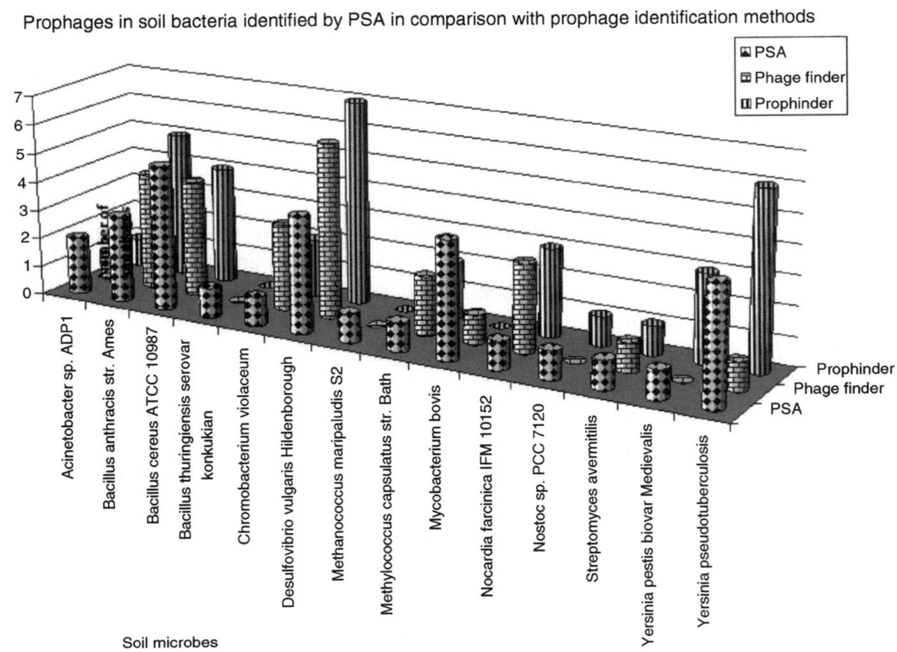

Fig. 5.3 Prophages in soil bacteria detected by PSA compared with other methods

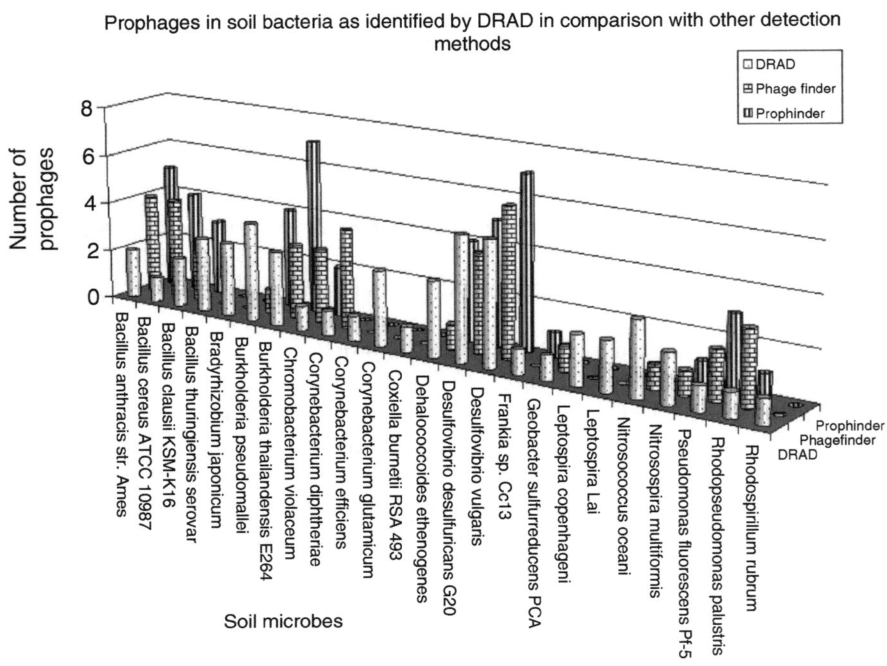

Fig. 5.4 Prophages in soil bacteria detected by DRAD compared with other methods

resolvase, integrase, capsid, transposase, putative glycosyl transferase, fimbrial, and pilus proteins. In the genomes of *Nitrosococcus oceani ATCC 19707*, *Geobacter sulfurreducens PCA*, and *Corynebacterium diphtheriae NCTC 13129*, Prophinder detects no prophages. In *Frankia*, Phagefinder program reports no prophage locus.

5.2.2 Impact of Prophages in Soil Bacteria

5.2.2.1 Organization of Prophages in Selected Soil Bacteria

Screening 200 bacterial genomes resulted in detection of 30 prophage-like elements by PSA method and 52 elements by DRAD method. A diversity of genomic organization exists for the prophage elements found in soil bacteria genomes. The first category of prophages, which are relatively scarce detected by both methods, is very cryptic with genome size less than 5 kb encoding only less than ten ORFs (Group 1). The second category is medial with genome size between 15 and 40 kb (Group 2). The third class are extremely huge with size ranging from 50 to 100 kb (Group 3). Figure 5.5 represents the distribution of prophages groups. Thus, the predominant distribution of prophages ranges between 15 and 40 kb genome size.

Among the soil bacterial genomes scanned for prophages by PSA method, *Bacillus cereus* ATCC 10987 was found to harbor highest number of prophages with a total of five. However, out of 15 genomes with prophage hits, 8 genomes harbor only one prophage. With DRAD, two genomes *Desulfovibrio desulfuricans*

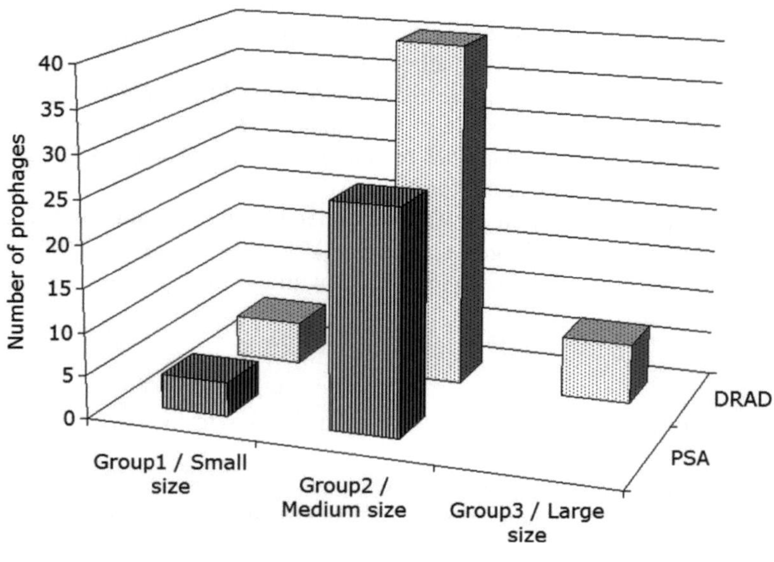

Fig. 5.5 Prophage distribution

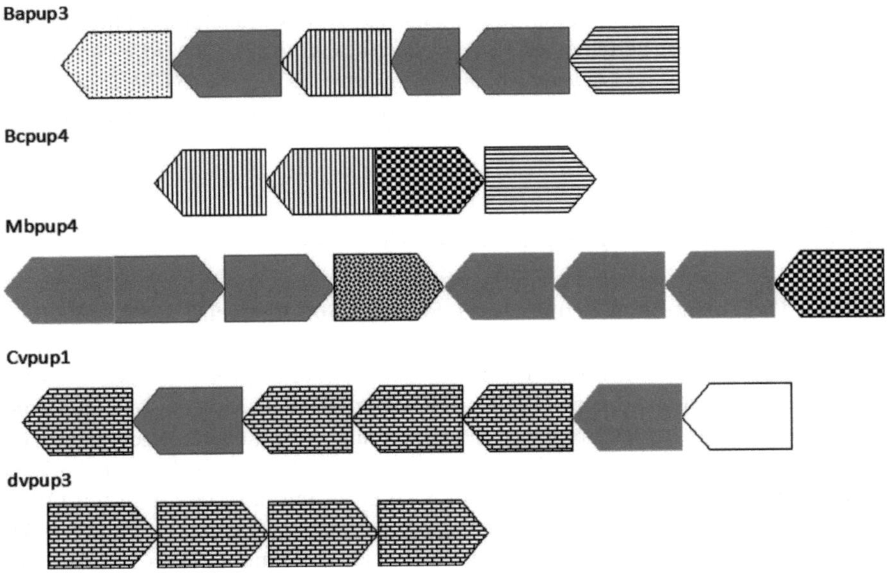

Fig. 5.6 Gene arrangements of cryptic prophages detected by PSA. Genes are not drawn to scale. *Gray* indicates hypothetical, *dotted* – terminase, *checks* – transcriptional regulators, *vertical lines* recombinase, *horizontal lines* – exonuclease, *white* – prophage protein, *bricks* – tail head structural proteins, *black dots* – functional host proteins, *fully black* – lysis proteins, *black circles* – restriction modification systems, *wavy* – nonphage protein, and *crosslines* – transposase

G20 and *Desulfovibrio vulgaris* subsp. *vulgaris str. Hildenborough* harbor five prophages each. From the 24 genomes with prophages, 11 genomes have only one prophage detected by DRAD method.

Among the cryptic prophages detected by PSA, three different gene organization was observed. As indicated in Fig. 5.6, prophages Mbpup4 and Bcpup4 have transcriptional regulators, Bapup3 has recombinase alone. Prophages Cvpup1 and Dvpup3 are typical GTAs with tail proteins.

Cryptic prophages detected by DRAD either had the integration machinery genes (Frankiapup1) or lysis genes (bcerpup1) or only transposase (Mbov2) in their genome. Burpsepup2 was a typical GTA with tail fiber proteins and integration genes (Fig. 5.7).

5.2.2.2 Prophage Encoded Gene Clusters in Soil Bacteria

In this section, prophage encoded gene clusters in selected genomes will be discussed. The protein organization of prophages identified in three categories soil bacteria was considered, which include pathogenic, nitrogen fixing, and extremophilic genomes.

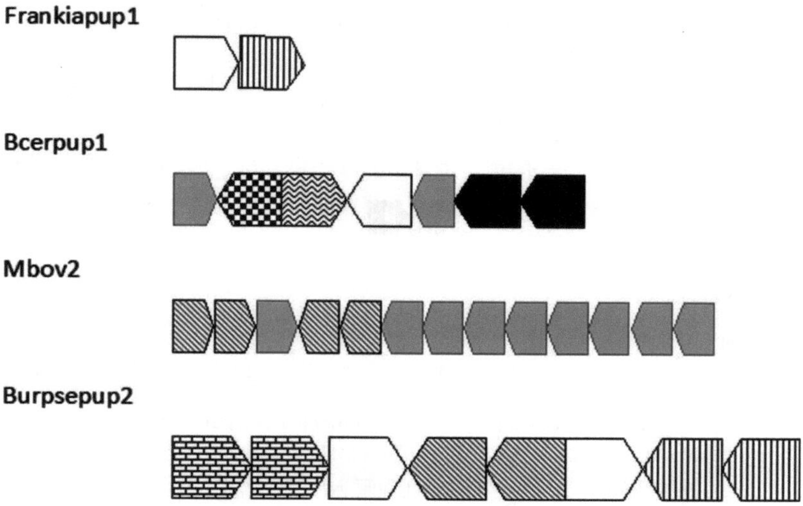

Fig. 5.7 Gene arrangements of cryptic prophages detected by DRAD. Genes are not drawn to scale. *Gray* indicates hypothetical, *dotted* – terminase, *checks* – transcriptional regulators, *vertical lines* recombinase, *horizontal lines* – exonuclease, *white* – prophage protein, *bricks* – tail head structural proteins, *black dots* – functional host proteins, *fully black* – lysis proteins, *black circles* – restriction modification systems, *wavy* – nonphage protein, and *crosslines* – transposase

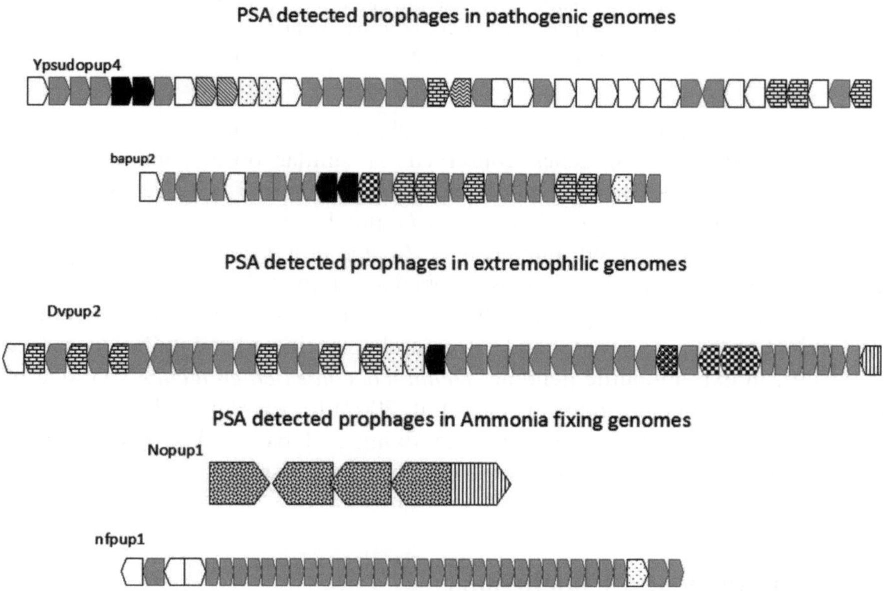

Fig. 5.8 PSA identified prophage gene clusters. Genes are not drawn to scale. *Gray* indicates hypothetical, *dotted* – terminase, *checks* – transcriptional regulators, *vertical lines* recombinase, *horizontal lines* – exonuclease, *white* – prophage protein, *bricks* – tail head structural proteins, *black dots* – functional host proteins, *fully black* – lysis proteins, *black circles* – restriction modification systems, *wavy* – nonphage protein, and *crosslines* – transposase

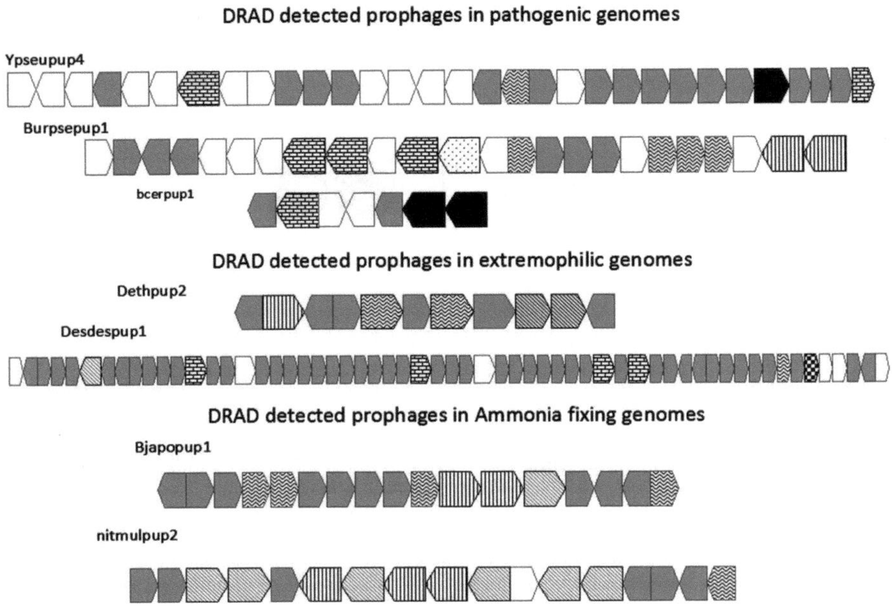

Fig. 5.9 DRAD identified prophage gene clusters. Genes are not drawn to scale. *Gray* indicates hypothetical, *dotted* – terminase, *checks* – transcriptional regulators, *vertical lines* recombinase, *horizontal lines* – exonuclease, *white* – prophage protein, *bricks* – tail head structural proteins, *black dots* – functional host proteins, *fully black* – lysis proteins, *black circles* – restriction modification systems, *wavy* – nonphage protein, and *crosslines* – transposase

In pathogenic soil bacteria, *Yersinia* prophage ypsudopup2 has lysis module, terminase, and structural genes conserved. A similar observation is seen in *B anthracis* prophage Bapup2. In Dvpup2, integration and lysis genes are conserved. In nitrogen fixing bacterium, *Nocardia* prophage Nfpup1 integration genes and terminase are conserved. Interestingly, *Nostoc* prophage encodes nitrogenase genes (Fig. 5.8). All the above hits were obtained from PSA.

DRAD detected prophages in pathogenic genomes mostly have integration–recombination, lysis genes, and structural proteins (head–tail) conserved (Fig. 5.9). In extremophilic genome, *Dehalococcoides ethenogenes* 195, one of the key protein (dehydogenase) (Fung et al. 2007) for extremophilic behavior is located in prophage locus. In ammonia fixing soil bacterial prophages, only integration and recombination genes are seen to be conserved.

5.3 Summary and Conclusions

With increasing loads of bacterial genome sequencing, data is available for comparative genomics and for understanding laterally transferred gene elements. A set of soil bacterial genomes and proteomes were scanned for the presence of prophage loci by DRAD and PSA methods. Analysis of prophages in soil bacteria helped identify interesting aspects. However, the contribution of prophages for the host survival in soil could not be correlated. Among the 30 prophage loci identified by PSA, *Nostoc* prophage encodes nitrogenase genes which are required for nitrogen fixation. Although other loci encode phage proteins and virulence factors, they do not have critical genes that could contribute for the host to survive in a soil environment. The same trend is seen with DRAD as well as the other methods Phagefinder and Prophinder. However, it should be noted that irrespective of the method, all prophage loci do harbor hypothetical proteins. Detailed analysis of these proteins needs to be addressed to arrive at their contribution to host phenotype.

Acknowledgments Bioinformatics facilities provided by Department of Biotechnology, Government of India under Centre of Excellence. CSIR and UGC RFSMS for fellowship to KVS.

References

Abebe HM, Sadowsky MJ, Kinkle BK, Schmidt EL (1992) Lysogeny in *Bradyrhizobium japonicum* and its effect on soybean nodulation. Appl Environ Microbiol 58:3360–3366

Appunu C, Dhar B (2008) Isolation and symbiotic characteristics of two Tn5-derived phage-resistant *Bradyrhizobium japonicum* strains that nodulate soybean. Curr Microbiol 57:212–217

Arber W (1991) Elements in microbial evolution. J Mol Evol 33:4–12

Ashelford KE, Day MJ, Fry JC (2003) Elevated abundance of bacteriophage infecting bacteria in soil. Appl Environ Microbiol 69:285–289

Baba T, Takeuchi F, Kuroda M, Yuzawa H, Aoki K et al (2002) Genome and virulence determinants of high virulence community-acquired MRSA. Lancet 359:1819–1827

Banks DJ, Beres SB, Musser JM (2002) The fundamental contribution to GAS evolution, genome diversification and strain emergence. Trends Microbiol 10:515–521

Barbe V, Vallenet D, Fonknechten N, Kreimeyer A, Oztas S et al (2004) Unique features revealed by the genome sequence of *Acinetobacter* sp. ADP1, a versatile and naturally transformation competent bacterium. Nucleic Acids Res 32:5766–5779

Blaisdell BE, Campbell AM, Karlin S (1996) Similarities and dissimilarities of phage genomes. Proc Natl Acad Sci USA 93:5854–5859

Bourhy P, Frangeul L, Couve E, Glaser P, Saint Girons I et al (2005) Complete nucleotide sequence of the LE1 prophage from the spirochete *Leptospira biflexa* and characterization of its replication and partition functions. J Bacteriol 187:3931–3940

Boyd EF, Brussow H (2002) Common themes among bacteriophage-encoded virulence factors and diversity among the bacteriophages involved. Trends Microbiol 10:521–529

Boyd EF, Davis BM, Hochhut B (2001) Bacteriophage-bacteriophage interactions in the evolution of pathogenic bacteria. Trends Microbiol 9:137–144

Brussow H (2007) Bacteria between protists and phages: from antipredation strategies to the evolution of pathogenicity. Mol Microbiol 65:583–589

Brussow H, Canchaya C, Hardt WD (2004) Phages and the evolution of bacterial pathogens: from genomic rearrangements to lysogenic conversion. Microbiol Mol Biol Rev 68:560–602

Canchaya C, Fournous G, Brussow H (2004) The impact of prophages on bacterial chromosomes. Mol Microbiol 53:9–18

Canchaya C, Proux C, Fournous G, Bruttin A, Brussow H (2003) Prophage genomics. Microbiol Mol Biol Rev 67:238–276, table

Casjens S (2003) Prophages and bacterial genomics: what have we learned so far? Mol Microbiol 49:277–300

Chitra D, Archana P (2002) Horizontal gene transfer and bacterial diversity. J Biosci 27:27–33

Collyn F, Fukushima H, Carnoy C, Simonet M, Vincent P (2005) Linkage of the horizontally acquired ypm and pil genes in *Yersinia pseudotuberculosis*. Infect Immun 73:2556–2558

de Almeida R, Trevilato PB, Bartoleti LA, Proenca-Modena JL, Hanna ES et al (2004) Bacteriophages and insertion sequences of *Chromobacterium violaceum* ATCC 12472. Genet Mol Res 3:76–84

DeShazer D (2004) Genomic diversity of *Burkholderia pseudomallei* clinical isolates: subtractive hybridization reveals a *Burkholderia mallei*-specific prophage in *B. pseudomallei* 1026b. J Bacteriol 186:3938–3950

Fierer N, Bradford MA, Jackson RB (2007) Toward an ecological classification of soil bacteria. Ecology 88:1354–1364

Fouts DE (2006) Phage_Finder: automated identification and classification of prophage regions in complete bacterial genome sequences. Nucleic Acids Res 34:5839–5851

Frunzke J, Bramkamp M, Schweitzer JE, Bott M (2008) Population heterogeneity in *Corynebacterium glutamicum* ATCC 13032 caused by prophage CGP3. J Bacteriol 190:5111–5119

Fung JM, Morris RM, Adrian L, Zinder SH (2007) Expression of reductive dehalogenase genes in *Dehalococcoides ethenogenes* strain 195 growing on tetrachloroethene, trichloroethene, or 2, 3-dichlorophenol. Appl Environ Microbiol 73:4439–4445

Garcia-Vallve S, Guzman E, Montero MA, Romeu A (2003) HGT-DB: a database of putative horizontally transferred genes in prokaryotic complete genomes. Nucleic Acids Res 31:187–189

Ghosh D, Roy K, Williamson KE, White DC, Wommack KE et al (2008) Prevalence of lysogeny among soil bacteria and presence of 16S rRNA and trzN genes in viral-community DNA. Appl Environ Microbiol 74:495–502

Gonzalez MD, Lichtensteiger CA, Caughlan R, Vimr ER (2002) Conserved filamentous prophage in *Escherichia coli* O18:K1:H7 and *Yersinia pestis biovar orientalis*. J Bacteriol 184:6050–6055

Hacker J, Carniel E (2001) Ecological fitness, genomic islands and bacterial pathogenicity. A Darwinian view of the evolution of microbes. EMBO Rep 2:376–381

Hsiao W, Wan I, Jones SJ, Brinkman FS (2003) IslandPath: aiding detection of genomic islands in prokaryotes. Bioinformatics 19:418–420

Ishikawa J, Yamashita A, Mikami Y, Hoshino Y, Kurita H et al (2004) The complete genomic sequence of *Nocardia farcinica* IFM 10152. Proc Natl Acad Sci USA 101:14925–14930

Johnson DR, Brodie EL, Hubbard AE, Andersen GL, Zinder SH et al (2008) Temporal transcriptomic microarray analysis of "*Dehalococcoides ethenogenes*" strain 195 during the transition into stationary phase. Appl Environ Microbiol 74:2864–2872

Kalinowski J, Bathe B, Bartels D, Bischoff N, Bott M et al (2003) The complete *Corynebacterium glutamicum* ATCC 13032 genome sequence and its impact on the production of L-aspartate-derived amino acids and vitamins. J Biotechnol 104:5–25

Karlin S (1998) Global dinucleotide signatures and analysis of genomic heterogeneity. Curr Opin Microbiol 1:598–610

Leplae R, Hebrant A, Wodak SJ, Toussaint A (2004) ACLAME: a CLAssification of Mobile genetic Elements. Nucleic Acids Res 32:D45–D49

Lima-Mendez G, Van Helden J, Toussaint A, Leplae R (2008) Prophinder: a computational tool for prophage prediction in prokaryotic genomes. Bioinformatics 24:863–865

Mantri Y, Williams KP (2004) Islander: a database of integrative islands in prokaryotic genomes, the associated integrases and their DNA site specificities. Nucleic Acids Res 32:D55–D58

Mavrodi DV, Loper JE, Paulsen IT, Thomashow LS (2009) Mobile genetic elements in the genome of the beneficial rhizobacterium *Pseudomonas fluorescens* Pf-5. BMC Microbiol 9:8

Maymo-Gatell X (2005) Fundamental things apply: the case of *Dehalococcoides ethenogenes*. Int Microbiol 8:137–140

Mehta P, Casjens S, Krishnaswamy S (2004) Analysis of the lambdoid prophage element e14 in the *E. coli* K-12 genome. BMC Microbiol 4:4

Moreau S, Leret V, Le Marrec C, Varangot H, Ayache M et al (1995) Prophage distribution in coryneform bacteria. Res Microbiol 146:493–505

Mural RJ, Friedman DI (1974) Isolation and characterization of a temperate bacteriophage specific for *Rhodopseudomonas spheroides*. J Virol 14:1288–1292

Norton JM, Klotz MG, Stein LY, Arp DJ, Bottomley PJ et al (2008) Complete genome sequence of *Nitrosospira multiformis*, an ammonia-oxidizing bacterium from the soil environment. Appl Environ Microbiol 74:3559–3572

Ou HY, Chen LL, Lonnen J, Chaudhuri RR, Thani AB et al (2006) A novel strategy for the identification of genomic islands by comparative analysis of the contents and contexts of tRNA sites in closely related bacteria. Nucleic Acids Res 34:e3

Pourcel C, Salvignol G, Vergnaud G (2005) CRISPR elements in *Yersinia pestis* acquire new repeats by preferential uptake of bacteriophage DNA, and provide additional tools for evolutionary studies. Microbiology 151:653–663

Qin JH, Zhang Q, Zhang ZM, Zhong Y, Yang Y et al (2008) Identification of a novel prophage-like gene cluster actively expressed in both virulent and avirulent strains of *Leptospira interrogans serovar Lai*. Infect Immun 76:2411–2419

Rao GV, Mehta P, Srividhya KV, Krishnaswamy S (2005) A protein similarity approach for detecting prophage regions in bacterial genomes. Genome Biol 6:p11

Ravel J, Fraser CM (2005) Genomics at the genus scale. Trends Microbiol 13:95–97

Regeard C, Maillard J, Dufraigne C, Deschavanne P, Holliger C (2005) Indications for acquisition of reductive dehalogenase genes through horizontal gene transfer by *Dehalococcoides ethenogenes* strain 195. Appl Environ Microbiol 71:2955–2961

Rohwer F, Edwards R (2002) The phage proteomic tree: a genome-based taxonomy for phage. J Bacteriol 184:4529–4535

Schloss PD, Handelsman J (2006) Toward a census of bacteria in soil. PLoS Comput Biol 2:e92

Seshadri R, Adrian L, Fouts DE, Eisen JA, Phillippy AM et al (2005) Genome sequence of the PCE-dechlorinating bacterium *Dehalococcoides ethenogenes*. Science 307:105–108

Smith MJ, Jeddeloh JA (2005) DNA methylation in lysogens of pathogenic *Burkholderia* spp. requires prophage induction and is restricted to excised phage DNA. J Bacteriol 187:1196–1200

Srividhya KV, Alaguraj V, Poornima G, Kumar D, Singh GP et al (2007) Identification of prophages in bacterial genomes by dinucleotide relative abundance difference. PLoS ONE 2: e1193

Srividhya KV, Geeta VR, Raghavenderan L, Preeti M, Jaime P et al (2006) Database and comparative identification of prophages. LNCIS 344:863–868

Summer EJ, Gonzalez CF, Bomer M, Carlile T, Embry A et al (2006) Divergence and mosaicism among virulent soil phages of the *Burkholderia cepacia* complex. J Bacteriol 188:255–268

Tinsley CR, Bille E, Nassif X (2006) Bacteriophages and pathogenicity: more than just providing a toxin? Microbes Infect 8:1365–1371

Tuanyok A, Leadem BR, Auerbach RK, Beckstrom-Sternberg SM, Beckstrom-Sternberg JS et al (2008) Genomic islands from five strains of *Burkholderia pseudomallei*. BMC Genomics 9:566

Verheust C, Jensen G, Mahillon J (2003) pGIL01, a linear tectiviral plasmid prophage originating from *Bacillus thuringiensis serovar israelensis*. Microbiology 149:2083–2092

Wang PW, Chu L, Guttman DS (2004) Complete sequence and evolutionary genomic analysis of the *Pseudomonas aeruginosa* transposable bacteriophage D3112. J Bacteriol 186:400–410

Ward N, Larsen O, Sakwa J, Bruseth L, Khouri H et al (2004) Genomic insights into methanotrophy: the complete genome sequence of *Methylococcus capsulatus* (Bath). PLoS Biol 2:e303

Williamson KE, Radosevich M, Smith DW, Wommack KE (2007) Incidence of lysogeny within temperate and extreme soil environments. Environ Microbiol 9:2563–2574

Williamson KE, Radosevich M, Wommack KE (2005) Abundance and diversity of viruses in six Delaware soils. Appl Environ Microbiol 71:3119–3125
Williamson KE, Schnitker JB, Radosevich M, Smith DW, Wommack KE (2008) Cultivation-based assessment of lysogeny among soil bacteria. Microb Ecol 56:437–447
Woods DE, Jeddeloh JA, Fritz DL, DeShazer D (2002) *Burkholderia thailandensis* E125 harbors a temperate bacteriophage specific for *Burkholderia mallei*. J Bacteriol 184:4003–4017
Yoon SH, Hur CG, Kang HY, Kim YH, Oh TK et al (2005) A computational approach for identifying pathogenicity islands in prokaryotic genomes. BMC Bioinformatics 6:184

Chapter 6
Back to the Soil: Retroviruses and Transposons

Omar Bagasra and D. Gene Pace

6.1 Soil Bacteria and Associated Retroelements

From structural, functional, and taxonomical perspectives, soil bacteria are an impressively diverse group. We know that they vary from free-living bacteria (that procure their own food independent of a host) to single fungi capable of extending their growth over large distances of multiple square kilometers. Still, we know relatively little about soil bacteria because of the difficulties associated with their cultivation (Banfield and Young 2009; Harris 2009). Only very limited number of species have been classified because less than 1% grow easily on agar plates. Unfortunately, at present most soil microbes simply cannot be cultivated, given present methodological challenges and limited capabilities (Harris 2009). Consequently, scientists depend on indirect analytical methodologies, mainly biochemical markers, as well as on measures of the metabolic activity in either entire soil microbial communities or selected segments of such communities. Scientists have yet to establish whether the microbial communities of the soil simply reflect general ecological patterns or if they could play major roles in soil restoration projects. Research has underscored the essential functions that microorganisms play in soil quality. This is particularly evident in the important areas of the cycling of essential nutrients, the decomposition of organic materials, the regulation of the productive capacity of plant life, in the dynamics of soil microbial community considered holistically (Ochman and Raghavan 2009), and in the structural aspects that accompany the generation of soil (Muotri et al. 2007). The results of

O. Bagasra (✉)
Department of Biology, South Carolina Center for Biotechnology, Claflin University, 400 Magnolia Street, Orangeburg, SC 29115, USA
e-mail: obagasra@claflin.edu
D.G. Pace
Department of English and Foreign Languages, South Carolina Center for Biotechnology, Claflin University, 400 Magnolia Street, Orangeburg, SC 29115, USA

G. Witzany (ed.), *Biocommunication in Soil Microorganisms*, Soil Biology 23,
DOI 10.1007/978-3-642-14512-4_6, © Springer-Verlag Berlin Heidelberg 2011

contemporary research will be the building blocks for future ecological restoration programs.

In spite of their potential for soil improvement, soil microbes have been largely overlooked in ecological restoration until fairly recently, and solid research into the interplay between plants and microorganisms has commenced yielding meaningful findings vis-à-vis both ecosystems ranging from highly degraded to fundamentally undisturbed (Harris 2009). The latest restoration-related research into the roles of microbes has branched off in two principal directions: investigation that describes conditions and target locations in the ecosystem, and research that focuses on system manipulation. This second area stresses the creative manipulation of system components to facilitate more rapid arrival at desired systematic states through overcoming challenges (so-called "biotic barriers") posed by paucity of "mutualists" and other positive components or by the presence of "invasive" plants and other negative influences (Jordan et al. 2002).

Studies may be divided into two categories: those that clarify mechanisms on natural sites and those that focus on restoration sites. At times, the two research areas are not mutually exclusive. For example, it is challenging to classify research on a long abandoned field that is currently undergoing restoration to produce grassland that is rich in species previously common in the region. Restoration theorist and practitioners sometimes quibble over how far real restoration should go. Should the ultimate aim be mature and ecologically rich forest land (as found in Northern Europe, for instance), or is grassland abounding in native species a more pragmatic approach in the modern world. The utilization of research results based on studies of communities of soil microbes has long been used to judge the ecological soundness of soil dominant ecosystems. Contributions in this field give solid evidence that shows a positive correlation between (1) the intentional decrease of intensive site usage to achieve greater ecological homeostasis and (2) an improved fungal–bacterial biomass ratio as a product of the introduction of more sophisticated organic material into soil matrices accompanied by a reduction of physical perturbations. Further improvement in this important ratio is achieved by enhanced scrub growth and general forest development. As systems shift from primitive to more complex successional stages, corresponding shifts in the flow of energy and other resources from root channels of energy to fungal channels ensues (Banfield and Young 2009). Scholarship in this field suggests two important conclusions. First, microbial communities are not independent microcosms; they are dependent on the community above ground level, and second, they serve as a valuable barometer of the management success of restoration projects (Ochman and Raghavan 2009).

6.1.1 Insertion Sequences

The most basic of transposable element forms, which are part of practically every bacterial genome, are the insertion sequences (ISs) (Jordan et al. 2002). Although these ISs lack the genetic encoding capacity to directly alter the bacterial host phenotype, they have escaped evolutionary elimination. Consequently, it appears probable that scholars have concluded that ISs can hardly be simplistically labeled mere "selfish" DNA that have avoided elimination through natural selection because of the more dynamic replication that results from transposition. Such transposition typically entails structural alterations in the DNA themselves that engender to the mutation creation in the form of deletions, insertions, translocations, and inversions, which collected exert phenotypic influences. IS insertion may prompt genetic disruption and, through upstream promoter formation, may also promote the activation of adjacent cryptic genes, notably their activation. Consequently, in bacterial genomes, the most prominently recombinogenic role is played by ISs. In fact, their presence substantially contributes to the assurance that these genomes become influenced to varied structural rearrangements, as opposed to becoming structures that are monolithic and static. These elements therefore contribute significantly to increased variability and, in turn, to host capacity vis-à-vis both evolution and adaptation. Moreover, ISs often are found on bacteriophages and plasmids, which can then promote their continued propagation via the process of lateral transfer among diverse bacterial populations. There are ISs capable of creating composite transposons (Tns). These Tns possess the capacity to carry diverse phenotypic traits, including the ability to effectively use various carbon materials, and to resist heavy metals and antibiotics. Thus far, researchers have isolated over 800 elements of ISs from over 200 prokaryotic species, counting those that derive from either Archea or bacteria. It was sequencing projects that facilitated the identification of most of the known transposable elements. Still, most of these still await experimental confirmation.

6.1.2 Beneficial Effects of Horizontal Transfer of Genes

Among the ways by which both ISs and TEs may alter and exert beneficial effects on the bacteria of the soil is through horizontal transfer as it relates to genetic material. Genes that are encoded on phage, plasmids, and other mobile genetic agents may be horizontally transferred, along with their accompanying "hitchhikers" (e.g., integrons, ISs, conjugative elements, transposons, and integrative elements). This process of horizontal transfer promotes the rapid genomic diversification of microorganisms, which subsequently impacts their fundamental character in the critical areas of pathogenicity, ecology, metabolism, and physiology. Prokaryotic genome research analysis demonstrates that the horizontal transfer of genes is a significant factor that contributes to the genomic innovation of microbes. A sizeable quantity of genetic data is deleted, or inserted, via the mechanisms associated with horizontal gene transfer. This is because of the significantly dynamic nature of microbial genomes. Therefore, the transfer of genes horizontally and the immense quantity of mobile elements enbue microbial communities with extensive capabilities to react expeditiously to ecological settings and take advantage of novel environmental niches. The growth, and even the survival, of soil microbes is seriously threatened by radionuclide pollution, by metallic substances, and by other contaminants, regardless of the source of the contamination, including aquifers. These contaminants lack the transformative capacity of organic xenobiotic pollutants to undergo benign transformation or biodegradation that would render them nontoxic (Harris 2009). As this chapter stresses, TEs and associated genetic elements play a fundamental role in positive soil transformation.

6.1.3 Abundance of Transposons in Soil Bacteria

Recent scholarship that has focused on *paracocci* (*Proteobacteria*) has gone beyond matters of basic identification and further characterization to an examination of the patterns of functionality of associated transposable elements. These elements, in terms of their physiology, include some of the most adaptable bacteria and contribute functionally in a variety of growth modes. During the first decade of the twentieth century, significant taxonomic alterations have occurred in the genus *Paracoccus*. The isolation of several novel species has been a side benefit of this research, as has the reevaluation of previously identified species (Jordan et al. 2002; Harris 2009). The recent research brings the total species count for the genus to 17. To this point, the single identified transposable element associated with these bacteria is IS*1248* of *P. denitrificans* PdX13, which is in turn classified as a member of the family IS*5*.

Abundant information, however, exists for the ISs that accompany a related genus known as *Rhizobium* (http://www.is.biotoul.fr/is.html). In an important study by Bartosik et al. 2003, these scholars provide a lucid analysis of the facultative chemolitoautotroph *P.pantotrophus* and its associated transposable elements. Under controlled aerobic conditions, it can grow as an electron donor in the presence of sulfide, thiosulfate, or molecular hydrogen. Moreover, each of the trio of substrates may be utilized to promote mixotrophic growth and development. Varied organic compounds support the growth and development of these bacteria, both aerobically and anaerobically (capable of denitrification). Impressive physiological heterogeneity is observed in strains of *P. pantotrophus*, as seen, for example, the ability of some to experience methylotrophic growth. The substantial variability, as well as high plasticity, of the species under consideration is indicated, for instance, by their observed appearance within a strain known as ATCC 35512 (which lacks the capacity to utilize methanol) to produce spontaneous mutants capable of flourishing *with* methanol as their only carbon and energy source. There is presently no conclusive evidence that ties the existence of transposable elements to fundamental phenotypic paracocci characteristics; nevertheless, we cannot safely rule out that possibility that their physiological heterogeneity may be a consequence of diverse transposition events.

In the transposable-element-rich species *P. Pantotrophus*, four transposable elements abound in species and they represent four distinctive families that have been identified: IS*256*, IS*5*, IS*66*, and Tn*3*. (IS*256* includes nucleotide sequence IS*Ppa1*; IS*5* includes nucleotides IS*Ppa2*, IS*Ppa3*, and IS*Ppa4*; IS*66* sequences IS*Ppa5*; and Tn*3* sequences Tn*3434*, and Tn*5393*). In the genetic organization of identified ISs that of IS*Ppa1*, IS*Ppa3*, and IS*Ppa4* are representative. These nucleotide sequences "encode a single ORF for a transposase protein with a family-specific DDE motif." Moreover, "the structure of IS*Ppa2* (IS*247* group of IS*5* family) differs, since it encodes two overlapping ORFs as well as the conserved frameshift motif, which is likely to promote the generation of a fusion protein (ORF1 plus ORF2) as a result of programmed translational frameshifting" (Bartosik et al. 2003). The *ISPpa2* fusion protein can allegedly encode the DDE motif, which (because it is transposase-specific) firmly implies that the ribosomal frameshifting process may play a fundamental regulatory role with regard to this element's transposition. To date, transframe protein production has been observed only for members pertaining to families IS*1* and IS*3*. However, IS*Ppa5* apparently facilitates transposase gene expression through a totally distinct regulatory mechanism. This particular element not only encodes the narrowly overlapping ORF sets ORF1 and ORF2 but also ORF3 and ORF4. For all IS*66* family members, there was analogous overlap. However, IS*Ppa5*, as well as its associated relatives, is not capable of encoding the previously mentioned frameshifting motif. This is significant because it suggests the inability of transframe fusion transposase generation. Consequently, it is likely that sets "ORF1 and ORF2 of all these elements (as well as ORF3 and ORF4 of IS*Ppa5*) may be produced by a translational coupling mechanism" (Bartosik et al. 2003; Han et al. 2001), and that "ORF3 of IS*Ppa5* (although it carries a putative DDE motif), unlike other bacterial transposases, have a relatively low isoelectric point (pI 6.75) and does not encode an HTH DNA binding motif" (Bartosik et al. 2003). It is important to note that on the one hand, ORF1 (pI 8.77) has been found to encode an HTH motif, yet on the other hand, ORF2 (the *IS66* family foremost conserved protein) encodes pI 10.41, which is a basic protein. Bartosik, Schacka, and Baj "speculate that the proteins encoded by IS*66*-like elements may form multiprotein complexes with transposase activity," and argue that "this is in agreement with the results of mutational analysis of IS*679* (identified in plasmid pB171 of enteropathogenic *E. coli* B171), which proved that all three IS*679*-encoded proteins (homologous to ORF1 to ORF3 of IS*Ppa5*) are necessary for transposition" (Han et al. 2001). Future research is needed into the roles that ORF4 of nucleotide sequence IS*Ppa5* plays. Bartosik, Schacka, and Baj identified two Tns (novel cryptic, Tn*3434* and Tn*5393*) in *P. pantotrophus*, along with the ISs; the two belong to the family Tn*3*. These two Tns carry a module that is streptomycin resistant. Earlier research into *Tn5393* showed that this Tn is very common among various gram-negative bacteria, in agricultural habitats in which streptomycin had been utilized. *Tn5393* experiences an elevated transposition rate into an entrapment vector, which could help account for its broad distribution in numerous environmental isolates. The significant upshot of these findings is that soil bacteria function effectively as ecologically beneficent reservoirs that promote antibiotic resistance determinants in the biosphere.

Bartosik and his coresearchers employed hybridization analysis to analyze the distribution of elements they had identified, and also the copy number of each. Such findings, however, to not automatically prove that each copy of a designated element, whether IS or Tn, is either identical to others. The sole transposable elements that would be strain-specific "were IS*Ppa1*, Tn*3434* of *P. pantotrophus* DSM 11072, and Tn*5393* of LMD 82.5, as well as IS*Ppa4* of DSM 65 (residing on plasmid pHG16-a)," which these investigators suggested may demonstrate acquisition by lateral transfer events that are of recent occurrence. The most common of the elements that they tested to be found in paracocci were IS*Ppa2*, which they found to be related strongly to IS*1248*. The cross-hybridization capacity of IS*Ppa2* and IS*1248* allowed the Bartosik team to discern these elements through hybridazation methodology, which functions well because they are substantially identical (87%) at the level of nucleotide sequences. Sequences similar to IS*Ppa2* have been found to be present in DSM 65 (pHG16-a) and DSM 11073 (pKLW1) plasmids, which invites the interpretation that their dissemination may well be the result of lateral transfer. There is a good likelihood that the insertion, on a preferential basis, of ISs within TAG and TAA, which function as putative stop codons, could assist in the frequency of minimization for transpositionally motivated disruption of DNA coding regions. This could conceivably render the more widespread element dissemination more likely. A process of hybridization coupled with partial sequencing has demonstrated that IS*Ppa5* is the only element known to be trapped by DSM 11073, an entrapment vector. Bartosik and his coauthors admitted that the cause for the atypically "high frequency of transposition of this element in this strain (103 higher than in DSM 11072)" is not yet known, and maintained that "it cannot be excluded, that its transposition might be stimulated by a host-encoded factor or factors or DNA sequence or sequences adjacent to the insertion site (e.g., transcriptional activation of IS*Ppa5*-encoded genes by a foreign promoter or promoters)."

The majority of the identified transposable elements seem to be of chromosomal origin. Each of the paracoccal strains that the Bartosik team evaluated carry megaplasmids, which in genetic terms are extrachromosomal elements that are characterized by their resistance to purification through the standard methodology involving alkaline lysis. It is still an open question whether or not hitherto identified transposable elements reside within a plasmid with an elevated molecular weight or not. Still, the preliminary finding of Bartosik, Sochacka, and Baj are intriguing and inviting in terms of their reasoned conclusions and foundational information for further research related to transposable elements associated with each of the species belonging to the *Paracoccus* genus.

6.2 Origin of Transposons

6.2.1 Single Stranded RNA to Double-Stranded Nucleic Acid

Primitive forms of life are believed to be RNA in nature; therefore, it follows that the invaders of host RNA genomes were also RNA-based. These were self-replicating nucleic acids; thus, the intruders evolved in a manner that allowed them to integrate into RNA nucleic acids (Doolittle and Sapienza 1980; Szathmáry 1999). During early evolutionary stages, it was challenging to separate host from parasite. Therefore, the mixing of genomes presumably caused enormous radiation of these early life forms and allowed an unlimited diversity of life forms.

This developmental phase subsequently evolved into unicellular and later multicellular life forms and the related radiation of speciation. As this process of speciation matured further, the invading microlife forms did likewise and came to be organized into unique genomes, even as they evolved novel invasion strategies that could threaten ever larger and complex genetic structures. This Darwinian evolution of host–parasite enmity, this perennial defense and counterdefence chess game, results in the proliferation of new species of both hosts and parasites. As the organisms developed the means to oppose the integration capacities of invading transposons and retroelements through expression of endogenized pieces of nucleic acids from the invaders themselves as well as specialized proteins developed into well defined defenses, the opposing parasites also devise methods to neutralize host defenses (Amemiya et al. 2007; Bai et al. 2008; Conley et al. 2008a, b; Heimberg et al. 2008; Jordan et al. 2003; Piriyapongsa and Jordan 2008; Shabalina and Koonin 2008).

6.2.2 Pretransposons and Archea

To date, there is an absence of fossil evidence to validate the reality of primitive agents known as pretransposons. These pretransposons are believed to have penetrated primeval RNA life forms, which were presumably merely naked RNA capable of self-replication. These RNA continuously enlarged, extending their presence and expanding their diversification through pretransposons over eons of time. Although still limited, we have needed to depend upon, and found our scientific assumption on, extent evidence in the form of extant Archeal life forms vis-à-vis transposons and other REs (Brouns et al. 2008). In a significant study, Tang et al. (2005) have noted, in hyperthermophile Archea-*S. solfataricus*, the presence of antisense RNA. From this archaeon *S. solfataricus*, they produced a specialized cDNA library and were able to identify 57 new small noncoding RNA (ncRNA) possibilities and validated their expression through Northern blot analysis. The greater part of these ncRNAs fell into either the antisense or antisense-box RNA categories, in which the antisense-box RNA display only partial complementarity to specific RNA targets under examination. They observed that the most preeminent antisense RNA group was amenable to transcription in an orientation opposite to that of the transposase genes, which involved encoded via insertion elements (transposons). In this manner, they postulate the antisense RNAs are potentially responsible for the regulation of insertion element transposition through the inhibition of the expression of transposase mRNA. Unexpectedly, the antisense

RNAs were also found to possess RNAs that complemented either tRNAs or sRNAs (small-nucleolar-like RNAs).

In the case of the antisense-box ncRNAs, it proved possible to relegate most of them to a small RNA class that specifies sites of 2′-*O*-methylation on either rRNAs or tRNAs (also called C/D sRNA). It was postulated that five C/D sRNAs in this particular group would target methylation in 13 distinct tRNAs at six different sites, which would indicate a far-reaching impact of these specific sRNA species vis-à-vis the alteration of tRNA in Archaea. Yet another set of antisense-box RNAs, one without the customary C/D sRNA patterns, was predicted to target different regions (3′-untranslated) of selected mRNAs, which suggests the ancient nature of the fundamental mechanisms of miRNA/RNAi, which stretch perhaps as much as 3.5 billion years into the past. Moreover, an ncRNA that failed to register antisense elements was discovered to have been transcribed from a particular repeat unit associated with a specific cluster, in *S. solfataricus*, of small repeats (regularly spaced), something that suggests the causal involvement of *S. solfataricus* in replicon partitioning. The previously described research was the initial published evidence, in an archael species, of antisense RNAs that are stably expressed, thus introducing the possibility that antisense-rooted mechanisms play an important and widespread role in gene expression in Archaea (Amemiya et al. 2007).

6.2.3 Birth of Intracellular or Molecular Immunity

We maintain that this primeval evolutionary phase gave birth to molecular immunity or intracellular small-RNA host defenses (Bagasra 2005, 2008; Bagasra and Pace 2010). What was initially a simple defense system based on homology recognition eventually gave birth to exceptionally specialized defense systems that we now see in the form of microRNAs (miRNA)- and RNAi-based immunity (Bagasra 1999). This initial system was founded on evolution of a singular molecular-based recognition system capable of distinguishing the similarities between integrated self DNA (noncoding nucleic acid sequences that were originally derived from the parasitic retroelements) and the genetic fragments of invading agents. Consequently, double-stranded small RNAs, expressed from previously integrated retroelements, developed the capacity to disable the reintegration and or the replication of numerous intracellular genetic parasites, including DNA and RNA viruses, retroviruses, transposons, and retrotransposons (Bagasra 1999). Because the genetic mutations and recombination within and between intracellular agents were consistently creating potentially newer iterations of genetic invaders, consistent with Darwinian evolutionary theories, molecular immunity came to be a constantly evolving phenomenon, which periodically hitting bumps on the road to self-preservation. One recent bump that *Homo sapiens* have hit is the arrival of HIV-1/AIDS (Bagasra 1999). However, throughout the evolution of life forms, comparable bumps have threatened, and even eliminated, many life forms, among them large chimpanzee colonies in the recent past (Caswell et al. 2008).

Our current understanding of the evolutionary history of retroelements (RE) incorporations in the higher life forms is just beginning to develop. However, it appears that as early as the emergence of Archea, REs were being incorporated into the host genome that served as intracellular defense for survival (Bagasra 1999). As evolution progressed, many of the REs became extinct due to their inability to find suitable hosts, and the incorporated REs genes were co-opted to serve other gene regularity functions. REs have profoundly affected the evolution of prokaryotic and eukaryotic forms. The evidence of such evolutionary events can be seen in the presence of ~50% gene sequences of genomes for most contemporary life forms that share genetic similarities to TEs and REs or their remnants (Kidwell and Lisch 2001).

Amid this scene of miRNA defenses and invading REs, the eternal host–parasite struggle-life blossomed into innumerable fauna and flora while defensive means against parasites did likewise. The prime directive of speciation is maintenance of genome integrity but this could not be achieved without symbiosis of TEs and REs with the evolving host genomes (Bagasra 1999, 2006). First among the successful defense systems were apparently prokaryotic forms that used restriction enzymes to destroy foreign DNA while protecting their own by methylation and miRNA (Brouns et al. 2008) indicate that bacteria began to defend themselves against REs as RE fragments formed genomic regions, clusters of regularly interspaced short palindromic repeats (CRISPRs). CRISPRs provide heritable memory of infections. The CRISPR region in Escherichia coli is transcribed, and the CRISPR-associated gene called casE promotes cleavage of the transcript into small ~57-nucleotide CRISPR-RNAs (crRNAs).

6.2.4 *Self vs. Nonself and REs*

This molecular pattern recognition system pioneered in distinguishing between self and nonself (Brouns et al. 2008). As DNA size increased and as prokaryotes evolved into eukaryotes, gene regulation mushroomed, and protection of self-DNA through restriction enzymes or CRISPRs became difficult, therefore, emergence of another more sophisticated layer of defense became necessary for host survival (Bagasra 1999). These old defenses lacked effectiveness against REs and certain pathogenic viruses that have modified their genes to evade the existing defenses. Many life forms accommodated REs rather than fight them. This gave birth to "molecular immunity" (MI) (dsRNA-based bio-defensive systems), where hairpin RE pieces were expressed as double-stranded (ds) microRNAs, or small interfering RNAs (siRNAs), to bind homologous sequences of invading REs, split them via DICER-like (DCL) enzymatic systems, or block integration through triplex-formation (TF). This MI has their origins in Archea and Prokaryotes.

At this point, it is helpful to note that the first immunity to develop was molecular immunity based on small RNAs. These existed among RNA life forms of the most rudimentary forms and later developed into a DsRNA-based defense system in more sophisticated, advanced life forms in which basic and defining genetic information was not stored in RNA but rather in DNA (Bagasra 1999; Bagasra and Amjad 1997, 2000), and that this is still the most prevalent immunity and constitutes the primary defense against intracellular invaders. Molecular immunity is found in every life form, no matter how primitive (Bagasra 1999; Bagasra and Amjad 1997, 2000). This immunity, primarily derived from retroelements, has profoundly influenced the evaluation of prokaryotic and eukaryotic life forms (Siomi and Siomi 2008). Researchers have observed the existence of genome sequences at the ~50% level for most contemporary life forms that display genetic resemblance to either retroelements or transposable elements (Conley et al. 2008a, b; Piriyapongsa and Jordan 2008). Out of this miniature drama in which miRNA defensive systems did battle against retroelements intruders, countless flora and fauna emerged (Lisch 2002).

As time passed, many life forms chose accommodation with retroelements rather than biological combat (Brouns et al 2008; Horvath and Barrangou 2010), a pattern that played an intermediary role for "molecular immunity" (MI), with its defensive system made up of expression of small fragments of noncoding genetic fragments drawn from the previously incorporated double-stranded RNA of retroelements. Molecular immunity appears to have origins in both Prokaryotes and Archea.

6.3 Endogenous Retroviruses from Soil to Mammals: Protective Lessons

Endogenous retroviruses heavily colonize vertebrate genomes, which share approximately 50% of their genomic DNA. These Endogenous retroviruses have emerged from host cell retroviral infections via evolutionary progression, which permits the permanent integration of viral genomes into host DNAs and facilitates multigenerational transmission (Lisch 2002). Endogenous retroviruses obstruct replication cycles of exogenous pathogenic retroviruses that are transmitted horizontally. We hypothesize that endogenous retroviruses have afforded protection to hosts against pathogenic retroviruses that share genetic sequences are similar to those of the integrated viruses. Researchers (Arnaud et al. 2007) have recently characterized the molecular virology and evolutionary history of the Jaagsiekte sheep retrovirus (endogenous beta-retroviruses, enJSRVs) and pointed out the crucial function of integrated retroviral genes in the struggle to oppose exogenous retroviruses. These scholars found that (1) two loci from enJSRV, which had invaded the host genome prior to speciation within sheep (i.e., genus Ovis) approximately three million years ago had acquired, following integration, a defective and mutated viral protein that was able to block exogenous retroviruses; (2) both transdominant enJSRV loci had became permanently established in the host genome by at least the time of sheep domestication (i.e., 10,000 years ago); (3) the intrusion of endogenous JSRV/enJSRV retroviruses persists to this day; and (4) there are new (<200 years ago) viruses that elude the transdominant enJSRV loci. Hence, momentous virus–host combat goes on; hosts counter retroviral infections via endogenization, and the judicious selection of endogenous retroviruses affords molecular defense.

As life forms grew increasingly complex, increasingly sophisticated parasites also emerged; some evaded integration into the host DNA; rather, they replicated beyond the confines of the nuclear system while cannibalizing raw matter from the host and from associated synthetic machinery (Dunn et al. 2008). To find viable protection against these novel RNA and DNA viruses became paramount. Consequently, small interfering RNAs, as well as triplex-forming microRNAs (Bagasra and Amjad 1997; Bagasra 1999, 2006) countered by challenging, and effectively interfering, with the cycles of viral replication (Bagasra 1999; Bagasra and Amjad 1997, 2000; Hakim et al. 2008; Dunn et al 2008; Matzke et al 2000; Dooner and Weil 2007; Buchon and Vaury 2006). Hosts attempted genetic editing (i.e., APOBEG3G enzymes: (Ulenga et al. 2008). For their part, viruses responded with miRNAs to disable host miRNAs (Walker and Goulder 2000). A balance resulted, and host genomes continued to accommodate more and more retroelements until about half of their genomes had REs; these seem to have provided the requisite number of small dsRNA permutations to block retroelements invasions (Bagasra 1999; Bagasra and Amjad 1997, 2000; Horvath and Barrangou 2010). Increasing retroelements numbers called for the control, without damaging host replication, of numerous potentially active endogenous retroelements as well as coding genes. In fact, the present mechanism for effecting mammalian gene regulation resembles a huge orchestra, one that promulgates life's musical scores through complex patterns of synchronic balance. Early in their progression, hosts evolved sizeable numbers of retroelements and also co-opted endogenized RE-miRNAs for both internal and external regulation (Agudelo-Romero et al. 2008; Schramke and Allshire 2003). Moreover, as life forms accumulated multiple organs and layers, cellular differentiation and specialization led to selected gene expression in cell types on a differentiated basic. This, in turn, prompted the expression of noncoding genes at differential stages (Taganov et al. 2007). Meanwhile, evolution led to resource conservation, and life forms developed cell surface receptors for purposes of expression for gathering resources and for differentiation. Accompanying these developments was the invitation of specific retroviruses, and other viruses as well, to invade cells in specific ways and at specific levels of development. For example, many pathogens primarily invade immunologically naïve host with much more vigor than the older more immunologically experienced host. Therefore, childhood infections of numerous origins are more devastating for humans at early age than later life. However, some of these infections are purely due to lack of preexisting immunity but others are related to receptor expression levels on the susceptible cell types. Poliovirus is a glaring example that infects the peripheral nerves of young human host who express a relatively large number of polio-specific receptor on their peripheral neurons than what is expressed in an adult, making children more susceptible to polio infection and paralytic poliomyelitis (Furesz and Levenbook 1998; Karttunen et al. 2003). As for miRNAs, they were differentially expressed in a manner that allowed them to check invasion through miRNA arrays (Navon et al.

2009; Schmeier et al. 2009; Taganov et al. 2007) and the differential expression may allow certain microbes to invade certain cell types due to absence of complementary miRNAs in these cell (Taganov et al. 2007).

Each host cell developed the means to block retroelements and to curtail other viral entry (Bagasra 1999; Bagasra and Amjad 1997, 2000). However, to operate properly required that surface receptors routinely regulate nutrition and systematically communicate signals (Buchon and Vaury 2006; Agudelo-Romero et al. 2008). Retroelements and viruses targeted the most vulnerable parts of host cells. The contest still continues, as observed in both the negative and positive utilization of entry routes for such menaces to life on the planet as SIVs, HIV-1, and human herpesvirus-6 and -7. CD4 molecules are important players in the contest; they are critical for the proper functioning of CD4+ T cells, as well as other immune cells (Karp and Auwaerter 2007; Lisco et al. 2007; Lusso et al. 1994; Frenkel et al. 1990; Ryt-Hansen et al. 2006). There are cases (e.g., *Caenorhabditis elegans*) in which miRNA deterrence has gained such ascendancy as to prevent invasion by any natural virus (Schott et al. 2005). When it comes to unnatural viruses, however, *C. elegans* are vulnerable to unnatural viruses, including human varieties that prove deadly in vitro (Schott et al. 2005). These worms have not developed the necessary miRNAs to quell unnatural infections (Bagasra 1999; Bagasra and Amjad 1997, 2000; Pandrea et al. 2008; Schott et al. 2005).

6.4 Innate Immunity and Development of Immunity Based on Pattern Recognition

As the multicellular life forms developed more sophisticated means of communications and signaling pathways, the pathogens also evolved to exploit the surface receptors (Bagasra 1999; Bagasra and Amjad 1997, 2000; Brouns et al. 2008; Horvath and Barrangou 2010). Therefore, organisms developed another layer of defense system that was based on pattern recognition. This so-called innate immune system lacked the fine specificity that is so precise in case of nucleic acid homology dependent immune defense that has so well documented in the form of RNAi or miRNAs, but it has an uncanny ability to recognize self vs. nonself based on recognition of repeating patterns of molecules on the surface of invading agents.

The surface of microorganisms is generally covered with repeating patterns of molecular structures; their nucleic acids likewise display predictable patterns. Bacterial DNA, for example, contains unmethylated repeats of dinucleotides CpG. Viruses nearly always require dsRNA in their life cycles. A singular immune system developed that could recognize so-called foreign pattern recognition in the invading agents through pattern-recognition receptors (PRR: Allavena et al. 2004). One such receptor capable of pattern recognition is mannose-binding lectin or MBL(Palm and Medzhitov 2009). MBL and a number of other protein receptors discern particular sugar patterns pathogenic surfaces while simultaneously recognizing that these particular patterns are not evident in the host. This protein exists in human blood plasma, as a free protein it participates in the activation of a complement, which constitutes yet another portion of the innate immune system, forms a crucial link between various immune defense layers and levels. Through PRR, MBL can sense invading pathogens and accurately distinguish them from self (Allavena et al. 2004; Palm and Medzhitov 2009).

The initial line of systemic defense in the host is, naturally, dependent on those protective gears that check invader entry. Examples include the antimicrobial enzymes of lacrimal glands, keratinized skin, respiratory epithelia, etc. Once the invading life forms reach the host target cells, the host cell receptors play a crucial role and the differential expression of receptors may decide if the invaders can get inside the cells. Once inside the cells, the first immune systems to counter the invading microorganisms are those that are perpetually ready to resist an invader. The PRR system systematically discriminates between self and nonself through a scanning process that analyzes the differences in patterns on the surface of newcomers and judges their similarity or difference vis-à-vis the patterns found on self. When necessary, they then attack and destroy the harmful intruders. Looking at it from a practical perspective, the surfaces of many microorganisms bear repeating three-dimensional patterns of mannose, a sugar that is present on the surface of many microorganisms in a repeated fashion and with a specific orientation. This 3D pattern that is sensed by MBL, which performs a binding action, and activates a complement cascade that pokes holes in membranes of the invaders, which causes them first to leak and then to perish (Helm 2004).

6.5 Evolution of Antitransposon Resistance in Bacteria to Large Mammals

A basic understanding of immunological evolution suggests a parallel with other instances of natural selection in the form of "survival of the fittest." In fact, Darwin's theoretical interpretations provided new viewpoints through which early immunologists could organize their research. Through the eyes of Darwin's theory, they saw that a battle for survival was raging between the early RNA retroelements and bacteria pathogens and humans. Their adaptations of Darwin's ideas led immunologists to see that immune defense involved systematic complexity, including natural selection of proteins. Evolutionary immunology molded thinking substantially beginning in the latter part of the nineteenth century, as researchers increasingly speculated about the origins of the complex system of human immunological defenses, both intracellular and extracellular.

All life forms including Archea, bacteria, unicellular prokaryotes and eukaryotes, and multicellular organisms consistently possess the capacity to differentiate self from nonself, a characteristic that facilitates preservation from attacks by pathogens or threats to required nutrients. The mechanisms that evolved in the self were countered by a parallel evolutionary process by nonself organisms. The ensuing battle, therefore, involved elements of both predictability and change, as defensive and offensive alterations emerged in a dynamic, ever-changing contest. Although original unicellular molecules were RNA, as time passed, multicellular organisms came to prevail. Consistent with the adversarial arrangement seen everywhere in the natural life, DNA emerged as the principle molecular guardian of genetic memory and species, even as parasites developed their own novel strategies of attack. The capacity of the molecular immune system to function as a molecular pattern recognition system was of vital significance. Such a recognition system utilized self vs. nonself logic to differentiate nucleic acid sequences (Bagasra 1999; Bagasra and Amjad 1997, 2000; Horvath and Barrangou 2010). It thus exercised a developed capacity to counter retroelements through the implementation of detection mechanisms involving homologous nucleic acid. Recent evidence backs the position that the first, and most solid, immune mechanism to materialize among unicellular life forms was a sequence recognition system that was molecularly based (Bagasra 1999; Bagasra and Amjad 1997, 2000; Brouns et al. 2008; Horvath and Barrangou 2010). Primeval parasites, retroelements and transposons (Bagasra 1999; Bagasra and Amjad 1997, 2000), were exceptionally versatile life forms that were imbued with the ability to attack host genomes, easily entering and leaving the genetic material of the threatened host. Evidence suggests that these retroelements, in their early formative stages, may have encouraged extensive biological diversity (Bagasra 1999; Bagasra and Amjad 1997, 2000; Brouns et al. 2008; Horvath and Barrangou 2010). In those initial evolutionary phases, the means to effectively block the integration of retrotransposons may not have been fully functioning, the diversification of life forms demanded the preservation of speciation; the host, therefore, promoted molecular defense through systematically co-opting retroelements (Bagasra 1999; Bagasra and Amjad 1997, 2000; Brouns et al. 2008; Horvath and Barrangou 2010). The expression of integrated genetic sequences as endogenous small RNA was significant because these small RNA identified homologous motifs and subsequently checked their integration. The radiation of life forms in unicellular organisms, which adapted into more complex multicellular varieties, was accompanied by a parallel sophistication in molecular defense systems, which became at once more diversified and more intricate. Logic would not sustain the position that with the passage of time, multicellular life forms abandoned the primary defense system that had developed and simply proceeded on to some other system. Actually, the system based on small RNA became increasingly diverse, and original retroelements played fundamental roles in the warding off of viruses, in the regulation of cell replication and endogenous genes and in the area of chromosomal modeling (Obbard et al. 2009; Zamore 2006).

Research in recent decades has led modern immunologists to conclude that immunological protection may have begun when relatively simple organisms – soil bacteria – with a single cell began to employ gene-disabling molecules as well as toxic peptides to block the intent of microorganisms to promote their own reproduction by exploiting them as raw material. Support for this position is found to this day both in the simplest of eukaryotes and in animals of greater complexity. Moreover, with the evolution of multicellular creatures came the dedication of specialized cells, including phagocytic cells, to differentiated tasks (Bagasra 1999; Bagasra and Amjad 1997, 2000; Brouns et al. 2008; Horvath and Barrangou 2010).

The cellular specialization and compartmentalization of multicellular biological life was paralleled by parasitical adaptation to more effectively consume raw material in extracellular compartments of hosts. Such compartments are of strategic importance because they nourish a number of vital systems, including the vascular and nervous systems. Creative host adaptation was necessary to counter creative attack strategies. Consequently, hosts developed novel approaches to halt parasite attacks because the small RNA- and miRNA-based molecular immune system was incapable of reaching outside cells to quell threatening parasites. Among the pathogens involved in the shifting-sand struggle for control of host material were fungi, parasites, and bacteria that were capable of surviving, and even thriving, in such liquid body substances as interstitial, spinal, gut lumen, blood, and pericardial fluids.

At present, higher organisms posses immune systems with a broad variety of multiple defenses, including a far-reaching molecular defense system that is small dsRNA-based, and versatile protein receptors that distinguish common characteristics of perilous pathogens that have been universally implanted into animal genomes. This protective "innate" immunity that provides first-line defensive protection for animals, relies on molecules and cells that hasten to infection sites (Fig. 6.1). Such inflammatory reaction – according to comparative research into sea squirts, earthworms, and sponges – suggests an ancient response pattern that dates to the very origins of life. Many researchers believe that the characteristics of the innate immune response system appear in the incipient stages of evolutionary adaptation. Sponges that belong to the phylum Porfera, and other basic multicellular organisms, grow in colonies of relatively undifferentiated cells. The functions of food gathering, waste product disposal, and host defense are all carried out by the same type of cell. Nevertheless, these organisms can detect the encroachment of cells from another colony and kill the invaders. With the arrival in the lower invertebrates of circulatory systems, organ development, and multiple body layers, the functions of host nutrition and host defense are carried out by separate cell types. Cells in the circulation of lower invertebrates and prevertebrates are specialized for detecting infected cells via the expression of a small number of pattern recognition molecules (PRMs) of relatively broad specificity. The primitive phagocytes began to develop the PRMs that can distinguish self from nonself and engulf the nonself by utilizing these PRMs. Lectins, antimicrobial molecules, and at least some complement or complement-like components are also present in lower invertebrates and prevertebrates (Schott et al. 2005). The range of PRMs expands as one proceeds higher up the prevertebrate and higher invertebrate evolutionary branches (Schluter and Marchalonis 2003). More sophisticated phagocytes patrol the body and engulf invaders, and large arrays of antimicrobial peptides and proteins are produced. No true classical immunity – where lymphocytes, antibodies, or MHC molecules play any role in host defense can be seen in either the lower

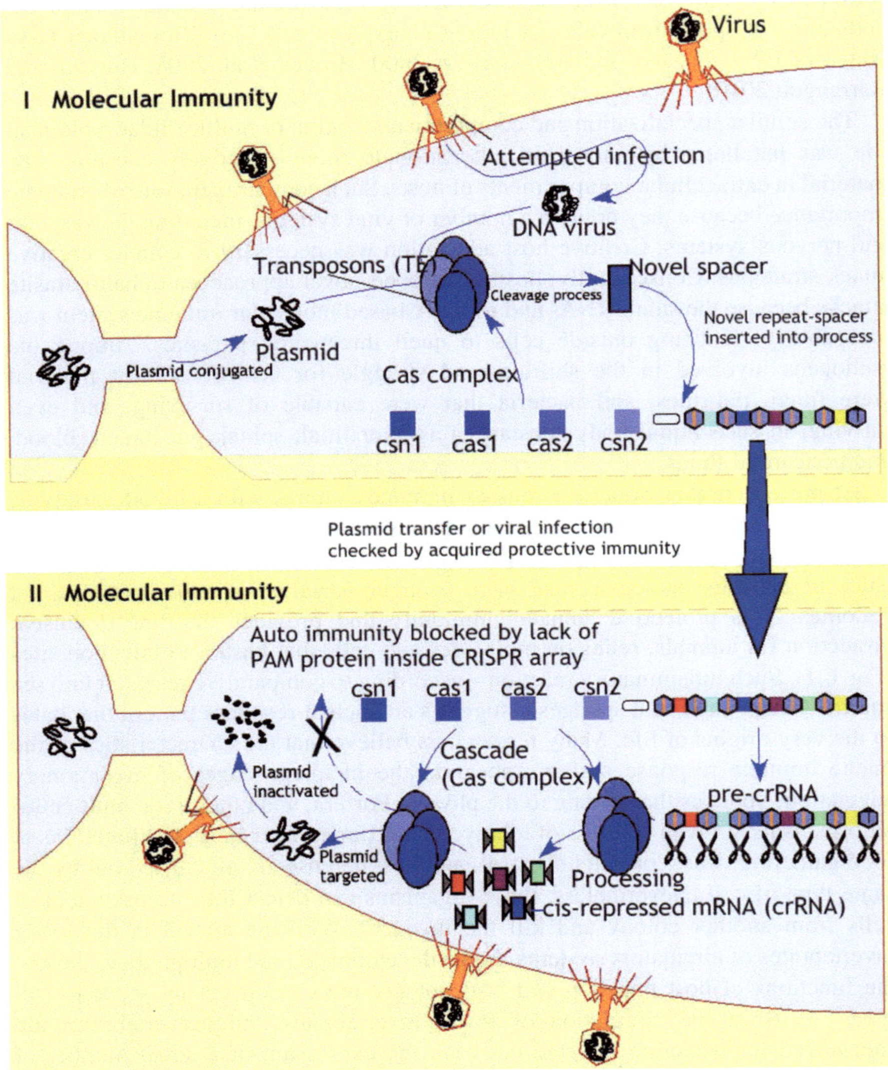

Fig. 6.1 Outline of the CRISPR defense system, with the evolution of molecular immunity process following exposure to invading DNA from transposons or plasmids (A), and the CRISPR-encoded molecular immunity mechanism of action, to inactivate invading DNA (B). Repeats are represented as diamonds, spacers as rectangles (adopted from Barrangou 2010)

or higher invertebrates or in prevertebrates (Kidwell and Lisch 2001). Thus, it seems that adaptive immunity was not needed for the survival of these species. These organisms have a limited habitat range, relatively short life spans (short reproductive cycle) and large reproductive capacities. The classical immune repertoire is absent and other primitive layers are predominant (Bagasra 1999; Bagasra and Amjad 1997, 2000), as must be true when the genes encoding defense molecules do not undergo somatic recombination and are "hardwired" in the germline. However, this small repertoire of PRMs and phagocytic cells is sufficient for survival because a limited habitat means that the range of pathogens encountered is generally narrow. A short span means the total number of pathogens encountered is relatively low, and a large reproductive capacity means that the huge numbers of offspring produced offset the loss of substantial numbers of them to pathogens. Vertebrate species show enhanced anatomical complexity accompanied by increased mobility, such that these animals very often wander over great distances. In addition, vertebrates have longer life spans that correlate with an increased time to reach reproductive maturity. As a result, vertebrates frequently encounter a large number and wide variety of pathogens prior to successful reproduction. Vertebrates also produce fewer offspring than either invertebrates or prevertebrates, such that severe losses to pathogens could threaten the species as a whole. These evolutionary pressures are thought to have promoted the development of adaptive immunity in vertebrates (Amemiya et al. 2007; Helm 2004). Although no conventional antibody can be detected in the lowest vertebrates (such as the jawless fish, Agnatha), primitive gut-associate lymphoid tissues (GALT) is present. Cartilaginous fish like the sharks have hinged jaws, making them better predators, able to take advantage of a broader range of nutritional opportunities. However, with such a diet comes an increased chance of internal injury and/or infection (Schluter and Marchalonis 2003). Even with innate immunity in place, a strictly germline-encoded repertoire of nonself recognition molecules would not be sufficiently diverse to counter all the pathogens such vertebrates meet. Cartilaginous fish are thus the first organisms in which there exists a mechanism to somatically diversify immune system genes and expand the immune repertoire. A distinct thymus and spleen and true lymphocytes expressing forms of immunoglobulins (Igs) and T-cell receptor (TCR) molecules are present, with IgM and nonmammalian antibody isotypes being produced. The terminal complement components foreign cell lysis are also first seen in the cartilaginous fish. Vertebrates like amphibians that move from the sea to the land require additional host defense mechanisms to cope with the varying environments. These animals have the limbs necessary to move on land and sophisticated vascular systems containing multiple types of circulating cells. Skins designed to shield the exposed animal from the sun's harmful rays are present, providing a physical barrier against pathogens. Bone marrow-like tissue serves as a source of distinct T and B cells, and lymphoid tissues of increased complexity and wider distribution are present. In the reptiles, there are advanced lymphoid tissues plus eggs in which the young develop in a self-contained aqueous system enclosed by multiple membranes. This innovation frees these cold-blooded animals from having to return to the water to reproduce and increase barrier protection for the offspring. However, the cold-bloodedness of the amphibians and reptiles affects their immune responses, causing measurable seasonal variations in the proliferative capacity of lymphocytes and the production of cytokines. The warm-bloodedness of birds and mammals removes this variation and allows these animals to forage and hunt at night when coldblooded vertebrates are less active.

However, with a permanently warm body comes an environment favoring pathogen growth. To counter the onslaught, birds and mammals have highly differentiated and structured lymphoid tissues, complete with distinct germinal centers and lymph nodes. Cell-mediated and humoral responses are optimally coordinated and controlled to deliver the most efficient response possible with the least amount of collateral damage to the host. It is these responses that would be the subject of our overview below.

The Adaptive or the classical immune system is a new comer in the defense realm of the host. It is only 300 MY old, appearing first in the Jawed fish-in sharks (Schluter and Marchalonis 2003). Our current paradigm is based on this new system. Here we would explore the genesis of this new immune system which is called by some the Big Bang of Immunity. Let us see how and where this first appeared?

6.6 The Big Bang of Immunity

As it is described in the earlier sections that various forms of immune systems have evolved as part of the evolutionary process that started from Archea and still persist in soil bacteria to the higher vertebrates. However, in what has been called the "big bang of immunology," most vertebrates later evolved a rather more sophisticated form of immunity, in which white blood cells exquisitely targeted to destroy "extracellular" microorganisms and invading DNA introduced by a virus. This unusual immunity first appeared in a jawed fish creature (Pradeu 2009).

6.6.1 Adaptive Immunity and TEs

The most surprising aspect of this new immunity was that it came into existence due to transposons. It may seem ironic that retroelements endowed vertebrates with the keys to a new microbial defense, but it illustrates that retroelements and other microscopic and submicroscopic life forms have shaped the evolution of animals for billions of years. Indeed, a few researchers now suggest that immune systems evolved as much to manage and exploit beneficial microbes as to fend off nasty ones. As we have proposed in our previous writings that scientists should soon have a more detailed view of immune evolution as they decipher the genomes of more invertebrates and vertebrates and tally up the defensive weapons shared by the various branches of life (Bagasra 1999; Bagasra and Amjad 1997; Bagasra and Amjad 2000).

It was only shortly after *On the Origin of Species* was published in 1859 that infectious diseases were discovered and became a compelling example of a Darwinian struggle – humans pitted against pathogens. The life-long works of Luis Pasteur and Robert Koch were to uncover the damaging effects of microbes. The enormous work of these two pioneers in microbiology subsequently led to the discovery of the classical immunity and unfortunately, it lead to the development of dogma that this is and was the only form of defense man has. All the paradigm of immune system were developed around this basic dogma. To understand that contest, immunology emerged in the late nineteenth century as the science of host defense. Soon, scientists were fighting over the importance of two competing defense mechanisms: the humoral system of antibodies in the blood vs. mobile amoeba-like cells known as phagocytes. The first forwarded by Paul Ehrlich as "the magic bullet" theory and the second championed Russian Elie Metchnikoff who was the father of cell mediated or phagocytic theory (de Kruif 1926).

A half-century later, another major intellectual advance within immunology bore the fingerprints of Darwinian evolution. As Darwin's theory of evolution held that a large amount of variation exists among individuals in a species and that species can adapt to new circumstances because evolution weeds out the less fit, favoring variants that improve reproduction and survival. A brilliant immunologist Frank Macfarlane Burnet (1959) forwarded a hypothesis that how the body forms its antibodies, the pathogen-binding molecules. Burnet proposed that the lymphocyte was the key evolutionary player being selected within the body. Those white blood cells making antibodies that react to the body's own tissues would be deleted, whereas one whose antibodies recognized a pathogen would survive and indeed be stimulated to expand greatly in number. The Clonal Selection hypotheses – as it is called now – could not answer all the mysteries about antibody formation. Although Burnet's idea assumed a large variation in preexisting antibodies, immunologists in the 1960s and '70s realized that animals could generate distinct antibodies to almost any protein or other molecular feature of a microbe. In fact, the vertebrate immune system could raise antibodies specific even to man-made molecules not found in nature. Given the prevailing phenomenon, which became known as the generation of diversity, or GOD, problem dogma that behind every protein there was a specific gene, immunologists were at a loss to explain how 5×10^{17} genes can exist in lymphocytes (it should be remembered that most of the life forms carry less than 40,000 genes).

6.6.2 Generation of Deversity and Somatic Recombination

In the late 1970s, in work that would earn him a Nobel Prize, Susumu Tonegawa of the MIT showed that B cells can produce such a vast array of antibodies thanks to a complicated process called VDJ recombination. A maturing B cell starts with dozens to hundreds of three classes of gene segments – the V's, D's, and J's – and as it develops, the cell excises all but one of each class (Sompayrac 2008). The surviving V, D, and J then get splices together into a DNA sequence that encodes an antibody unique to each mature B cell. (The other key player in the adaptive system, the T cell, also bypasses the one gene–one protein hurdle and similarly recombines gene segments to create distinct cell-surface receptors for pathogens.). It is akin to language where 27 alphabets can generate millions of words. However, here – one can image – that three different languages are being used, each with unique numbers of alphabets – each representing V, D, and J.

The elucidation of VDJ recombination gradually exposed immunology's big bang. By 1990, investigators at the Whitehead Institute for Biomedical Research in Cambridge had identified two genes essential to VDJ recombination, recombination-activating genes or *RAG1* and *RAG2* (Schluter and Marchalonis 2003). Sharks and all the other jawed vertebrates with classical immunity or adaptive immunity have these genes, but all the evidence at the time indicated that hagfish, lampreys, and invertebrates didn't. So, where did *RAG1* and *RAG2* come from? The most surprising part of this discovery was that the two genes are located immediately next to each other (Sompayrac 2008). Scientists wondered if whether the pair had once been part of a DNA recombination system in fungi or viruses that got incorporated into vertebrates. As immunologists teased out what the proteins encoded by the two did, they realized the molecules are the scissors and knitting needles that cut out all but one V, D, and J and stitch those remaining three gene segments together. In 1995, Craig Thompson (1995) was first to propose that the DNA now encoding *RAG1* and *RAG2* was once a mobile genetic element – a transposon. As we have discussed previously that transposons are some of the earliest evolution makers and are responsible for the radiation and diversity of life (Sompayrac 2008; Thompson 1995). These agents can cut themselves out of one DNA sequence and stick themselves back in another, so immunologists could envision those skills being co-opted to recombine V, D, and J gene segments. In this "transposon hypothesis," Thompson suggested that at some point after jawed and jawless vertebrates split into two branches, about 450 million years ago, a transposon invaded the former lineage, perhaps brought in by a retrovirus that infected a germ cell and hence the enzymes that would ultimately provide adaptive immunity, by creating diverse antibodies and T cell receptors, were now in place and could mutate into that new role (Sompayrac 2008).

6.6.3 *RAG1/RAG2* Dilemma

Many research teams began trying to verify the transposon hypothesis. In 1998, two groups independently showed that the enzymes encoded by *RAG1* and *RAG2* could, in addition to cutting out DNA sequences, actually insert one stretch of DNA into another (Pradeu 2009; Sompayrac 2008). The recent evidence suggests that the *RAG1-RAG2* transposon may have entered sea urchins and vertebrates independently. Whatever the actual evolutionary pathway that led to the very complex vertebrate adaptive system, it was surely a gradual progression that co-opted many preexisting immune mechanisms into more robust systems (Pradeu 2009; Sompayrac 2008).

We will explore the bases of its emergence and describe it in reasonable detail so the readers can determine that why this new immune system is not evolved to protect us against retroelements but its primary purpose is to protect us against extracellular parasites. Until now the readers might have recognized several sophisticated immune defense system that have evolved throughout the evolution and are still persistent in higher vertebrates including primates like man. Each of these systems has their specialized utility and emerged as the result of evolutionary necessity – to support the survival of the fittest. All through evolution on obvious fact is always glaring us on the face is that every time a new system of defense emerges, it is always the result of accommodation and coexistence. The great radiation of the single-celled life forms most likely was the result of accommodating the parasitic retrotransposons and eventually co-opting them to preserve the species. As a result over 50% of the man's DNA is made of retroelements and 8% is retroviral in origin (Bagasra 1999). Similarly, over 90% of bacteria and other microorganisms that inhabits our bodies are "beneficial" life forms co-opted to protect us from the harmful pathogenic ones. The reason that we make antibodies to HIV upon infection is not because this particular response is protective against this type of pathogens but because this response is purely a classical immune response and not necessarily protective against retroelements like HIV (Bagasra and Pace 2010).

6.6.4 Classical Immunity Barrowed Its Model from Molecular Immunity

The classical immune system has evolved for a particular purpose. It evolved to solve a particular problem. The teleology of the classical immune system is to protect the body from threats posed by toxic substances and extracellular pathogens and to do so in a way that minimizes harm to the body and ensures its continued functioning. The term extracellular pathogen embraces a plethora of inimical micro-organisms, such as bacteria, parasites, viruses, and fungi that constantly assault the body. These pathogens are the source of many diseases and ailments, for example, pneumonia is caused by bacteria, and influenza is caused by viruses, and malaria is caused by parasites. It should be clarified here that as opposed to retroelements, transposons, and retroviruses that can have to integrate – invade the host genomes to survive and reproduce these kinds of parasites have evolved to replicate either in the cytoplasm of the host cells or in the body fluids of the multicellular organisms. As we have discussed in the preceding section that the retroelements are essentially defended by miRNAs, however, these other pathogens are essentially quelled by the adaptive immune system. Of course, it does not mean that mRNAs would not be involved in defeating the intracellular pathogens that replicate inside the cell – in the cytoplasm. This would be true in case of viruses and there is ample evidence to show that miRNAs are involved in defense against almost all human viruses. However, the range of miRNA diminishes significantly in case of extracellular pathogens or in case of intracellular nonviral organisms like mycobacterium, listeria, and brucella to name a new variety of intracellular bacteria. These pathogens have their own protective coats inside the host cells that protect them from miRNAs. It should be realized that in evolutionary terms there has to be certain degree of overlap between various immune defenses and when investigators are attempting to prepare vaccine against HIV using classical immunity and see results in terms of antibodies and cell-mediated immunity it does not necessarily mean that the result represent real protective mechanisms. On the contrary, the classical immunity will response to any epitope that they see as foreign, but it does not mean that it is a productive response. We know that neither the antibodies against HIV nor the cellular immunity has any protective effect. Therefore, it should be clear that we should look at the real anti-HIV (antiretroviral) immunity. There are two major factors that one has to keep in mind when evaluating protective responses against a specific microbe; (1) host genetic makeup related to the invading microbe, and (2) microbe's degree of virulence. These are two opposing forces eventually decide "survival of the fittest" for the host and the microbe.

Unchecked replicating pathogens can lead to a rapid demise of the host. There are two aspects to the problem that the CI faces: the identification or *detection* of pathogens, and the efficient *elimination* of those pathogens while minimizing harm to the body, from both pathogens and the classical immune system itself. The detection problem is often described as that of distinguishing "self" from "nonself" (which are elements of the body and pathogens/toxins, respectively). However, many foreign micro-organisms are not harmful, and an immune response to eliminate them may damage the body. In these cases, it would be healthier not to respond, so it would be more accurate to say that the problem faced by the classical immune system is that of distinguishing between *harmful* nonself and everything else. As a matter of fact over 90% of the microbes found in and on human body are "beneficial microbes and the host need not response to it. Once pathogens have been detected, the classical immune system must eliminate them. Different pathogens have to be eliminated in different ways, and by different the components of the immune. However, the fact that the classical immune system does react to "harmless" micro-organisms is essential for an immunization system that accomplishes this; they are called *effectors*. The elimination problem facing the classical immune system is that of choosing the right effectors for the particular kind of pathogen to be eliminated.

6.7 Summary and Conclusion

By now the readers might have recognized several sophisticated immune defense system that have evolved throughout the evolution and are still persistent in higher vertebrates including primates like man. All of the host defense systems have their origin in Archea and bacterial life forms. Each of these systems has their specialized utility and emerged as the result of evolutionary necessity – to support the survival of the fittest. All through evolution on obvious fact is always glaring us on the face is that every time a new system of defense emerges it is always the result of accommodation and coexistence. The great radiation of the single-celled life forms most likely was the result of accommodating the parasitic retro-elements and eventually co-opting them to preserve the species. As a result over 50% of the man's DNA is made of retroelements and 8% is retroviral in origin. Similarly, over 90% of bacteria and other microorganisms that inhabits our bodies are "beneficial" life forms co-opted to protect us from the harmful pathogenic ones.

Acknowledgment We would like to thank Robert T. Pace for his art work for Figure 1. This work was partially supported by a grant from the US Department of Energy (grant #DE-EM0000479).

Appendix

Abbreviation	Term	Meaning
Ago	Argonaute protein	The catalytic core of RISC that binds short RNAs and, in many cases, displays RNase H-like mRNA-cleaving activity. Key domains include the PAZ (Piwi-Argonaute-Zwille) and Piwi domains. Ago is named after an *Arabidopsis* developmental mutant that resembles the tentacles of a paper nautilus (Argonautidae)
Dcr	Dicer protein	The RNase-III family ribonuclease that cleaves dsRNA into short RNAs. Key domains include a helicase C-terminal domain, dsRNA-binding domains, a PAZ domain and two RNase-III domains. Named for its "dicing" activity
dsRNA	Double-stranded RNA	Small double-stranded RNAs are the basic molecules involved in gene silencing, gene regulation and viral defense
endo-siRNA	Endogenous siRNA	A short RNA (other than an miRNA) that is derived from the host genome, rather than an exogenous source (e.g. a virus)
IS	Insertional sequences	The most basic of transposable element forms, which are part of practically every bacterial genome, are the insertion sequences
miRNA	microRNA	Single-stranded RNAs of 21–22 nt, derived from short fold-back hairpins (pre-miRNAs) and involved in translational control
piRNA	Piwi-interacting RNA	Single-stranded RNAs of 24–29 nt that function in complex with Piwi family Argonaute proteins in the animal germ line
rasiRNA	Repeat-associated siRNA	Short RNAs derived from repeat sequences, such as TEs, sometimes considered a subclass of piRNAs in *Drosophila*
RdRp	RNA-dependent RNA polymerase	RNA polymerase directed by RNA, especially eukaryotic polymerases involved in the amplification of RNAi in nematodes and plants

(*continued*)

Abbreviation	Term	Meaning
RISC	RNA-induced silencing complex	The complex comprising Argonaute, a short RNA, and several other proteins, which mediates RNAi through sequence-specific complementarity
RNAi	RNA interference	The class of processes that use short single-stranded RNA molecules in complex with an Argonaute protein to bind complementary nucleic acids and modify their action and/or processing
siRNA	Short interfering RNA	Single-stranded RNAs of 20–30 nt involved in RNAi (especially those not classed as microRNAs)
ncRNA	Small noncoding RNA	Small noncoding RNAs play a crucial role in molecular immunity based on sequence homology
SRNA	Small nucleolar-like RNA	The pre-RNA that are several hundred to thousand or so bp long and contain hairpin structure. These molecules are cut into smaller size and exported to cytoplasm by Exportin-5. Subsequently they take the shape of siRNA
TE	Transposable element	A stretch of DNA capable of moving around the genome, either by excision (cut-and-paste transposons) or through an RNA intermediate (retro-elements)
UTR	Untranslated region	Nonprotein-coding regions at the 5′ and 3′ ends of an mRNA
viRNA	Viral RNA	siRNAs derived from viral sequences
VSR	Viral suppressor of RNAi	Any viral gene that inhibits host RNAi function

References

Agudelo-Romero P, Carbonell P, Perez-Amador MA, Elena SF (2008) Virus adaptation by manipulation of host's gene expression. PLoS ONE 3:e2397

Allavena P, Chieppa M, Monti P, Piemonti L (2004) From pattern recognition receptor to regulator of homeostasis: the double-faced macrophage mannose receptor. Crit Rev Immunol 24 (3):179–192

Amemiya CT, Saha NR, Zapata A (2007) Evolution and development of immunological structures in the Lamprey. Curr Opin Immunol 19(5):535–541

Arnaud F, Caporale M, Varela M, Biek R, Chessa B, Alberti A, Golder M, Mura M, Zhang YP, Yu L, Pereira F, Demartini JC, Leymaster K, Spencer TE, Palmarini M (2007) A paradigm for virus-host coevolution: sequential counter-adaptations between endogenous and exogenous retroviruses. PLoS Pathog 3:e170

Bagasra O (2008) Role of mutually homologous microRNAs in HIV-1 inhibition. 33rd FEBS Congress & 11th IUBMB Conference, Athens, Greece, PP5A-1

Bagasra O (2006) A unified concept of HIV-1 latency. Expert Opin Biol Ther 6:1135–1149

Bagasra O (2005) RNAi as an anti-viral therapy. Expert Opin Biol Ther 5(11):1463–1474

Bagasra O (1999) HIV and molecular immunity: prospect for AIDS vaccine. Eaton publishing, Natick, MA, USA

Bagasra O, Pace G (2010) A new perspective on HIV vaccine design: a view point. Ibnosina J Med and Biomed Sci 2(1):1–13

Bagasra O, Amjad M (2000) Protection against retroviruses is owing to a different form of immunity: a RNA-based molecular immunity hypothesis. Appl Immunohistochem Mol Morphol 8:133–146

Bagasra O, Amjad M (1997) Natural immunity against HIV-1: prospect for AIDS vaccine. Front Biosci 2:387–402

Bagasra O, Stir AE, Pirisi-Creek L, Creek KE, Bagasra AU, Glenn N, Lee JS (2006) Role of Micro-RNAs in regulation of lentiviral latency and persistence. Appl Immunohistochem Mol Morphol 14:276–290

Bai Y, Casola C, Betrán E (2008) Evolutionary origin of regulatory regions of retrogenes in Drosophila. BMC Genomics 9:240–250

Banfield JF, Young M (2009) Microbiology. Variety – the splice of life – in microbial communities. Science 326:1198–1199

Bartosik D, Sochacka M, Baj J (2003) Identification and characterization of transposable elements of Paracoccus pantotrophus. J Bacteriol 185(13):3753–3763

Brouns SJ, Jore MM, Lundgren M, Westra ER, Slijkhuis RJ, Snijders AP, Dickman MJ, Makarova KS, Koonin EV, van der Oost J (2008) Small CRISPR RNAs guide antiviral defense in prokaryotes. Science 321(5891):922–923

Buchon N, Vaury C (2006) RNAi: a defensive RNA-silencing against viruses and transposable elements. Heredity 96(2):195–202

Burnet FM (1959) The clonal selection theory of acquired immunity. Vanderbilt University Press, TN

Caswell JL, Mallick S, Richter DJ, Neubauer J, Schirmer C, Gnerre S, Reich D (2008) Analysis of chimpanzee history based on genome sequence alignments. PLoS Genetics 4:e10000057

Conley AB, Piriyapongsa J, Jordan IK (2008a) Retroviral promoters in the human genome. Bioinformatics 24:1563–1567

Conley AB, Miller WJ, Jordan IK (2008b) Human *cis* natural antisense transcripts initiated by transposable elements. Trends Genet 24:53–56

de Kruif P (1926) Microbe hunters. Harcourt, Brace and Company, New York

Doolittle RF, Sapienza C (1980) Selfish genes, the phenotypic paradigm, and genome evolution. Nature 284:601–603

Dooner HK, Weil CF (2007) Give-and-take: interactions between DNA transposons and their host plant genomes. Curr Opin Genet Dev 17:486–492

Dunn CW, Hejnol A, Matus DQ, Pang K, Browne WE, Smith SA, Seaver E, Rouse GW, Obst M, Edgecombe GD, Sørensen MV, Haddock SH, Schmidt-Rhaesa A, Okusu A, Kristensen RM, Wheeler WC, Martindale MQ, Giribet G (2008) Broad phylogenomic sampling improves resolution of the animal tree of life. Nature 452:745–749

Frenkel N, Schirmer EC, Wyatt LS, Katsafanas G, Roffman E, Danovich RM, June CH (1990) Isolation of a new herpesvirus from human CD4+ T cells. Proc Natl Acad Sci USA 87:748–752

Furesz J, Levenbook I (1998) New assays for the quality control of live oral poliovirus vaccine. Acta Microbiol Immunol Hung 45(3–4):391–399

Hakim ST, Alsayari M, McLean DC, Saleem S, Addanki KC, Aggarwal M, Mahalingam K, Bagasra O (2008) A large number of the human microRNAs target lentiviruses, retroviruses, and endogenous retroviruses. BBRC 369:357–362

Han CG, Shiga Y, Tobe T, Sasakawa C, Ohtsubo E (2001) Structural and functional characterization of IS*679* and IS*66*-family elements. J Bacteriol 183:4296–4304

Harris J (2009) Soil microbial communities and restoration ecology: facilitators or followers? Science 325:573–574

Helm T (2004) Basic immunology: a primer. Minn Med 87(5):40–44

Heimberg AM, Sempere LF, Moy VN, Donoghue PC, Peterson KJ (2008) MicroRNAs and the advent of vertebrate morphological complexity. Proc Natl Acad Sci USA 105:2946–2950

Horvath P, Barrangou R (2010) CRISPR/Cas, the immune system of bacteria and archaea. Science 327(5962):167–170

Jordan IK, Rogozin IB, Wolf YI, Koonin EV (2002) Microevolutionary genomics of bacteria. Theor Popul Biol 61(4):435–447

Jordan IK, Rogozin IB, Glazko GV, Koonin EV (2003) Origin of a substantial fraction of human regulatory sequences from transposable elements. Trends Genet 19(2):68–72

Karp CL, Auwaerter PG (2007) Coinfection with HIV and tropical infectious diseases. II. Helminthic, fungal, bacterial and viral pathogens. Clin Infect Dis 45:1214–1220

Karttunen A, Pöyry T, Vaarala O, Ilonen J, Hovi T, Roivainen M, Hyypiä T (2003) Variation in enterovirus receptor genes. J Med Virol 70(1):99–108

Kidwell MG, Lisch DR (2001) Perspective: transposable elements, parasitic DNA, genome evolution. Evolution 55:1–24

Lisch D (2002) Mutator transposons. Trends Plant Sci 7:498–504

Lisco A, Grivel JC, Biancotto A, Vanpouille C, Origgi F, Malnati MS, Schols D, Lusso P, Margolis LB (2007) Viral interactions in human lymphoid tissue: human herpesvirus 7 suppresses the replication of CCR5-tropic human immunodeficiency virus type 1 via CD4 modulation. J Virol 81:708–717

Lusso P, Secchiero P, Crowley RW, Garzino-Demo A, Berneman ZN, Gallo RC (1994) CD4 is a critical component of the receptor for human herpesvirus 7: Interference with human immunodeficiency virus. Proc Natl Acad Sci USA 91:3872–3876

Matzke MA, Mette MF, Matzke AJ (2000) Transgene silencing by the host genome defense: implications for the evolution of epigenetic control mechanisms in plants and vertebrates. Plant Mol Biol 43:401–415

Muotri AR, Marchetto MC, Coufal NG, Gage FH (2007) The necessary junk: new functions for transposable elements. Hum Mol Genet 16(Review Issue 2):R159–R167

Navon R, Wang H, Steinfeld I, Tsalenko A, Ben-Dor A, Yakhini Z (2009) Novel rank-based statistical methods reveal microRNAs with differential expression in multiple cancer types. PLoS One 4(11):e8003

Obbard DJ, Gordon KH, Buck AH, Jiggins FM (2009) The evolution of RNAi as a defence against viruses and transposable elements. Philos Trans R Soc Lond B Biol Sci 364:99–115

Ochman H, Raghavan R (2009) Systems biology. Excavating the functional landscape of bacterial cells. Science 326:1200–1201

Palm NW, Medzhitov R (2009) Pattern recognition receptors and control of adaptive immunity. Immunol Rev 227(1):221–233

Pandrea I, Sodora DL, Silvestri G, Apetrei C (2008) Into the wild: simian immunodeficiency virus (SIV) infection in natural hosts. Trends Immunol 29:419–428

Piriyapongsa J, Jordan IK (2008) Dual coding of siRNAs and miRNAs by plant transposable elements. RNA 14:814–821

Pradeu T (2009) Immune system: "Big Bang" in question. Science 325(5939):393

Ryt-Hansen R, Katzenstein TL, Gerstoft J, Eugen-Olsen J (2006) No influence of GB virus C on disease progression in a Danish cohort of HIV-infected men. AIDS Res Hum Retroviruses 22:496–498

Schluter SF, Marchalonis JJ (2003) Cloning of shark RAG2 and characterization of the RAG1/RAG2 gene locus. FASEB J 17:470–472

Schott DH, Cureton DK, Whelan SP, Hunter CP (2005) An antiviral role for the RNA interference machinery in Caenorhabditis elegans. Proc Natl Acad Sci USA 102:18420–18424

Schmeier S, MacPherson CR, Essack M, Kaur M, Schaefer U, Suzuki H, Hayashizaki Y, Bajic VB (2009) Deciphering the transcriptional circuitry of microRNA genes expressed during human monocytic differentiation. BMC Genomics 10:595

Schramke V, Allshire R (2003) Hairpin RNAs and retrotransposon LTRs effect RNAi and chromatin-based gene silencing. Science 301:1069–1074

Shabalina SA, Koonin EV (2008) Origins and evolution of eukaryotic RNA interference. Trends Ecol Evol 23(10):578–587

Siomi MC, Siomi H (2008) Characterization of endogenous human Argonautes and their miRNA partners in RNA silencing. Nucleic Acids Symp Ser (Oxf) 52:59–60

Sompayrac LM (2008) How the immune system works, 3rd edn. Wiley-Blackwel, Malden, MA

Szathmáry E (1999) The origin of the genetic code: amino acids as cofactors in an RNA world. Trends Genet 15:223–229

Taganov KD, Boldin MP, Baltimore D (2007) MicroRNAs and immunity: tiny players in a big field. Immunity 26:133–137

Tang TH, Polacek N, Zywicki M, Huber H, Brugger K, Garrett R, Bachellerie JP, Hüttenhofer A (2005) Identification of novel non-coding RNAs as potential antisense regulators in the archaeon *Sulfolobus solfataricus*. Mol Microbiol 55(2):469–481
Thompson CB (1995) New insights into V(D)J recombination and its role in the evolution of the immune system. Immunity 3(5):531–539
Ulenga NK, Sarr AD, Hamel D, Sankale JL, Mboup S, Kanki PJ (2008) The level of APOBEC3G (hA3G)-related G-to-A mutations does not correlate with viral load in HIV type 1-infected individuals. AIDS Res Human Ret 24:1285–1290
Walker BD, Goulder PJ (2000) AIDS. Escape from the immune system. Nature 407:313–314
Zamore PD (2006) RNA interference: big applause for silencing in Stockholm. Cell 127: 1083–1086

Chapter 7
Ubiquitous Bacteriophage Hosts in Rice Paddy Soil

Makoto Kimura, Guanghua Wang, Natsuko Nakayama, and Susumu Asakawa

7.1 Introduction

Viruses are the most abundant biological entities in marine and freshwater environments. Many studies have indicated the ecological importance of viruses in global biogeochemical nutrient cycles (Weinbauer 2004). Viruses are recognised as the greatest genomic reservoirs in marine and freshwater environments due to their huge abundance and diversity (Weinbauer and Rassoulzadegan 2004; Frost et al. 2005; Paul and Sullivan 2005). This understanding has led to a resurgence of interest in virus ecology (Mann 2005).

Bacteriophages (phages) are estimated to comprise the majority of viruses in aquatic environments from circumstantial evidence such as the greater abundance of bacteria than other planktonic hosts, the predominance of viruses within the virioplankton with phage-sized genomes and the ability of changes in bacterial abundance to predict changes in viral abundance (Wommack and Colwell 2000).

The diversity of biological entities is evaluated from their phylogenetic diversity, but until recently, it was common knowledge among virus taxonomists that there is no single gene that is commonly present in all virus genomes. However, recent developments in viral genomics revealed that comparable genomic information is preserved among viral subsets and can be used as a powerful tool for the phylogenetic classification of viruses and the evaluation of viral diversity in the environment (Weinbauer and Rassoulzadegan 2004). The most popular candidate

M. Kimura (✉), N. Nakayama, and S. Asakawa
Graduate School of Bioagricultural Sciences, Nagoya University, Furocho, Chikusa, Nagoya 464-8601, Japan
e-mail: kimuram@agr.nagoya-u.ac.jp
G. Wang
Northeast Institute of Geography and Agroecology, Chinese Academy of Sciences, Harbin 150081, China
e-mail: guanghuawang@hotmail.com

G. Witzany (ed.), *Biocommunication in Soil Microorganisms*, Soil Biology 23,
DOI 10.1007/978-3-642-14512-4_7, © Springer-Verlag Berlin Heidelberg 2011

gene for the evaluation of viral diversity is the capsid gene, *g23*, for the T4-type phage subset in *Myoviridae*. Tétart et al. (2001) and Desplats and Krisch (2003) grouped T4-type phage members into T-evens, PseudoT-evens, SchizoT-evens and ExoT-evens, with increasing divergence from coliphage T4 as determined from the sequence comparison of the tail and capsid genes *g18*, *g19* and *g23*. Filée et al. (2005) elucidated that the majority of the sequences of *g23* PCR products from diverse marine environments belonged to five previously uncharacterised groups (Marine Groups I to V).

Although viruses may be important components of the microbial world in soil, to date the ecology of soil viruses has been studied little at the community level (Kimura et al. 2008).

The paddy field is a unique agro-ecosystem, where the soil is maintained under flooded conditions during most periods of rice cultivation. In this chapter, the seasonal variation in the abundance of virus-like particles (VLPs) and bacteria in the floodwater of a rice field was first measured. Then, as phages comprised the majority of viral communities in the floodwater, the abundance of phages of common heterotrophic bacteria in the floodwater was surveyed. The frequency of phage-infected bacterial cells in the floodwater indicated the importance of phage communities in the microbial food web in rice fields. Furthermore, the lysogeny of oligotrophic soil bacteria was evaluated for understanding the impact of phages on autochthonous soil bacteria. From the analysis of *g23* clones in rice field soils in Japan and Northeast China, many novel *g23* gene sequences were detected, which suggested the presence of soil-specific phage communities. These findings indicated the importance of viruses as a genomic reservoir and in the nutrient cycle in rice fields.

7.2 Abundance and Morphology of Viruses in the Floodwater

Viruses are the most abundant biological entities in the floodwater of rice fields. Viruses in the floodwater have been enumerated for four plots in Anjo City under a long-term fertiliser trial since 1925: a plot without fertiliser (NF plot), a plot with chemical fertilisers (CF plot), a plot with chemical fertilisers and lime (CFCa plot) and a plot treated with chemical fertilisers, lime and compost (CM plot) (Nakayama et al. 2007a). Enumeration of viruses was performed by epifluorescence microscopy (EFM) after staining with SYBR Green I. As all the viral particles that were counted by EFM were not confirmed to be viruses, enumerated viruses were labelled as VLPs.

VLP abundance in the floodwater during rice cultivation period ranged from 5.6×10^6 to 1.2×10^9 VLPs mL^{-1} (mean abundance of 1.5×10^8 VLPs mL^{-1}; Fig. 7.1). This value fluctuated by 50 to 200 times in the plot fields. VLP in the irrigation water ranged from 5.1×10^6 to 1.1×10^7 VLPs mL^{-1} (Fig. 7.1), which was smaller than that in the floodwater. Smaller VLP abundance in the irrigation water than that in the floodwater indicated that viruses in the floodwater were not derived directly from the irrigation water but occurred in the floodwater. The high

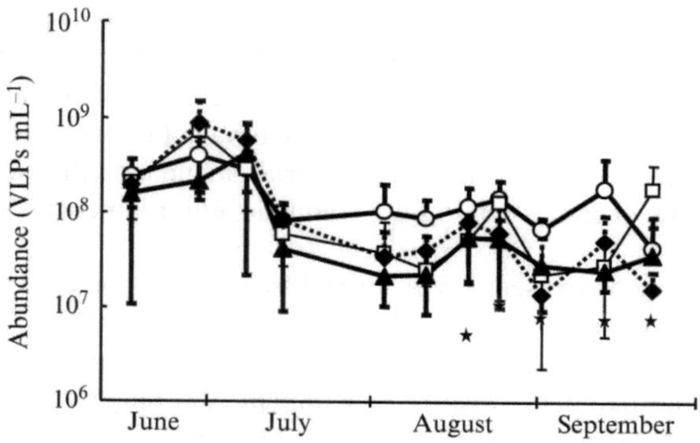

Fig. 7.1 Seasonal variations in abundance of virus-like particles (VLPs) in the water samples from the NF plot (*open circle*), CF plot (*open square*), CFCa plot (*filled diamond*), CM plot (*filled triangle*) and the irrigation water (*filled star*) during the rice cultivation period

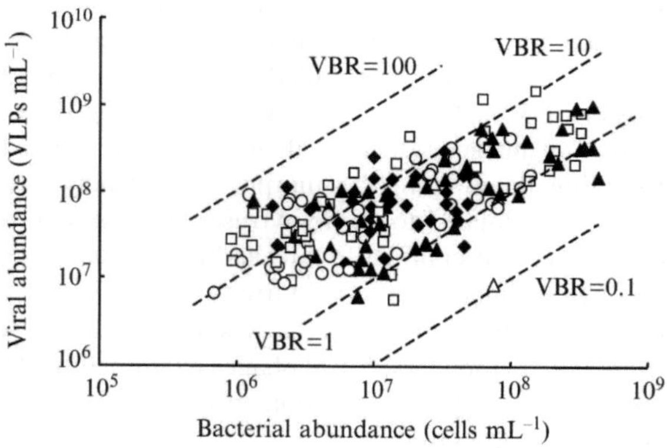

Fig. 7.2 Seasonal variations in virus-to-bacterium ratios (VBRs) in the floodwater from the NF plot (*open circle*), CF plot (*open square*), CFCa plot (*filled diamond*) and CM plot (*filled triangle*) during the rice cultivation period. The correlation coefficient between the abundance of VLPs and bacteria was 0.608 (significant at 1% level)

VLP abundance in the floodwater suggested that it is an aquatic environment abundant in viruses in comparison with marine, estuarine and freshwater environments, where VLP abundance is generally less than 10^7 VLPs mL^{-1} (Wommack and Colwell 2000; Weinbauer 2004).

The virus-to-bacterium ratio (VBR) in the floodwater ranged from 0.11 to 72. There was no significant difference in VBRs among the plots. VLP abundance showed significant correlation with bacterial abundance for every plot ($P < 0.01$: Fig. 7.2). In contrast to the general findings that VBRs fall generally between 3 and 25 in aquatic environments (Fuhrman 1999; Wommack and Colwell 2000; Weinbauer 2004), 35 and 6% of the floodwater samples showed lower and higher

VBRs, respectively, than the range in aquatic environments (Fig. 7.2). Although there was no significant correlation between VLP abundance and VBR, seasonal variation in VBR correlated significantly with the change of bacterial abundance, and VBR increased with the decrease in bacterial abundance ($P < 0.01$). The inverse relationship between VBR and bacterial abundance that was observed in the present study occurs commonly in aquatic environments (Wommack and Colwell 2000).

The morphology of viruses was examined by transmission electron microscopy (TEM) for the same floodwater samples (Nakayama et al. 2007c). The capsid size distribution was monophasic, and the median values fell into the 50–60 or 60–70 nm fraction. More than half of the viruses fell within the size ranges from either 40–50 or 50–60 nm to either 60–70 or 70–80 nm for most samples. Thus, most capsid sizes of viral communities in the floodwater were distributed within narrow ranges. The predominance of viruses with the size classes of <100 nm is attributed to the predominance of phages in viral communities because the average capsid size for viruses of eukaryotic algae is reported to be 152 nm (Van Etten et al. 1991). Ackermann and DuBow (1987) also summarised that phages with isometric capsids have a clear preference for the 55–65-nm range for 180 phages of various bacteria.

The most common capsid forms were tailed or non-tailed, isometric icosahedral form with a diameter of 50–70 nm irrespective of sampling dates and plots, and other forms were too rare and sporadic to evaluate the difference in diversity among the plots and sampling dates from the morphology (Nakayama et al. 2007c). Dominant viruses were estimated to be myoviruses (with a long contractile tail), siphoviruses (with a long non-contractile tail) and podoviruses (with a short non-contractile tail). Rough estimation of tailed viruses by TEM accounted for 2–54% of the total in this study (data not shown). As Ackermann (2001) summarised that 96% of phages were tailed among ca. 5,100 phages examined by TEM, the sonication process used in our sample preparation might have caused tail loss in a considerable proportion of the viruses.

7.3 Abundance of Bacteriophages of Common Heterotrophic Bacteria in the Floodwater

Lytic phages of 18 heterotrophic bacterial strains belonging to a large variety of phylogenetic taxa were enumerated by subjecting the same floodwater samples in Sect. 7.2 (Nakayama et al. 2007b). The bacterial strains were isolated from the floodwater of a nearby rice field. Table 7.1 shows the seasonal variations in abundance (in plaque forming units (PFU) mL^{-1}) of phages that are infectious to the bacterial strains by grouping them into a *Cytophaga-Flexibacter-Bacteroides* (CFB) group, α-, β- and γ-Proteobacteria, Actinobacteria and Firmicutes. In general, phage abundance in the floodwater was highest on 15 June, and it decreased with time until the midseason drainage (from 15 July to 25 July). The abundance tended to be higher after the mid-season drainage than before it ($P < 0.01$), except

Table 7.1 The seasonal variations in abundance of phages infectious to the bacterial strains

Bacterial strain	June 15	June 28	July 7	July 14	August 2	August 10	August 18	August 24	September 1	September 13	September 22
α-Proteobacteria											
Rhizobium 061238*	aaaa	aabb	aaaa	aaaa	aaaa	aaaa	aaaa	dcba	cbbd	bccd	abaa
Rhodobacter 072807	aaad	acad	aaab	aaaa	aaaa	aacb	abaa	cddb	baaa	abac	aaaa
Sphingomonas 070409	eeee	acda	eeae	eeee	baaa	babb	aaba	aabc	abbb	dddd	aaaa
Sphingomonas 072822	baaa	aaaa	aaaa	aaaa	abaa	bcbb	abab	bbdb	aaca	aaaa	aaaa
Sphingomonas 072833*	bccc	aaaa	aaaa	bbab	aaaa	aaaa	aaaa	bcda	cbba	aaab	aaaa
Sphingomonas 081631	aaaa	aaaa	bbba	aaba	aabb	aaaa	aaaa	aaac	bbcb	aaca	aaaa
β-Proteobacteria											
Ralstonia 050629	cccd	aaaa	babb	aaba	ccca	bcac	aaaa	aaba	bccd	bbbc	ccbc
Ralstonia 050603*	acaa	aaaa	abab	bbaa	baaa	aaaa	aaab	aaaa	aaaa	bbbb	aaaa
γ-Proteobacteria											
Pseudomonas 050644*	daaa	aaaa	bbab	aabb	aaaa	aaaa	aaaa	bbba	aaaa	bdbb	aaaa
Enterobacter 061235*	ecdc	abac	bbca	aaba	bbcb	edde	acba	bcdb	dbdc	bdbc	bcaa
CFB group											
Chryseobacterium 052302*	eddc	aaaa	abab	aaaa	aaaa	abab	aaaa	abac	abbb	aaaa	dbdd
Cytophaga 052318*	ccca	aaac	bbbb	aaaa	eeee	eddd	bbba	eebe	cccc	daad	bbab
Flavobacterium 070416	aaaa	aaaa	aaaa	aaaa	aaaa	aaba	aaaa	acca	cbba	baab	aaaa
Pedobacter 081642*	cccc	aaaa	bbbb	aaaa	aaaa	caba	aaaa	abbc	bbab	abba	bbbb
Firmicutes											
Bacillus 061217*	aaab	aaab	bbbb	aaaa	babb	aaaa	aaaa	ccee	bbca	cdbd	bbdc
Actinobacteria											
Streptomyces 061220*	abac	aaab	aaaa	aaaa	cdba	aaaa	aaaa	aaaa	cabb	cbab	aaaa
Microbacterium 070464*	eddd	acaa	aaaa	abbb	cbbb	baaa	eddd	aaab	bebb	aaaa	cbdb
Agromyces 061242*	cccc	eeee	aaaa	aaaa	aaaa	aaaa	aaaa	aaaa	aabb	aaaa	aaaa

Quartette letters indicate the phage abundance in the NoF plot, CF plot, CFCa plot and CM plot, respectively, from the left to the right, and a, b, c, d and e in the quartettes show the phage abundances of $<1\ ml^{-1}$, $1–10\ ml^{-1}$, $10–10^2\ ml^{-1}$, $10^2–10^3\ ml^{-1}$ and $>10^3\ ml^{-1}$, respectively. The strain name with the symbol *asterisk* indicates the bacterial strains that contained prophage(s) in their genomes for *Sphingomonas* sp. 070409 and *Agromyces* sp. 061242. There was no clear preference among the plots for phages of the respective bacteria. There was no floodwater sample that contained less than a phage per mL infectious to either of the tested strains. The effect of heavy rain events on phage abundance was not observed.

Phages were detected with an abundance of more than 4800 PFU mL^{-1} in 22 cases (Table 7.1). Phage abundances greater than 10^3 PFU mL^{-1} were observed 33 times for eight bacterial strains, and phages of *Enterobacter* sp. 061235 and *Cytophaga* sp. 052318 were detected at more than one PFU mL^{-1} for 33 and 32 samples, respectively, out of 44 total samples (Table 7.1). Thus, culturable bacteria were the hosts of many phages in the floodwater of the rice field, notwithstanding the small numbers of phages of culturable heterotrophic bacteria present in aquatic environments (Wommack and Colwell 2000).

Phages of *Sphingomonas* spp. in the floodwater seemed to be strain-dependent, especially before the midseason drainage, and they were numerous for *Sphingomonas* sp. 070409 but small for *Sphingomonas* spp. 072822 and 081631 (Table 7.1). Specificity of phage infectivity at the strain level was also observed between *Ralstonia* spp. 050629 and 050503.

Prophage induction with mitomycin C was observed for 12 of 18 bacterial strains (Table 7.1). Although the induced prophages were siphoviruses for *Rhizobium* sp. 061238, *Enterobacter* sp. 061235 and *Chryseobacterium* sp. 052302, and a myovirus for *Pedobacter* sp. 081642, there was no difference in the proportion of phage-detected floodwater samples between bacterial strains with and without prophages.

It is generally held that lysogenic bacteria are immune to superinfection by the same or related phage types (Marsh and Wellington 1994; Mann 2003). However, *Pedobacter* sp. 081642, *Microbacterium* sp. 070464 and *Agromyces* sp. 061242, having isometric icosahedral prophages, did not show any difference in the proportion of phage-detected floodwater samples from the other bacterial strains. Superinfection immunity might not have been the factor determining the proportion of phage-detected floodwater samples for the respective bacterial strains in this study.

7.4 Morphology and Host Range of *Sphingomonas*/*Novosphingobium* Phages

Members of the *Sphingomonas*-related genera (*Sphingomonas*, *Sphingobium*, *Novosphingobium* and *Sphingopyxis*) were dominant in bacterial isolates obtained from the floodwater (Nakayama et al. 2006). Fifty-eight *Sphingomonas*/*Novosphingobium* phages were isolated from the floodwater and examined with regard to their host ranges in relation to the phylogenetic positions of hosting bacteria. Strains 070409, 072822 and 081631 were their original hosts, and the phages were obtained by purifying the plaques in Sect. 7.3 (Nakayama et al. 2009c).

All of the phages had an isometric or elongated icosahedral capsid with a long and thin noncontractile tail (diameter 8–9 nm; Table 7.2). The morphology of the tails indicated that all of them belong to the family *Siphoviridae* in the order

Table 7.2 Morphology, genome size, host range and phylogenetic affiliation of *g23* genes of respective isolated phages (Nakayama et al. 2009a)

Host bacterium	Phages isolated[a]	Capsid size (nm)	Tail length (nm)	Genome size (kb)	Infection type[b]	Affiliation of *g23* gene[c]
070409	N-AFCM0615-3	55 × 55	190	60	2	I
	N-AFCa0615-4	50 × 50	170	60	2	I
	N-AFCa0615-5	55 × 55	205	60	2	II
	N-AFCa0615-6	50 × 50	180	60	1	III**
	N-AFNF0628-7	145 × 60	240	160	2	II
	N-AFCF0628-8	125 × 55	170	160	2	I
	N-AFCF0628-9	165 × 50	240	160	2	IV
	N-AFCM0628-10	160 × 75	240	160	2	II
	N-AFCa0628-11	155 × 50	180	160	2	II
	N-AFCa0628-12	150 × 70	280	160	2	V
	N-AFNF0707-13	140 × 60	235	160	2	II
	N-AFCF0707-14	140 × 55	240	160	2	III
	N-AFCF0707-15	155 × 70	240	160	2	IV
	N-AFCF0707-16	140 × 55	215	160	2	VI*
	N-AFCa0707-17	140 × 70	220	160	2	V
	N-AFCa0707-18	150 × 60	250	160	2	I
	N-AFCF0714-51	150 × 60	220	160	2	III
	N-AFCM0714-52	140 × 50	245	160	2	III
	N-AFNF0802-19	155 × 60	230	160	2	III
	N-AFNF0802-20	160 × 55	200	160	2	III
	N-AFCF0802-21	140 × 50	270	160	2	IV
	N-AFCF0802-22	160 × 60	250	160	2	V
	N-AFCa0802-23	60 × 60	220	40	2	VI
	N-AFCa0802-24	60 × 60	240	40	2	III
	N-AFCa0802-56	100 × 50	220	100	2	I
	N-AFCa0802-57	120 × 60	240	100	2	III
	N-AFNF0810-25	160 × 60	180	160	2	III
	N-AFCF0810-26	130 × 55	260	160	2	VI*
	N-AFCM0810-27	130 × 55	200	160	2	VI
	N-AFCM0810-28	170 × 60	280	160	2	IV
	N-AFCa0810-29	120 × 60	200	160	2	I
	N-AFCa0810-30	130 × 55	230	160	2	III
	N-AFNF0824-31	150 × 50	220	160	2	I
	N-AFCF0824-32	135 × 75	220	160	2	IV
	N-AFCM0824-33	140 × 50	200	160	2	III
	N-AFCa0824-34	130 × 55	200	160	2	VI
	N-AFCa0824-35	125 × 55	210	160	2	IV
	N-AFCa0824-36	110 × 55	200	160	2	II
	N-AFCF0901-37	130 × 60	170	160	2	I
	N-AFCM0901-40	125 × 50	240	160	2	IV
	N-AFCa0901-41	110 × 50	220	160	2	IV
	N-AFCa0901-42	160 × 65	205	160	2	V
	N-AFNF0913-43	160 × 65	220	160	1	IV
	N-AFCM0913-47	150 × 50	220	160	2	III**
	N-AFNF0922-55	150 × 55	190	160	2	III
072822	S-AFNF0615-61	60 × 60	170	60	3	VI*
	S-AFCF0802-69	60 × 60	120	60	3	VI*
	S-AFNF0810-70	45 × 45	140	60	3	VI*

(*continued*)

Table 7.2 (continued)

Host bacterium	Phages isolated[a]	Capsid size (nm)	Tail length (nm)	Genome size (kb)	Infection type[b]	Affiliation of *g23* gene[c]
	S-AFCa0824-71	80 × 80	160	160	3	VI*
	S-AFCa0901-72	80 × 80	250	160	3	VI
081631	N-AFCa0707-83	60 × 60	170	60	6	VI*
	N-AFCa0714-84	70 × 70	150	60	6	VI*
	N-AFCa0802-85	70 × 70	190	60	6	VI*
	N-AFCM0802-86	60 × 60	170	60	6	VI*
	N-AFCF0901-89	60 × 60	150	60	5	VI*
	N-AFCa0901-90	60 × 60	150	60	5	VI*
	N-AFCM0901-91	60 × 60	190	60	5	VI
	N-AFCa0913-92	60 × 60	190	60	4	VI*

Symbols *asterisk* and *double asterisk* indicate the identical sequences among the members at the nucleotide level

[a]The first letter N or S shows the genus name of bacterium from which the phage was isolated (N; *Novosphingomonas*, S; *Sphingomonas*)

[b]See Fig. 7.3 for the infection type of respective phages

[c]Roman figures indicate the group numbers in the Paddy Group

Caulovirales (Fauquet et al. 2005). The genomes were double-stranded DNA with sizes of 40, 60, 100 or 160 kb. The observation of only four types of genomic sizes in the isolated phages did not imply that the phage communities in the floodwater were simple, as evidenced in Sect. 7.7.

Host ranges of the phages were grouped into six infection types (Fig. 7.3), summarised in Table 7.2. Phages N-AFCa0615-6 and N-AFNF0913-43 lysed only strain 070409, the host strain used for isolation of those phages. In contrast, seven phages induced lysis in ten of sixteen bacterial strains (Table 7.2), and the host range covered three genera, including the genus *Porphyrobacter* for four phages (N-AFCa0707-83, N-AFCa0714-84, N-AFCa0802-85 and N-AFCM0802-86).

The host range was primarily determined by the host strain used for the isolation of these phages (Table 7.2). It was very narrow for the phages of strain 070409 and broad for those of strain 081631, covering three genera. It is notable that the five phages that were isolated from strain 072822 had 60- and 160-kb genomes and belonged to the Type 3 group (Table 7.2), indicating that host range is determined by the host strain used for isolation. In addition, the phages belonging to infection types 3 to 6 (Fig. 7.3) infected two or three bacterial genera, but they could not infect some bacterial strains of the same genera with which they were isolated. This indicates that phylogenetic relatedness to the host bacterium was not a decisive factor in determining the infectious ability of the phage.

Prophage induction with mitomycin C was observed for strains 072806, 072814 and 072833. Many phages infected strain 072833, while no phage could infect strains 072814 and 072806 (Fig. 7.3). Thus, the superinfection immunity by prophages was not seen in strain 072833, as observed previously.

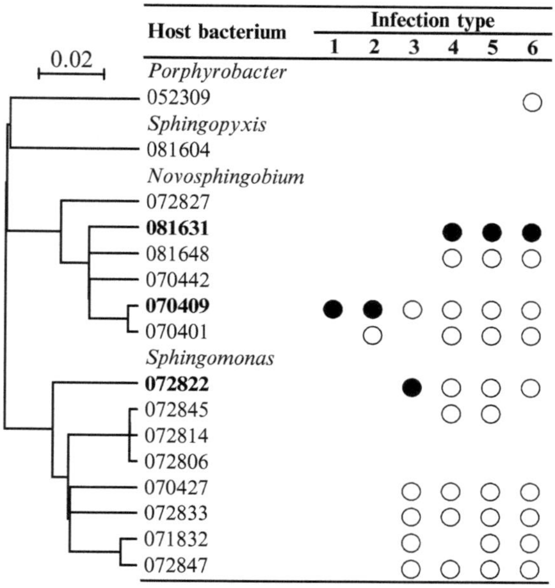

Fig. 7.3 Host range pattern of phages isolated in this study. Symbols *filled circle* and *open circle* indicate the bacterium from which the phage was isolated and a bacterium which the phage can infect, respectively. The phylogenetic relations among the hosts are shown at the *left side*. The *scale bar* represents the amount of nucleotide substitutions per residue

7.5 Frequency of Phage-Infected Bacterial Cells in the Floodwater

Weinbauer (2004) reviewed the virus-mediated carbon flow in pelagic oceans, where 6–26% of the carbon fixed by primary producers enters into the dissolved organic carbon (DOC) pool via virus-induced lysis at different trophic levels. The situation is similar in freshwater environments. Fischer and Velimirov (2002) estimated that, in a eutrophic oxbow lake of the Danube River, the contribution of phage lysis to bacterial mortality ranges from 56 to 63%, while the grazing of heterotrophic nanoflagellates accounted for an average of only 5% of the bacterial mortality.

The bacterial mortality in the floodwater due to phage lysis was evaluated by measuring the frequency of phage-infected bacterial cells with a transmission electron microscope (TEM) (Nakayama et al. 2009b). Floodwater samples were the same samples used in Sect. 7.2. The frequency of visibly infected cells (FVIC) was calculated from the proportion of phage-infected cells in the total number of cells examined. The FVIC in the CM plot was larger, ranging from 2.4 to 3.6% (average 3.0%), than that in the NPKCa plot, which ranged from 1.6 to 2.9% (average 2.0%) (Table 7.3). The higher FVIC in the CM plot was due, on one hand, to the larger proportion of phage-infected cells in long-rod members and, on the other hand, to the increase in the proportion of short rods with a relatively larger fraction of phage-infected cells (Table 7.3).

Table 7.3 Morphological distribution of bacteria, frequency of visibly infected cells (FVIC), frequency of phage-infected cells (FIV) and fraction of bacterial mortality from phage lysis (FMPL)

	Site	Long rods	Short rods	Cocci	Total	FVIC (%)[a]	FIC (%)[a]	FMPL (%)[a]
No-compost plot								
15 June	Right	134(2)	5(0)	11(0)	150(2)	1.8 ± 0.8	12.0 ± 5.3	15.1 ± 5.9
	Left	122(2)	11(1)	16(1)	149(4)			
	Counter	123(1)	14(0)	9(1)	146(2)			
7 July	Right	135(4)	11(0)	2(0)	148(4)	2.9 ± 1.1	19.0 ± 7.5	27.3 ± 8.6
	Left	131(5)	7(0)	7(1)	145(6)			
	Counter	130(3)	8(0)	12(0)	150(3)			
10 August	Right	137(1)	7(1)	6(0)	150(2)	1.6 ± 0.4	10.5 ± 2.7	12.8 ± 2.8
	Left	137(1)	11(0)	2(1)	150(2)			
	Counter	136(3)	7(0)	7(0)	150(3)			
24 August	Right	142(3)	6(0)	2(0)	150(3)	1.6 ± 0.4	10.5 ± 2.7	12.8 ± 2.8
	Left	132(1)	15(1)	3(0)	150(2)			
	Counter	130(1)	16(1)	4(0)	150(2)			
13 September	Right	133(4)	11(0)	6(0)	150(4)	2.2 ± 0.4	14.7 ± 2.7	19.3 ± 2.8
	Left	125(3)	11(0)	14(0)	150(3)			
	Counter	115(2)	17(1)	18(0)	150(3)			
Total		1962(36)	157(5)	119(4)	2238(45)	2.0 ± 0.8	13.4 ± 5.2	17.2 ± 5.8
Compost plot								
15 June	Right	99(2)	39(2)	12(0)	150(4)	3.6 ± 0.8	22.4 ± 5.3	34.8 ± 5.9
	Left	101(3)	34(3)	15(0)	150(6)			
	Counter	89(5)	42(0)	19(1)	150(6)			
7 July	Right	121(3)	20(0)	9(0)	150(3)	2.4 ± 0.4	16.0 ± 2.7	21.7 ± 2.8
	Left	106(3)	27(1)	17(0)	150(4)			
	Counter	121(4)	18(0)	11(0)	150(4)			
10 August	Right	87(2)	53(0)	10(1)	150(3)	2.7 ± 1.2	17.3 ± 7.9	24.2 ± 9.1
	Left	100(1)	47(2)	3(0)	150(3)			
	Counter	103(4)	42(2)	5(0)	150(6)			
24 August	Right	113(4)	37(1)	0(0)	150(5)	2.7 ± 0.7	17.8 ± 4.7	25.0 ± 5.1
	Left	97(4)	34(0)	7(0)	138(4)			
	Counter	128(3)	18(0)	4(0)	150(3)			
13 September	Right	118(2)	27(3)	5(0)	150(5)	3.6 ± 1.1	22.5 ± 7.2	35.0 ± 8.2
	Left	107(4)	36(2)	5(1)	148(7)			
	Counter	112(4)	29(0)	9(0)	150(4)			
Total		1602(48)	503(16)	131(3)	2236(67)	3.0 ± 0.9	19.3 ± 5.8	27.9 ± 6.5

[a]The values are mean ± standard deviation

By TEM, phage particles are recognisable in the infected cells only in the late stage of the lytic cycle. Therefore, the frequency of phage-infected cells (FIC) was estimated from the FVIC by considering the fraction of the eclipse period in the latent period (Proctor et al. 1993; Binder 1999). The FIC ranged from 10.5 to 19.0% (average 13.4%) for the NPKCa plot and from 16.0 to 22.5% (average 19.3%) for the CM plot (Table 7.3). According to the non-linear steady-state model of Binder (1999), the fraction of bacterial mortality from phage lysis (FMPL) ranged from 12.8 to 27.3% (average 17.2%) for the NPKCa plot and from 21.7 to 35.0% (average 27.9%) for the CM plot (Table 7.3). The frequency of phage lysis in floodwater was within the frequency ranges observed in other aquatic environments (Weinbauer 2004). The impact of bacterial mortality from phage lysis seemed to be large enough to redirect the microbial food web and induce the change and succession of bacterial communities in the floodwater since phages are, in general, strictly host-specific. Bacterial abundance in the floodwater fluctuated by two orders of magnitude during the rice cultivation period in Sect. 7.2 (Nakayama et al. 2007a). Large fluctuations of phage and bacterial abundances might have been due partly to the bacterial mortality from phage lysis.

7.6 Comparison of Lysogeny Between Copiotrophic and Oligotrophic Bacteria

The production and survival of virulent phages require frequent, successful host–phage encounters to exceed the rates of phage destruction and inactivation in the environment, whereas temperate phages in the lysogenic state are independent of host cell density and physiological state (Weinbauer and Suttle 1996). Lysogeny is an effective strategy for phages to maintain their populations where a host population is too small and/or hosts are inactive for phage populations to pursue a lytic infection life cycle (Mann 2003; Weinbauer et al. 2003).

Microorganisms in soil are generally in the dormant state because the soil environment is poor in substrates for their growth. Lysogeny seems to be a logical choice as a survival strategy for phages in a soil environment. Lysogeny may also be favourable for host bacteria because lysogenic bacteria are immune to superinfection by the same or related types of phages (Mann 2003; Marsh and Wellington 1994). However, lysogenic hosts are burdened with the synthesis of extra prophage DNA requiring additional substrates (Marsh and Wellington 1994). Lysogeny may have pros and cons for the host bacteria, depending on their adaptation/survival strategy in soils (Kimura et al. 2008). The lysogenesis of cells induced by mitomycin C was compared among 34 copiotrophic and 20 oligotrophic bacterial isolates in different phylogenetic positions from a rice field soil near Sendai City (unpublished data). Prophage induction by mitomycin C was also compared among the isolates of different phylogenetic positions. Here, bacteria capable of growth on 100-fold diluted nutrient broth (DNB) but not on single-strength nutrient broth (NB) are defined to be oligotrophic.

According to the colony-forming curves (CFCs), the isolates were grouped into four CFC groups based on visible colony development during incubation periods of 0–30 h, 31–51 h, 52–114 h and 115–265 h as Groups I, II, III and IV, respectively (Gorlach et al. 1994).

Prophage induction by mitomycin C was observed for 13 out of 54 strains (24%): i.e., 6 out of 20 oligotrophs (30%) and 7 out of 34 copiotrophs (21%). There was no significant difference in the frequency of lysogeny between oligotrophs and copiotrophs (Table 7.4, column 6). All of the induced virions were filamentous with the lengths of 550–2,600 nm and diameters of 10–15 nm, except for the virion of α-Proteobacteria S23328a (Table 7.4, column 6). The floodwater bacteria used in Sect. 7.3 were copiotrophs belonging to CFC Group I. The frequency of their lysogeny was 67% (Table 7.1).

The frequency of lysogeny was significantly higher for α-proteobacterial strains (46%) than for High G+C Gram-positive strains (12%), and it was also higher for strains belonging to CFC Groups II and IV (43–50%) than for those belonging to CFC Group I (5%, $P < 0.05$; Table 7.1). Significantly, smaller proportions of lysogeny for CFC Group I and High G+C Gram-positive strains were partly due to the absence of lysogenic *Arthrobacter* strains belonging to CFC Group I (12 strains). Table 7.4 also indicates that the frequency of lysogeny was dependent on the species. Prophages were detected from three out of four *Mesorhizobium loti* strains and two out of three *Bacillus thuringiensis* strains, whereas none of nine *Arthrobacter oryzae* strains was lysogenic. The frequency of lysogeny of floodwater *Sphingomonas*-related strains in Sect. 7.4 was 19%. Williamson et al. (2007) conducted induction assays with mitomycin C on Antarctic soils and temperate Delaware soils and found inducible fractions ranging from 4 to 20% and from 22 to 68%, respectively. The lysogenic level of the bacterial strains in this study (24%) was intermediate within the levels of bacterial populations in those soils.

All of the virions were estimated to belong to *Inoviridae* from the filamentous form, harbouring single-stranded DNA and a sensitivity to nuclease S1 (Fauquet et al. 2005). As most temperate phages isolated from marine bacteria were *Caudovirales* belonging to the family *Myoviridae* or *Siphoviridae* (Dillon and Parry 2008; Jiang and Paul 1994), prophages belonging to the family *Inoviridae* seemed to be characteristic of the soil bacteria.

Lysogenic strains belonging to *Mesorhizobium*, *Bacillus*, and *Arthrobacter* grew slower than or at a similar rate to the non-lysogenic strains, although *Flavobacterium* showed the adverse tendency (Table 7.4). The burden of additional prophage DNA synthesis seemed to be a factor determining the growth rate of bacterial strains.

However, the greater tendency of lysogeny in oligotrophic strains (30%; 6 out of 20 strains) than in copiotrophic strains (21%; 7 out of 34 strains) could not be explained ecologically from the superinfection immunity against the same or related-type virulent phages because most of viruses in aquatic environments, including the floodwater of rice field, were phages with a roughly isometric capsid (Børsheim 1993; Nakayama et al. 2007a). Genomic analysis of prophages and their

Table 7.4 CFC grouping and growth on NB media of bacterial strains and morphology and nucleic acid type of their lysogenic phages (Nakayama et al. unpublished)

Strain name	Accession no.[a]	CFC[b] group	DNB[c] Bact.	Closest relative (similarity %)	Prophage[d]	Size(nm)		Type of genome
						Length	Width	
α-Proteobacteria								
S23321	D84604	III	DNB	*Bradyrhizobium betae* (99.7)	–			
S24556	D84644	IV	DNB	*Bradyrhizobium yuanmingense* (99.4)	filamentous	900–2500	10–15	ssDNA
S24543	AB495348	IV	DNB	*Bradyrhizobium elkanii* (99.1)	–			
S24554a	AB495349	IV	NB	*Mesorhizobium loti* (99.8)	filamentous	630–1010	10–15	ssDNA
S23423	D84623	III	NB	*Mesorhizobium loti* (99.9)	filamentous	800–1000	10–15	ssDNA
S23442	D84631	III	NB	*Mesorhizobium loti* (99.9)	–			
S24538	D84639	IV	DNB	*Mesorhizobium loti* (99.9)	filamentous	1160–1300	10–15	ssDNA
S23436	D84627	III	DNB	*Erythromicrobium ramosum* (94.8)	filamentous	1000	10–15	ssDNA
S23328a	AB495350	III	DNB	*Sphingomonas eaquatilis* (97.5)	filamentous	660–960	15–20	ssDNA
S23322	D84605	III	DNB	*Sphingomonas echinoides* (97.6)	–			
S23301	D84598	III	DNB	*Novosphingobium stygium* (98.2)	–			
S23306	D84601	III	DNB	*Novosphingobium stygium* (98.1)	–			
S23435	D84626	III	DNB	*Novosphingobium pentaromativorans* (95.9)	–			
β-Proteobacteria								
S21012a	AB495351	I	DNB	*Duganella zoogloeoides* (98.3)	filamentous	640–2600	10–15	ssDNA
S24561	D84645	IV	DNB	*Variovorax soli* (97.5)	–			
S23408	D84617	III	DNB	*Variovorax boronicumulans* (97.6)	–			
S21104	D84572	I	DNB	*Janthinobacterium agaricidamnosum* (97.7)	–			
S21124	D84576	I	DNB	*Janthinobacterium agaricidamnosum* (97.6)	–			
S22204	D84582	II	DNB	*Janthinobacterium agaricidamnosum* (97.5)	–			
S22229	D84590	II	NB	*Janthinobacterium agaricidamnosum* (97.7)	–			
γ-Proteobacteria								
S21027	D84568	I	NB	*Pseudomonas lini* (99.5)	–			
CFB bacteria								
S22202a	AB496407	II	DNB	*Flavobacterium pectinovorum* (98.2)	filamentous	1000–1900	10–15	ssDNA
S24524a	AB496408	IV	DNB	*Flavobacterium pectinovorum* (98.2)	–			
Low G+C Gram-positive bacteria								
S21001	D84560	I	NB	*Bacillus acidiceler* (99.7)	–			
S21018	D84565	I	NB	*Bacillus acidiceler* (99.7)	–			
S21032	D84570	I	NB	*Bacillus megaterium* (99.6)	–			
S23402	D84614	III	NB	*Bacillus thuringiensis* (99.7)	–			

(*continued*)

Table 7.4 (continued)

Strain name	Accession no.[a]	CFC[b] group	DNB[c] Bact.	Closest relative (similarity %)	Prophage[d]	Size(nm)		Type of genome
						Length	Width	
S23440	D84630	III	NB	*Bacillus thuringiensis* (99.7)	filamentous	710–1030	10–15	ssDNA
S23335a	AB496409	III	NB	*Bacillus thuringiensis* (99.8)	filamentous	1090–1150	10–15	ssDNA
High G+C Gram-positive bacteria								
S23316	D84602	III	NB	*Agromyces ramosus* (99.0)	–			
S21004	D84562	I	NB	*Arthrobacter humicola* (99.9)	–			
S22236	AB496410	II	NB	*Arthrobacter humicola* (99.9)	filamentous	550–650	10–15	ssDNA
S21003	D84561	I	NB	*Arthrobacter oryzae* (99.7)	–			
S21011	D84563	I	NB	*Arthrobacter oryzae* (99.9)	–			
S21020	D84566	I	NB	*Arthrobacter oryzae* (99.9)	–			
S21022	D84567	I	NB	*Arthrobacter oryzae* (99.9)	–			
S21106	D84573	I	NB	*Arthrobacter oryzae* (99.8)	–			
S21114	D84575	I	NB	*Arthrobacter oryzae* (99.9)	–			
S21130	D84578	I	NB	*Arthrobacter oryzae* (99.9)	–			
S21135	AB496411	I	NB	*Arthrobacter oryzae* (99.1)	–			
S22227	D84588	I	NB	*Arthrobacter oryzae* (99.9)	–			
S22232	D84591	I	NB	*Arthrobacter oxydans* (99.1)	–			
S22237	D84595	I	NB	*Arthrobacter oxydans* (99.2)	–			
S22242	D84596	II	NB	*Arthrobacter sulfonivorans* (98.6)	filamentous	1000–1100	10–15	ssDNA
S23334a	AB496412	III	NB	*Actinoplanes digitatis* (98.7)	–			
S23338	D84612	III	NB	*Oryzihumus leptocrescens* (96.2)	–			
S23437	D84628	III	NB	*Oryzihumus leptocrescens* (96.6)	–			
S23409	D84618	III	NB	*Phycicococcus dokdonensis* (99.7)	–			
S23422	D84622	III	NB	*Phycicococcus dokdonensis* (99.7)	filamentous	1020–2260	10–15	ssDNA
S23430	D84624	III	NB	*Phycicococcus dokdonensis* (99.7)	–			
S22243	D84597	II	NB	*Terrabacter terrae* (98.8)	–			
S23303	D84599	III	NB	*Terracoccus luteus* (98.0)	–			
S24526	D84638	IV	DNB	*Leifsonia ginsengi* (96.8)	–			
S23405	D84616	III	DNB	*Nocardioides kribbensis* (96.9)	–			

[a]16S rRNA, partial sequence
[b]Groups I to IV according to the colony forming curve (Gorlach et al. 1994)
[c]NB and DNB indicate the bacterium that can and cannot grow on one-strength DB medium, respectively
[d]–indicates the absence of prophage in the chromosome

roles in the lysogenic state might be integral to understanding the life of oligotrophs in soil.

7.7 Characteristics and Diversity of Phages in Rice Field Soils – Estimation by *g23* Gene Sequences of T4-Type Phages

To evaluate the phylogenetic characteristics and diversity of phages in rice field soils, *g23* genes of T4-type phages, *Myoviridae*, were PCR-amplified and sequenced from DNA extracted from rice field soils in Japan and Northeast China (Fig. 7.4). The *g23* genes in the floodwater were also studied by using the same

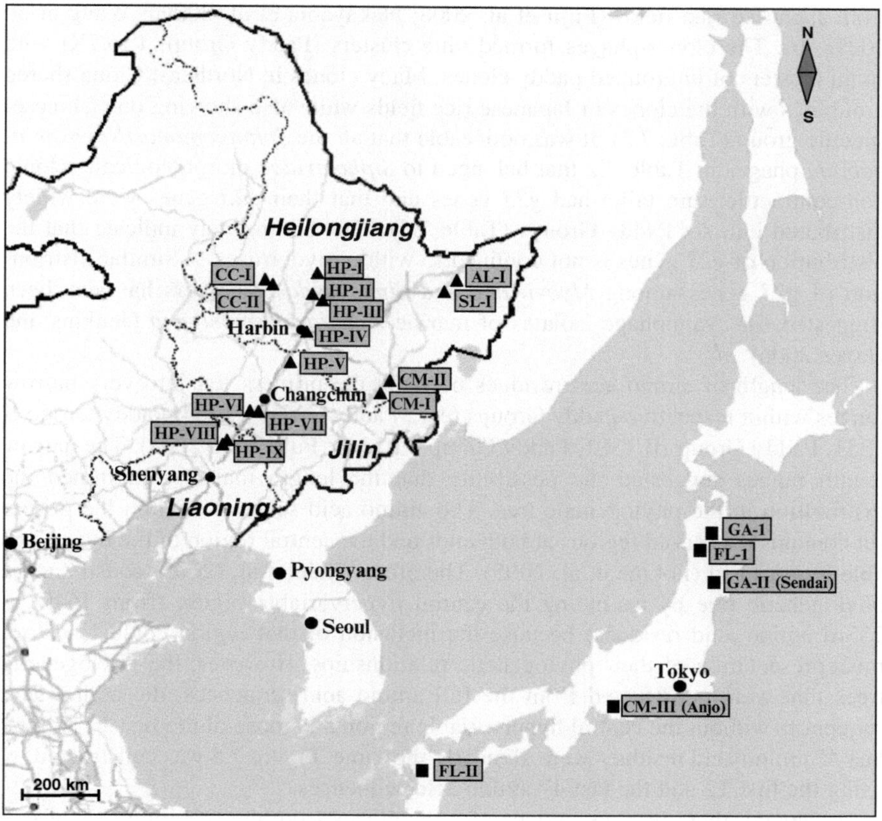

Fig. 7.4 Locations of study rice fields in Japan and Northeast China designated by the soil types. *AL* Albic Luvisols, *CC* Calcic Chernozems, *CM* Cambisols, *FL* Fluvisols, *GA* Gleyic Andisols, *HP* Haptic Phaeozems, *SL* Solonchaks

floodwater samples as in Sect. 7.2 and the phages isolated from *Sphingomonas* and *Novosphingobium* in Sect. 7.4. The primers used for the *g23* amplification were MZIA1bis and MZIA6 (Filée et al. 2005). Denaturing gradient gel electrophoresis (DGGE) was conducted on the PCR products to efficiently obtain different *g23* clones.

Deduced amino acid residues of the partial *g23* fragments were aligned first together with the representatives of T-evens, PseudoT-evens, SchizoT-evens and ExoT-evens as well as marine *g23* clones obtained by Filée et al. (2005). The majority of soil *g23* sequences were distantly related to the T-evens sequences and those of marine origin (Table 7.5) and formed several independent clusters (Fujii et al. 2008; Nakayama et al. 2009a, c; Wang et al. 2009a, b). T4-type phage communities in rice fields consisted of previously uncharacterised members phylogenetically distant from those found in marine environments.

Figure 7.5 shows the neighbour-joining phylogenetic tree giving the relationships of *g23* amino acid sequences from Chinese *g23* clones with those obtained from Japanese rice fields (Fujii et al. 2008; Nakayama et al. 2009a; Wang et al. 2009a, b). The clones/phages formed nine clusters (Paddy Groups I to IX) with small clusters of ungrouped paddy clones. Many clones in Northeast China shared groupings with the clones in Japanese rice fields while also showing one Chinese-specific group (Table 7.5). It was noticeable that all the *Sphingomonas*/*Novosphingobium* phages in Table 7.2 that belonged to *Siphoviridae* morphologically (long, non-contractile, thin tails) had *g23* genes and that their *g23* genes were widely distributed into six Paddy Groups (Table 7.5). This finding may indicate that the distribution of *g23* genes is not confined to within myoviruses. A similar distribution of *g23* genes among *Myoviridae* and *Siphoviridae* members has also been suggested for cyanophage isolates of marine *Nodularia spumigena* (Jenkins and Hayes 2006).

The length of amino acid residues between the primers fell into very narrow ranges within respective Paddy Groups (e.g., Paddy Group I 1264, Paddy Group II 1351, Paddy Group III 1381, Paddy Group VI 1344; Fujii et al. 2008). The narrow length ranges suggested the possibility that the length mainly determined the distribution in the phylogenetic tree. The amino acid sequence within the primer set contains conserved regions at both ends and the central region of the hypervariable loop capsid (Fokine et al. 2005). Therefore, Filée et al. (2005) constructed a phylogenetic tree by excluding the central hypervariable region (from 164th to 233rd amino acid residues) because the inclusion of that region could lead to a misrepresentation of their phylogenetic relationships. However, the phylogenetic trees that were constructed from the full amino acid sequences, the amino acid sequences without the central hypervariable region and those of the first 12 and the last 47 amino acid residues were all nearly the same. Figure 7.5 was constructed by using the first 12 and the last 47 amino acid sequences.

In general, there was no tendency for the clones from the specific types of soil and fertilisers, the stage during rice cultivation, or the location, at least within Japan or Northeast China, to belong to the specific Paddy Groups, indicating that they were not the major factors in determining T4-type phage communities.

Table 7.5 Grouping of *g23* clones that were obtained from various rice fields in Japan and Northeast China

	Total	PG-I	PG-II	PG-III	PG-IV	PG-V	PG-VI	PG-VII	PG-VIII	PG-IX	UG	MG	T-evens	bp range[a]	
Anjo	17	3	0	0	2	1	5	0	0	0	0	2	4	366–564	Jia et al. (2007)
Daisen	39	7	6	4	3	4	10	0	1	0	4	0	0	372–594	Fujii et al. (2008)
Kuroishi, Anjo, Chikugo	56	5	5	6	4	7	7	0	4	6	5	0	7	300–549	Wang et al. (2009a)
Northeast China	53	7	0	0	0	15	0	9	4	6	8	4	0	348–450	Wang et al. (2009b)
Anjo	58	8	6	13	9	4	18	0	0	0	0	0	0	371–479	Nakayama et al. (2009a)
Anjo	40	6	5	6	6	1	3	0	5	1	1	2	4	345–531	Nakayama et al. (2009b)
Nagakute, Ichinomiya	97	12	0	9	0	16	6	0	14	22	12	3	3	357–531	Wang et al. (2009c)
Sum	360	48	22	38	24	48	49	9	28	35	30	11	18		

PG Paddy Group, *UG* Ungrouped, *MG* Marine Group, T-evens: T-, PseudoT-, SchizoT- and ExoT-evens

[a]It shows the range of the length of nucleotide sequences between the primers

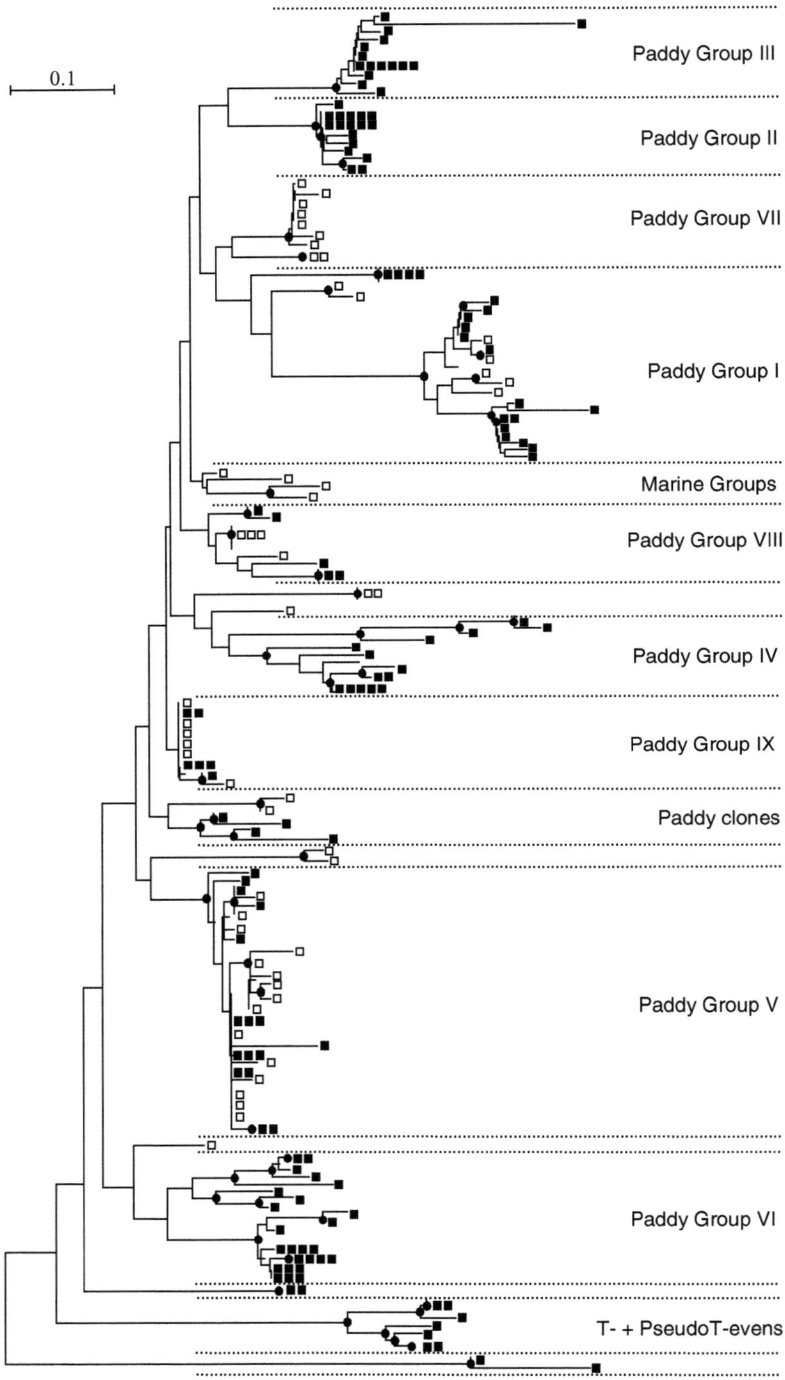
0.1
Paddy Group III
Paddy Group II
Paddy Group VII
Paddy Group I
Marine Groups
Paddy Group VIII
Paddy Group IV
Paddy Group IX
Paddy clones
Paddy Group V
Paddy Group VI
T- + PseudoT-evens

Short and Suttle (2005) reported that nearly identical (>99%) bacteriophage *g20* sequences were recovered from marine and freshwater environments, suggesting that either closely related hosts and the viruses infecting them were distributed widely across environments, or that horizontal gene exchange occurs among phage communities in very different environments. Similar clues suggesting the horizontal gene exchange/transfer were obtained in the *g23* sequence analysis. First, the identical *g23* sequences at the nucleotide level were observed in two cases among the phages having genome sizes of 60 and 160 kb, as well as among some phages with isometric and elongated capsids in the Anjo (CM-III) rice field. Furthermore, identical *g23* sequences at the nucleotide level were also observed between the Anjo (CM-III) field and the FL-I field (Nakayama et al. 2009a). In addition, several cases of identical *g23* sequences at the nucleotide level were observed among the clones obtained from the Anjo (CM-III) and GA-I rice fields in Japan and from the CC-I, HP-II, HP-V and HP-VII rice fields in Northeast China (Nakayama et al. 2009a; Wang et al. 2009a, b).

The unrooted phylogenetic tree demonstrated that the majority of *g23* fragments from rice fields (soils and floodwater) formed several clusters independent from those derived from marine environments (Fig. 7.6). The marine groups were distributed narrowly in the tree, which indicated that *g23* genes in the paddy fields are more divergent in comparison with those in the marine environment.

7.8 Changes in Major Capsid Genes (*g23*) of T4-Type Bacteriophages with Soil Depth

Microbial communities in soil are well known to change with soil depth (i.e., Fierer et al. 2003; Agnelli et al. 2004). The soil depth profiles of T4-type phage communities were studied in two Japanese rice fields near Anjo City, and *g23* clones in soil DNA extracts to a depth of 1 m were examined (Wang et al. 2009c).

Finer and coarser soil layers were stratified in the subsoil in both soil profiles, and the coarse (loamy coarse sand) layers contained less organic matter and showed smaller CEC than the finer layers. Rice roots were observed to have developed densely in the A and B layers (the surface 20 cm in the Nagakute field and 30 cm in the Ichinomiya field).

Similar banding patterns of DGGE fingerprints were observed between the A and B layers (termed the rooting layers) and among the layers below them

Fig. 7.5 Neighbour-joining phylogenetic tree showing the relationships of *g23* amino acid sequences in Chinese rice fields with those obtained from Japanese rice fields (Fujii et al. 2008; Nakayama et al. 2009a; Wang et al. 2009a, b) by using the conserved regions in the *g23* clones. The symbol *filled circle* indicates internal nodes with at least 50% bootstrap support. Clones obtained from Chinese and Japanese rice fields are shown with the symbols *open square* and *filled square* (one symbol per clone), respectively

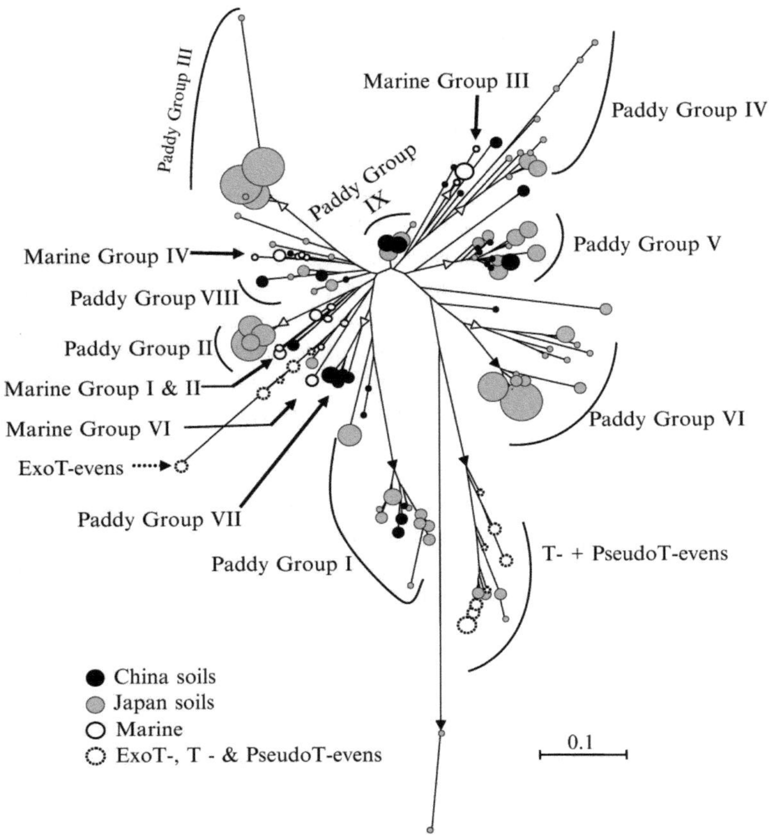

Fig. 7.6 Unrooted phylogenetic tree comparing *g23* sequences obtained from Japanese and Chinese rice fields with those of T-evens, PseudoT-evens, SchizoT-evens, ExoT-evens and marine origins. The *closed* and *open triangles* indicate internal nodes with at least 90% and 50% bootstrap supports, respectively. The size of *circles* at the end of branches is proportional to the number of clones/phages, and the *smallest* and *largest circles* represent one and seven clones/phages, respectively. The *scale bar* represents the number of amino acid substitutions per residue

(termed the subsoil layers) for both soil profiles. There was no difference in the banding patterns between coarse textured layers and fine ones in the subsoil for both soil profiles. Ninety-seven *g23* clones were sequenced in total, among which 28 and 23 clones were from the rooting layers and 21 and 25 clones from the subsoil layers in the Nagakute and Ichinomiya profiles, respectively.

As shown in Fig. 7.7, the majority of *g23* clones formed several clusters independent from those of enterobacteria and marine origins. Generally, *g23* clones in the rooting layers and in the subsoil layers formed their own clusters away from each other (Fig. 7.7). In addition, *g23* clones in the rooting layers formed different clusters between the Nagakute and Ichinomiya profiles, while such differences were not observed between the profiles in the subsoil layers (Fig. 7.7). Thus, T4-type

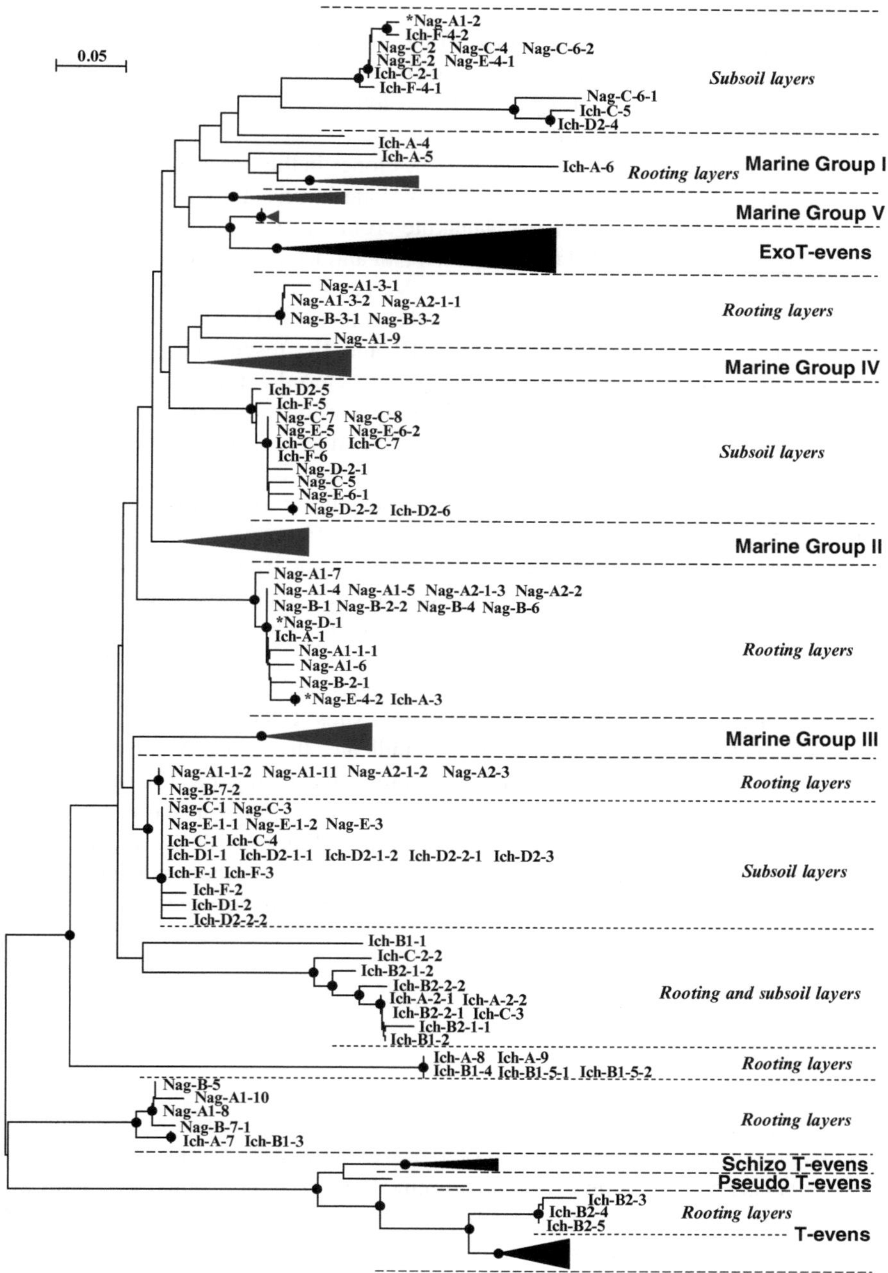

Fig. 7.7 Neighbour-joining phylogenetic tree showing the relationships of *g23* amino acid sequences in the soil layers with those of enterophages and of phages of marine origin (Filée et al. 2005). Clone names with *asterisk* indicate that they are an exception in the rooting or subsoil layers. The symbol *filled circle* indicates internal nodes with at least 50% bootstrap support. The *scale bar* represents the abundance of amino acid substitutions per residue

phage communities of rooting layers were different from those of subsoil layers. The development of rice roots seemed to be an important factor determining the T4-type phage community in the surface soil layers. Since both subsoils consisted of coarse or fine soil layers, it is reasonable to hypothesise that soil texture did not affect T4-type phage communities in both profiles.

The phylogenetic analysis revealed that the clones in the soil layers distributed to Paddy Groups I, III, V, VI, VIII and IX. No clone in this study fell within Paddy Group II, IV and VII clusters. In the Nagakute profile, *g23* clones obtained from the rooting layers belonged to Paddy Groups I, V, VI and IX, and those from the subsoil layers were grouped into Paddy Groups I, V, VIII and IX. In the Ichinomiya profile, the clones from the rooting layers and the subsoil layers belonged to Paddy Groups III, V and VI and Paddy Groups I, III, VIII and IX, respectively. This finding suggested that root development may change eubacterial communities in the rooting layers, with resultant changes in *g23* composition (T4-type phage communities). Several clones were identical at the nucleotide level with the clones in the AL-I, HP-II, HP-V and HP-VII rice fields (Wang et al. 2009c).

In the evolutionary history of *g23* genes in the environment, two hypotheses are conceivable: one being that a common ancestral *g23* gene primitively spread to marine and terrestrial environments with later adaptation by exchange, deletion and addition of amino acid moieties in the gene. The other hypothesis is that many *g23* genes emerged separately in different environments and evolved independently afterwards. The *g23* genes in all the clones and phages obtained from marine and soil environments as well as from enterophages possess the conserved regions, with the same length at both ends between the primer sites (Filée et al. 2005; Fujii et al. 2008; Wang et al. 2009a, b), which strongly indicates that *g23* genes in marine and soil environments were descended from a common ancestral gene. The time of their spread to soil and marine environments might be old enough in the evolutionary history of *g23* genes to experience independent evolution, with the adaptation of *g23* genes to their respective environments resulting in the sequence changes of the conserved regions and the changes in length and structure between the conserved regions.

7.9 Conclusions

Viruses are the most abundant biological entities in rice fields, as found elsewhere, and phages comprised the majority among viral communities. Floodwater phages of *Sphingomonas* and *Novosphingobium*, dominant bacterial members in the floodwater, were exclusively siphoviruses with various host ranges. The high frequency of phage-infected bacterial cells in the floodwater indicated that the bacterial mortality from phage lysis could be significant enough to redirect the microbial food web and induce the change and succession of the bacterial communities. There were no significant differences in the frequency of lysogeny between oligotrophs and copiotrophs obtained from rice field soil, although slower growth due to the burden of additional prophage DNA synthesis was indicated among the strains at the genus or species level. Although lysogeny occurred commonly in rice fields, the effect of superinfection immunity was not observed, probably due to the phylogenetic differences in prophages from the virulent phages in soil environment. Thus, the hosts of phages are ubiquitous in soil environments. The majority of the *g23* sequences of T4-type phages in rice fields were distantly related to the T-evens sequences and those of marine origins, which indicated that T4-type phage communities in rice fields consist of previously uncharacterised members phylogenetically distant from those in marine environments. The occurrence of horizontal gene exchange/transfer was suggested by the presence of identical *g23* sequences in distant rice fields (Japan and Northeast China). The *g23* genes in paddy fields seem to have diverged more compared with marine *g23* genes. These results indicated the ecological importance of viruses in nutrient cycles in soils. The many uncharacterised *g23* genes and their large diversification strongly suggested that soil environments are great genomic reservoirs of viruses.

References

Ackermann HW (2001) Frequency of morphological phage descriptions in the year 2000. Arch Virol 146:843–857

Ackermann HW, DuBow MS (1987) Viruses of prokaryotes, vol. 2. Natural groups of bacteriophages. CRC Press Inc, Boca Raton, FL

Agnelli A, Ascher J, Corti G, Ceccherini MT, Nannipieri P, Pietramellara G (2004) Distribution of microbial communities in a forest soil profile investigaged by microbial biomass, soil respiration and DGGE of total and extracellular DNA. Soil Biol Biochem 36:859–868

Binder B (1999) Reconsidering the relationship between virally induced bacterial mortality and frequency of infected cells. Aquat Microb Ecol 18:207–215

Børsheim KY (1993) Native marine bacteriophages. FEMS Microbiol Ecol 102:141–159

Desplats C, Krisch HM (2003) The diversity and evolution of the T4-type bacteriophages. Res Microbiol 154:259–267

Dillon A, Parry JD (2008) Characterization of temperate cyanophages active against freshwater phycocyanin-rich *Synechococcus* species. Freshwater Biol 53:1253–1261

Fauquet CM, Mayo MA, Maniloff J, Desselberger U, Ball KA (eds) (2005) Virus taxonomy: eighth report of the international committee on taxonomy of viruses. Elsevier Academic Press, London

Fierer N, Schimel JP, Holden PA (2003) Variations in microbial community composition through two soil depth profiles. Soil Biol Biochem 35:167–176

Filée J, Tétart F, Suttle CA, Krisch HM (2005) Marine T4-type bacteriophages, a ubiquitous component of the dark matter of the biosphere. Proc Natl Acad Sci USA 102:12471–12476

Fischer U, Velimirov B (2002) High control of bacterial production by viruses in a eutrophic oxbow lake. Aquat Microb Ecol 27:1–12

Fokine A, Leiman PG, Shneider MM, Ahvazi B, Boeshans KM, Steven AC, Black LW, Mesyanzhinov VV, Rossmann MG (2005) Structural and functional similarities between the capsid proteins of bacteriophages T4 and HK97 point to a common ancestry. Proc Natl Acad Sci USA 102:7163–7168

Frost LS, Leplae R, Summers AO, Toussaint A (2005) Mobile genetic elements: the agents of open source evolution. Nat Rev Microbiol 3:722–732

Fuhrman JA (1999) Marine viruses and their biogeochemical and ecological effects. Nature 399:541–548

Fujii T, Nakayama N, Nishida M, Sekiya H, Kato N, Asakawa S, Kimura M (2008) Novel capsid genes (*g23*) of T4-type bacteriophages in a Japanese paddy field. Soil Biol Biochem 40:1049–1058

Gorlach K, Shingaki R, Morisaki H, Hattori T (1994) Construction of eco-collection of paddy field soil bacteria for population analysis. J Gen Appl Microbiol 40:509–517

Jenkins CA, Hayes PK (2006) Diversity of cyanophages infecting the heterocystous filamentous cyanobacterium *Nodularia* isolated from the brackish Baltic Sea. J Mar Biol Ass UK 86:529–536

Jia Z, Ishihara R, Nakajima Y, Asakawa S, Kimura M (2007) Molecular characterization of T4-type bacteriophages in a rice field. Environ Microbiol 9:1091–1096

Jiang SC, Paul JH (1994) Seasonal and diel abundance of viruses and occurrence of lysogeny/ bacteriocinogeny in the marine environment. Mar Ecol Prog Ser 104:163–172

Kimura M, Jia Z, Nakayama N, Asakawa S (2008) Ecology of viruses in soils: past, present, and future perspectives. Soil Sci Plant Nutr 53:1–32

Mann NH (2003) Phages of the marine cyanobacterial picophytoplankton. FEMS Microbiol Rev 27:17–34

Mann NH (2005) The third age of phage. PLoS Biol 3:753–755

Marsh P, Wellington EMH (1994) Phage–host interactions in soil. FEMS Microbiol Ecol 15:99–108

Nakayama N, Asakawa S, Kimura M (2009a) Comparison of *g23* gene sequence diversity between *Novosphingobium* and *Sphingomonas* phages and phage communities in the floodwater of a Japanese paddy field. Soil Biol Biochem 41:179–185

Nakayama N, Asakawa S, Kimura M (2009b) Frequency of phage-infected bacterial cells in the floodwater of a Japanese paddy field. Soil Biol Biochem 41:186–191

Nakayama N, Okabe A, Toyota K, Kimura M, Asakawa S (2006) Phylogenetic distribution of bacteria isolated from the floodwater of a Japanese paddy field. Soil Sci Plant Nutr 52:305–312

Nakayama N, Okumura M, Inoue K, Asakawa S, Kimura M (2007a) Seasonal variations in abundances of virus-like particles and bacteria in the floodwater of a Japanese paddy field. Soil Sci Plant Nutr 53:420–429

Nakayama N, Okumura M, Inoue K, Asakawa S, Kimura M (2007b) Abundance of bacteriophages of common heterotrophic bacteria in the floodwater of a Japanese paddy field. Soil Sci Plant Nutr 53:595–605

Nakayama N, Okumura M, Inoue K, Asakawa S, Kimura M (2007c) Morphological analysis of viral communities in the floodwater of a Japanese paddy field. Soil Biol Biochem 39:3187–3190

Nakayama N, Tsuge T, Asakawa S, Kimura M (2009c) Morphology, host range and phylogenetic diversity of *Sphingomonas* phages in the floodwater of a Japanese paddy field. Soil Sci Plant Nutr 55:53–64

Paul JH, Sullivan MS (2005) Marine phage genomics: what have we learned? Curr Opin Biotech 16:299–307

Proctor LM, Okubo A, Furhman JA (1993) Calibrating estimates of phage-induced mortality in marine bacteria: ultrastructural studies of marine bacteriophage development from one-step growth experiments. Microb Ecol 25:161–182

Short CM, Suttle CA (2005) Nearly identical bacteriophage structural gene sequences are widely distributed in both marine and freshwater environments. Appl Environ Microbiol 71:480–486

Tétart F, Desplats C, Kutateladze M, Monod C, Ackermann HW, Krisch HM (2001) Phylogeny of the major head and tail genes of the wide-ranging T4-type bacteriophages. J Bacteriol 183:358–366

Van Etten JL, Lane KC, Meints RH (1991) Viruses and viruslike particles of eukaryotic algae. Microbiol Rev 55:586–620

Wang G, Hayashi M, Saito M, Tsuchiya K, Asakawa S, Kimura M (2009a) Survey of major capsid genes (*g23*) of T4-type bacteriophages in Japanese paddy field soils. Soil Biol Biochem 41:13–20

Wang G, Jin J, Asakawa S, Kimura M (2009b) Survey of major capsid genes (*g23*) of T4-type bacteriophages in rice fields in Northeast China. Soil Biol Biochem 41:423–427

Wang G, Murase J, Taki K, Ohashi Y, Yoshikawa N, Asakawa S, Kimura M (2009c) Changes in major capsid genes (*g23*) of T4-type bacteriophages with soil depth in two Japanese rice fields. Biol Fertil Soils 45:521–529

Weinbauer MG (2004) Ecology of prokaryotic viruses. FEMS Microbiol Rev 28:127–181

Weinbauer MG, Brettar I, Hofle M (2003) Lysogeny and virus-induced mortality of bacterioplankton in surface, deep, and anoxic waters. Limnol Oceanogr 48:1457–1465

Weinbauer MG, Rassoulzadegan F (2004) Are viruses driving microbial diversification and diversity? Environ Microbiol 6:1–11

Weinbauer MG, Suttle CA (1996) Potential significance of lysogeny to bacteriophage production and bacterial mortality in coastal waters of the Gulf of Mexico. Appl Environ Microbiol 62:4374–4380

Williamson KE, Radosevich M, Smith DW, Wommack KE (2007) Incidence of lysogeny within temperate and extreme soil environments. Environ Microbiol 9:2563–2574

Wommack KE, Colwell RR (2000) Virioplankton: viruses in aquatic ecosystems. Microbiol Mol Biol Rev 64:69–114

Chapter 8
Phage Biopesticides and Soil Bacteria: Multilayered and Complex Interactions

Antonet M. Svircev, Susan M. Lehman, Peter Sholberg, Dwayne Roach, and Alan J. Castle

8.1 Phages as Biopesticides

Control of plant pathogens in agriculture faces many unique challenges. These include development of pathogen resistance to conventional pesticides, absence of resistant plant host material, the requirement for environmentally friendly alternatives to traditional pesticide, and the gradual regulatory removal of traditional chemicals used for plant pathogen control. The resurgence in research on the development of bacteriophage-based control programs is evident in the inspection of current literature. Debate continues on the suitability of phages as biological control agents (BCAs) (Stewart 2001). Recently, phages have been tested as BCAs or biopesticides for the control of bacterial spot of tomato (Flaherty et al. 2000; Balogh et al. 2003; Obradovic et al. 2004; Obradovic et al. 2005; Jones et al. 2005; Balogh et al. 2005; Iriarte et al. 2007), bacterial wilt of tomato (Kumar et al. 2006; Yamada et al. 2007); bacterial blight of geranium (Flaherty et al. 2001), citrus canker (Balogh et al. 2008), fire blight in pear and apple (Schnabel et al. 1999; Schnabel and Jones 2001; Svircev et al. 2002a; Gill et al. 2003; Svircev et al. 2006; Lehman 2007), soft-rot of calla lilies (Svircev et al. 2002b; Ravensdale et al. 2007), leaf blight of onion (Lang et al. 2007), seed treatments (Goyer 2005), and bacterial

A.M. Svircev (✉)
Agriculture and Agri-Food, Victoria Ave. North, Vineland Station, ON L0R 2G0, Canada
e-mail: Antonet.Svircev@agr.gc.ca
S.M. Lehman
Division of Healthcare Quality Promotion, Centers for Disease Control and Prevention, 1600 Clifton Rd NE, Mail Stop C-16, Atlanta, GA 30333, USA
P. Sholberg
Agriculture and Agri-Food Canada, Pacific Agri-Food Research Centre, Summerland, BC V0H 1Z0, Canada
D. Roach and A.J. Castle
Department of Biological Science, Brock University, 500, Glenridge Avenue St. Catharines, ON L2S 3A1, Canada

G. Witzany (ed.), *Biocommunication in Soil Microorganisms*, Soil Biology 23,
DOI 10.1007/978-3-642-14512-4_8, © Springer-Verlag Berlin Heidelberg 2011

spot of peach (Zaccardelli et al. 1992; Saccardi et al. 1993). Jones et al. (2007) provide a thorough and up-to-date review on the use of bacteriophages as agricultural biopesticides. In addition, the authors discuss the advantages, challenges, and novel approaches that are utilized for the integration of phages into modern diseases control programs.

8.1.1 Aerial Application of Phage Biopesticides and Impact on Soil Ecology

The aerial surfaces of plants provide a unique habitat for bacterial microorganisms. The complex interactions between these organisms as a whole are only beginning to be examined and studied (Andrews and Harris 2000). In the apple and pear orchard, *Erwinia amylovora* is the pathogen responsible for a necrotic wilt disease commonly named fire blight (Vanneste 2000). Fire blight biological control programs have been developed and implemented with commercially available bacterial BCAs (Johnson and Stockwell 2000). While *E. amylovora* is commonly present in the aerial portions of the canopy, it has been isolated from the orchard soil and a soil-inhabiting microarthropod (Hildebrand et al. 2001). The authors postulate that the insect may be responsible for reducing the pathogen population in the soil. Little is understood on the impact of *Erwinia* spp. phages on the removal of *E. amylovora* from soil. However, majority of the phages isolated by Gill et al. (2003) originated from the soil in the immediate vicinity of active fire blight infections.

The aerial application of phage biopesticide for the control of the fire blight pathogen occurs in the spring when the blossoms are fully open (Lehman 2007). Biopesticides are commonly applied at 0–25%, 25–50%, and 75–100% bloom, to help establish the BCA populations ahead of the pathogen and prevent the ingression of the pathogen *E. amylovora* into the host via the flower cup or hypanthium. In the phage-based biopesticide program, the common orchard epiphyte *Pantoea agglomerans* performs as a phage "carrier" and incubator. The carrier and phage are simultaneously applied to the opened blossoms at multiplicity of infection (MOI=1) or 1:1 ratio of bacteria:phage at 10^8 CFU and PFU/ml. Real-time PCR technology was used to monitor in situ the phage, carrier, and the artificially introduced pathogen populations during field trials (Lehman 2007). Disease control was obtained in in vivo bioassays and field trials when phages were able to decrease the population of the pathogen below 10^4 CFU/ml (Lehman 2007).

The impact of the aerial application of phages on the orchard soil ecology is poorly understood. Biological control programs generally focus on understanding the interaction(s) between the biological agent and the pathogen ignoring the more complex interactions of pathogen–epiphyte–BCA–plant-associated microorganisms on aerial and soil surfaces. Studying the impact of the phage biopesticide on the soil ecosystem is hampered by the lack of specific technologies. The techniques available for the study of phages in soil ecosystems are further discussed later in this chapter. We cannot distinguish between the applied phage-carrier and the indigenous phage populations. However, molecular techniques that follow overall populations of phage, *P. agglomerans* and *E. amylovora* have been developed (Lehman 2007) and serve an important role in following the population dynamics in the flower and, potentially, in the orchard soil. The yearly application of phage and/or bacterial biologicals may alter the ecology of the orchard soil but, to date, there are no reported studies that study directly the impact of phage biopesticides on the soil microbial communities.

8.2 Phage Biopesticides in Greenhouse Soils: Control of *Pectobacterium carotovorum*

The use of bacteriophages to control disease in greenhouse soil mixtures creates unique challenges since the plant pathogen and biopesticide interaction takes place on soil particles in an arid-aqueous environment. Soils and water solutes, including fertilizers, affect the survival of phage biopesticides and their interactions with the pathogen. In this section, we look at the use of bacteriophages that inhibit *Pectobacterium carotovorum* soft-rot as a model system for the biocontrol of greenhouse diseases and the problems that are encountered during the development of such a biocontrol system.

Pectobacterium carotovorum subsp. *carotovorum* (*Pcc*, Jones 1901) Hauben et al. 1999 (= *Erwinia carotovora* subsp. *carotovora*) and its relatives, most notably *P. carotovorum* subsp. *atroseptica*, are the causative agents of soft-rot disease of numerous plant species. These bacteria are common in surface waters (Harrison et al. 1987), are distributed worldwide, and in the case of *Pcc* show little host specificity (Pérombelon and Kelmon 1980). These characteristics contribute to significant crop damage in a broad range of host species.

Ravensdale et al. (2007) were the first group to assess the feasibility of using bacteriophages as BCAs for soft-rot caused by *Pcc*. Fourteen bacteriophage isolates, classified by morphology to the order *Caudovirales*, have host ranges specific to *Pcc* isolates from calla lilies. *Pcc* is a devastating pathogen of calla lilies (*Zantedeschia* spp.) and can cause anywhere from <3% to complete loss of greenhouse crops (Blom and Brown 1999).

Prior to greenhouse trials, Ravensdale et al. (2007) tested phage survival in preplant-treatment and fertilizer solutions. Phages were completely inactivated by 48 h in three preplant-treatment solutions. The mechanism of inactivation is unknown but may be attributed to deleterious effects of copper (Sagripanti 1992) or nonionic surfactants (Chattopadhyay et al. 2002). Fertilizer solutions did not inhibit the phages, while the choice of water source was critical for the phage-based biopesticide. Reverse osmosis removes an inhibitory agent in tap water, most likely chlorine (Berg et al. 1989) or a cation scrubbed from piping. Later work on bacteriophages that attack *Erwinia amylovora* showed that EDTA amendment increased phage survival in sterile tap water thus supporting the "cation inhibitor" hypothesis (Svircev and Castle, unpublished data).

Three tests for the effects of the phages on *Pcc* populations were reported by Ravensdale et al. (2007). These tests included inhibition in liquid media, on tissue plugs from calla lily tubers and on intact calla lilies grown in the greenhouse. The tests in liquid media included phage and *Pcc* in nutrient broth (Ravensdale et al. 2007), in fertilizer solution containing Fe-EDTA or in fertilizer solution lacking Fe-EDTA. The phages reduced *Pcc* populations in nutrient broth and fertilizer lacking Fe-EDTA but were ineffective in the presence of Fe-EDTA. The authors proposed two plausible mechanisms for this inhibitory effect. First, bacteriophage replication is dependent upon the activity of bacteriophage-encoded ribonucleotide reductase, an iron-dependent enzyme. An iron chelator such as EDTA could reduce effective, unbound iron concentrations within the host cell thereby inhibiting bacteriophage propagation (Romeo et al. 2001). Alternatively, free EDTA could bind other essential cations such as magnesium or calcium that stabilize many phages.

Several bacteriophages, alone or in mixtures, reduce soft-rot of calla lilies by up to 70% in greenhouse trials. These tests indicated that bacteriophages show promise as biocontrol agents and warrant further study. Several routes for optimization of biocontrol efficacy could be followed. For example, MOI of 100 was used in these studies. Higher MOIs may give better control by allowing greater diffusion of phages within the soil. Bacteriophages can diffuse up to 10 cm in porous soils and can persist at relatively stable concentrations for several weeks (Assadian et al. 2005). Lower effective MOIs may be achieved with phages that are difficult to amplify but nevertheless give promising biocontrol results in laboratory conditions. Testing biocontrol efficacy for crops in different soils would determine the overall applicability of these agents. Different soils also generate a new set of problems to be addressed, including the effects of various microbial populations on bacteriophage propagation and survival and sorption of applied phages to particulate matter (Chattopadhyay et al. 2002).

The Ravensdale et al. (2007) study gives a glimpse of the broad spectrum of factors that should be considered when developing a new bacteriophage-based biocontrol agent for bacterial diseases of greenhouse and, by extension, field crops. One must test all potential factors such as preplant treatment, fertilizer, water source, or surfactants that may come in contact with the agent. Soil type, delivery vehicle, multiplicity of infection, and timing of application should also be considered for optimal results. Despite this complexity, however, useable biocontrol agents may be realized even at intermediate developmental stages.

8.3 Phages and Rhizobacteria

Soil and plant-associated phages have been reviewed before with different focuses each time. Brüssow and Kutter (2005a) stated that the work is in its infancy compared with phage research in marine environments. They see the plant rhizosphere as an important area of study for bacteriophages in the soil environment especially regarding bacterial populations and gene transfer. The review by Kimura et al. (2008) is the most up-to-date on strictly soil-associated phages and provides a good ecological summary with references of all that is currently known about phages in the soil environment. The review by Weinbauer (2004) is an all encompassing ecological view of prokaryotic phages and presents a detailed picture of phage evolution, their interactions between one another and their bacterial hosts, and their potentially huge influence on food webs. Three other reviews that should be mentioned in relation to soil-associated bacteriophages are the reviews by Marsh and Wellington (1994) on phage–host interactions in soil, the review on plant interactions by Gill and Abedon (2003) and the recent review on plant disease control by Jones et al. (2007).

8.3.1 Effect of Soil on Phages

The soil has a profound effect on the microflora that lives in it and indirectly on the phages that parasitize these organisms. Some important factors that affect viral inactivation and infectivity are temperature, pH, clay type, organic matter, heavy metals, acid pollutants, aerobicity, ionic strength (Assadian et al. 2005), and moisture content (Kimura et al. 2008). In general, lower temperatures resulted in longer periods of survival, longer latent periods, and reduced burst size of bacteriophages. Phages were not found in soils with a pH value lower than 6.1 despite the presence of potentially susceptible acidophilic hosts in these soils (Sykes et al. 1981). This could be because they are hydrophilic and their adsorption depends on the electrostatic properties of the soil surface which largely depends on pH. A variety of factors influence the binding of phages to soil particles. Kimura et al. (2008) list them as (1) type of clay minerals; (2) cation exchange capacity of the soil; (3) soil-associated and dissolved organic matter; (4) soil pH; (5) ionic strength and its constituent; and (6) type of phage. These parameters interact in negatively charged phages with soils having both negatively and positively charged sites. The net result depends on soil pH and ionic strength and its constituents that determine the binding force between phage and soil. This is influenced by organic matter which weakens the electrostatic binding of phages to soils. The ionic environment of the phage affects life cycle and survival. Divalent cations such as magnesium and/or calcium appear to have a positive affect whereas sodium has a variable affect.

Bacteriophages can move horizontally and vertically in soil for several meters even though soils are efficient adsorbers of them. In soil, desiccation and adsorption to soil colloids could be major reasons for phage destruction. On the other hand, some phages adsorb to colloids together with their hosts allowing the phage to find its host (Weinbauer 2004). An experiment in which host cells and a particular phage were introduced into the soil showed that host cells were lysed in the presence of added nutrients, and that the phage was more abundant in fertilized soils (Smit et al. 1996). Encapsulation of the host cells in alginate beads inhibited lysis by the phage in soil. The lysogenic lifestyle is a distinct advantage for bacteriophages in soil where long periods of host inactivity can be survived by residence within host populations (Marsh and Wellington 1994).

8.3.2 Phages in the Rhizosphere

The rhizosphere is that portion of the soil which is subject to the influence of the plant root system and supports greater microbial activity than soil more distant from roots (Katznelson et al. 1948). It is a unique zone exerting a powerful stimulation on soil microorganisms which varies with type, variety, age, and vigor of the plant and the type, treatment, and moisture content of the soil in which it grows. An important factor in the study of rhizosphere microorganisms is the rhizosphere:soil (R:S) ratio or the number of organisms in the rhizosphere soil divided by the number in the soil at a distance from the root. This ratio is of fundamental importance for evaluating the influence of soil type, treatment, and other factors on the root surface microflora (Katznelson et al. 1948). Phage can attack bacteria directly associated with the rhizosphere (Gracia-Garza et al. 2004). Phage ecology in the rhizosphere is thought to approximate solid-phase growth in laboratory media to some extent. It is speculated that if the soil is not disturbed, bacterial microcolonies within the rhizosphere display periods of boom or bust with regard to phage attack that is influenced by microcolony size which increases the likelihood of phage-microcolony encounter. Infection of one bacterium within a localized bacterial clone could result in the destruction of part or all of a genetically homogenous bacterial microcolony. However, bacteria have several ways to thwart such an attack such as variation in life cycle, motility, and sequestration away such as infection of a root nodule.

The challenge of using phages to control plant pathogens in the rhizosphere has been summarized by Jones et al. (2007) (Hagens and Offerhaus 2008). They listed several factors that can hinder success of disease control in the rhizosphere. The rate of diffusion through soil is low and changes as a function of available free water. Phages can become trapped in biofilms, adsorbed to clay particles, and inactivated by low soil pH. For these reasons, only a low number of viable phages are available to lyse target bacteria and a high population is needed to insure bacterial lysis. In soils, there are “hot spots” for microorganisms where they proliferate very actively (Kimura et al. 2008). These are the habitats around plant roots or the rhizosphere where the lytic life may be favorable to phages because the host turnover rate is short enough. The number of host cells required for bacteriophages to increase has been studied in natural ecosystems. Bacteriophages required a host cell population of at least 100–1,000 per g of soil in order to multiply (Germida 1986). Wiggins and Alexander (1985) found that the minimum density of a strain of *Bacillus subtilis* to increase the number of a specific bacteriophage was 3×10^4 CFU/ml. They concluded that phages do not affect the number of bacteria in environments where the density of the host species is below 10^4 CFU/ml. Research on the soil rhizosphere in specific crops has been conducted in barley, wheat, and sugar beets. The earliest research was on the rhizosphere soil of sugarbeets by Stephens et al. (1987). They found that a key factor in decline of a beneficial fluorescent pseudomonas strain were bacteriophages and phages antagonistic toward *Pseudomonas* strain B2/6 that was present in 43% of the soils from the major sugarbeet growing regions of Ireland. Five bacteriophages infecting *Pseudomonas fluorescens* and *P. putida* were isolated from barley rhizosphere soil (Campbell et al. 1995). Four of the phages belonged to the Myoviridae family with large isometrical heads on contractile tails and had complex protein and DNA profiles. The ecological importance of these phages could be their slow multiplication rates suggesting a possible mechanism of balanced phage–host coexistance in the rhizosphere. Slow multiplication might reflect complexity in formation of large bacteriophages. Free *Pseudomonas* spp. bacterial hosts in the rhizosphere of wheat were only slightly affected by the addition of bacteriophages, while cells escaping from alginate beads were effectively lysed (Blom and Brown 1999). It is hypothesized that the cells escaping from beads might be in a more active metabolic state allowing phage infection to occur. Phages could potentially infect cells from beads preventing them from colonizing wheat roots.

8.3.3 Effect of Phages on Root Nodulation

Root-nodule bacteria of leguminous plants are classified into two genera: *Rhizobium* and *Bradyrhizobium*. Inoculation of soybean with *B. japonicum* is often unsuccessful because the inoculum strains do not nodulate soybeans in the presence of indigenous strains. Studies have shown that it is possible to reduce nodulation with indigenous strains by amending the soil with a bacteriophage specific for the indigenous strain. Nodulation was increased from 48 to 82% by coating the seed with a phage and *B. japonicum* (Basit et al. 1992). Therefore, the elimination of a single strain of rhizobia from soil enhanced nodule occupancy by the inoculum strain in this case. Unfortunately soils often contain numerous strains of rhizobia, so phage coating of seed only has limited value. Nitrogen fixation by bacteria is influenced by phage infection. Kleczkowska (1957) found that as long as the phage is present, phage-resistant mutants are also present that may be more effective at nitrogen fixation. Novikova et al. (1993) provided evidence using phages specific to *R. loti* that *Rhizobium* strains nodulating *Astragalus*, *Hedysarum*, *Glycyrrhiza*, and *Ononis* plant species are related to each other.

8.3.4 Effect of Phages on Yield and Disease Control

Pseudomonas fluorescens, a plant growth promoting rhizobacterium (PGPR), is optimally infected by bacteriophages at 26°C (Sillankorva et al. 2006). The prevalence of bacteriophages in PGPR bacteria was investigated in four bacterial species belonging to the genus *Azospirillum* (Harrison et al. 1987). The study showed that there were many phages present because 11 strains out of 24 released phage particles. Moreover, each type of bacteriophage seemed to be associated with a specific bacterial species because only "big" phages were found for *A. brasilense* and "small" phages for *A. lipoferum* strains. A successful use of PGPR bacteria resistant to phage was given by Suslow (1986) with his patented technique in which yield of root crops such as potatoes, sugar beets, and radishes are increased. In this process, bacteriophage-resistant strains are applied to seeds or root pieces at concentrations of 10^5 to 10^9 cells/ml in an acceptable carrier medium. The technique is not without problems because the soil presents several obstacles to the successful use of bacteriophages-resistant strains. The use of phages for disease control is an expanding area of plant protection but a major problem has been the development of bacterial strains resistant to the phage. A patented process was developed to prevent occurrence of phage-resistant mutants (Hagens and Offerhaus 2008). Mixtures of mutant phages are prepared that are able to lyse bacterial strains that are resistant to the parent phage, while still capable of lysing the wild-type bacterium. This gives them an extended host compared with the parent strain and has been shown to be an effective strategy for phage application and disease control.

8.4 Lysogeny and Soil–Phage Interaction

Lysogenic replication by temperate bacteriophages involves the suppression of lytic functions by down-regulating specific gene expression to establish a quiescent state inside the host (Campbell 2006). The bacteriophage genome (Canchaya et al. 2004) is stably maintained as an integrated part of the bacterial chromosome or as an autonomous extrachromosomal element. Prophage replication is coordinated with host-genome replication and is present in the progeny of the lysogenic parent bacterium. It is latent and so its presence does not promote cell death or the production of bacteriophage virions. Reversion to the lytic functions occurs at a certain frequency in growing lysogenic populations, although prophage induction is usually caused by environmental signals that cause physiological stress to the host cell. The stress causes loss of expression of the repressor protein(s), subsequent reinstatement of lytic functions, and release of bacteriophage progeny (Campbell 2006).

8.4.1 Lysogeny as a Bacteriophage Survival Mechanism

While little is known about the factors leading to the establishment of lysogeny in nature, it is generally believed that lysogeny provides a refuge for temperate bacteriophages when conditions are unfavorable for robust replication or when host abundance is low (Marsh and Wellington 1994; Weinbauer 2004). A study of the interactions between *Bacillus subtilis* and its bacteriophages in soil ecosystems is consistent with this idea that lysogeny increases bacteriophage survival

(Pantastico-Caldas et al. 1992). At equilibrium, both temperate and virulent bacteriophages were much less abundant than the bacterial host. The temperate bacteriophage did not reduce the equilibrium host titre while the virulent bacteriophage reduced the titre tenfold as compared with soil lacking phage. The authors suggested that the dynamics of this system was the result of the acidic soil, which caused a rapid and permanent inactivation of free bacteriophages. This inactivation drives selection for temperate bacteriophages capable of forming a lysogenic association. Pantastico-Caldas et al. (1992) suggest that a temperate life cycle would be commonplace in harsh soil environments. Another study of soil ecosystems also has shown correlations between host-cell density, nutrient availability, and the frequency of temperate bacteriophages (Ashelford et al. 1999). The authors determined that in a sugar beet rhizosphere a high level of *Serratia* temperate bacteriophages occurred during periods of high host-cell density and elevated metabolic activity early in the growing season. Virion titres were much lower when cell densities were low due to nutrient depletion late in the growing season. These data suggest that lytic replication by temperate bacteriophages occurs during times of rich resources in the rhizosphere microbial community, and lysogeny is favored when host growth is limited.

When considering bacteriophage as BCAs, one question to be addressed concerns their artificial predominance in a particular soil microenvironment. In virulent bacteriophage populations, the only route for cell-to-cell transmission of genetic material is through horizontal gene transfer (HGT) by generalized or abortive transduction. Lysogenic populations, through lysogeny, are capable of cell-to-cell transmission by various forms of specialized transduction where host genes are excised along with the prophage and carried to another host. Furthermore, lysogenic infections can result in lysogenic conversion of the host where the expression of novel genes on the bacteriophage genome alters the host phenotype by the addition or loss of various characteristics (Herron 2004; Brüssow and Kutter 2005b). This phenomenon can provide the host with new phenotypic characteristics conferring pathogenicity or enhanced virulence addition of toxins.

8.4.2 Phage Gene Transfer in Soil

There are two mechanisms by which bacteriophages can mediate bacterial gene transfer, phage conversion, and transduction. Phage conversion occurs when the phenotype of the host changes due to a gene within the genome of a temperate phage and has been the most studied aspect of phage-mediated gene transfer (Herron 2004). Transduction is the phage-mediated gene transfer between a donor and a recipient host cell followed by expression of the genetic traits in the progeny of the recipient (Weinbauer 2004). Only a small fraction of generalized transducing bacteriophages have been characterized presumably because most are not culturable. Sander and Schmieger (2001) developed a host-independent method to detect these phages, and the method is being used to estimate the contribution of generalized transduction to HGT.

Generalized transduction was observed for a number of phages among strains of *R. meliloti* and it was thought that rhizobiophages capable of specialized transduction could be useful for genetic studies in *B. japonicum* (Abebe et al. 1992). Integration of phage V into the genome of a phage-resistant isolate was accompanied by the inability of that isolate to nodulate soybean plants. The fact that the phage V integrated into the genome of some strains offers hope for the development of a *B. japonicum*-specific transducing system that would greatly facilitate genetic studies. *R. leguminosarum* bacteriophages were isolated from a field where survival of a genetically modified host had been monitored for several years (Mendum et al. 2001). The authors found evidence of infrequent generalized transduction of a plasmid-located gene for neomycin resistance. It is thought that a small proportion of phage particles enclosed fragments of the host genome, although the maximum length of host DNA that can be packaged is not known. The presence of sequences from both virulent and temperate phage in indigenous bacteria indicates that phage–bacteria interactions occur in soil. The authors believe that where phage and susceptible bacteria coincide, such as in the rhizosphere, infection will occur to make gene transduction possible. The virulent phages could provide a reservoir of bacterial genes in conditions where the host might not survive. Polymerase-chain reaction (PCR) studies with phage-specific primers show promise for future studies on the ecology of phages in soil.

8.5 Detection of Phages in Soil Systems

Direct observation of phage populations is conducted for many different purposes. Metagenomic studies require the isolation of large amounts of genomic phage DNA, without regard to individual viral particles. Ecological studies often involve the recovery of infectious particles and can be either qualitative (presence/absence) or quantitative (O'Brien and Lindow 1989). A growing interest in phages as biopesticides has increased the need for tools that allow investigators to monitor the population dynamics of their therapeutically applied phages over the course of treatment and then to trace the environmental fate of those phages. Ecological and biopesticide studies are increasingly looking toward molecular genetic methods of phage enumeration, but these will not replace the direct recovery of viable virions in all situations.

The great diversity of soil types, and the wide range of physical and chemical properties observed in even one soil type, makes it difficult to establish one, or even a few, standard methods for detecting phages in soil systems. This is particularly true if quantitative detection is desired, since accurate enumeration is so easily inhibited by soil chemistry and by direct interactions of phages with soil particles and other components of the soil environment. The following discussion considers some of the challenges and uses of various approaches to detecting and enumerating phage communities in soil. These topics are mostly considered in the context of phage biopesticides, but are equally applicable to studies of general soil ecology and of viral persistence and transport in wastewater- and biosolids-amended soils.

8.5.1 Isolation of Phage Particles

8.5.1.1 Enrichment Methods

Whole phages can be isolated from soil using either enrichment techniques or direct recovery of existing virions. Enrichment methods rely on the basic principle that phages will replicate if they have access to a susceptible host in a suitable environment. If even one viable phage is present in a sample, it should be detectable in far greater numbers following enrichment on an appropriate bacterial host culture. In reality, the efficiency of enrichment can be greatly affected by the enrichment conditions. Infective phages may be temporarily unavailable to host bacteria as a result of interactions with charged substances; the nutritional status of the host cell can alter expression of the cell surface molecules that are required for phage adsorption; the energetic state of the host cell can affect the ability of the phage genome to replicate within an infected host; and factors such as pH, temperature, and the availability of cofactors such as Ca^{2+} or Mg^{2+} can also influence the efficiency of phage replication and thus which phages are detected (Guttman et al. 2005). The choice of host culture will also influence which phages are enriched. An environmental soil sample can be reasonably expected to contain multiple phage types. Obviously, the choice of a host species influences which phages are recovered, but within a host species, the number and type of strains used will favor the preferential enrichment and recovery of some phages over others. For example, the use of multiple host strains vs. a single host strain preferentially enriches phages having broader host ranges (Jensen et al. 1998; Gill et al. 2003; Lehman 2007).

E. amylovora phages are easily isolated from soil beneath rosaceous hosts exhibiting signs of an active fire blight infection (Erskine 1973; Schnabel and Jones 2001; Gill et al. 2003) but are not normally recovered from soil beneath healthy trees that have not been treated with phages, even when using enrichment techniques (Ritchie and Klos 1977; Ritchie 1978; Ritchie and Klos 1979; A. M. Svircev, unpublished data). Similarly, free *Bacillus* phages were recovered from environmental soil samples in very low numbers until the population of endogenous bacterial hosts was increased by adding rich media (Tan and Reanney 1976). In both cases, the presence of actively growing host cells and high concentrations of multiple phage strains were recovered. If the potential for future population growth of a therapeutically applied phages is the main consideration, then enrichment methods also have the benefit of detecting phages that are present in the soil environment within lysogenic, pseudolysogenic, or poorly replicating cells such as those in stationary phase or biofilms [for a comprehensive discussion of these particular phage–host relationships see Abedon (2008)].

Basic enrichment methods have been well described (Adams 1959; Carlson 2005; Van Twest and Kropinski 2009).Van Twest and Kropinski (2009) report substantially better recovery from soil samples that are allowed to dry before adding enrichment media. Traditionally, enrichment has been limited to phage–host systems for which, at a minimum, the host can be isolated and grown in pure culture. This is an important restriction, since it has been estimated that as few as 1–5% of endogenous soil bacteria are currently culturable (Torsvik et al. 1990).

8.5.1.2 Elution Methods

Recovery of phages from soil is hindered by adsorption of phages to soil particles. Adsorption is mediated by pH-dependent electrostatic interactions (Burge and Enkiri 1978a, b; Taylor et al. 1981; Dowd et al. 1998) and seems to be greater for phages with longer tails (Williamson et al. 2003; Ashelford et al. 2003). For a detailed review of the principles of viral adsorption to soil, see Kimura et al. (2008).

Several detailed studies of elution techniques have been published, but tend to lack the type of truly systematic approach that is needed to ascertain which particular factors influenced the relative success of a particular method. In some studies, different sample handling procedures have been used with each elution medium (Hu 1998), or each medium has been used at a different pH (Williamson et al. 2003), making it difficult to determine which factor is responsible for differential recovery efficacies. It should also be noted that some studies report pH of the elution medium prior to use, while others report the pH of the soil–eluent mixture. Perhaps the most useful single study of phage elution was conducted by Lanning and Williams (1982). They tested recovery of actinophages from multiple soil types using media at a fixed pH, and then examined the effect of pH on recovery using the best-performing medium. Based largely on the results of Lanning and Williams (1982), we have compared the elution of *E. amylovora* phages from sandy orchard soil using five media, all at pH 8.0: nutrient broth, nutrient broth with 0.1% egg albumin, 250 mM glycine, 10 mM tetrasodium pyrophosphate, and 10 mM potassium phosphate buffer amended with sodium chloride and magnesium chloride (Lehman 2007). In this case, no significant difference in phage elution efficiency was observed. Proteinaceous substances and egg albumin in particular have been found to aid the release of phages from soil in some cases (Lanning and Williams 1982) but not all (Lehman 2007). Since adsorption is mediated by pH-dependent electrostatic processes (Taylor et al. 1981; Dowd et al. 1998), it is possible that there is no detectable difference between elution media at a near-optimal pH, but that at nonoptimal pH levels, the specific types of charged species in each medium are differentially successful in disrupting the interactions between phages and soil particles.

Apart from the chemical factors affecting phage recovery from soil, the physical treatment of the sample also appears to be important. Actinophage recovery by Lanning and Williams (1982) was generally more efficient when samples were agitated using a reciprocal flask shaker vs. an orbital shaker or magnetic stirrer.

Sonication is fairly common practice for viral recovery from marine and freshwater sediments (Danovaro et al. 2001; Leroy et al. 2008) and does not appear to reduce the viability of myoviruses (Fu et al. 2009). Guzmán et al. (2007) compared the effects of homogenization procedures, centrifugation, and filtration on recovery of coliphages and F-specific RNA phages. Centrifugation and filtration reduced bacterial contamination without reducing phage viable counts, but unlike in the Lanning and Williams (1982) study, no one method proved superior. Despite the variation in methodology among all of these studies, three common themes emerge: the efficacy of any given elution technique varies with soil type, is lower for phages with long tails, and tends to be better when using eluents at slightly basic pHs.

8.5.2 Direct Detection by Microscopy

Transmission electron microscopy (TEM) is commonly used to directly count phages in soils (Borsheim et al. 1990; Weinbauer and Suttle 1997; Ashelford et al. 2003; Williamson et al. 2005; Yamada et al. 2007; Srinivasiah et al. 2008; Swanson et al. 2009). The reported populations of virus-like particles (VLPs) were at least 350-fold higher than those estimated from viable plaque counts (Ashelford et al. 2003). However, the ecological impact of these extra VLPs is questionable since some of these samples did not yield any viable phages even after enrichment on multiple hosts. Epifluorescence microscopy (EFM) may be employed to assess phage abundance in soils. Danovaro et al. (2001) successfully applied EFM to study viral abundance in marine sediments.

EFM is more sensitive than TEM when applied to marine water samples (Hennes and Suttle 1995; Weinbauer and Suttle 1997), but may be less specific since the lower magnification reveals approximate dimensions rather than detailed particle morphology (Borsheim et al. 1990; Brussaard et al. 1996). The use of high concentrations of SYBR Green may improve threshold-based differentiation of viruses from extremely small bacteria (Danovaro et al. 2001). The sensitivity of microscopic detection methods is also impacted by the same factors that limit phage elution from soil, since viruses attached to soil particles cannot be resolved, which necessitates a preparatory phage elution step. Using a combination of techniques, Swanson et al. (2009) estimated $0.87–1.1 \times 10^9$ virions/g dry agricultural soil.

8.5.3 Direct Detection of Biopesticides by Molecular Methods

In the case of phage-based biopesticides, there is a clear need to monitor the populations of at least one phage and one host over time, and often there will be multiple types or strains of each. The population dynamics of the phages, their target, and any alternative bacterial hosts can confirm that disease control is attributable to phage action or can suggest reasons for treatment failure (Lehman 2007). Even if the therapeutic outcome of interest occurs on aerial plant tissue, the surrounding soil should be expected to act as both a sink for applied phages (due to inactivation and adsorption) and as a protected reservoir (since not all phages washed down into soil will be inactivated or otherwise lost). In following the fate of the phage biopesticide, there may sometimes only be a need to know whether a phage persists, in which case enrichment and subsequent identification of a particular phage type is sufficient. However, there is often interest in the more detailed ecology of surviving phages. Two such cases have been presented in this chapter: when phage biopesticides are being applied to subsurface plant tissues such as tubers, making biopesticide efficacy directly dependent on phage interactions with the rhizosphere; and when a nonpathogen that also supports replication of the therapeutic phages is present in abundance, as in the case of *P. agglomerans* and phages that target *E. amylovora*. Quantitative tracking of phage biopesticides may also become a regulatory issue that must be addressed to register a phage-based biopesticide.

This type of tracking is a challenge for culture-based detection methods, even if only two or three different phages are being used to target a single host species since true therapeutic cocktails are likely to contain 4–6 phages in the mixture. (Markoishvili et al. 2002; Guenther et al. 2009). Plaque morphologies often cannot be used to reliably distinguish among phage types, and there is no phage plating equivalent to selective bacterial culture media. Rather than attempting to separate phage types within a mixture, it is far easier to apply DNA detection methods such as PCR. Endpoint PCR can be used for detection, real-time PCR for quantification, and reverse transcriptase steps can be incorporated into either if the target is an RNA virus or a transcription product.

When a phage enrichment step is not desirable, PCR is performed after community DNA extraction or the elution of phage particles. However, successful extraction or elution does not necessarily equate to successful detection and quantification, since soil contains many substances that are known to inhibit PCR, and that tend to be extracted or washed from the soil along with the recovered phages. As little as 10 ng humic acid can inhibit a conventional, endpoint PCR reaction (Tsai and Olsen 1992). It has been postulated that phenolic moities in humic substances react with, and covalently bind to, DNA and protein, preventing the necessary interactions between the polymerase and the target DNA, or between primers and target DNA (Young et al. 1993). Humic acids may also interfere with the fluorescence processes upon which real-time PCR depends by quenching fluorescence of SYBR Green, Hoescht 33258, and PicoGreen complexed to DNA (Bachoon et al. 2001; Zipper et al. 2003).

A variety of metal ions can inhibit PCR, apparently by interfering with the binding and activity of the polymerase enzyme. Calcium ions in milk can interfere with PCR amplification (Bickley et al. 1996). Iron and other heavy metals are generally present in soil, whether in high levels as pollutants or levels appropriate for micronutrition of plants, and are known PCR inhibitors (Wilson 1997; Ogram 1998; Hao et al. 2002). Any substances that sequester Mg^{2+} will also inhibit the DNA polymerase enzyme, which requires the ion as a cofactor and is sensitive to changes in its concentration (Satsangi et al. 1994; Wilson 1997). Polyamines

(Ahokas and Erkkila 1993), phenol (Katcher and Schwartz 1994), and plant polysaccharides (Demeke and Adams 1992) can inhibit amplification by directly affecting the DNA polymerase. Foulds et al. (2002) were able to remove PCR inhibitors by washing *E. coli* cells collected from environmental water samples with EDTA, a metal ion chelator, prior to DNA extraction. Extensive work has also been done to develop methods of removing these inhibitors in the course of extracting total community DNA from soil or soil eluates (Zhou et al. 1996; Sjöstedt et al. 1997; Miller et al. 1999; Desai and Madamwar 2006).

There are many studies describing DNA extraction techniques that can be used, with varying success, to remove soil-derived PCR inhibitors. The Tth and Tfl polymerases have been shown to be much more resistant than Taq polymerase to inhibitors that directly affect the DNA polymerase (Katcher and Schwartz 1994; Wiedbrauk et al. 1995) and may offer some improvement in DNA amplification from soil extracts. The SDS-based method of Zhou et al. (1996) was used successfully with several soil types. The Ultraclean Soil DNA Isolation kit from MO BIO Laboratories has been used to extract total bacterial DNA for real-time PCR detection of *E. coli* O157:H7 (Ibekwe et al. 2002), though Desai and Madamwar (2006) describe a protocol that removes metallic and organic inhibitors more efficiently.

Few studies bother to mention attempts to amplify microbial DNA directly from soil eluate, and those that do, report consistent failure (Sjöstedt et al. 1997; Lehman 2007). Some commonly used eluents that help dissociate phages from soil particles are not optimal for direct PCR, and any that may help chelate soil-derived PCR inhibitors will also tend to chelate cofactors required for PCR. For example, tetrasodium pyrophosphate has been used by soil scientists to dissolve organic matter and extract metals bound to humic substances (McKeague et al. 1971; Manninen et al. 1996), but it reduces the sensitivity of real-time PCR (Lehman 2007), presumably by sequestering Mg^{2+} ions. The addition of small amounts of EDTA to soil eluate, followed by ultrafiltration with buffer replacement, may remove most soil-derived PCR inhibitors, but the overall detection efficiency is still dominated by the elution process (Lehman 2007).

8.6 Summary

Phages can have strong influences on the performance of microbial food webs, microbe diversity, and biogeochemical cycles in various environments, although many specific details on the mechanisms of these influences are lacking (Weinbauer 2004). We are at the verge of understanding the influence of phages on links between ecosystem stability, functioning, and diversity. With respect to phages effects in soils, circumstantial evidence suggests that these viruses play important roles in biogeochemical nutrient cycles and as genomic reservoirs similar to those in the sea (Kimura et al. 2008). Phages may regulate host populations by lysis, but this regulation is probably extremely limited in soil where populations have periods of inactivity (Marsh and Wellington 1994). Changes in environmental parameters such as moisture content, temperature, pH, and aerobicity frequently fluctuate because of weather and field management. These changes have a direct impact on microbial activities and may induce lysogenic changes in soil bacteria. Lysogeny can influence host bacteria in two ways, by permitting survival and in rare circumstances in soils mediating HGT. Lysogeny represents a compromise between hosts and phage where both parties are granted advantages in terms of improved survival capabilities in return for reduced abundance.

In regard to applied biocontrol of soil-borne plant diseases, bacteriophages have great potential because they are widely present, are self-replicating, can be targeted against specific bacterial receptors, are nontoxic to eukaryotes, and are specific to bacteria (Hagens and Offerhaus 2008).

Our ability to study phage ecology in soil systems generally lags behind our ability to study phages in water systems. For phage-based biopesticides, the principle value of good phage detection methods is to permit a quantitative assessment of phage activity and to explain why a treatment is successful or not. This is especially important when one considers how many of the substances that we use to fertilize agricultural soils are known to impact phage survival and activity (Ravensdale et al. 2007; Kimura et al. 2008; Bouzari et al. 2008). DNA-based methods of phage detection and quantification are generally the most sensitive. These require some specific sample processing to remove soil-derived enzymatic and fluorescence inhibitors, but with elution techniques yielding only 40–60% recovery in most cases (Lanning and Williams 1982; Danovaro et al. 2001; Lehman 2007), only the most abundant phage types will be reliably detected by elution-dependent methods. This makes DNA extraction directly from soil, followed by PCR-based quantification, the most effective technique currently available for quantitative studies of soil phage communities.

To date, limited information is available on the large-scale impact of bacteriophages applied as biopesticides on the soil ecosystem. While many phages that are used as biopestides have originated from soil (Svircev et al. 2002b; Gill et al. 2003; Goyer 2005; Svircev et al. 2006; Jones et al. 2007; Ravensdale et al. 2007), determining the environmental fate of the phages has been restricted to laboratory and greenhouse. Future developments in molecular detection technologies may permit study of biopesticide-related phages in agricultural ecosystems and should lead to greater elucidation of the impact and importance of phages on bacterial activities in soils in general.

References

Abebe HM, Sadowsky MJ, Kinkle BK, Schmidt EL (1992) Lysogeny in *Bradyrhizobium japonicum* and its effect on soybean nodulation. Appl Environ Microbiol 58:3360–3366

Abedon ST (2008) Phage, ecology, and evolution. In: Abedon ST (ed) Bacteriophage ecology. Advances in molecular and cellular microbiology, vol 15. Cambridge University Press, Cambridge UK, pp 1–28

Adams MH (1959) Bacteriophages. Interscience Publishers, New York

Ahokas H, Erkkila MJ (1993) Interference of PCR amplification by the polyamines, spermine and spermidine. PCR Methods Appl 3:65–68

Andrews JH, Harris RF (2000) The ecology and biogeography of microorganisms on plant surfaces. Ann Rev Phytopathol 38:145–180

Ashelford KE, Day MJ, Bailey MJ, Lilley AK, Fry JC (1999) In situ population dynamics of bacterial viruses in a terrestrial environment. Appl Environ Microbiol 65:169–174

Ashelford KE, Day MJ, Fry JC (2003) Elavated abundance of bacteriophage infecting bacteria in soil. Appl Environ Microbiol 69:285–289

Assadian NW, Giovanni GD, Enciso J, Iglesias J, Lindemann W (2005) The transport of water-borne solutes and bacteriophage in soil subirrigated with a wastewater blend. Agric Ecosyst Environ 111:279–291

Bachoon DS, Otero E, Hodson RE (2001) Effects of humic substances on fluorometric DNA quantification and DNA hybridization. J Microbiol Methods 47:73–82

Balogh B, Canteros BI, Stall RE, Jones JB (2008) Control of citrus canker and citrus bacterial spot with bacteriophages. Plant Dis 92:1048–1052

Balogh B, Jones JB, Momol MT, Olson SM (2005) Persistence of bacteriophages as biocontrol agents in the tomato canopy. Acta Hortic 695:299–302

Balogh B, Jones JB, Momol MT, Olson SM, Obradovic A, King P, Jackson LE (2003) Improved efficacy of newly formulated bacteriophages for management of bacterial spot on tomato. Plant Dis 87:949–954

Basit HA, Angle JS, Salem S, Gewaily EM (1992) Phage coating of soybean seed reduces nodulation by indigenous soil bradyrhizobia. Can J Microbiol 38:1264–1269

Berg G, Sanjaghsaz H, Wangwongwatana S (1989) Potentiation of the virucidal effectiveness of free chlorine by substances in drinking water. Appl Environ Microbiol 55:390–393

Bickley J, Short SK, McDowell DG, Parkes HC (1996) Polymerase chain reaction (PCR) detection of *Listeria monocytogenes* in diluted milk and reversal of PCR inhibition caused by calcium ions. Lett Appl Microbiol 22:153–158

Blom TJ, Brown W (1999) Preplant copper-based compounds reduce *Erwinia* soft rot on calla lilies. HortTechnol 9:56–59

Borsheim KY, Bratbak G, Heldal M (1990) Enumeration and biomass estimation of planktonic bacteria and viruses by transmission electron microscopy. Appl Environ Microbiol 56:352–356

Bouzari M, Emtiazi G, Mehdipour-Moghaddam MJ (2008) The effects of heavy metals and chelating agents on phage development and enumeration of *Rhizobium* by phage counting in different soils. Am Eurasian J Agric Environ Sci 3:420–424

Brussaard CPD, Kempers RS, Kop A, Riegman R, Heldal M (1996) Virus-like particles in a summer bloom of *Emiliana huleyi* in the North Sea. Aquat Microb Ecol 10:105–113

Brüssow H, Kutter E (2005a) Phage ecology. In: Kutter E, Sulakvelidze A (eds) Bacteriophages: biology and applications. CRC Press, Boca Raton, pp 129–163

Brüssow H, Kutter E (2005b) Genomics and evolution of tailed phages. In: Kutter E, Sulakvelidze A (eds) Bacteriophages: biology and applications. CRC Press, Boca Raton, pp 91–128

Burge WD, Enkiri NK (1978a) Virus adsorption by five soils. J Environ Qual 7:73–76

Burge WD, Enkiri NK (1978b) Adsorption kinetics of bacteriophage P29103000X-174 on soil. J Environ Qual 7:536–541

Campbell A (2006) General aspects of lysogeny. In: Calendar R (ed) The bacteriophages, 2nd edn. Oxford University Press, New York

Campbell JIA, Albrechtsen M, Sorensen J (1995) Large *Pseudomonas* phages isolated from barley rhizosphere. FEMS Microbiol Ecol 18:63–74

Canchaya C, Fournous G, Brüssow H (2004) The impact of prophages on bacterial chromosomes. Mol Microbiol 53:9–18

Carlson K (2005) Working with bacteriophages: common techniques and methodological approaches. In: Kutter E, Sulakvelidze A (eds) Bacteriophages: biology and applications. CRC Press, Boca Raton, pp 437–487

Chattopadhyay D, Chattopadhyay S, Lyon WG, Wilson JT (2002) Effect of surfactants on the survival and sorption of viruses. Environ Sci Technol 36:4017–4024

Danovaro R, Dell'Anno A, Trucco A, Serresi M, Vanucci S (2001) Determination of virus abundance in marine sediments. Appl Environ Microbiol 67:1384–1387

Demeke T, Adams R (1992) The effects of plant polysaccharides and buffer additives on PCR. BioTechniques 12:333–334

Desai C, Madamwar D (2006) Extraction of inhibitor-free metagenomic DNA from polluted sediments, compatible with molecular diversity analysis using adsorption and ion-exchange treatments. Bioresour Technol 98:761–768

Dowd SE, Pillai SD, SookYun W, Corapcioglu MY (1998) Delineating the specific influence of virus isoelectric point and size on virus adsorption and transport through sandy soils. Appl Environ Microbiol 64:405–410

Erskine JM (1973) Characteristics of *Erwinia amylovora* bacteriophage and its possible role in the epidemiology of fire blight. Can J Microbiol 19:837–845

Flaherty JE, Harbaugh BK, Jones JB, Somodi GC, Jackson LE (2001) H-mutant bacteriophages as a potential biocontrol of bacterial blight of geranium. HortSci 36:98–100

Flaherty JE, Jones JB, Harbaugh BK, Somodi GC, Jackson LE (2000) Control of bacterial spot on tomato in the greenhouse and field with h-mutant bacteriophages. HortSci 35:882–884

Foulds IV, Granacki A, Xiao C, Krull UJ, Castle AJ, Horton PJ (2002) Quantification of microcystin-producing cyanobacteria and *E. coli* in water by 5′-nuclease PCR. J Appl Microbiol 93:825–834

Fu W, Forster T, Meyer O, Curtin JJ, Lehman SM, Donlan RM (2009) Bacteriophage cocktail for the prevention of biofilm formation by *Pseudomonas aeruginosa* on catheters in an in vitro model system. Antimicrob Agents Chemother. doi:10.1128/AAC.00669-09

Germida JJ (1986) Population dynamics of *Azospirillum brasilense* and its bacteriophage in soil. Plant Soil 90:117–128

Gill J, Abedon ST (2003) Bacteriophage ecology and plants. APSnet http://www.apsnet.org/online/feature/phages/ Cited: July 2009

Gill JJ, Svircev AM, Smith R, Castle AJ (2003) Bacteriophages of *Erwinia amylovora*. Appl Environ Microbiol 69:2133–2138

Goyer C (2005) Isolation and characterization of phages Stsc1 and Stsc3 infecting *Streptomyces scabiei* and their potential as biocontrol agents. Can J Plant Pathol 27:210–216

Gracia-Garza JA, Blom TJ, Brown W, Roberts DP, Schneider K, Freisen M, Gombert D (2004) Increased incidence of *Erwinia* soft-rot on calla lilies in the presence of phosphorous. Eur J Plant Pathol 110:293–298

Guenther S, Huwyler D, Richard S, Loessner MJ (2009) Virulent bacteriophage for efficient biocontrol of *Listeria monocytogenes* in ready-to-eat foods. Appl Environ Microbiol 75:93–100

Guzmán C, Jofre J, Blanch AR, Luana F (2007) Development of a feasible method to extract somatic coliphages from sludge, soil and treated biowaste virol Meth 144:41–48

Guttman B, Raya R, Kutter E (2005) Basic phage biology. CRC Press, Boca Raton, FL

Hagens S, Offerhaus ML (2008) Bacteriophages – new weapons for food safety. Food Technol Chicago 62:46

Hao Y, Dick WA, Tuovinen OH (2002) PCR amplification of 16S rDNA sequences in Fe-rich sediment of coal refuse drainage. Biotechnol Lett 24:1049–1053

Harrison MD, Franc DG, Maddox DA, Michaud JE, McCater-Zorner NJ (1987) Presence of *Erwinia carotovora* in surface water in North America. J Appl Bacteriol 62:565–570

Hennes JE, Suttle CA (1995) Direct counts of viruses in natural waters and laboratory cultures by epifluorescence microscopy. Limnol Oceanogr 40:1050–1055

Herron, PR (2004) Phage ecology and genetic exchange in soil. In: Kowalchuk GA, Bruijn, FJD, Head IM, Akkermans ADL and Elsas JDV (eds) Vol. II, Molecular microbial ecology manual, pp 1173–1183

Hildebrand M, Tebbe CC, Geider K (2001) Survival studies with the fire blight pathogen *Erwinia amylovora* in soil and in a soil-inhabiting insect. J Phytopath 149:635–639

Hu TL (1998) A comparison of two methods to recover phages from soil samples. Bioresour Technol 65:167–169

Ibekwe AM, Watt PM, Grieve CM, Sharma VK, Lyons SR (2002) Multiplex fluorogenic real-time PCR for detection and quantification of *Escherichia coli* O157:H7 in dairy wastewater wetlands. Appl Environ Microbiol 68:4853–4862

Iriarte FB, Balogh B, Momol MT, Smith LM, Wilson M, Jones JB (2007) Factors affecting survival of bacteriophage on tomato leaf surfaces. Appl Environ Microbiol 73:1704–1711

Jensen EC, Schrader HS, Rieland B, Thompson TL, Lee KW, Nickerson KW, Kokjohn TA (1998) Prevalence of broad-host-range lytic bacteriophages of *Sphaerotilus natans*, *Escherichia coli*, and *Pseudomonas aeruginosa*. Appl Environ Microbiol 64:575–580

Johnson KB, Stockwell VO (2000) Biological control of fire blight. In: Vanneste JL (ed) Fire blight: the disease and its causative agent, *Erwinia amylovora*. CaB International, Wallingford, Oxon, pp 319–337

Jones JB, Jackson LE, Balogh B, Obradovic A, Iriarte FB, Momol MT (2007) Bacteriophages for plant disease control. Ann Rev Phytopathol 45:245–262

Jones JB, Momol MT, Obradovic A, Balogh B, Olson SM (2005) Bacterial spot management on tomatoes. Vol. 695 Acta Hortic 119–124

Katcher HL, Schwartz I (1994) A distinctive property of *Tth* DNA polymerase: enzymatic amplification in the presence of phenol. BioTechniques 16:84–92

Katznelson H, Lochhead AG, Timonin MI (1948) Soil microorganisms and the rhizosphere. Bot Rev 14:543–587

Kleczkowska J (1957) A study of the distribution and the effects of bacteriophage of root nodule bacteria in soil. Can J Microbiol 3:171–180

Kimura M, Jia Z-J, Nakayama N, Asakawa S (2008) Ecology of viruses in soils: past, present and future perspectives. Soil Sci Plant Nut 54:1–32

Kumar MKP, Khan ANA, Reddy CNL, Basavarajappa MP, Venkataravanappa V, Basha Z (2006) Biological control of bacterial wilt of tomato caused by *Ralstonia solanacearum*, race-1, biovar-III. J Plant Dis Sci 1:176–181

Lang JM, Gent DH, Schwartz HF (2007) Management of *Xanthomonas* leaf blight of onion with bacteriophages and a plant activator. Plant Dis 91:871–878

Lanning S, Williams ST (1982) Methods for the direct isolation and enumeration of actinophages in soil. J Gen Microbiol 128:2063–2071

Lehman SM (2007) Development of a bacteriophage-based biopesticide for fire blight. PhD thesis, Brock University

Leroy M, Prigent MDM, Confalonieri F, Dubow M (2008) Bacteriophage morphotype and genome diversity in Seine River sediment. Freshw Biol 53:1176–1185

Manninen A, Hall GEM, Vaive JE, MacLaurin AI (1996) Analytical aspects of the application of sodium pyrophosphate reagent in the specific extraction of the labile organic component of humus and soils. J Geochem Explor 56:23–36

Markoishvili K, Tsitlanadze G, Katsarava R, Morris JGJ, Sulakvelidze A (2002) A novel sustained-release matrix based on biodegradable polyester amides impregnated with bacteriophages and an antibiotic shows promise in management of infected venous stasis ulcers and other poorly healing wounds. Int J Dermatol 41:453–458

Marsh P, Wellington EMH (1994) Phage–host interactions in soil. FEMS Microbiol Ecol 15:99–107

McKeague JA, Brydon JE, Miles NM (1971) Soil Sci. Soc Am Proc 35:33–38

Mendum TA, Clark IM, Hirsch PR (2001) Characterization of two novel *Rhizobium leguminosarum* bacteriophages from a field release site of genetically-modified rhizobia. Antonie Van Leeuwenhoek 79:189–197

Miller DN, Bryant JE, Madsen EL, Ghiorse WC (1999) Evaluation and optimization of DNA extraction and purification procedures for soil and sediment samples. Appl Environ Microbiol 65:4715–4724

Novikova NI, Pavlova EA, Limeshchenko EV (1993) Phage sensitivity and host range of *Rhizobium* strains isolated from root nodules of temperate legumes. Plant Soil 151:45–53

O'Brien RD, Lindow SE (1989) Effet of plant species and environmental conditions on epiphytic population sizes of *Pseudomonas syringae* and other bacteria. Phytopathology 79:619–627

Obradovic A, Jones JB, Momol MT, Balogh B, Olson SM (2004) Management of tomato bacterial spot in the field by foliar applications of bacteriophages and SAR inducers. Plant Dis 88:736–740

Obradovic A, Jones JB, Momol MT, Olson SM, Jackson LE, Balogh B, Guven K, Iriarte FB (2005) Integration of biological control agents and systemic acquired resistance inducers against bacterial spot on tomato. Plant Dis 89:712–716

Ogram A (1998) Isolation of nucleic acids from environmental samples. In: Burlage RS, Atlas R, Stahl D, Geesey G, Sayler G (eds) Techniques in microbial ecology. Oxford University Press, Oxford UK, pp 273–288

Pantastico-Caldas M, Duncan KE, Istock CA, Bell JA (1992) Population dynamics of bacteriophage and *Bacillus subtilis* in soil. Ecology 73:1888–1902

Pérombelon MCM, Kelmon A (1980) Ecology of the soft rot *Erwinias*. Ann Rev Phytopathol 18:361–387

Ravensdale M, Blom TJ, Gracia Garza JA, Svircev AM, Smith RJ (2007) Bacteriophages and the control of *Erwinia carotovora* subsp. *carotovora*. Can J Plant Pathol 29:121–130

Ritchie DF (1978) Bacteriophages of *Erwinia amyovora*: Their isolation, distribution, characterization, and possible involvement in the etiology and epidemiology of fire blight. PhD thesis, Michigan State University

Ritchie DF, Klos EJ (1977) Isolation of *Erwinia amyolovora* bacteriophage from aerial parts of apple trees. Phytopathology 67:101–104

Ritchie DF, Klos EJ (1979) Some properties of *Erwinia amylovora* bacteriophages. Phytopathology 69:1078–1083

Romeo AM, Christen L, Niles EG, Kosman DJ (2001) Intracellular chelation of iron by bipyridyl inhibits DNA virus replication. J Biol Chem 24:301–308

Saccardi A, Gambin E, Zaccardelli M, Barone G, Mazzucchi U (1993) *Xanthomonas campestris* pv. *pruni* control trials with phage treatments on peach in the orchard. Phytopathol Mediterr 32:206–210

Sagripanti J-L (1992) Metal-based formulations with high microbiocidal activity. Appl Environ Microbiol 58:3157–3162

Sander M, Schmieger H (2001) Method for host-independent detection of generalized transducing bacteriophages in natural habitats. Appl Environ Microbiol 67:1490–1493

Satsangi J, Jewell DP, Welsh K, Bunce M, Bell JI (1994) Effect of heparin on polymerase chain reaction. Lancet 343:1509–1510

Schnabel EL, Fernando WGD, Meyer MP, Jones AL, Jackson LE (1999) Bacteriophage of *Erwinia amylovora* and their potential for biocontrol. Acta Hortic 489:649–653

Schnabel EL, Jones AL (2001) Isolation and characterization of five *Erwinia amylovora* bacteriophages and assessment of phage resistance in strains of *Erwinia amylovora*. Appl Environ Microbiol 67:59–64

Sillankorva S, Oliveira R, Viera MJ, Sutherland I, Azeredo J (2006) *Pseudomonas fluorescens* infection by bacteriophage øS1: the influence of temperature, host growth phase and media. FEMS Microbiol Lett 241:13–20

Sjöstedt A, Eriksson U, Ramisse V, Garrigue H (1997) Detection of *Bacillus anthracis* spores in soil by PCR. FEMS Microbiol Ecol 23:159–168

Smit E, Wolters AC, Lee H, Trevors JT, van Elsas JD (1996) Interactions between a genetically marked *Pseudomonas fluorescens* strain and bacteriophage FR2f in soil: effects of nutrients, alginate encapsulation, and the wheat rhizosphere. Microb Ecol 31:125–140

Srinivasiah S, Bhavsar J, Thapar K, Liles M, Schoenfeld T, Wommack KE (2008) Phages across the biosphere: contrasts of viruses in soil and aquatic environments. Res Microbiol 159:349–357

Stephens PM, O'Sullivan M, Gara F (1987) Effect of bacteriophages on colonization of sugar beet roots by *Pseudomonas* spp. Appl Environ Microbiol 53:1164–1167

Stewart A (2001) Commercial biocontrol – reality or fantasy? Australas. Plant Pathol 30:127–131

Suslow TV (1986) Bacteriophage-resistant plant growth promoting rhizobactria. USA Patent 4584274

Svircev AM, Gill JJ, Sholberg P (2002a) *Erwinia amylovora* (Burrill) Winslow, Broadhurst, Buchanan, Krumwiede, Rogers and Smith, fire blight (*Enterobacteriaceae*). In: Mason PG, Huber JT (eds) Biological control programmes in Canada. CaBI Publishing, Wallingford UK, pp 448–451

Svircev AM, Lehman SM, Kim W-S, Barszcz E, Schneider KE, Castle AJ (2006) Control of the fire blight pathogen with bacteriophages. In: Proceedings of the 1st International Symposium on Biological control of Bacterial Plant Diseases. Zeller W and Ullrich C (eds) Arno Brynda, Berlin Germany, pp 259–261

Svircev AM, Smith R, Gracia-Garza JA, Gill JJ, Schneider K (2002b) Biocontrol of *Erwinia* with bacteriophages. Bull OILB/SROP 25:10, p. 139–142

Swanson MM, Fraser G, Daniell TJ, Torrance L, Gregory PJ, Taliansky M (2009) Viruses in soils: morphological diversity and abundance in the rhizosphere. Ann Appl Biol 155:51–60

Sykes IK, Lanning S, St W (1981) The effect of pH on soil actinophage. Microbiol 122:271–280

Tan JSH, Reanney DC (1976) Interactions between bacteriophages and bacteria in soil. Soil Biol Biochem 8:145–150

Taylor DH, Moore RS, Sturman LS (1981) Influence of pH and electrolyte composition on adsorption of poliovirus by soils and minerals. Appl Environ Microbiol 42:976–984

Torsvik V, Goksřyr J, Daae FL (1990) High diversity in DNA of soil bacteria. Appl Environ Microbiol 56:782–787

Tsai YL, Olsen B (1992) Rapid method for separation of bacterial DNA from humic substances in sediments for polymerase chain reaction. Appl Environ Microbiol 58:2292–2295

Vanneste JL (ed) (2000) Fire blight: the disease and its causative agent, *Erwinia amylovora*. CaB International, Wallingford, UK

Van Twest R, Kropinski AM (2009) Bacteriophage enrichment from water and soil. In: Clokie MRJ, Kropinski AM (eds) Bacteriophages, methods and protocols, Vol. 1: Isolation, characterization and interactions. Methods in molecular biology, Walker JM (series ed). Humana Press part of Springer Science + Business Media, Totowa, NJ

Weinbauer MG (2004) Ecology of prokaryotic viruses. FEMS Microbiol Rev 28:127–181

Weinbauer MG, Suttle CA (1997) Comparison of epifluorescence and transmission electron microscopy for counting viruses in natural marine waters. Aquat Microb Ecol 13:225–232

Wiedbrauk DL, Werner JC, Drevon AM (1995) Inhibition of PCR by aqueous and vitreous fluids. J Clin Microbiol 33:2643–2646

Wiggins BA, Alexander M (1985) Minimum bacterial density for bacteriophage replication: implications for significance of bacteriophages in natural ecosystems. Appl Environ Microbiol 49:19–23

Williamson KE, Radosevich M, Wommack KE (2005) Abundance and diversity of viruses in six Delaware soils. Appl Environ Microbiol 71:3119–3125

Williamson KE, Wommack KE, Radosevich M (2003) Sampling natural viral communities from soil for culture-independent analyses. Appl Environ Microbiol 69:6628–6633

Wilson IG (1997) Inhibition and facilitation of nucleic acid amplification. Appl Environ Microbiol 63:3741–3751

Yamada T, Kawasaki T, Nagata S, Fujiwara A, Usami S, Fujie M (2007) New bacteriophages that infect the phytopathogen Ralstonia solanacearum. Microbiology 153:2630–2639

Young CC, Burghoff RL, Keim LG, Minak-Bernero V, Lute JR, Hinton SM (1993) Polyvinylpyrrolidone-agarose gel electrophoresis purification of polymerase chain reaction-amplifiable DNA from soils. Appl Environ Microbiol 59:1971–1974

Zaccardelli M, Saccardi A, Gambin E, Mazzucchi U (1992) *Xanthomonas campestris* pv. *pruni* bacteriophages on peach trees and their potential use for biological control. Phytopathol mediterr 31:133–140

Zhou J, Bruns MA, Tiedje JM (1996) DNA recovery from soils of diverse composition. Appl Environ Microbiol 62:316–322

Zipper H, Buta C, Lämmle K, Brunner H, Bernhagen J, Vitzthum F (2003) Mechanisms underlying the impact of humic acids on DNA quantification by SYBR Green I and consequences for the analysis of soils and aquatic sediments. Nucleic Acids Res 31:e39

Chapter 9
Interactions Between Bacteriophage DinoHI and a Network of Integrated Elements Which Control Virulence in *Dichelobacter nodosus*, the Causative Agent of Ovine Footrot

Brian F. Cheetham, Gabrielle Whittle, Michael Ting, and Margaret E. Katz

9.1 Introduction

Dichelobacter nodosus is an anaerobic, non-spore-forming, non-flagellated gram-negative rod (Egerton et al. 1988) with terminal knob-like swellings and polar type 4 *N*-methylphenylalanine fimbriae (Dalrymple and Mattick 1987). It has been assigned to the family *Cardiobacteriaceae* in the gamma subdivision of the *Proteobacteria* (Dewhirst et al. 1990). *D. nodosus* is the principal causative agent of footrot (Beveridge 1941), a mixed bacterial infection of the hooves of sheep, goats, deer and cattle. The genome of one strain of *D. nodosus* has been sequenced (Myers et al. 2007) and is very small (1.3 Mb) with an unusually small proportion of genes devoted to regulation. We have identified seven mobile genetic elements which integrate into the *D. nodosus* genome and have proposed that the virulence of this bacterial pathogen is modulated by these integrated genetic elements (Katz et al. 1991; Katz et al. 1994; Cheetham et al. 1995; Bloomfield et al. 1997; Whittle et al. 1999; Cheetham et al. 2008; Tanjung et al. 2009). We discuss here evidence for interactions between these mobile genetic elements and possible mechanisms for coordinate control.

9.2 Transmission of Ovine Footrot

D. nodosus is free-living and may be cultured under anaerobic conditions in the laboratory. The bacteria may persist and multiply for months or years in footrot lesions in hooves of infected animals. Although the cells die rapidly away from the host due to oxygen sensitivity, they may survive in soil or faeces or on the pasture

B.F. Cheetham (✉), G. Whittle, M. Ting, and M.E. Katz
Molecular and Cellular Biology, University of New England, Armidale, NSW 2351, Australia
e-mail: bcheetham51@gmail.com

G. Witzany (ed.), *Biocommunication in Soil Microorganisms*, Soil Biology 23,
DOI 10.1007/978-3-642-14512-4_9,

for 4 to 5 days (Laing and Egerton 1978). This period of survival in the soil is crucial for the transmission of footrot, since infective material is transferred from exposed lesions on the feet of one animal to the soil or pasture, thereby contaminating the feet of previously unaffected animals. Transmission is favoured by warm, moist conditions and long pasture, which may damage the interdigital skin and facilitate penetration of the bacteria. Sheep tend to walk in single file, thus contributing to the spread of footrot (Stewart 1989). Footrot infections in sheep start in the interdigital cleft of the hoof and progressively spread down the inside wall of the claw and across the sole. This eventually results in severe underrunning and separation of the hoof from the underlying epidermis, causing severe pain and lameness which interferes with the ability of the sheep to graze, leading to loss of body condition and reduced wool production.

9.2.1 Virulence of D. nodosus

Different strains of *D. nodosus* cause disease of differing severity, ranging from benign to virulent. Benign footrot results in only mild lameness, which usually heals spontaneously when dry environmental conditions return. By contrast, virulent footrot spreads rapidly within a flock of sheep and results in severe lameness. Extracellular proteases produced by virulent strains are more thermostable than those produced by benign strains (Depiazzi and Rood 1984) and have greater elastolytic activity (Stewart 1979) which may aid in penetration of the hoof. In addition, virulent strains show greater twitching motility, generated by polar fimbriae (Depiazzi and Richards 1985) and twitching motility is essential for virulence (Kennan et al. 2001; Han et al. 2008).

9.3 Mobile Genetic Elements in the *D. nodosus* Genome

In an attempt to identify genes with a role in virulence, DNA sequences present in the virulent strain, A198, but absent from the benign strain, C305, were isolated (Katz et al. 1991). This work led to the identification of the *intA* (formerly *vap*) element (Katz et al. 1992; Katz et al. 1994; Cheetham et al. 1995) and the virulence-related locus, *vrl* (Haring et al. 1995; Billington et al. 1999). Further analysis of the integration site for the *intA* element in different strains led to the isolation of the *intB* (Bloomfield et al. 1997), *intC* (Whittle et al. 1999) and *intD* elements (Tanjung et al. 2009). Bacteriophage induction experiments led to the isolation of the bacteriophage DinoHI (Cheetham et al. 2008) and analysis of sequences adjacent to the integrated bacteriophage indicated the presence of another integrated genetic element, designated element X. The properties of these integrated genetic elements are discussed in detail below.

9.3.1 The intA *Element*

The *intA* element (Cheetham et al. 1995) is found in most virulent strains, but is also found in about 30% of benign strains. It consists of an integrase gene, *intA*, together with *toxA* and a series of *vap* (virulence-associated protein) genes, *vapA-vapH* (Fig. 9.1). Although these genes were isolated as potential virulence genes, they show no similarity to genes involved in virulence in other bacteria. Instead, *vapB-vapH* are similar to genes involved in the replication or maintenance of plasmids or bacteriophages and the *vapA/toxA* operon encodes a plasmid addiction system (Bloomfield et al. 1997) highly related to the *higB/higA* system from the killer plasmid Rts1 of *E. coli* (Tian et al. 1996).

In the prototype virulent strain A198 there are three copies, or partial copies, of the *intA* element, designated *vap* regions 1, 2 and 3 (Fig. 9.1). *Vap* region 1 is integrated into *tRNA-ser*$_{GCU}$ immediately downstream from *csrA* (formerly *glpA*, (Whittle et al. 1999) with the integrase gene, *intA1* (Myers et al. 2007) located about 200 nucleotides downstream from the end of *tRNA-ser*. *Vap* region 2 is integrated into a different tRNA-ser gene, *tRNA-ser*$_{GGA}$, immediately downstream from *pnpA*. *Vap* region 3 is located adjacent to *vap* region 1. A 19-nt sequence from the ends of both the tRNA-ser genes is duplicated at the ends of *vap* regions 2 and 3 and forms the attachment (*att*) site for integration. In most strains studied, the *intA* element, if present, is found integrated at one or both of these sites. However, in strain AC3577, the *intA* element is maintained stably as a plasmid (Billington et al. 1996).

9.3.2 The intB *Element*

The *intB* element (Bloomfield et al. 1997) is found in all strains tested, both benign and virulent. It is located immediately after *vap* region 3 in strain A198 and contains an integrase gene, *intB*, followed by *regA*, which has high similarity to the repressor proteins from several bacteriophages. This is followed by three genes of unknown function, *gepA*, *gepB* and *gepC* and then by genes encoding a potential sulphate transporter. The limits of the *intB* element have not been determined as the attachment sites have not been identified. Furthermore, Southern blot analysis showed that *gepA* was absent and *regA* was not directly adjacent to *gepB* in many strains. It is likely that the *intB* element at the end of *vap* region 3 in strain A198 has been disrupted. In this strain, but not some others, *intB* is a pseudogene.

A partial copy of *intB* is found at the end of *vap* region 2 in several strains. In some strains which lack the *intA* element, *intB* is located immediately downstream from either *tRNA-ser*$_{GCU}$ or *tRNA-ser*$_{GGA}$. Thus, the *intA* and *intB* elements have a common integration site and tandem integrations are possible.

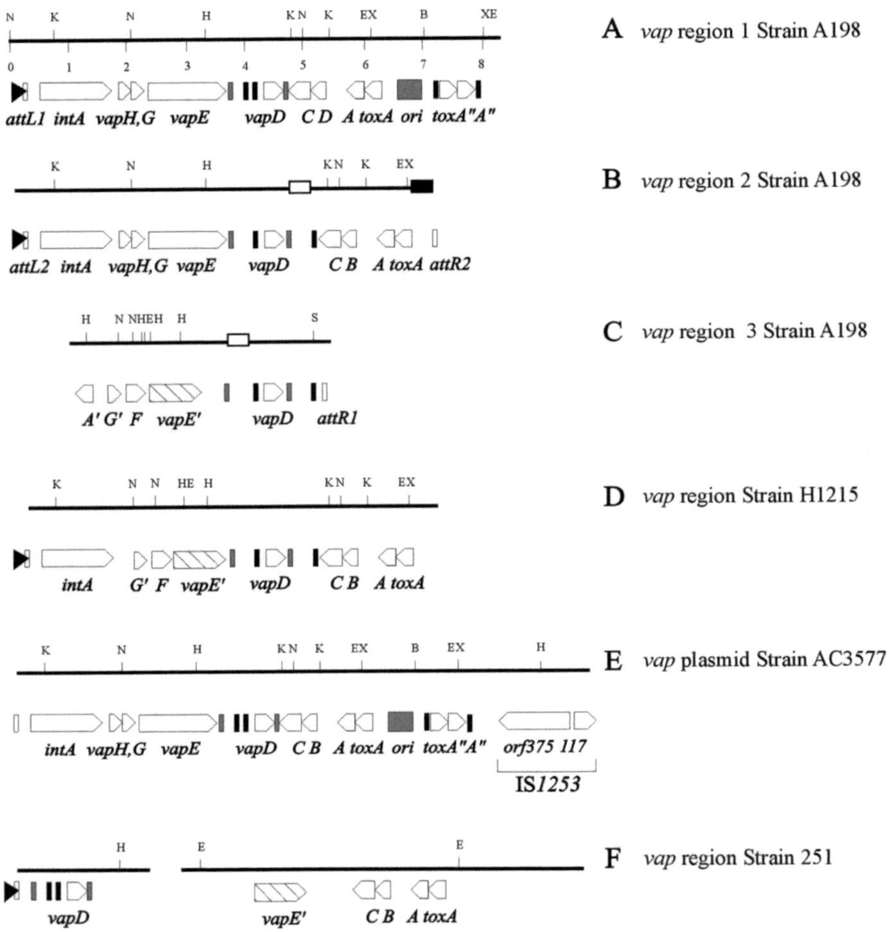

Fig. 9.1 Different forms of the *intA* element. The numbers show the distance in kb from the leftmost *Nru*I site in *vap* region 1, strain A198. The restriction sites shown are *Bam*HI (B), *Eco*RI (E), *Hin*dIII (H), *Kpn*I (K), *Nru*I (N), *Sac*I (S) and *Xho*I (X). Not all restriction sites are shown for strain 251. The major potential genes are shown by *open arrows*. The 19 bp *att* sites are indicated by small *open boxes*, 103 bp repeats or partial copies are indicated by small *shaded boxes*, 102 bp repeats or partial copies are indicated by small *solid boxes* and the putative origin of replication is indicated by a large *shaded box*. The tRNA-ser genes are indicated by *solid triangles*. DNA sequences found in *vap* region 2 but not in *vap* region 1 are indicated on the *scale line* for *vap* region 2. The complete DNA sequences of *vap* regions 1 and 3 have been determined. The maps of the remaining strains are based on partial DNA sequencing, Southern blotting and PCR experiments. The maps A–E have been published previously (Cheetham et al. 1999)

9.3.3 The intC Element

The *intC* element (Whittle et al. 1999) is present in most strains, both benign and virulent. It has been fully characterised from the benign strain C305 and contains an integrase gene, *intC*, whose predicted product has 55% amino acid similarity with *intA*, followed by two genes of unknown function, *orf242* and *orf171* and then genes highly related to *vapGH* from the *intA* element (Fig. 9.2). These are followed by an insertion sequence, IS*1253*, which is also found on the *vap* plasmid (Billington et al. 1996). However, as with the *intB* element, Southern blot analysis of the *intC* element from different strains suggests that the *intC* element in strain C305 has been disrupted and may have lost genes. A longer form of the *intC* element which does not contain the insertion sequence is present in some strains. The integration site for the *intC* element is *tRNA-ser*$_{GCU}$. No strains with the *intC* element integrated into *tRNA-ser*$_{GGA}$ have been identified.

9.3.4 *The* intD *Element*

The *intD* element (Tanjung et al. 2009) is found in only about 20% of strains, almost all of which are benign. It is 32 kb in size and contains an integrase gene, *intD*, genes related to *vapGH* of the *intA* element, a gene related to *orf242* and *orf171*of the *intC* element, together with genes encoding a putative type IV secretion system and a mobilisation region (Fig. 9.2). Thus, it has the features of an integrated conjugative plasmid. In all strains examined so far, the *intD* element is integrated into *tRNA-ser*$_{GGA}$.

The integrase genes *intA*, *intB*, *intC* and *intD* encode integrases with about 50% amino acid identity. These integrases are generally more highly related to each other than to other integrases in the databases, but have diverged a long time ago.

9.3.5 *The Virulence-Related Locus,* vrl

The *vrl* (Haring et al. 1995; Billington et al. 1999) is a 27-kb region which has many features of an integrated genetic element, but lacks an integrase gene. It is found in most virulent strains and is absent from most benign strains. As with the *intA*, *intB*,

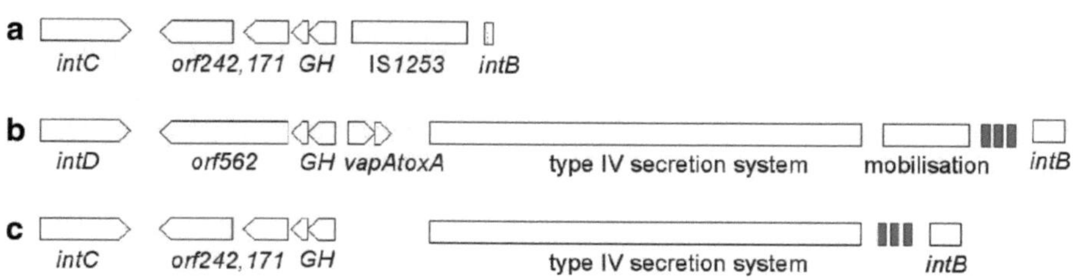

Fig. 9.2 The *intC* and *intD* elements of *D. nodosus*. (**a**) The *intC* element from strain C305. (**b**) The *intD* element from strain 819. (**c**) The proposed hybrid *intC*/*intD* element. The diagram is not drawn to scale. *Grey boxes* represent 102 bp repeated sequences. *GH* refers to the *vapG* and *vapH* genes

intC and *intD* elements, no genes with similarities to known bacterial virulence determinants were identified. The integration site for the *vrl* is *ssrA*, which encodes a potential regulatory 10Sa RNA molecule.

9.3.6 The Bacteriophage DinoHI

To investigate whether any of the integrated genetic elements were prophages, *D. nodosus* strains were treated with agents known to induce prophage excision. These experiments did not induce any of the known integrated genetic elements, but resulted in the isolation of a novel bacteriophage, DinoHI (Cheetham et al. 2008) from *D. nodosus* strain H1215. The DinoHI genome is 40 kb in size and contains an integrase gene, *intP*, which has approximately 32% amino acid identity with the integrases from the *intA*, *intB*, *intC* and *intD* elements. Although it is found predominantly in virulent strains, no genes with similarity to genes associated with virulence in bacteria were identified. The integration site for DinoHI is distinct from the integration sites for the *vrl* and the *intA* elements.

9.3.7 Element X

Analysis of sequences adjacent to DinoHI in strain H1215 indicates the presence of another integrated element, which has been designated element X (Cheetham et al. 2008). A 4 kb segment of DNA from one end of this element has been sequenced and shown to encode a type I restriction-modification system.

9.4 A Model for the Control of Virulence by Integrated Genetic Elements

The distribution of integrated genetic elements in *D. nodosus* strains is not random. In particular, the *intA* element and the *vrl* are preferentially associated with virulent strains. However, neither of these genetic elements encodes proteins with a direct role in bacterial virulence. Analysis of the integration sites for these elements has revealed three genes with a potential role as global regulators of virulence, and we have proposed that the integrated genetic elements modulate the expression of these putative global virulence regulators (Whittle et al. 1999).

9.4.1 CsrA

The *tRNA-ser*$_{GCU}$ gene is immediately downstream from *csrA* in the *D. nodosus* genome. CsrA and the closely related protein RsmA are virulence repressors in

Salmonella enterica (Lawhon et al. 2003), *Helicobacter pylori* (Barnard et al. 2004), *Legionella pneumophila* (Molofsky and Swanson 2003) and *Erwinia carotovora* (Mukherjee et al. 1996). Virulent strains of *D. nodosus* have either *intA* or *intC* integrated next to *csrA*. Furthermore, the loss of the *intC* element, which was integrated next to *csrA* in *D. nodosus* strain 1311, resulted in a strain, 1311A, with the *intB* element next to *csrA*, which had lost protease thermostability, a virulence characteristic.

We used reverse-transcriptase PCR to show that *csrA* transcripts extend into the adjacent integrated element (Fig. 9.3). RNA was prepared from four *D. nodosus* strains which have *intA*, *intB* or *intC* next to *csrA*. A forward primer (primer 1) from within the *csrA* coding region was used in conjunction with reverse primers located within *csrA* (primer 2), within the adjacent tRNA gene (primer 3), and at three locations within the *intA* element (primers 4–6). Primer 4 is located in a sequence which is conserved between the *intA* and *intC*, but not the *intB*, elements. These experiments showed that *csrA* is expressed in all strains, and that the transcripts

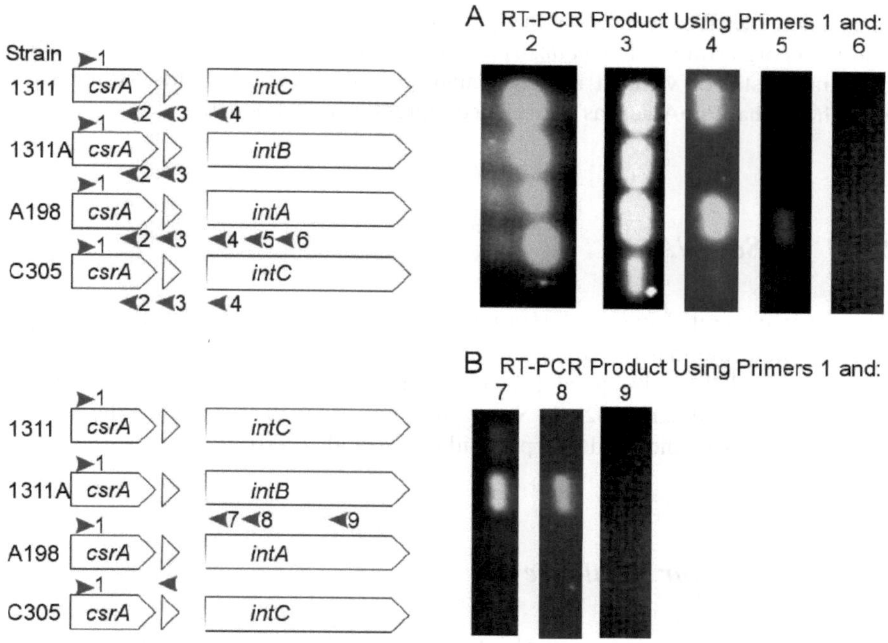

Fig. 9.3 Transcription of *csrA* extends into the adjacent integrated element. Reverse transcriptase PCR using forward primer 1 with reverse primers 2, 3, 4, 5, 6, 7, 8 or 9 was carried out on RNA prepared from *D. nodosus* strains 1311, 1311A, A198 and C305, which have *intC*, *intB*, *intA* or *intC*, respectively, next to *csrA*. The positions to which the primers bind are shown as numbered *arrowheads*. Primer 4 is complementary to a region of identical sequence in the *intA* and *intC* elements. The tRNA gene is shown as a *triangle*. Sections of agarose gels showing products of different sizes for the different primer combinations are shown in panels A and B. No products were detected when reverse transcriptase was omitted from the reaction

include the tRNA gene and extend into the *intA* and *intC* elements (Fig. 9.3a). Using a different reverse primer complementary to a sequence in the *intB*, but not *intA* or *intC* elements, we showed that transcripts also extend into the *intB* element (Fig. 9.3b). Thus, the 3′ ends of *csrA* transcripts are different in strains with different integrated elements at this locus. This may alter transcript stability or the capacity of the mRNA to be translated, thereby altering the expression of this putative virulence regulator.

9.4.2 PnpA

The second integration site for the *intA* element is *tRNA-ser*$_{GGA}$, which is located immediately downstream from *pnpA*. The *pnpA* product, polynucleotide phosphorylase, is a global regulator of virulence in *S. enterica* (Clements et al. 2002). Almost all virulent strains of *D. nodosus*, which have been tested, have *intA* at this integration site. We have shown that partial knockouts of polynucleotide phosphorylase in benign strains of *D. nodosus* result in increased twitching motility, which is a characteristic of virulent strains (Palanisamy et al., 2010). This supports the hypothesis that PnpA acts as a virulence repressor in *D. nodosus*.

9.4.3 10Sa RNA

The *vrl* is integrated into *ssrA* (Haring et al. 1995) which encodes 10Sa RNA, a tmRNA molecule that is involved in the removal of ribosomes stalled on mRNA molecules and the subsequent proteolytic degradation of the resultant peptides (Withey and Friedman 2002). In *S. enterica*, mutations in *ssrA* reduce the expression of genes from the virulence plasmid (Julio et al. 2000).

9.4.4 Model for Virulence

Comparison of the arrangement of integrated genetic elements has shown that, in general, virulent strains have *intA* or *intC* next to *csrA*, *intA* next to *pnpA* and the *vrl* next to *ssrA* (Fig. 9.4). Benign strains may have other integrated genetic elements next to *csrA* or *pnpA* or may lack the *vrl*. We have proposed that these genetic elements modulate the expression of *csrA*, *pnpA* and *ssrA*, which encode virulence regulators. The regulation of virulence by integrated genetic elements in *D. nodosus* may compensate for the dearth of regulatory genes in the very small genome (Myers et al. 2007).

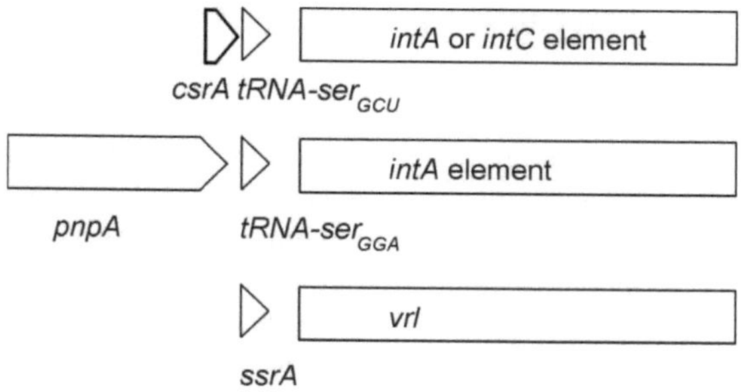

Fig. 9.4 Model for virulence of *D. nodosus*. Virulent strains have either the *intA* or *intC* element integrated into *tRNA-ser*$_{GCU}$, the *intA* element integrated into *tRNA-ser*$_{GGA}$ and the *vrl* integrated into *ssrA*. Benign strains differ in at least one integration site

9.5 Different Forms of the *intA* Element

There is considerable variation in the *intA* element, both within and between strains (Fig. 9.1). In strain A198, *vap* region 1 contains *intA*, a complete suite of *vap* genes (*vapA-H*), *toxA*, and a second copy of the plasmid addiction system, *vapA″/toxA″*. *Vap* region 2 contains the same genes with the exception of *vapA″/toxA″*, but has a small insertion between *vapD* and *vapC* and a short additional sequence after *toxA*. *Vap* region 3 lacks *toxA* and *vapA-C* and has *vapE′* in the place of *vapE*. VapE′ has 67% amino acid identity with VapE. The plasmid form of the *intA* element found in strain AC3577 is identical to *vap* region 1, with the addition of the insertion sequence IS*1253*. In some strains, such as strain H1215, a complete set of *vap* genes is present, but *vapE′* is found instead of *vapE*. Other strains, such as strain 251, have been identified which lack *intA* and have *vapD* immediately downstream from *intA* (Tanjung, unpublished).

Alternative forms of the *intA* element may have arisen by insertions or deletions after the primary integration event. However, analysis of sequences from the *intA* element shows that these events are not simple. For example, strain 251 lacks *intA* and has *vapD* immediately downstream from the integration site. This arrangement could be generated by integration of the *intA* element, followed by the deletion of sequences between the integration site and *vapD*. However, DNA sequencing of the region between the integration site and *vapD* showed that it contains an additional sequence which is found after *toxA* in *vap* region 2. Strain 251 contains *vapE′*, which would normally be located between *vapD* and *intA*, and also contains *vapA*, *vapB* and *vapC*. The exact arrangement of these genes is unknown as Southern blot analysis does not link *vapD* with any of the other *vap* genes in this strain (Fig. 9.1). It is possible that *vapA-C* and *vapE′* are integrated at a different site.

We have identified two families of repeated sequences within *vap* regions 1 and 3 (Cheetham et al. 1995) and have proposed that these sequences may play a role in DNA rearrangements within the *vap* regions. Comparison of DNA sequences between *vap* regions 1, 2 and 3 of strain A198 and the *vap* plasmid of strain AC3577 has shown that these repeats (or partial copies of the repeats) are often located at boundaries where the sequences diverge (Cheetham and Katz 1995; Billington et al. 1996).

9.6 Interactions Between the Integrated Genetic Elements

While the *intA* element is usually integrated into the genome, it may replicate as a plasmid (Billington et al. 1996). The *intD* element has the features of a conjugative plasmid, and excision of the *intD* element has been seen in one strain (Tanjung et al. 2009). Loss of the *intC* element has also been observed (Whittle et al. 1999), so excision of this integrated element is possible. The bacteriophage DinoHI can be stably maintained in the integrated form, but may be induced to replicate by ultraviolet light (Cheetham et al. 2008). The *intB* element and the *vrl* have some bacteriophage-like features but have only been found in the integrated state. Features of these integrated elements which suggest that their replication may be co-ordinately controlled are discussed below.

9.6.1 vapGH *and Bacteriophage Immunity*

The *vapGH* region is located immediately downstream from *intA* in *vap* regions 1 and 2 of strain A198. Related sequences are found on both the *intC* (Whittle et al. 1999) and *intD* (Tanjung et al. 2009) elements (Fig. 9.5). Sequence analysis suggests that *vapH* is not translated since there is no likely Shine-Dalgarno sequence upstream of the start codon. Blast searches revealed that *vapH* and *vapG* are related to *orf*179 of *S. flexneri*, which is in turn related to *orf199* of bacteriophage P4 (Fig. 9.6). During lysogeny in bacteriophage P4, there is constitutive transcription from the promoter P_{LE} (Fig. 9.6), which is located 36 nt downstream from the start codon for *orf199*. Within these transcripts are four

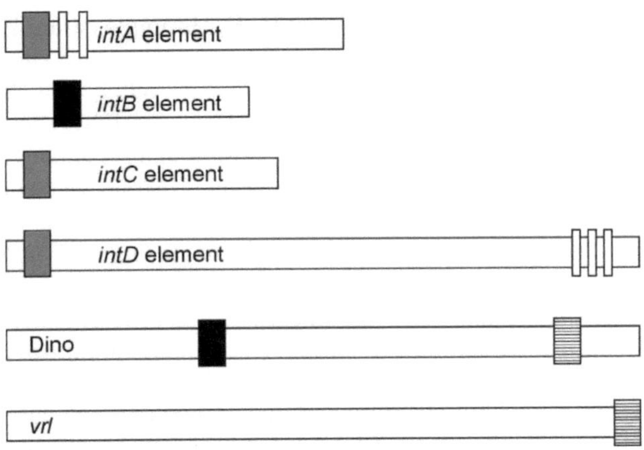

Fig. 9.5 Location of regulatory sequences shared by integrated elements of *D. nodosus*. *Grey boxes* – *vapGH*, *white boxes* – 102 bp repeats, *black boxes* – *regA*, *hatched boxes* – DinoHI packaging site. The figure is not drawn to scale

a

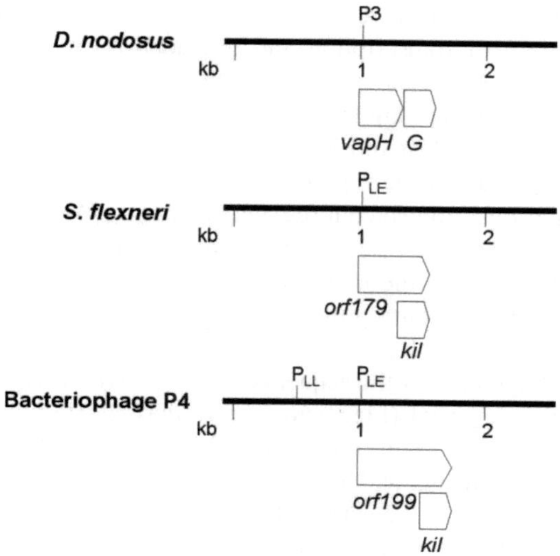

b

```
   orf199                                         -35           P_LE         -10
P4 ATGATGATGCCCGCTCTGCAAAAACTCCCTTTTTCTGGCTTGCCTTTTTCCGGCATTTGCGGATATAGTT  50
                         orf179
Sf -CGGTGATGAAAGGAAAGAGAAAATGTTGAATGTAGCTATTGAAAACCAGAACGGG-TGGAATTATAGTG  48
                          vapH
Dn ATTTAAGGAGATTATTCAATAAAAATGAAGACTTATCTATTGACAAGCCAAGAGCTTTTAGCCTATAGTC  50

                                 seqA
P4 TTTCCGCTGCCGCAAAATCGGCAGCCGGGCGTGAGAACCCGAGCAATTCACAGGCGACATATGACGCGCC 100
Sf CACCTGCACCTCATAAAACGGGTGCCGGGATTGCTACCCCGACGATGACTACTGCGCATAACCGCGCTCA  98
Dn CACTCGCTACAGTAAGCACTGTAGTCGGGATTGGAACCCCGTTGATACGAAAGCGTGTACAGCGCTTATT 100

P4 ACGCGTCTTTTTTTGTGTCGCAATCAATGCCACAGAGCGCCAGATTATGGTGTGGTGTGCGGGGCAGCCT 150
Sf GGCGGTTTTTTTATGCGTC------AAGCACAGCCACATTCAGATTATGGTGGGGCGTGCAGGGCAGCCG 142
Dn AAGCCATTTTTTATG-----CAA--AATTTGTCTATGCAAGCATACGTGGGAGAACCGAAAGGT--TCGC 141

                                 seqB                                seqC
P4 CCGGGCTGGCCGGTATCCTGTGAAGCCGGTATTCTCACCCCCGTCTGGGCTATCGCCATCGAGCGTGAGA 200
Sf CAAGGCTGGCCGGTTTCGGTAGTCACCGGTTGTAGCAACCCTGTGCGTCTCACCACCCATGAGATTGCTA 192
Dn --CGTTTCTTTCGTAGCGGTAGTTCCAATCTCGCGTATGCTGTCGCCCTCTTATTTGGAACTAATGGCGG 189

                                seqC'          kil
P4 ACTCCGGCGATAGCGTATTTGCTATT-CACAGGAGGCTGCCATCATGGCTACAACCCTTACCCCCTCACA 250
Sf CCTCTGGTGGTGAGTCATTTAAACTGACTATCGAGGCTGCCATCATGGCTACTATCCTTACCCTTTCTCA 242
                                               vapG
Dn CAGTCAATCCAAAATTAC-----------GAAAGGAGACTGCGAATGTCTAATTCATTATCGCTGACTAT 232
```

Fig. 9.6 (**a**) Similarities between the *vapGH* region of *D. nodosus*, *orf179* of *S. flexneri* and the immunity region of bacteriophage P4. The location of promoters P3, P_{LE} and P_{LL} is shown. (**b**) Alignment of DNA sequences from the immunity region from bacteriophage P4 (P4) containing *orf199* with regions containing *orf179* from *Shigella flexneri* (Sf) and *vapH* and *vapG* from

sequences designated *seqA*, *seqB*, *seqC* and *seqC′* which form secondary structures resulting in premature termination of transcription and degradation by RNases (Ghisotti et al. 1992; Deho et al. 1992; Lindqvist et al. 1993). Transcription of *orf199* can be initiated from the upstream promoter P_{LL} (Fig. 9.3), but translation of *orf199* and the in-frame gene *kil* (Fig. 9.3) is blocked by binding of the transcripts from the P_{LE} promoter to the *seqB* and *seqC* regions. This system maintains the lysogenic state and prevents superinfection since it prevents incoming P4 particles from expressing genes required for the lytic cycle and replication. In bacteriophage P4, the *kil*-encoded protein is involved in the inhibition of cell division (Ghisotti et al. 1992) and only expressed from P_{LL} during the lytic cycle.

In *D. nodosus*, there is a promoter, P3 (Fig. 9.6) located 16 nt downstream from the start codon for *vapH*. Alignment of the sequences of *vapH* with *orf179* of *S. flexneri* and *orf199* of bacteriophage P4 (Fig. 9.6b) shows that *vapH* contains sequences equivalent to *seqA*, *seqB*, *seqC* and *seqC′*, located in approximately the same positions and capable of forming secondary structures. The predicted start codon for *vapG* is in the same position as the start codons for *kil*in *S. flexneri* and bacteriophage P4. Thus, the *vapGH* region of *D. nodosus* closely resembles the *orf199* region of bacteriophage P4, and *vapH* transcripts may prevent the translation of *vapG* from integrated elements, possibly to maintain the integrated state. Although *vapG* does not share sequence homology with *kil*, it may perform a similar function. It is of interest that only one strain of *D. nodosus* has been identified in which the *intA* element is stably maintained as a plasmid and in this strain, AC3577, there is a deletion of six nucleotides together with 18 base substitution mutations within *vapH*.

The *vapGH* system is found on the *intA*, *intC* and *intD* elements. Since numerous strains contain both the *intA* and the *intC* elements, the *intA* and *intD* elements, the *intC* and *intD* elements, or multiple copies of the *intA* element, the *vapGH* system does not appear to affect superinfection.

9.6.2 *A Repressor Gene Common to Bacteriophage DinoHI and the* intB *Element*

The *intB* element contains a gene, *regA*, which is related to several bacteriophage repressor proteins, including the cI repressor from lambdoid bacteriophages (Bloomfield et al. 1997). The DinoHI genome (Cheetham et al. 2008) contains

---

Fig. 9.6 (continued) D. nodosus (Dn). Start codons for *orf199*, *orf179*, *vapH*, *kil* and *vapG* are *underlined* and *bold*. Sequences that can form *seqA*, *seqB*, *seqC* and *seqC′* which interact to form stem-loop structures are in *bold*. The consensus sequence for the P_{LE} promoter of bacteriophage P4 and the corresponding promoter regions from *S. flexneri* and *D. nodosus*, respectively, are *underlined*. Numbers on the *right hand side* show the nucleotide number. Accession numbers are as follows: P4 (MYP4CG), Sf (Z23101) and Dn (L31763)

a highly related gene, *regA2* (Fig. 9.5) which has 97% DNA sequence identity over 701 nt, resulting in a protein product which is identical for 231 of the 232 amino acids of the *intB* RegA. However, the start codon is absent in the DinoHI sequence. Instead, there is another potential start codon upstream which would result in an extra 17 amino acids at the N-terminus for the DinoHI RegA protein.

The likely function of RegA is to maintain the *intB* element and the DinoHI prophage in the integrated state. The high level of amino acid identity between RegA and RegA2 suggests that the two proteins may be functionally interchangeable, i.e. these repressor proteins would maintain both these genetic elements in the integrated state. Thus, excision of bacteriophage DinoHI and the *intB* element could be co-ordinately controlled.

9.6.3 Interactions Between the vrl, *DinoHI and the* intA *Element*

A 1.8 kb DNA sequence from the DinoHI genome which contains the linear ends is also found at the end of the *vrl* (Cheetham et al. 2008). Thus, strains which contain both DinoHI and the *vrl* have two copies of this DNA sequence. This sequence is used for packaging of the DinoHI genome into the DinoHI bacteriophage particle. The presence of this sequence at the ends of the *vrl* suggests that the *vrl* may be transferred between *D. nodosus* strains by transduction.

The *vrl* lacks an integrase gene, but all strains which contain the *vrl* contain the *intA* element, suggesting that the *intA* integrase may be used for integration of the *vrl* (Billington et al. 1996).

9.6.4 Mobilisation of the intA *Element by the* intD *Mobilisation Cassette*

The *intD* element encodes a type IV secretion system with a probable role in DNA transport during conjugation (Tanjung et al. 2009). A putative origin of replication, *oriV*, and an adjacent origin of transfer, *oriT*, have also been identified. Thus, the *intD* element appears to be self-transmissible. The *intD* element contains a mobilisation region which may be involved in the transfer of smaller plasmids or other non-self-transmissible elements. Mobilisation regions typically encode proteins that function in the nicking of the *oriT* region, the subsequent piloting of the 5′ end of the nicked strand into the recipient cell and recircularisation and priming of the complementary strand for replication (Boyd et al. 1989). Four potential genes clustered at the end of the *intD* element, which contains *oriT*, share amino acid similarity with mobilisation genes. The *oriT* sequence from the *intD* element is also found on the *intA* element, next to the origin of replication (Cheetham et al. 1995). This strongly suggests that the *intA* element can be mobilised and transferred to other *D. nodosus* strains by the *intD* element.

9.6.5 *Common Repeated Sequences on the* intA *and* intD *Elements*

A 102 bp repeat is found at several locations on the *intA* element (Cheetham et al. 1995) and may have a role in rearrangements since copies or partial copies of this sequence are found at boundaries where there is sequence divergence in different forms of the *intA* element (Sect. 9.5). The *intD* element contains three copies of this 102 bp repeat, suggesting that it may be possible to exchange sequences between these two elements.

9.6.6 *Relationships Between the* intC *and* intD *Elements*

Southern blot analysis of a range of *D. nodosus* strains has shown the presence of large segments of the *intD* element in strains which lack the integrase gene, *intD* (Tanjung et al. 2009). This raises the possibility that sequences from the *intD* element may be associated with a different integrated element in other strains (Fig. 9.2). All these strains contained *intC*, and loss of the *intC* element in one of these, strain 1311, resulted in concomitant loss of the segments of the *intD* element. Thus, it seems likely that the genes from the *intD* element are found on the *intC* element in some strains. As described in Sect. 9.3.3, the *intC* element has only been fully characterised from strain C305 where it may have been disrupted by an insertion sequence. The strains which lack *intD* but contain segments of the *intD* element are those strains which appear to have a longer form of the *intC* element.

9.7 Evolutionary Significance

The analysis of *D. nodosus* strains from different sheep flocks in Australia has shown a great deal of genetic diversity. In addition, flocks of sheep or even individuals within the flock may be infected with multiple strains (Claxton et al. 1983), and it is not uncommon to isolate both benign and virulent strains from within the same flock. The persistence of both benign and virulent strains suggests that there may be evolutionary advantages under different conditions. Footrot infections are greatly influenced by environmental conditions. Under ideal conditions of moderate temperatures and high moisture levels, infections tend to be very severe and virulent strains may rapidly destroy the hoof, eventually resulting in death of the host. Under hot, dry conditions, benign footrot rapidly resolves and virulent strains may be more able to survive as isolated pockets of infection within the hoof. We have proposed that strains may switch between benign and virulent states by the excision, or acquisition and integration, of genetic elements, thereby altering the expression of three genes encoding virulence regulators (CsrA, PnpA and 10Sa RNA). This provides a mechanism for rapid and reversible adaptation of strains.

9.8 Conclusions

We have identified a series of genetic elements integrated into the *D. nodosus* genome and have proposed that these genetic elements modulate the expression of upstream genes encoding putative virulence regulators. The *intA*, *intB*, *intC* and *intD* elements integrate downstream from *csrA* while the *intA*, *intB* and *intD* elements may also integrate downstream from *pnpA*. The *vrl* integrates downstream from *ssrA*. The bacteriophage DinoHI does not appear to influence virulence directly, but may be responsible for the transfer of the *vrl* between strains. The *vapG*/*vapH* system may maintain the *intA*, *intC* and *intD* elements in the integrated state, while the putative repressor, RegA, may control both the *intB* element and DinoHI. Exchange of sequences between some of these genetic elements also appears likely. Thus, our model suggests that the integrated elements of *D. nodosus* control the expression of virulence determinants and that the elements themselves are controlled in a coordinated manner.

References

Barnard FM, Loughlin MF, Fainberg HP, Messenger MP, Ussery DW, Williams P, Jenks PJ (2004) Global regulation of virulence and the stress response by CsrA in the highly adapted human gastric pathogen *Helicobacter pylori*. Mol Microbiol 51:15–32

Beveridge WIB (1941) Footrot in sheep: a transmissible disease due to infection with *Fusiformis nodosus*. CSIRO Bull 140:1–40

Billington SJ, Huggins AS, Johanesen PA, Crellin PK, Cheung JK, Katz ME, Wright CL, Haring V, Rood JI (1999) Complete nucleotide sequence of the 27-kilobase virulence related locus (*vrl*) of *Dichelobacter nodosus*: evidence for extrachromosomal origin. Infect Immun 67:1277–1286

Billington SJ, Sinistaj M, Cheetham BF, Ayres A, Moses EK, Katz ME, Rood JI (1996) Identification of a native *Dichelobacter nodosus* plasmid and implications for the evolution of the *vap* regions. Gene 172:111–116

Bloomfield GA, Whittle G, McDonagh MB, Katz ME, Cheetham BF (1997) Analysis of sequences flanking the *vap* regions of *Dichelobacter nodosus*: evidence for multiple integration events, a killer system, and a new genetic element. Microbiology 143:553–562

Boyd AC, Archer JA, Sherratt DJ (1989) Characterization of the ColE1 mobilization region and its protein products. Mol Gen Genet 217:488–498

Cheetham BF, Katz ME (1995) A role for bacteriophages in the evolution and transfer of bacterial virulence determinants. Mol Microbiol 18:201–208

Cheetham BF, Parker D, Bloomfield GA, Shaw BE, Sutherland M, Hyman JA, Druitt J, Kennan RM, Rood JI, Katz ME (2008) Isolation of the bacteriophage DinoHI from *Dichelobacter nodosus* and its interactions with other integrated genetic elements. Open Microbiol J 2:1–9

Cheetham BF, Tattersall DB, Bloomfield GA, Rood JI, Katz ME (1995) Identification of a gene encoding a bacteriophage-related integrase in a *vap* region of the *Dichelobacter nodosus* genome. Gene 162:53–58

Cheetham BF, Whittle G, Katz ME (1999) Are the *vap* regions of the *Dichelobacter nodosus* genome pathogenicity islands? In: Kaper JB, Hacker J (eds) Pathogenicity islands. ASM Press, Washington, pp 203–218

Claxton PD, Ribeiro LA, Egerton JR (1983) Classification of *Bacteroides nodosus* by agglutination tests. Aust Vet J 60:331–334

Clements MO, Eriksson S, Thompson A, Lucchini S, Hinton JC, Normark S, Rhen M (2002) Polynucleotide phosphorylase is a global regulator of virulence and persistency in *Salmonella enterica*. Proc Natl Acad Sci USA 99:8784–8789

Dalrymple B, Mattick JS (1987) An analysis of the organization and evolution of type 4 fimbrial (MePhe) subunit proteins. J Mol Evol 25:261–269

Deho G, Zangrossi S, Sabbattini P, Sironi G, Ghisotti D (1992) Bacteriophage P4 immunity controlled by small RNAs via transcription termination. Mol Microbiol 6:3415–3425

Depiazzi LJ, Richards RB (1985) Motility in relation to virulence of *Bacteroides nodosus*. Vet Microbiol 10:107–116

Depiazzi LJ, Rood JI (1984) The thermostability of proteases from virulent and benign strains of *Bacteroides nodosus*. Vet Microbiol 9:227–236

Dewhirst FE, Paster BJ, La Fontaine S, Rood JI (1990) Transfer of *Kingella indologenes* (Snell and Lapage 1976) to the genus *Suttonella* gen. nov. as *Suttonella indologenes* comb. nov.; transfer of *Bacteroides nodosus* (Beveridge 1941) to the genus *Dichelobacter* gen. nov. as *Dichelobacter nodosus* comb. nov.; and assignment of the genera *Cardiobacterium*, *Dichelobacter*, and *Suttonella* to *Cardiobacteriaceae* fam. nov. in the gamma division of *Proteobacteria* on the basis of 16S rRNA sequence comparisons. Int J Syst Bacteriol 40:426–433

Egerton JR, Yong WK, Riffkin GG (1988) Footrot and foot abscess of ruminants. CRC Press, Florida

Ghisotti D, Chiaramonte R, Forti F, Zangrossi S, Sironi G, Deho G (1992) Genetic analysis of the immunity region of phage-plasmid P4. Mol Microbiol 6:3405–3413

Han X, Kennan RM, Davies JK, Reddacliff LA, Dhungyel OP, Whittington RJ, Turnbull L, Whitchurch CB, Rood JI (2008) Twitching motility is essential for virulence in *Dichelobacter nodosus*. J Bacteriol 190:3323–3335

Haring V, Billington SJ, Wright CL, Huggins AS, Katz ME, Rood JI (1995) Delineation of the virulence-related locus (*vrl*) of *Dichelobacter nodosus*. Microbiology 149:2081–2086

Julio SM, Heithoff DM, Mahan MJ (2000) *ssrA* (tmRNA) plays a role in *Salmonella enterica* Serovar Typhimurium pathogenesis. J Bacteriol 182:1558–1563

Katz ME, Howarth PM, Yong WK, Riffkin GG, Depiazzi LJ, Rood JI (1991) Identification of three gene regions associated with virulence in *Dichelobacter nodosus*, the causative agent of ovine footrot. J Gen Microbiol 137:2117–2124

Katz ME, Strugnell RA, Rood JI (1992) Molecular characterization of a genomic region associated with virulence in *Dichelobacter nodosus*. Infect Immun 60:4586–4592

Katz ME, Wright CL, Gartside TS, Cheetham BF, Doidge CV, Moses EK, Rood JI (1994) Genetic organization of the duplicated *vap* region of the *Dichelobacter nodosus* genome. J Bacteriol 176:2663–2669

Kennan RM, Dhungyel OP, Whittington RJ, Egerton JR, Rood JI (2001) The type IV fimbrial subunit gene (*fimA*) of *Dichelobacter nodosus* is essential for virulence, protease secretion, and natural competence. J Bacteriol 183:4451–4458

Laing EA, Egerton JR (1978) The occurrence, prevalence and transmission of *Bacteroides nodosus* infection in cattle. Res Vet Sci 24:300–304

Lawhon SD, Frye JG, Suyemoto M, Porwollik S, McClelland M, Altier C (2003) Global regulation by CsrA in *Salmonella typhimurium*. Mol Microbiol 48:1633–1645

Lindqvist BH, Deho G, Calendar R (1993) Mechanisms of genome propagation and helper exploitation by satellite phage P4. Microbiol Rev 57:683–702

Molofsky AB, Swanson MS (2003) *Legionella pneumophila* CsrA is a pivotal repressor of transmission traits and activator of replication. Mol Microbiol 50:445–461

Mukherjee A, Cui Y, Liu Y, Dumenyo CK, Chatterjee AK (1996) Global regulation in *Erwinia* species by *Erwinia carotovora rsmA*, a homologue of *Escherichia coli csrA*: repression of secondary metabolites, pathogenicity and hypersensitive reaction. Microbiology 142: 427–434

Myers GS, Parker D, Al Hasani K, Kennan RM, Seemann T, Ren Q, Badger JH, Selengut JD, Deboy RT, Tettelin H, Boyce JD, McCarl VP, Han X, Nelson WC, Madupu R, Mohamoud Y,

Holley T, Fedorova N, Khouri H, Bottomley SP, Whittington RJ, Adler B, Songer JG, Rood JI, Paulsen IT (2007) Genome sequence and identification of candidate vaccine antigens from the animal pathogen *Dichelobacter nodosus*. Nat Biotechnol 25:569–575

Palanisamy SKA, Fletcher C, Tanjung L, Katz ME and Cheetham BF (2010) Deletion of the C-terminus of polynucleotide phosphorylase increases twitching motility, a virulence characteristic of the anaerobic bacterial pathogen *Dichelobacter nodosus*. FEMS Microbiol Lett 302:39–45

Stewart DJ (1979) The role of elastase in the differentiation of *Bacteroides nodosus* infections in sheep and cattle. Res Vet Sci 27:99–105

Stewart DJ (1989) Footrot of sheep. In: Egerton JR, Yong WK, Rifkin GG (eds) Footrot and foot abscess of ruminants. CRC Press, Florida, pp 5–45

Tanjung LR, Whittle G, Shaw BE, Bloomfield GA, Katz ME, Cheetham BF (2009) The *intD* mobile genetic element from *Dichelobacter nodosus*, the causative agent of ovine footrot, is associated with the benign phenotype. Anaerobe 15(5):219–224

Tian QB, Ohnishi M, Tabuchi A, Terawaki Y (1996) A new plasmid-encoded killer gene system: cloning, sequencing, and analyzing *hig* locus of plasmid Rts1. Biochem Biophys Res Commun 220:280–284

Whittle G, Bloomfield GA, Katz ME, Cheetham BF (1999) The site-specific integration of genetic elements may modulate thermostable protease production, a virulence factor in *Dichelobacter nodosus*, the causative agent of ovine footrot. Microbiology 145:2845–2855

Withey JH, Friedman DI (2002) The biological roles of trans-translation. Curr Opin Microbiol 5:154–159

Chapter 10
Gene Network Holography of the Soil Bacterium *Bacillus subtilis*

Dalit Roth, Asaf Madi, Dror Y. Kenett, and Eshel Ben-Jacob

10.1 Introduction

The development of microarray chip technologies has made it possible to generate vast amounts of raw data regarding the genomic response of the soil bacteria, *Bacillus subtilis* (Caldwell et al. 2001; Fawcett et al. 2000; Hamoen et al. 2002; Molle et al. 2003; Serizawa et al. 2004) [for a review on microarray studies in *B. subtilis*, see (Kocabas et al. 2009)]. Advanced analysis methods have been devised to extract meaningful information from gene-expression data. Most of these methods focus on distinguishing between groups of subjects or identifying the most relevant genes that help to distinguish between these groups – marker genes that exhibit distinct up- or down-regulation.

Current gene-expression data analysis methodologies can be divided into two categories: supervised approaches that aim to determine genes that fit a predetermined pattern; and unsupervised approaches, which aim to characterize the components without a priori assumptions. Supervised methods are usually used to find individual genes such as the nearest neighbor approach (Golub et al. 1999) and/or multiple genes, such as decision trees (Quinlan 1992), neural networks (Rumelhart et al. 1986), and support vector machines (Furey et al. 2000; Hartigan 1975; Rumelhart et al. 1986). Unsupervised methods are usually based on cluster analysis (Everitt 1993; Hansen and Jaumard 1997; Hartigan 1975; Mirkin 1996). Several

D. Roth and A. Madi
Faculty of Medicine, Tel Aviv University, 69978 Tel Aviv, Israel

D.Y. Kenett,
School of Physics and Astronomy, Tel Aviv University, 69978 Tel Aviv, Israel

E. Ben-Jacob (✉)
School of Physics and Astronomy, Tel Aviv University, 69978 Tel Aviv, Israel
The Center for Theoretical and Biological Physics, University of California San Diego, La Jolla, CA 92093, USA
e-mail: eshel@tamar.tau.ac.il; eshelbj@gmail.com

G. Witzany (ed.), *Biocommunication in Soil Microorganisms*, Soil Biology 23,
DOI 10.1007/978-3-642-14512-4_10,

algorithmic techniques were previously used in clustering gene-expression data, including hierarchical clustering (Eisen et al. 1998), self organizing maps (Tamayo et al. 1999), K-means (Herwig et al. 1999), simulated annealing (Alon et al. 1999), and graph theoretic approaches: HCS (Hartuv and Shami 2000), CAST (Ben-Dor et al. 1999), and CLICK (Sharan et al. 2003).

Recently, we have presented the Genome Holography (GH) approach to the investigation and analysis of gene networks (Madi et al. 2008). This method is based on the Functional Holography method (FH), which was previously used to investigate many different complex systems, such as the brain (Baruchi and Ben-Jacob 2004), the immune system (Madi et al. 2009), and the stock market (Shapira et al. 2009). The FH method includes collective normalization of the correlations (according to the correlations of each gene with all the others). The matrices of normalized correlations are then analyzed using dimension reduction algorithms (here we use the Principal Component Algorithm – PCA) to keep the most relevant information. Next, to reveal functional motifs – functional relations between genes, the genes are located in a reduced three-dimensional (3D) space whose axes are the three leading principal vectors of the PCA. We note that projection on a low dimension space of principal vectors is common practice in clustering investigations. Apart from the projection on a 3D space, to retain putative relevant information that can be embedded in the higher dimensions, the genes in the reduced space are linked with lines colored according to the values of the correlations and for selected range of correlation values. Doing so makes it possible to decipher, for example, if distinct clusters in the 3D space are linked in the higher dimensions.

Here we present the approach by analyzing a database of gene expression of *B. subtilis* exposed to sublethal levels of antibiotics (Hutter et al. 2004). The gene-expression levels were monitored in response to 37 different kinds of antibiotics and at three time points after the exposure. We also note that, although the focus in this work on analyzing the matrices of correlations between genes (the gene correlation matrices), important information can also be extracted, in principle, by analyzing the matrices of correlations between the responses to the different antibiotics.

To test the capabilities of the GH method, we investigate its ability to identify, from the gene-expression data, the organization of genes into operons. These functional units (cliques) in the genome are composed of one or more genes cotranscribed into one polycistronic mRNA – a single mRNA molecule that codes for more than one protein. At the macroscopic scale, operons are organized as a network connected by regulators that control, as many other biological networks, joint biological functions and pathways. Recent advancement of experimental methods induced a rapid increase in the available detailed information about the various genes clustered in the operon system. However, knowledge is still lacking about the functional principles that govern the relationship between operon genes and external regulators. We will demonstrate that our analysis method can extract such information from gene-expression data. For example, when projecting the genes of each operon onto the 3D space according to their correlations with the other genes, they tend to form distinct clusters.

The strengths of the GH analysis of gene data will be further demonstrated by representing two more applications. First, the application of the GH analysis to investigate time development of gene regulatory networks under a specific stress will be introduced. Second, we will combine the FH analysis with the commonly used Minimal Spanning Tree (MST) analysis (Mantegna 1999; Mantegna and Stanley 2000; Tumminello et al. 2007; Xu et al. 2001). This allows a more complex yet informative 3D representation of the constructed MST.

Out of the gene database, genes belonging to three well known regulatory networks: sporulation, cannibalism, and competence were selected. These three networks are developmental pathways that maintain the survival of the bacteria under transition state stress conditions, such as nutrient depletion and high cell density (Serror and Sonenshein 1996; Serror et al. 2001; Sonenshein 2000). Other stresses that are also known to cause changes in those pathways are oxidative, osmotic, general stress conditions (Claverys et al. 2006; Dowds et al. 1987; Ruzal and Sanchez-Rivas 1998) and antibiotics (Atsuhiro et al. 2003; Jonas et al. 1990; Prudhomme et al. 2006; Rogers et al. 2007; Vazquez-Ramos and Mandelstam 1981).

Using this novel methodology, we show for a given network the internal gene interactions, in terms of their expression similarity. Furthermore, we demonstrate that these gene interactions change with time, which enabled to identify specific regulated genes that were significantly affected by the antibiotic stress. Finally, novel intra-gene network motifs that are found for a given antibiotic stress are presented.

10.2 Functional Holography Analysis of Gene Expression

The FH approach was first introduced by Baruchi et al. (Baruchi and Ben-Jacob 2004) for analysis of recorded human brain activity. The term hologram stands for "whole" – holo in Greek, plus "information" or "message" – gram in Greek. Implementing this methodology on gene expression makes possible to uncover hidden gene structural and functional motifs.

10.2.1 Holographic Presentation of the Genes

In Fig. 10.1, the normalized correlation matrix is presented, which efficiently sorts the genes according to the operons they belong to. In Fig. 10.2, we show the holographic presentation of the matrix of normalized correlations using the projection on the 3D space as described by Madi et al. (2008). Genes belonging to the same operon are given the same color. Lines colored according to the correlation values link pairs of genes with correlations above 0.7. It is possible to observe that

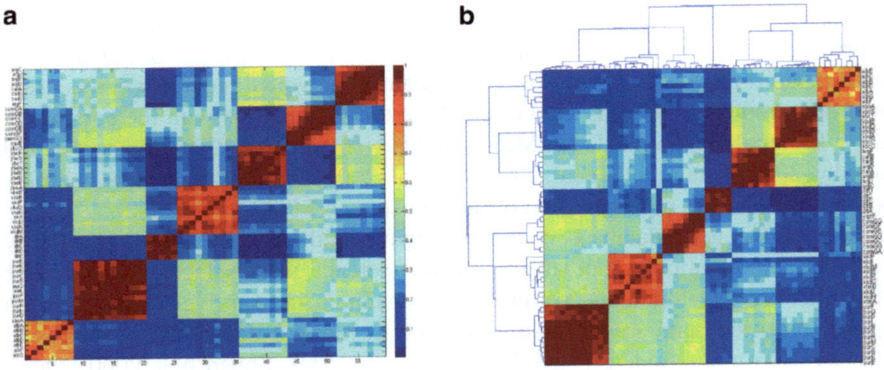

Fig. 10.1 Matrix of normalized correlations. (**a**) The supervised sorted matrix (according to the operons). (**b**) The unsupervised version, sorted by the dendrogram algorithm. Note that the dendrogram algorithm sorts the operons in a different order (7, 3, 1, 5, 2, 4, 6)

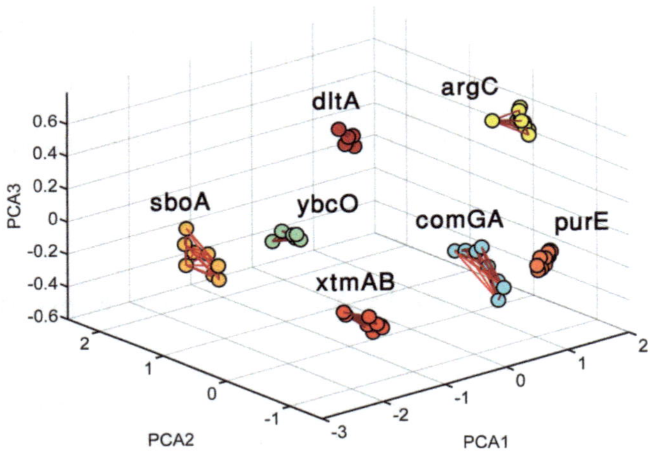

Fig. 10.2 Holographic presentation of the genes in the 3D space. The genes are located in the 3D space whose axes are the three leading principal components calculated by applying PCA to the matrix of normalized correlations. The links between pairs of genes are for correlations above 0.7

only genes belonging to the same operon are linked, implying that interoperon correlations are weaker than intra-operon correlations.

10.2.2 *Internal Structure of Gene Operons*

In this section, we focus on holographic zooming (Baruchi and Ben-Jacob 2004; Baruchi et al. 2006) analysis of subgroups of genes – the operons. The idea is to separately perform the collective normalization on the correlation matrix of the subgroup of genes and to calculate a new 3D space for this specific subgroup. A clear correspondence then surfaces between the genes functional relations and the known structures of the operons. The results are illustrated for specific operon – *pyrR* (Chander et al. 2005) that has a non trivial internal organization.

The *pyrR* operon (Chander et al. 2005) has a complex structure, as shown in Fig. 10.3a. It is composed of ten genes that are organized in three subunits: (1) The gene *pyrR* that acts as self-inhibitor of the operon as a whole and also acts as an inhibitor of the three subunits. (2) The gene *pyrP* is located downstream from the *pyrR* with a terminator segment in between. (3) The third subunit is composed of eight genes downstream from the *pyrP* with terminator and promoter segments in between. This operon is regulated by SigA and PurR.

The normalized correlation matrix of the *pyrR* operon genes, *pyrR*, *pyrP*, *pyrB*, *pyrC*, *pyrAA*, *pyrAB*, *pyrK*, *pyrD*, *pyrF* and *pyrE*, and the corresponding holographic presentation are shown in Fig. 10.3b, c. In Fig. 10.3c, it can clearly be seen that *pyrR* and *pyrP* are distinct from each other and from the eight genes of the third subunit of the operon. We also note that the holographic functional organization of the operon in the 3D space corresponds to the structural organization of the operon, as *pyrR* gene is linked only to the *pyrP*, which, in turn, is linked to the rest of the genes. The genomic scheme of the operon in Fig. 10.3a is consistent with these results, as *pyrR* and *pyrP* are separated from the rest of the genes in the operon by terminators. Furthermore, it was previously shown that PyrR is a protein that

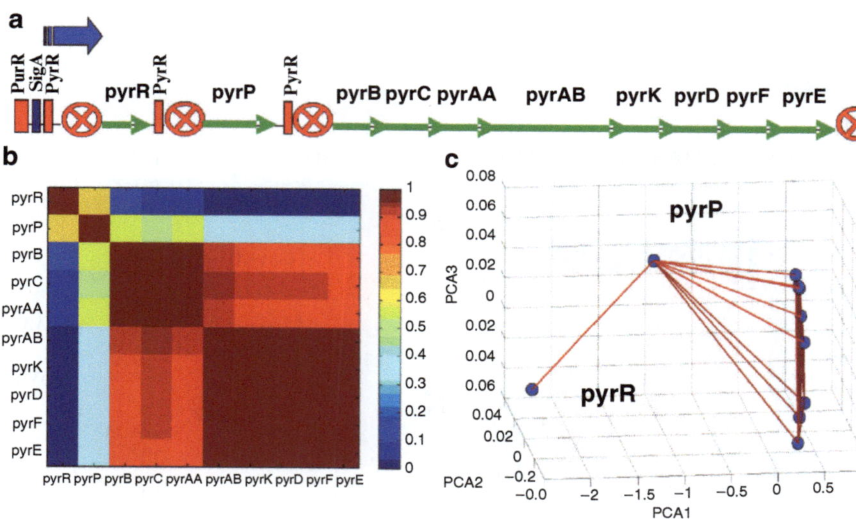

Fig. 10.3 The pyrR operon. (**a**) Schematic internal sequential structure presentation (Chander et al. 2005). Promoter represented by a *blue arrow*. Terminator represented by a *marked circle*. *Rectangles* represent regulation regions of *sigA*, *purR*, and *pyrR*. (**b**) Matrix of normalized correlation. Note the low correlations between *pyrR* to the rest of the genes, matching its function as a negative regulator. (**c**) FH holographic presentation, with correlation values above 0.7 marked by *lines*

regulates the expression of genes and operons of pyrimidine nucleotide biosynthesis (*pyr* genes) in many bacteria and specifically in B. subtilis. *pyrR* acts by binding to specific sequences on *pyr* mRNA, causing transcriptional attenuation when intracellular levels of uridine nucleotides are elevated (Chander et al. 2005).

Based on this demonstrated efficiency of the method, we proceeded to test its ability to predict/reveal unknown functional relations. To test this ability, we investigate the *spoVAA* operon (Azevedo et al. 1993). In Fig. 10.4a, we show the currently presumed internal structure of this operon. The matrix of normalized correlations and its holographic networks in the PCA space are shown in Fig. 10.4b, c, respectively. Inspecting these results, it was observed that *lysA* has weak correlations with the other genes in the operon. These results are somewhat unexpected since no terminator or regulation factors were found between *spoVAF* and *lysA* (Genbank L09228). Azevedo et al. found a 2.3 kb transcript originating about 1 kb upstream of the *lysA* start codon, suggesting that transcription of *spoVA* continues into the *lysA* gene. However, the *lysA* gene is also transcribed monocistronically as a 1.3-kb transcript. A possible explanation might be the existence of a regulation element, a terminator perhaps, between the *spoVAF* and *lysA* genes. Another possible explanation can also be the existence of an additional unknown pathway (through another gene) in which the *lysA* gene acts as a negative regulator of the spoVAA operon. LysA mediates the last step of the lysine biosynthesis (Rodionov et al. 2003). The lysine-mediated gene regulation in bacteria appears to operate via a unique RNA structural element similar to riboswitch that is involved in the regulation of purin biosynthesis (Mandal et al. 2003). The LYS element is

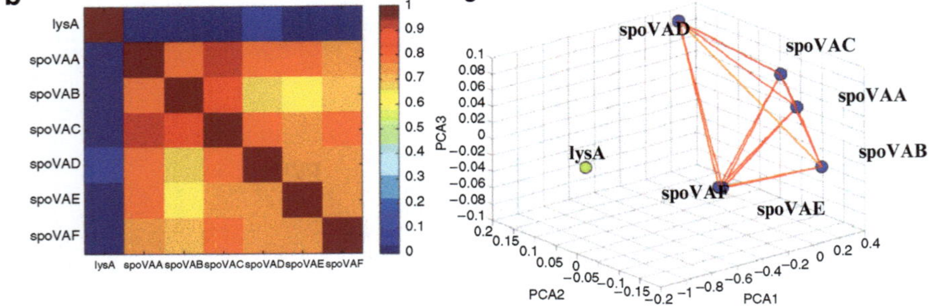

Fig. 10.4 The spoVAA operon (Azevedo et al. 1993). (**a**) Schematic internal sequential structure presentation. The promoter is represented by a *blue arrow*, the terminator by a *red crossed circle*, and the binding site of the activator by a *purple rectangle*. (**b**) Matrix of normalized correlations. (**c**) FH holographic presentation in 3D PCA space, where correlation values above 0.8 is shown in *lines*

characterized by its compact secondary structure with a number of conserved helices and extended regions of sequence conservation, which could be necessary for specific metabolite binding (Rodionov et al. 2003). Comparative genomic analysis predicted conserved RNA secondary structures in lysine metabolism genes such as *lysC* and *lysA*. Thus, our analysis supports the genomic prediction of a regulatory element adjacent to the *lysA* gene and transcription of lysA monocistronically.

10.3 Minimal Spanning Tree Analysis

The use of graph theory and network theory has become widespread over the past few years in the analysis of many physical (Donetti et al. 2005), biological (de Jong 2002; Ortega et al. 2008), and economic systems (Coronnello et al. 2005; Tumminello et al. 2007). In particular, graph theory can be useful for studying correlation-based systems.

The weighted adjacency matrix. The first step in order to utilize the analyses methods from network theory is to compute the adjacency matrix that corresponds to the correlation matrix. The adjacency matrix A is a binary matrix that was developed to describe the topology of the network connectivity. In the case of activity network such as correlations between stocks (Mantegna 1999) or synchronization between neuron firing (Fuchs et al. 2009), the function matrix (e.g., correlations or synchronizations) is first transformed into a distance matrix. This can be done in different ways; one of the more common ways is by using the ultrametric distance, first suggested by Mantegna et al. (Mantegna 1999; Mantegna and Stanley 2000). This is done by transforming the normalized correlation between two nodes, $Aff(i,j)$, into a distance by

$$D(i,j) = \sqrt{2[1 - Aff(i,j)]} \tag{10.1}$$

Note that high correlations correspond to short distances and vice versa. In a network that corresponds to a weighted adjacency matrix, all the nodes are connected (with weighted links between the nodes according to the distances) and hence the topological structure is that of a complete graph.

Minimum spanning tree. Since the relevant information can be obscured in the complete graph (West 2001), the idea is then to extract a subgraph, the minimum spanning tree (MST), from the complete network, based on the weights of the links in order to extract the most relevant information contained in the complete graph. We then focus on the topological structure of the sub graph, and make use of graph theory techniques to analyze this information (Albert and Barabási 2002; Graham and Hell 1985; Mantegna and Stanley 2000; Newman 2003).

Here we have chosen to use the Kruskal algorithm (Kruskal 1956) to compute the MST. This algorithm looks for a subset of the branches that forms a tree that includes every node, where the total weight of all the branches, namely the score that is derived from the correlation, in the tree is minimized. If the graph is not connected, then it finds a *minimum spanning forest* (a minimum spanning tree for each connected node). More specifically, the algorithm starts where every node considered a separate tree in the forest. Then, the algorithm considers each branch in turn, in order by increasing weight. If a branch connects two different trees, then it is added to the set of branches of the MST, and two trees connected by this branch are merged into a single tree on the other hand; if a branch connects two nodes in the same tree, then it is discarded. This is repeated until the forest is reduced into one tree.

To demonstrate the use of MST to analyze gene networks (see also Xu et al. 2001), we apply this method to study genes belonging to seven operons. This tree is presented in Fig. 10.5. We color-code the genes of each operon in the same color, and thus are able to observe that each operon is grouped together in the tree.

Finally, we combine the MST analysis with the FH analysis, from which we obtain the Functional Holography minimal Spanning Tree (FHMST). To achieve this, we first calculate the reduced PCA space for a given set of genes, and then project on it the gene connections, according to the results of the MST analysis.

The results of this method are demonstrated using the sporulation genes, discussed above. First, we perform the FH analysis on these genes, using the data for all three-time points for this example. Once we have created the FH space, we compute the MST for these genes, and from it obtains the connections between the genes. This is presented in Fig. 10.6.

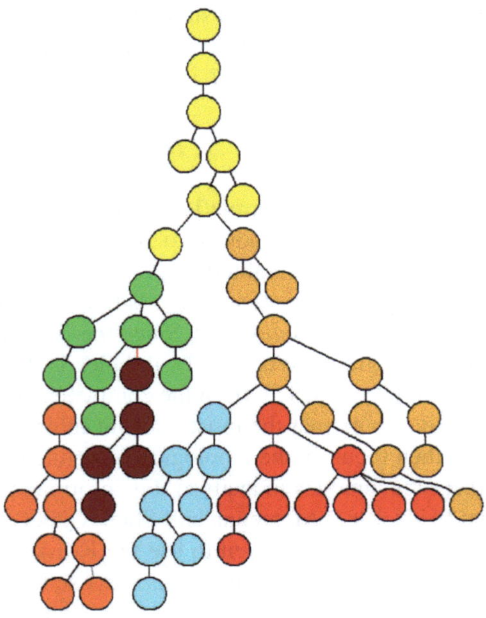

Fig. 10.5 MST analysis of gene operon data. We created the MST for genes belonging to seven operons and color-coded each operon according to Fig. 10.2. We observe that each operon is grouped in the tree

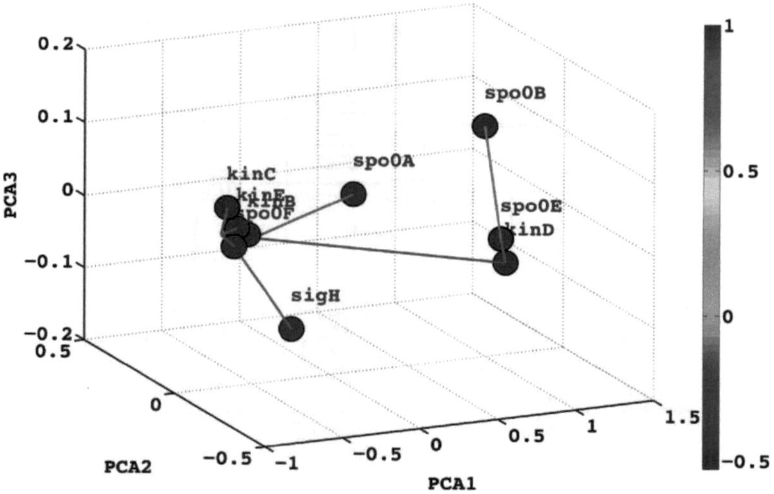

Fig. 10.6 Functional Holography Minimal Spanning Tree (FHMST) for the sporulation gene network

10.4 Time Progression of the Sporulation and Competence Networks

Sporulation is a multistage, developmental process that is responsible for the conversion of a growing cell into a dormant cell type known as the spore or endospore (Stragier and Losick 1996). Sporulation is not initiated automatically (deterministically) upon nutrient limitation, but instead it is the end result of a series of steps that might be described as cellular decisions regarding how to best cope with the stress (Veening et al. 2005).

The master regulator for entry into sporulation is the response regulator Spo0A (Fig. 10.7) (Hoch 1993). The activity of Spo0A is governed by a multicomponent phosphorelay, which consists of five histidine autokinases (KinA, KinB, KinC, KinD, and KinE), and two phosphorelay proteins (Spo0F and Spo0B) (Jiang et al. 2000), which are responsible for the phosphorelation of Spo0A (Burbulys et al. 1991). The level of phosphorylation of Spo0A~P is also influenced by dedicated phosphatases that remove phosphoryl groups from Spo0F~P and from Spo0A~P itself (Spo0E) (Grossman 1995; Parego et al. 1994).

The competence regulatory network, which acts as a stochastic switch, controls the escape rate into competence from the sporulation path and the exit time back from the competence state (Jiang et al. 2000). Competence is the ability of the bacteria to bind and take up exogenous DNA (Lorenz and Wackernagel 1994). In the center of the signal-transduction network of the competence system positioned, ComK (Fig. 10.7). *comK* is a positive auto-regulatory gene that regulate the onset of competence (Madi et al. 2008). Activation of the ComK transcription factor is

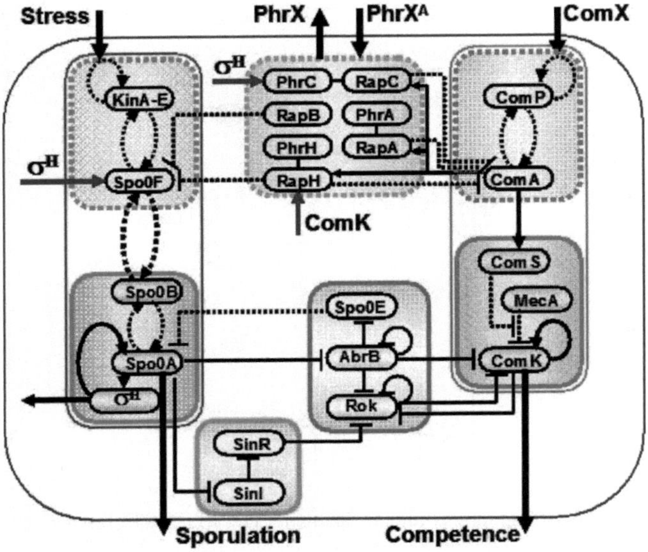

Fig. 10.7 The sporulation-competence signal transduction network. The network modules are described in details in the text. The sporulation module is shown on the *left* and the competence module is shown on the *right*. Each module is composed of a sensing unit and a regulatory unit. These are the KinA-E stress sensing and the Spo0A timer comprising the sporulation module and the Spo0P-ComA quorum sensing and the ComK switch comprising the competence module. The two main modules interact via the Rap communication and information processing module (*upper center*), the AbrB-Rok decision module (*middle center*) and the SinR-SinI commitment unit (*lower center*). The input and output signals are represented by the *wide solid lines* that cross the outer *black* (Schultz et al. 2009)

controlled by many genes and appears to be the step at which multiple physiological signals that affect competence are integrated (Grossman 1995).

10.4.1 Time Progress of Sporulation Initiation Gene Network

Here we focus on the sporulation initiation genes, which include *kinA-E*, *spo0F*, *sigH*, *spo0B*, *spo0A*, and *spo0E*. First, the normalized gene correlations were calculated for all time points (Fig. 10.8). In Fig. 10.9, the FH analysis is presented, for these genes for the three different time points, under the stress of all 37 antibiotics. The analysis of each time point is performed using the FH projection principle.

A close investigation of Fig. 10.9 leads to three important observations. First, in all three-time points we observe that the genes are separated into two main clusters that are negatively correlated. Second, we observe at the 80 min interval after exposure that the genes *spo0E* and *kinD* become more negatively correlated to the

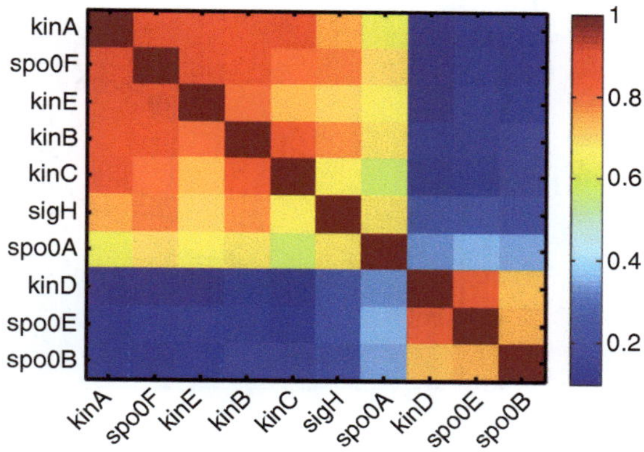

Fig. 10.8 Normalized correlation matrix for the sporulation initiation genes, for all three time points

Kin genes. Third, by comparing Fig. 10.9a, b, it is possible to observe that the gene *spo0A* significantly changes its location in the reduced 3D PCA space. In the first time point, 10 min after exposure, this gene has a special relationship to the remaining genes that organize as two separate clusters.

The gene that is affected the most by the different antibiotics is *spo0A*. The correlation of *spo0A* to the other genes changes throughout the three different time points. After 10 min, *spo0A* is situated in the PCA space between the two clusters and has positive correlation with *spo0E* from one cluster and *kinB*, *kinE*, and *spo0F* from the second cluster. After 40 and 80 min, *spo0A* becomes closer to the *Kin* cluster, *spo0F* and to *sigH* (Fig. 10.9). The transition of *spo0A* toward this latter cluster might indicate an ongoing process of synchronization in the sporulation system over the exposure time to antibiotics.

The separation of the genes into two groups under all antibiotics is quite surprising. Most of the genes mentioned above should have been correlated, leading to phosphorylation of Spo0A and entrance into sporulation. Spo0E is the only negative regulator in this network which dephosphorylate Spo0A ~ P. Here it was found that *kinD* and *spo0B* are also correlated to *spo0E* and have negative correlation to the other genes, negative correlation that becomes stronger as the exposure time increased (Fig. 10.9). More work is needed to clarify the negative correlation of *spo0B* to the other sporulation genes, especially with *spo0F*.

Fujita and Losick found that KinA, KinB, and KinC behave in a significantly different manner than KinD and KinE, with the former being capable of triggering sporulation (Fujita and Losick 2005). This provides a possible partial explanation to the negative correlation of *kinD* to the other *kin* genes. Furthermore, Hoch (Stephenson and Hoch 2002) has described the different expression time of the five *Kin* genes. However, while his work showed that *kinA*, *kinD*, and *kinE* are highly expressed during growth and at equal levels, we have not observed such high correlations between these three genes, rather only for the *kinA* and *kinE* genes.

In an attempt to resolve this issue, we considered analyzing the sporulation genes for only specific antibiotics, for the different exposure times. A number of studies

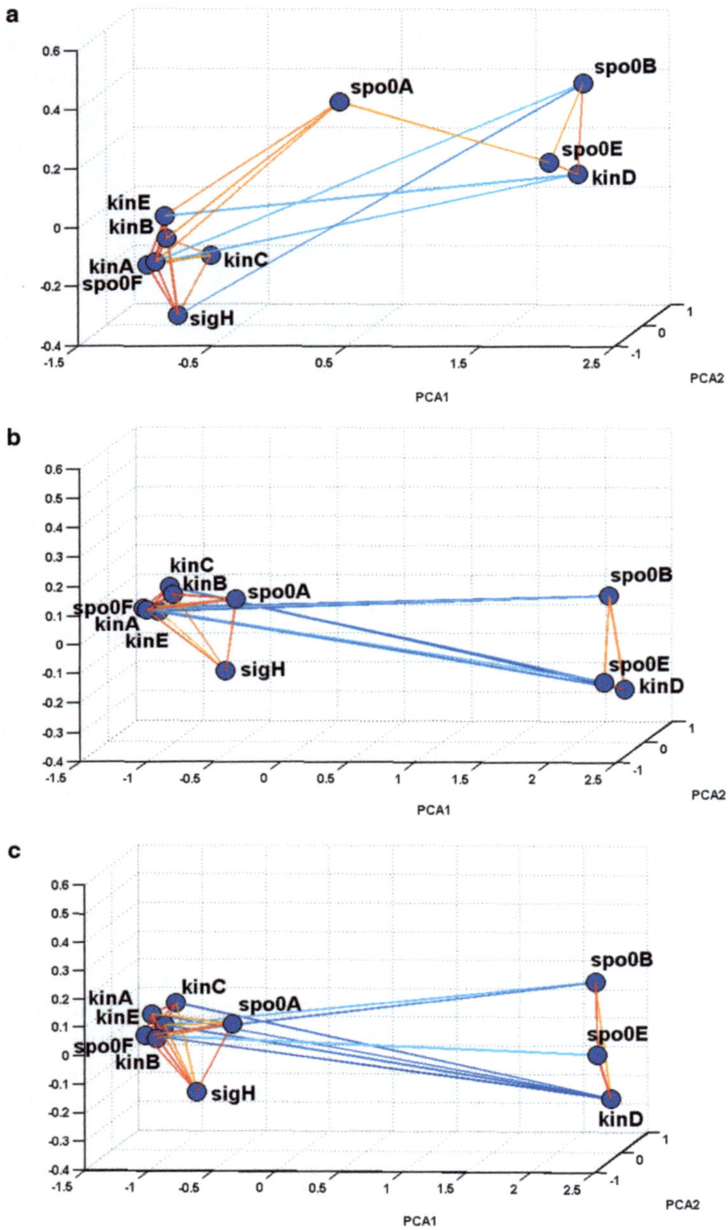

Fig. 10.9 Holographic network of sporulation initiation genes at three time points after exposure. (**a**) After 10 min of exposure, (**b**) after 40 min of exposure, and (**c**) after 80 min of exposure to all antibiotics. For visualization purposes, the correlation lines connecting the genes were set by correlation threshold greater than 0.4 and smaller than −0.4

have shown that some protein synthesis antibiotics, such as chloramphenicol, lincomycin, erythromycin, as well as antibiotics from other classes such as novobiocin, nalidixic acid, and penicillin, inhibit sporulation initiation (Atsuhiro et al. 2003; Jonas et al. 1990; Vazquez-Ramos and Mandelstam 1981). Thus, we analyzed the interaction between the genes under specific antibiotics that were predicted to have an effect on the expression of these genes and compared it to the interactions between the genes under all of the remaining antibiotics. The comparison of the FH analysis of the sporulation inhibition antibiotics to the noninhibiting antibiotics is presented in Fig. 10.11. Comparing the left panels of Fig. 10.11 to the right panels, it is possible to observe a different effect of these two groups of antibiotics. For the case of the sporulation inhibition antibiotics, the results of the FH analysis were similar to the ones obtained for all antibiotics (compare right panel of Fig. 10.11 to Fig. 10.9). In contrast, in the case of the noninhibiting antibiotics, all genes were clustered as one group with very high correlations between the genes (above 0.8).

We thus suggest that the high correlation within the sporulation network shows the synchronization of the genes under antibiotic stress. However, antibiotics that have a major effect on sporulation possibly influence this strong correlation. Since we know that most of the genes that lead to sporulation are correlated (*kinA, B, C, E, spo0A, sigH, sigF*), and that *spo0E* that has a negative regulation on sporulation has also negative correlation with those genes, this suggests that *kinD* and possibly *spo0B* are involved in negative regulation of the sporulation system under the effect of specific antibiotics.

10.4.2 Time Progress of Competence Gene Network Response

In this section, we illustrate the ability of the new approach to reveal additional dynamical motifs related to the time progress in the response of the two gene networks to the antibiotic stress. First, we examined the response of the competence network to 37 antibiotics (Fig. 10.12). The competence gene network studied here contains the genes *comK*, *Rok*, *abrB*, *degU*, and *comG* operon (Fig. 10.13). The FH analysis described above was repeated for these genes, separately for each time point following exposure (Fig. 10.14).

Studying Fig. 10.14, it is possible to see that the most important change is the formation of positive correlation between *degU* and late competence genes (*comK* and *comG* operon).

comK is an auto-regulatory gene in the center of the competence network. AbrB and Rok are repressors of *comK*, and DegU is a positive regulator. ComK is a positive regulator of the *comG* operon and a repressor of the *rok* gene (Fig. 10.13). Since *comG* operon is under a direct regulation of ComK and it triggers its transcription, the cluster of *comG* genes and *comK*, and the strong correlation between these genes at all time intervals is highly expected. Furthermore, ComK

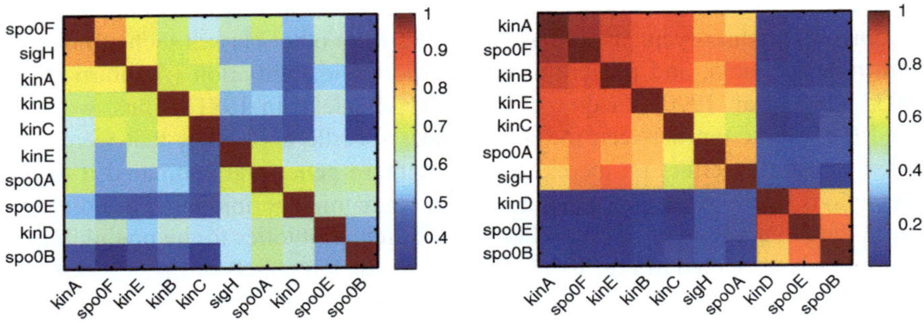

Fig. 10.10 Normalized correlation matrix for the sporulation initiation genes, for all three time points. The *right panel* represents the normalized correlations of the genes under specific antibiotics (chloramphenicol, clarithromycin, clindamycin, erythromycin, fusidic acid, neomycin, puromycin, spectinomycin, tetracyclin, novobiocin, nalidixic acid and penicillin). The *left pane* represent the normalized correlations of sporulation genes under 25 antibiotics from different classes, not including the former set

is the key regulator in triggering the late competence genes. When performing the FH analysis with all the competence genes (results not shown), *comK* was located in the cluster of all the late competence genes, showing the high correlation between this gene and the other competence genes. Note that even though *comG* operon is also negatively controlled by *comZ* (Ogura and Tanaka 2000), the correlation between *comK* and *comG* is not affected under stress of antibiotics.

DegU is one of *comK* positive regulators and was expected to have high correlation to *comK*. Nevertheless, in our findings we show that after 10 min of exposure, *degU* is correlated only with *abrB*. Correlation between *comK* and *comG* evolved over time under all antibiotics, but still the two genes are distant. Although DegU has a main role in *comK* regulation and competence, it was shown that the main function of DegU in the development of competence is to stimulate binding of ComK at the onset of competence development (Hamoen et al. 2000), and overproduction of ComK bypasses the need for DegU. ComK appears to be able to function without DegU if present at sufficiently high concentrations (Roggiani and Dubnau 1993). This might support our finding of the low correlation level between the two genes after 10 min. It has also been suggested in the past that unphosphorylated DegU is required for competence, whereas phosphorylated DegU activates the production of degradative enzymes (Dahl et al. 1992). Mutation that causes hyper phosphorylated form of DegU, by DegS, results in decreased competence, diminished motility, and glucose-insensitive sporulation. For this reason, we assume that correlation between *comK* and *degU* will not necessarily appear in the transcription level but they might exist in the phosphorylation level under non stressful conditions.

Nevertheless, the correlations between *degU* and late competence genes evolve over time under all antibiotics. It is possible that the influence of specific antibiotic that affect the *comK* network become dominant over time, as will be shown below.

The response of the competence to DNA topology ("topo") antibiotics antibiotics was examined, at the three exposure time points – 10, 40, and 80 min.

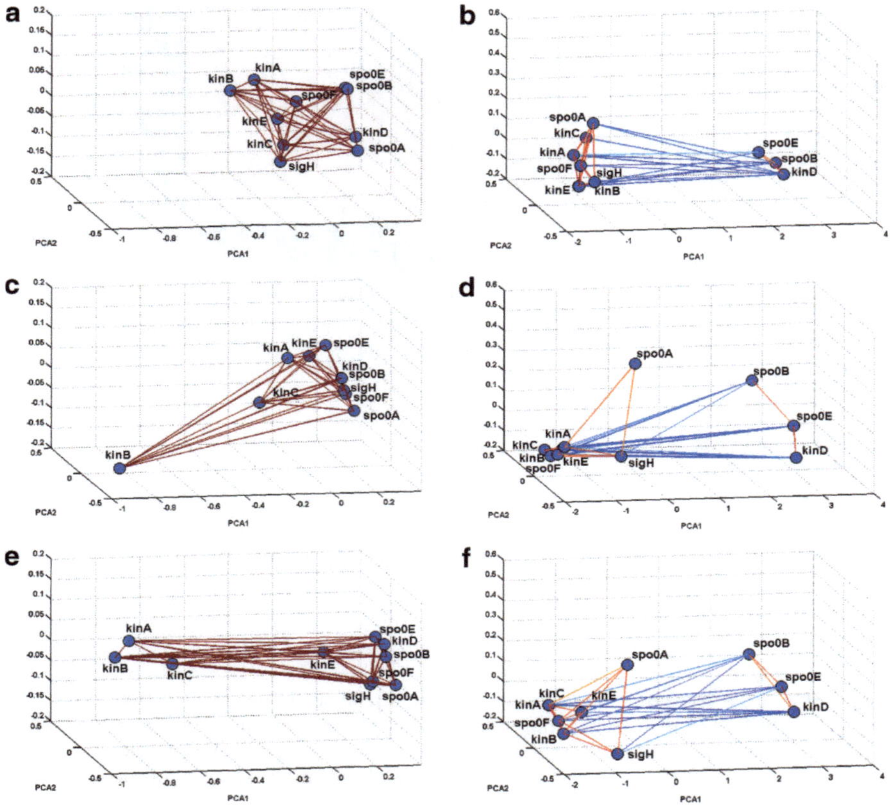

Fig. 10.11 Sporulation initiation genes under the effect of different antibiotics. The *right panels* represent the FH analysis of the genes under specific antibiotics (chloramphenicol, clarithromycin, clindamycin, erythromycin, fusidic acid, neomycin, puromycin, spectinomycin, tetracyclin, novobiocin, nalidixic acid, and penicillin). The *left panels* show the FH analysis of sporulation genes under 25 antibiotics from different classes, not including the former set, at 10 min of exposure (**a**, **b**), after 40 min of exposure (**c**, **d**), and after 80 min of exposure to all antibiotics (**e**, **f**). For visualization purposes, the correlation lines connecting the genes were set by correlation threshold greater than 0.4 and smaller than −0.4

In Fig. 10.15, we show the resulting holographic networks for the three time points after the exposure. Here the main change is detected between the 10 and 40 min time points. At 80 min after exposure, there is an additional effect, as *abrB* and *rok* become functionally more similar (closer in the 3D PCA space) and show higher correlations between them (Fig. 10.16).

The analysis of the response of competence genes to specific case of DNA topology antibiotics shows that *degU* had very high correlation to *comK* and *comG*. This increase in correlation can indicate the advance of synchronization of the competence system, as a result of exposure to antibiotics.

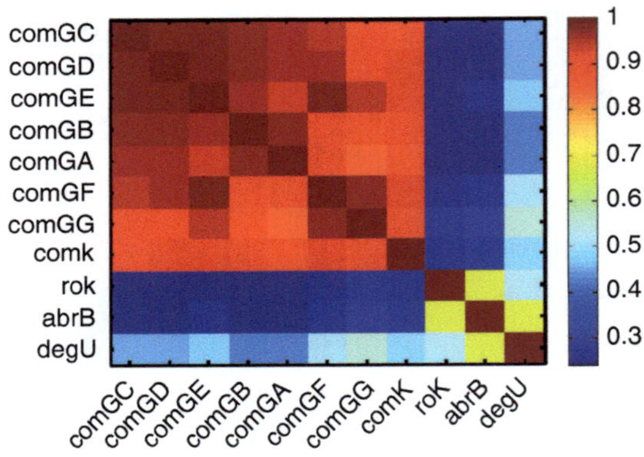

Fig. 10.12 Normalized correlation matrix for the competence genes, for all three time points

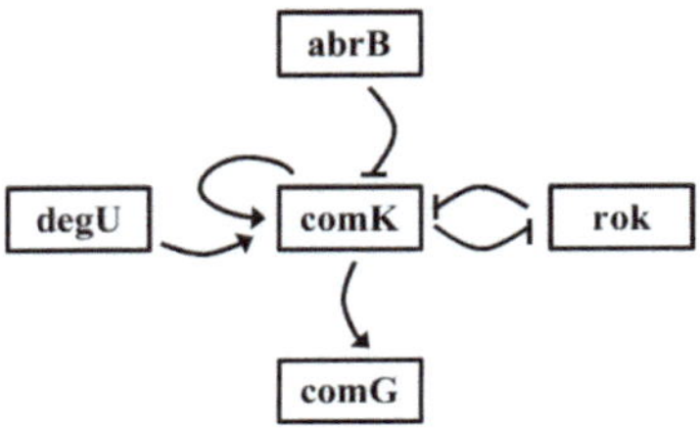

Fig. 10.13 A representation of the known regulation relations in our regulation gene network (Errington 1993; Grossman 1995; Stragier and Losick 1996). *Arrows* and *blunt arrows* represent positive and negative regulation, respectively. At the center of the chart lies the auto-regulatory gene *comK*. AbrB and Rok are repressors of *comK*, and DegU is a positive regulator. ComK is a positive regulator of the *comG* operon (comprised of the genes *comGA* to *comGG* and *yqze*) and a repressor of the *rok* gene

10.5 Time Progress of Cannibalism Gene Network

Cannibalism refers to behavior that was observed in *B. subtilis* cells in which the Spo0A-ON (sporulating) cells in the population trigger the lysis of nonsporulating bacteria (Spo0A-OFF cells) via the elaboration of a killing factor and a toxin (Gonzalez-Pastor et al. 2003). The Cannibalism network consists of two operons – *skf* and *sdp*. *skf* operon is involved in cannibalism and the production of an extracellular killing factor during sporulation (Fig. 10.17). This operon, through its products SkfE and SkfF, also confers resistance to the killing factor. The second operon, *sdp*, is controlled by SpoOA. SdpC is responsible for producing an extra-cellular factor that acts as a signaling protein among bacteria. SdpC strongly controls the transcription of a two-gene operon, *sdpR* and *sdpI*, located immediately downstream of the *sdp* operon. *sdpI-sdpR* use as an immune genes at the producing bacteria and also inhibit sporulation at other bacteria.

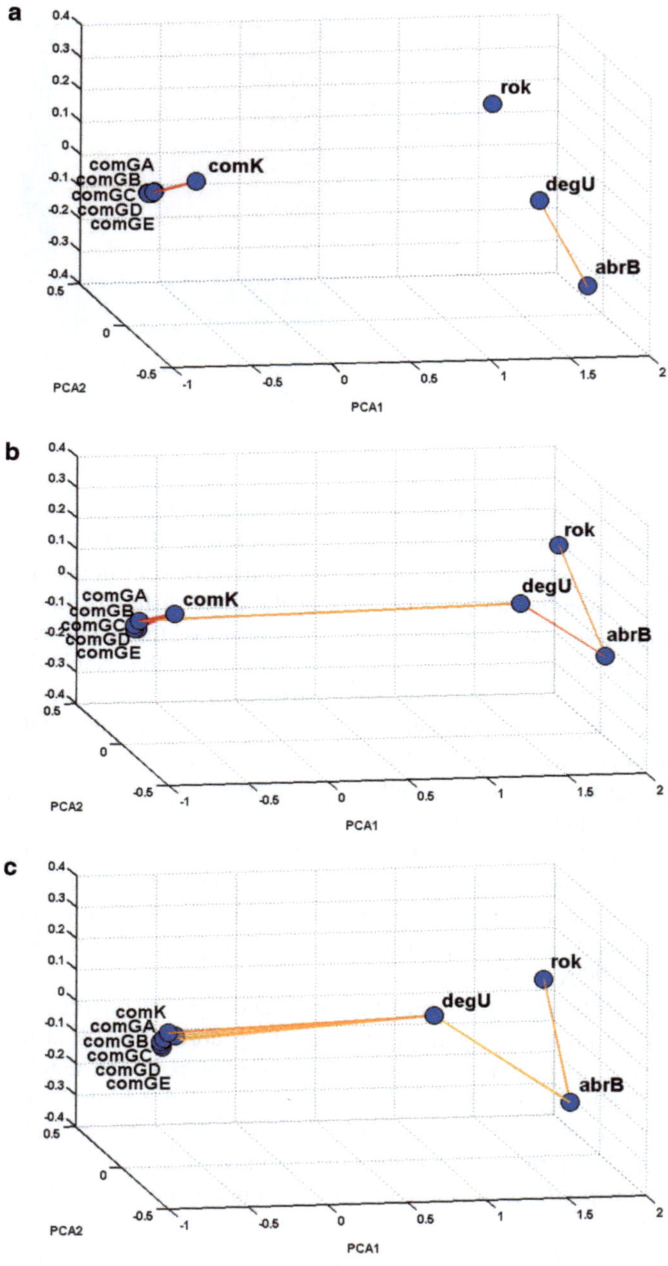

Fig. 10.14 Holographic network of competence genes at three time points after exposure. (**a**) After 10 min of exposure, (**b**) after 40 min of exposure, and (**c**) after 80 min of exposure to all antibiotics. For visualization purposes, the correlation lines connecting the genes were set by correlation threshold greater than 0.4 and smaller than −0.4

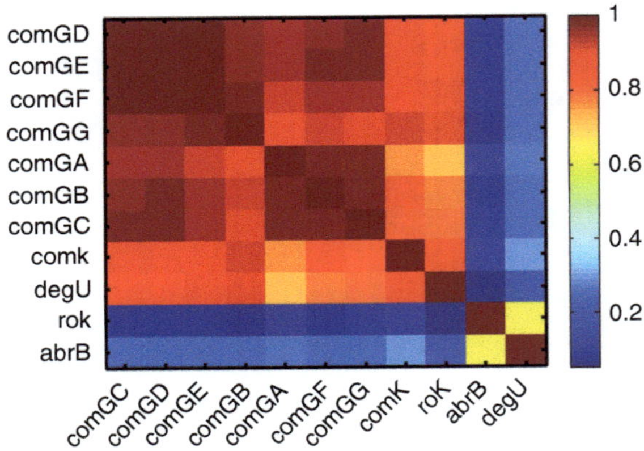

Fig. 10.15 Normalized correlation matrix for the competence genes, for all three time points, under exposure to the "topo" antibiotics

The cannibalism gene network includes the genes *skfA*, *skfB*, *skfC*, *skfD*, *skfE*, *skfF*, *skfG*, *skfH*, *sdpA*, *sdpB*, *sdpC*, *sdpR*, and *sdpI*. In Fig. 10.18, we show the normalized correlation matrix for these genes for all three-time points. In Fig. 10.19, we present the cannibalism gene network at the three different time points. It is possible to observe several changes that are a result of the exposure time to the antibiotics. First, *spo0E* becomes negative correlated to the cannibalism operons (*skf*, *sdp*) and *abrB*; Second, the two cannibalism operons become more correlated; third, *sigH* becomes more correlated to *spo0A*; and last, the gene *sdpI* dissociates from the cannibalism operons and becomes more correlated to *sdpR* and *spo0E*.

The cannibalism operons, *skf* and *sdp*, transcribe together at the beginning of sporulation. Although they are two separate operons, they cooperatively cause the lysis of Spo0A (off) cells. Our findings show that the two operons separate into two groups that are clustered together with high correlation. Moreover, the correlation becomes stronger as the exposure time to antibiotics grows. The remaining genes (*abrB*, *sigH*, *spo0A*, and *spo0E*) that participate in regulation of these operons are scattered in the PCA space.

Spo0A and abrB are mutual regulators of *skf* and *sdp*. The operons are positively regulated by Spo0A and negatively regulated by abrB (Rok is also a negative regulator of *sdp*, and PhoP is a positive regulator of *skf*). The regulation of *sdp* and *skf* by Spo0A is low-threshold activated, and under high-threshold *sdp* is repressed. This could provide a possible explanation to why *skf* and *sdp* do not have high correlation to *spo0A*. On the other hand, *sigH*, which induce sporulation, is very close to *spo0H*, as was already observed in the analysis of the sporulation gene network.

The gene *spo0E* becomes in time negatively correlated to the *sdp* genes. The function of the two cannibalism operons and Spo0E delays entry into sporulation. However, their role is different. Spo0E dephosphorylates Spo0A and thus maintains a low level of phosphorylated Spo0A. On the other hand, SdpC induces SdpR, which turns on the operons for ATP synthetase (atp), and for lipid catabolism

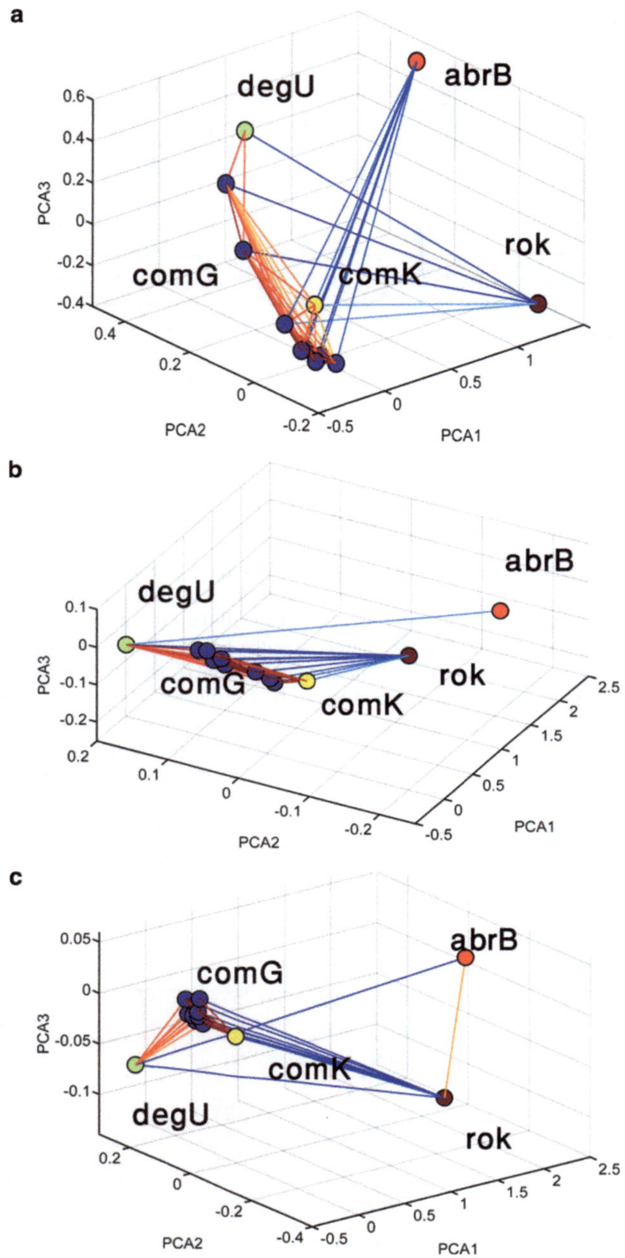

Fig. 10.16 Time progress of competence genes in response to "topo" antibiotics. The correlation matrices for the competence gene network for three time stages, projected on the joint FH holographic 3D of the PCA space for all time points (Fig. 10.15). (**a**) After 10 min of exposure, (**b**) after 40 min of exposure, and (**c**) after 80 min of exposure to antibiotics interfering with DNA topology

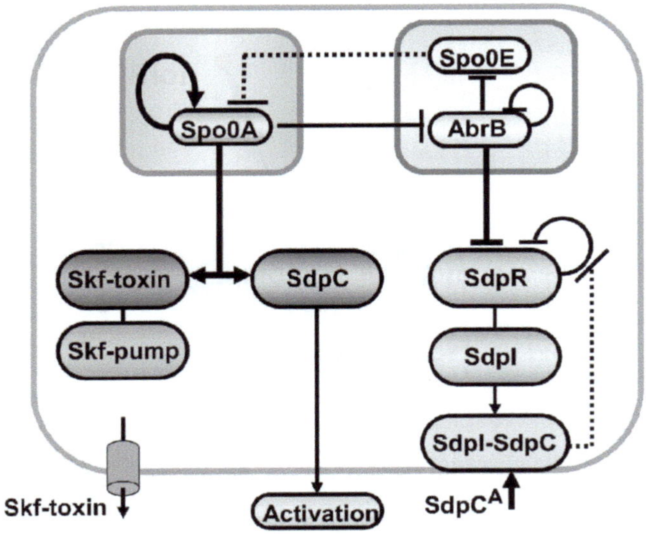

Fig. 10.17 The cannibalism regulatory network. One way of coping with transition state stress conditions in SpoOA-active cells is delaying sporulation by switching on two operons, *skf* and *sdp*. The *skf* operon is involved in the production of an extracellular killing factor (*skf*-toxin). Its other two products, SkfE and SkfF (*skf*-pump), antagonize the lethal action of the killing factor in the SpoOA-active cells. In contrast, cells containing inactive SpoOA are killed by the secreted killing factor, resulting in the release of nutrients that can be consumed by SpoOA-active cells. In both active and inactive SpoOA cells, a gene product of the *sdp* operon (SdpC) switches on expression of the transcription factor sdpR. This factor seems to delay sporulation in SpoOA-active cells, probably by activating lipid catabolism and ATP-producing enzymes (Engelberg-Kulka and Hazan 2003). Moreover, SdpC, which is a toxin, binds to SdpI, thereby triggering sequestration of SdpR at the membrane and activating *sdpRI* transcription. Increasing amounts of SdpI and SdpR molecules trap all SdpC molecules. Free SdpI cannot sequester SdpR, which remains in the cytoplasm, where it shuts off transcription of *sdpRI* (Stragier 2006)

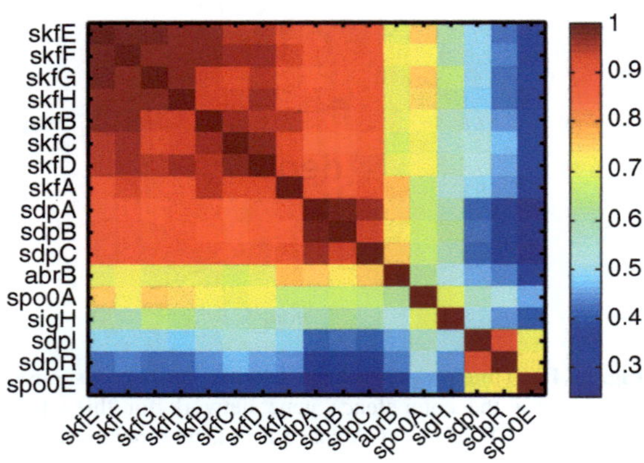

Fig. 10.18 Normalized correlation matrix for the cannibalism genes, for all three time points

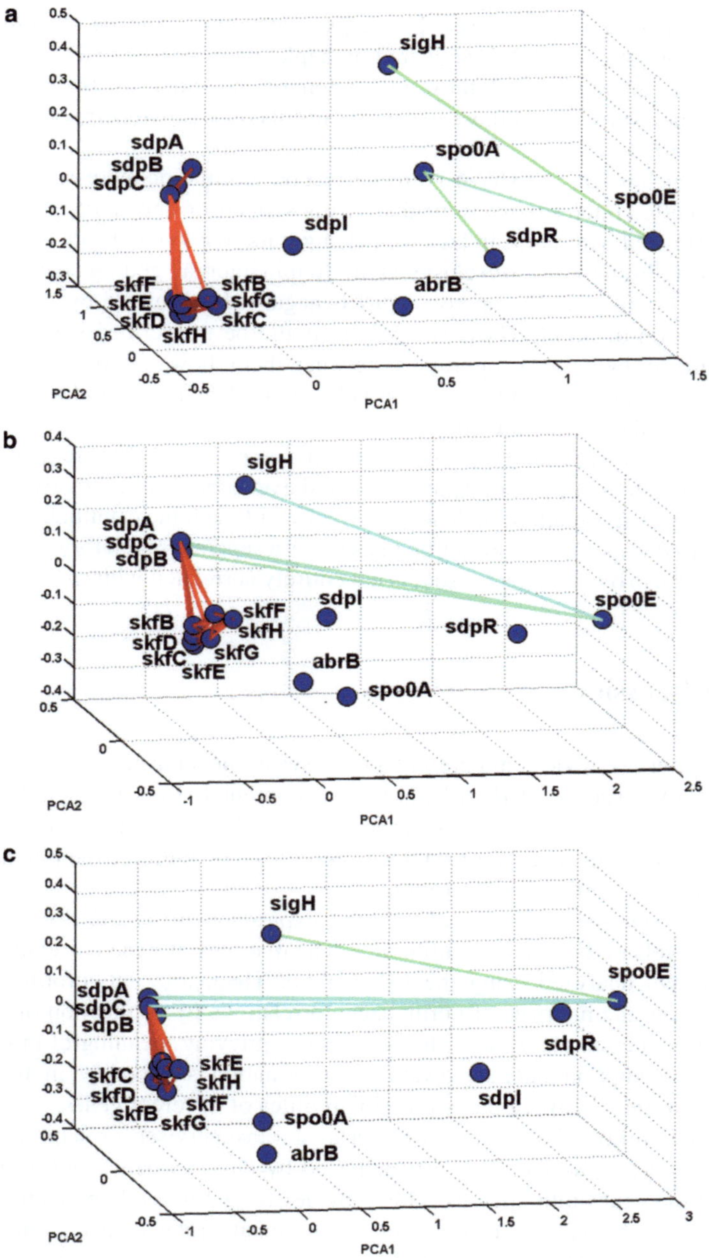

Fig. 10.19 Holographic network of cannibalism genes at three time points after exposure. (**a**) After 10 min of exposure, (**b**) after 40 min of exposure, and (**c**) after 80 min of exposure to all antibiotics. For visualization purposes, the correlation lines connecting the genes were set by correlation threshold greater than 0.8 and smaller than 0

enzymes (YusLKJ). The increase in energy production could be responsible for delaying sporulation, which is triggered by depletion of energy reserves. Since *skf* and *sdp* are activated by Spo0A, repression of the latter by Spo0E probably decreases the expression of those genes that results in negative correlation between *sdp* and *spo0E*.

sdpR, which is known to have a direct effect on sporulation repression, is located close to *spo0E*. Interestingly, *sdpI* that is transcribed on the same operon is not highly correlated with *sdpR*. In fact, *sdpI* changes its position in the PCA space over time, moving from the *skf*, *sdp* cluster at the 10 min interval, getting nearer to *spo0E* at the 80-min interval. The operon *sdpRI* is negatively regulated by AbrB and by auto-repression of SdpR. These two genes should be transcribed simultaneously, as there are no regulators or repressors between them; however, our findings show that they are not correlated (correlation below 0.4). This result could be due to an additional regulation that directly affects *sdpI*.

Once again, it was found that under antibiotic stress, genes that are related to the sporulation system become more synchronized over time and negatively correlated to repressors of sporulation, such as *spo0E*. Still, although the cannibalism network is related to the sporulation initiation phase, it is a separate regulatory network that is regulated also by other genes, so it is not entirely correlated with *spo0A* and *sigH*.

10.6 Discussion

In this chapter, we review and present a new, system-level analysis of the complex gene-network response of *B. subtilis* to environmental stress measured by DNA microarray chips. The method is based on the FH analysis that was originally developed for analyzing multichannels recordings of cultured neural networks activity and of recorded brain activity.

This method was used to analyze gene expression of *B. subtilis* exposed to sublethal levels of 37 different antibiotics. The matrices of gene correlations were computed and analyzed using the FH method. Then, relevant information was extracted from the matrices of normalized correlations by application of the PCA dimension reduction algorithm. The success in retrieving meaningful information proves the assumption that indeed valuable information is embedded in the correlations (similarities) between the expression profiles of different genes.

First, the ability of this method to act as an unsupervised or semi-supervised method was demonstrated. This was achieved by its successful identification and sorting of the genes into operons. This provides for a powerful application of this methodology, which is the identification of functionally related gene groups such as operons in an unsurprised way. In addition, this approach can also classify operons that are more cohesive from others.

Next, it was demonstrated that the approach can also be used to reveal the internal structure of the operons, thus relating the function (expression) to the form. A specific example was given, introducing how this method is able to deduce information about the existence of unknown structural motif in the case of the *spoVAA* operon. These results demonstrated that the method could be used as a prediction tool to reveal functional similarities of unknown genes or operons.

Next we focus on the ability of the GH method to present the dynamical (functional) correlation motifs in well-known gene regulatory networks. We show that in competence, sporulation, and cannibalism networks, our method is capable of identifying previously known positive and negative gene interactions. Examples of such positive gene interactions include the interactions between the sporulation initiation genes (*kinABCE*, *sigH*, *spo0F*, and *spo0A*), cannibalism operons (*skf*, *sdp*), and late competence genes (*comK*, *comG* operon), and the interactions of *spo0E* are an example of such known negative gene interactions. We also show that under specific antibiotics, this method is able to identify unknown regulations, as was found in the case of *kinD* and *spo0B* that were negatively correlated to other sporulation genes. Furthermore, using this tool to study development across time, we observed specific motifs that were affected as a result of the exposure to the antibiotics. For examples, *sdpIR* that became dissociated from the other cannibalism operons, and *kinD* that became negatively correlated to other sporulation genes. These two examples possibly suggest unknown regulation mechanisms that were uncovered by the FH methodology. Other examples discussed here include the approaching of *spo0A* to sporulation genes in the PCA space, and the increase of correlation over time of *degU* to late competence genes, which possibly indicates a synchronization of the sporulation and competence system under exposure to antibiotic stress.

The affect of different antibiotics on the sporulation and competence networks has been investigated thoroughly in the past (Atsuhiro et al. 2003; Prudhomme et al. 2006). Here we demonstrated the effect of antibiotic stress on the three regulatory networks. Indeed, it was observed that there was a clear effect of specific antibiotics, such as protein synthesis inhibitors, on the interaction between the sporulation genes and DNA topology antibiotics on competence network. These findings are in agreement with previous reports that found the inhibition of sporulation under this class of antibiotics (Madi et al. 2008).

Under all antibiotics, for all three networks, genes that lead to sporulation, cannibalism, and competence become more correlated over time. The regulators Spo0A, DegS, SigH, and AbrB govern several transition state pathways, thus the increase in correlation of these regulators to the network might indicate the synchronization and regulation of the system in prolonging response to antibiotics stress. The increase over time of intra gene network correlation under specific conditions may indicate the functionality of the network in relation to these conditions.

Finally, we integrate the FH methodology with the well-known MST method (Mantegna 1999; Mantegna and Stanley 2000; Tumminello et al. 2007; Xu et al. 2001). Using MST methodology to investigate gene-expression data has been performed in the past (Xu et al. 2001) and can be used as we have shown to identify specific groups of genes. Combining these two methods can provide additional information regarding the makeup of the gene network. We perform each analysis separately, and then use the links between genes calculated from the MST analysis to connect the genes in the reduced PCA space. This provides us with a simpler visualization of complex gene networks in a 3D in a sense that it sustains the information regarding the relationships between the genes given by the FH method while providing a smart well defined "filter" for the complexity of correlations information that we would like to present. Furthermore, it enabled us to make use of other network theory information methods, such as how central a gene is in the network (as shown qualitatively here).

In conclusion, we present here a new system-level analysis method for the investigation of gene-expression data. Possible applications of this methodology have been discussed in short, and some of the biological results obtained using this methodology were presented.

Acknowledgements We would like to thank Itai Baruchy for his help in applying the Functional holography method to the study of gene-expression data. We also thank Dr. Sharron Bransburg-Zabary, Yonatan Friedman, and Tamar Regev for their contribution to this project. This research has been supported in part by the Maugy-Glass Chair in Physics of Complex Systems and the Tauber Family Foundation at Tel Aviv University, by National Science Foundation-sponsored Center for Theoretical Biological Physics (CTBP) Grants PHY-0216576 and 0225630, and by the University of California at San Diego.

References

Albert A, Barabási AL (2002) Statistical mechanics of complex networks. Rev Mod Phys 47:47–97

Alon U, Barkai N, Notterman DA, Gish K, Ybarra S et al (1999) Broad patterns of gene expression revealed by clustering analysis of tumor and normal colon tissues probed by oligonucleotide arrays. Proc Natl Acad Sci USA 96:6745–6750

Atsuhiro A, Eiji K, Michihiro H, Mitsuo O (2003) Inhibition of *Bacillus subtilis* aprE expression by lincomycin at the posttranscriptional level through inhibition of ppGpp synthesis. J Biochem 134:691–697

Azevedo V, Sorokin A, Ehrlich SD, Serror P (1993) The transcriptional organization of the *Bacillus subtilis* 168 chromosome region between the spoVAF and serA genetic loci. Mol Microbiol 10:397–405

Baruchi I, Ben-Jacob E (2004) Functional holography of recorded neuronal networks activity. Neuroinformatics 2:333–352

Baruchi I, Grossman D, Volman V, Shein M, Hunter J et al (2006) Functional holography analysis: simplifying the complexity of dynamical networks. Chaos 16(1):015112

Ben-Dor A, Shamir R, Yakhini Z (1999) Clustering gene expression patterns. J Comput Biol 6:281–297

Burbulys D, Trach K, Hoch J (1991) Initiation of sporulation in *B. subtilis* is controlled by a multicomponent phosphorelay. Cell 8:545–552

Caldwell R, Sapolsky R, Weyler W, Maile RR, Causey SC et al (2001) Correlation between *Bacillus subtilis* scoC phenotype and gene expression determined using microarrays for transcriptome analysis. J Bacteriol 183:7329–7340

Chander P, Halbig KM, Miller JK, Fields CJ, Bonner HK et al (2005) Structure of the nucleotide complex of PyrR, the pyr attenuation protein from *Bacillus caldolyticus*, suggests dual regulation by pyrimidine and purine nucleotides. J Bacteriol 187:1773–1782

Claverys J, Prudhomme M, Martin B (2006) Induction of competence regulons as a general response to stress in gram-positive bacteria. Annu Rev Microbiol 60:451–475

Coronnello C, Tumminello M, Lillo F, Micciche S, Mantegna RN (2005) Sector identification in a set of stock return time series traded at the London Stock Exchange. Acta Phys Pol B 36:2653–2679

Dahl MK, Msadek T, Kunst F, Rapoport G (1992) The phosphorylation state of the DegU response regulator acts as a molecular switch allowing either degradative enzyme synthesis or expression of genetic competence in *Bacillus subtilis*. J Biol Chem 267:14509–14514

de Jong H (2002) Modelling and simulation of genetic regulatory systems: a literature review. J Comput Biol 9:67–103

Donetti L, Hurtado PI, Munoz MA (2005) Entangled networks, synchronization, and optimal network topology. Phys Rev Lett 95:188701

Dowds B, Murphy P, McConnell D (1987) Relationship among oxidative stress, growth cycle, and sporulation in *Bacillus subtilis*. J Bacteriol 169:5771–5775

Eisen M, Spellman P, Brown P (1998) Cluster analysis and display of genome-wide expression patterns. Proc Natl Acad Sci USA 95:14863–14868

Engelberg-Kulka H, Hazan R (2003) Microbiology. Cannibals defy starvation and avoid sporulation. Science 301:467–468

Errington J (1993) *Bacillus subtilis* sporulation: regulation of gene expression and control of morphogenesis. Microbiol Rev 57:1–33

Everitt BS (1993) Cluster analysis. Edward Arnold, London

Fawcett P, Eichenberger P, Losick R, Youngman P (2000) The transcriptional profile of early to middle sporulation in *Bacillus subtilis*. Proc Natl Acad Sci USA 97:8063–8068

Fuchs E, Ayali A, Ben-Jacob E, Boccaletti S (2009) Formation of synchronization cliques during development of modular neural networks. Phys Biol 6(3):036018

Fujita M, Losick R (2005) Evidence that entry into sporulation in *Bacillus subtilis* is governed by a gradual increase in the level and activity of the master regulator Spo0A. Genes Dev 19:2236–2244

Furey TS, Cristianini N, Duffy N, Bednarski DW, Schummer M et al (2000) Support vector machine classification and validation of cancer tissue samples using microarray expression data. Bioinformatics 16:906–914

Golub TR, Slonim DK, Tamayo P, Huard C, Gaasenbeek M et al (1999) Molecular classification of cancer: class discovery and class prediction by gene expression monitoring. Science 286:531–537

Gonzalez-Pastor JE, Hobbs EC, Losick R (2003) Cannibalism by sporulating bacteria. Science 301:510–513

Graham RL, Hell P (1985) On the history of the minimum spanning tree problem. IEEE Ann Hist Comput 7:43–57

Grossman A (1995) Genetic networks controlling the initiation of sporulation and the development of genetic competence in *Bacillus subtilis*. Annu Rev Genet 29:477–508

Hamoen LW, Smits WK, de Jong A, Holsappel S, Kuipers OP (2002) Improving the predictive value of the competence transcription factor (ComK) binding site in *Bacillus subtilis* using a genomic approach. Nucleic Acids Res 30:5517–5528

Hamoen LW, Van Werkhoven AF, Venema G, Dubnau D (2000) The pleiotropic response regulator DegU functions as a priming protein in competence development in *Bacillus subtilis*. Proc Natl Acad Sci USA 97:9246–9251

Hansen P, Jaumard B (1997) Cluster analysis and mathematical programming. Math Program 79:191–215

Hartigan JA (1975) Clustering algorithms. Wiley, New York

Hartuv E, Shami R (2000) A clustering algorithm based on graph connectivity. Inf Process Lett 76:175–181

Herwig R, Poustka AJ, Müller C, Bull C, Lehrach H et al (1999) Large-scale clustering of cDNA-fingerprinting data. Genome Res 9:1093–1105

Hoch J (1993) Regulation of the phosphorelay and the initiation of sporulation in *Bacillus subtilis*. Annu Rev Microbiol 47:441–465
Hutter B, Schaab C, Albrecht S, Borgmann M, Brunner NA et al (2004) Prediction of mechanisms of action of antibacterial compounds by gene expression profiling. Antimicrob Agents Chemother 48(8):2838–2844
Jiang M, Shao W, Perego M, Hoch J (2000) Multiple histidine kinases regulate entry into stationary phase and sporulation in *Bacillus subtilis*. Mol Microbiol 38:535–542
Jonas R, Holt S, Haldenwang W (1990) Effects of antibiotics on synthesis and persistence of sigma E in sporulating *Bacillus subtilis*. J Bacteriol 172:4616–4623
Kocabas P, Calik P, Calik G, Ozdamar TH (2009) Microarray studies in *Bacillus subtilis*. Biotechnol J 4:1012–1027
Kruskal JB (1956) On the shortest spanning subtree of a graph and the traveling salesman problem. Proc Am Math Soc 7:48–50
Lorenz M, Wackernagel W (1994) Bacterial gene transfer by natural genetic transformation in the environment. Microbiol Mol Biol Rev 58:563–602
Madi A, Friedman Y, Roth D, Regev T, Bransburg-Zabary S et al (2008) Genome holography: deciphering function-form motifs from gene expression data. PLoS ONE 3:e2708
Madi A, Hect I, Bransburg-Zabary S, Merbl Y, Zucker-Toledano M et al (2009) Organization of the autoantibody repertoire in healthy newborns and adults revealed by system level informatics of antigen microarray data. Proc Natl Acad Sci USA 106:14484–14489
Mandal M, Boese B, Barrick JE, Winkler WC, Breaker RR (2003) Riboswitches control fundamental biochemical pathways in *Bacillus subtilis* and other bacteria. Cell 113:577–586
Mantegna RN (1999) Hierarchical structure in financial markets. Eur Phys J B 11:193–197
Mantegna RN, Stanley HE (2000) An introduction to econophysics: correlation and complexity in finance. Cambridge University Press, Cambridge, UK
Mirkin B (1996) Mathematical classification and clustering. Kluwer Academic Publishing, Dordrecht, The Netherlands
Molle V, Fujita M, Jensen ST, Eichenberger P, Gonzalez-Pastor JE et al (2003) The Spo0A regulon of *Bacillus subtilis*. Mol Microbiol 50:1683–1701
Newman MEJ (2003) The structure and function of complex networks. SIAM Rev 45:167–256
Ogura M, Tanaka T (2000) *Bacillus subtilis* comZ (yjzA) negatively affects expression of comG but not comK. J Bacteriol 182:4992–4994
Ortega GJ, Sola RG, Pastor J (2008) Complex network analysis of human ECoG data. Neurosci Lett 147:129–133
Parego M, Hanstein C, Welsh KM, Djavakhishvili T, Glaser P et al (1994) Multiple protein-aspartate phosphatases provide a mechanism for the integration of diverse signals in the control of development in *B. subtilis*. Cell 79:1047–1055
Prudhomme M, Attaiech L, Sanchez G (2006) Antibiotic stress induces genetic transformability in the human pathogen Streptococcus pneumoniae. Science 313(5783):89–92
Quinlan J (1992) Programs for machine learning. Morgan Kaufmann, San Mateo, CA
Rodionov DA, Vitreschak AG, Mironov AA, Gelfand MS (2003) Regulation of lysine biosynthesis and transport genes in bacteria: yet another RNA riboswitch? Nucleic Acids Res 31:6748–6757
Rogers P, Liu T, Barker K, Hilliard G (2007) Gene expression profiling of the response of Streptococcus pneumoniae to penicillin. J Antimicrob Chemother 59:616–626
Roggiani M, Dubnau D (1993) ComA, a phosphorylated response regulator protein of *Bacillus subtilis*, binds to the promoter region of srfA. J Bacteriol 175:3182–3187
Rumelhart DE, Hinton GE, Williams RJ (1986) Learning internal representations by error propagation. In: Rumelhart DE, McClelland JL (eds) Parallel distributed processing, vol 1. MIT Press, Cambridge, MA, pp 318–362
Ruzal S, Sanchez-Rivas C (1998) In *Bacillus subtilis* DegU-P is a positive regulator of the osmotic response. Curr Microbiol 37:368–372
Schultz D, Wolynes PG, Ben Jacob E, Onuchic JN (2009) Deciding fate in adverse times: sporulation and competence in *Bacillus subtilis*. Proc Natl Acad Sci USA 106:21027–21034

Serizawa M, Yamamoto H, Yamaguchi H, Fujita Y, Kobayashi K et al (2004) Systematic analysis of SigD-regulated genes in *Bacillus subtilis* by DNA microarray and Northern blotting analyses. Gene 329:125–136

Serror P, Sonenshein A (1996) CodY is required for nutritional repression of *Bacillus subtilis* genetic competence. J bacteriol 178:5910–5915

Serror P, Wong K, Sonenshein A (2001) *Bacillus subtilis* CodY represses early-stationary-phase genes by sensing GTP levels. Genes Dev 15:1093–1103

Shapira Y, Kenett DY, Ben-Jacob E (2009) The index cohesive effect on stock market correlations. Eur Phys J B 72:657–669

Sharan R, Maron-Katz A, Shamir R (2003) CLICK and EXPANDER: a system for clustering and visualizing gene expression data. Bioinformatics 19:1787–1799

Sonenshein A (2000) Control of sporulation initiation in *Bacillus subtilis*. Curr Opin Microbiol 3:561–568

Stephenson K, Hoch JA (2002) Evolution of signalling in the sporulation phosporelay. Mol Microbiol 2:297–304

Stragier P (2006) To kill but not be killed: a delicate balance. Cell 124:461–463

Stragier P, Losick R (1996) Molecular genetics of sporulation in *Bacillus subtilis*. Annu Rev Genet 30:297–341

Tamayo P, Slonim D, Mesirov J, Zhu Q, Kitareewan S et al (1999) Interpreting patterns of gene expression with self-organizing maps. Proc Natl Acad Sci USA 96:2907–2912

Tumminello M, Coronnello C, Lillo F, Micciche S, Mantegna RN (2007) Spanning trees and bootstrap reliability estimation in correlation-based networks. Int J Bifurcat Chaos 17:2319–2329

Vazquez-Ramos J, Mandelstam J (1981) Inhibition of sporulation by DNA gyrase inhibitors. Microbiology 127:11–17

Veening JW, Hamoen LW, Kuipers OP (2005) Phosphatases modulate the bistable sporulation gene expression pattern in *Bacillus subtilis*. Mol Microbiol 56:1481–1494

West DB (2001) An introduction to graph theory. Prentice-Hall, Englewood Cliffs, NJ

Xu Y, Olman V, Xu D (2001) Minimum spanning trees for gene expression data clustering. Genome Inform 12:24–33

Chapter 11
Population and Comparative Genomics Inform Our Understanding of Bacterial Species Diversity in the Soil

Margaret A. Riley

11.1 Introduction

The diversity of soil bacteria is simply staggering. According to Torsvik and colleagues (Torsvik et al. 1990; Torsvik and Ovreas 2002), species diversity in soil samples is so high that our most powerful methods of estimation provide only the crudest measure of its magnitude. Nonetheless, many such estimates exist, and they suggest that a single gram of soil may contain over 10 billion microbial cells and more than 1,800 bacterial species (Torsvik and Ovreas 2002; Gans et al. 2005; Zhang et al. 2008). An equally compelling estimate is provided by Dykhuizen (Dykhuizen and Dean 2004), who examined levels of genetic diversity in soil bacteria and predicted that 30 g of forest soil contains over half a million species!

As our methods of empirically estimating bacterial diversity improve, so to do our methods of mathematically refining these numbers. For example, some analytical methods now take into account the fact that there are very few common soil species and untold numbers of rare species (Youssef and Elshahed 2009; Schloss and Handelsman 2006). These refinements push our estimates of bacterial diversity to over a million species per gram of soil (Gans et al. 2005). To put these numbers into perspective, similar studies of the human gastrointestinal tract predict a mere 400 distinct bacterial phylotypes (Rajilic-Stojanovic et al. 2007). This staggering level of species diversity is even more remarkable when one considers that the number of prokaryotes reported in the National Center for Biotechnology Information molecular database is only about 15,111 (Sayers et al. 2010). Clearly, the soil represents a vast reservoir of untapped bacterial diversity.

How we will classify this newfound diversity remains an open question. As molecular methods, such as whole genome sequencing, are more widely applied

M.A. Riley
University of Massachusetts, 611 N. Pleasant Street, Amherst, MA 01003, USA
e-mail: riley@bio.umass.edu

G. Witzany (ed.), *Biocommunication in Soil Microorganisms*, Soil Biology 23,
DOI 10.1007/978-3-642-14512-4_11,

to characterize bacterial diversity, our ability to make taxonomic sense of what we learn is severely challenged. The focus of this review is to explore how we can employ population and species level comparative genomics to provide a rational basis for identifying, and even naming, evolutionary "lineages." In essence, we want to know whether a functional and useful bacterial species concept emerges from the burgeoning genomic information overload.

11.2 Species Classification by 16s rRNA Comparison

Traditional bacterial species designations were based upon extensive phenotypic characterization of a large number of isolates. Although current methods now require the use of 16s rRNA sequence comparisons to identify the closest relatives of a proposed species, phenotype still remains the primary criterion by which species are identified (Rossello-Mora and Amann 2001).This phylo-phenetic bacterial species concept posits that a bacterial species is "a monophyletic and genomically coherent cluster of individual organisms that show a high degree of overall similarity with respect to many independent characteristics, and is diagnosable by a discriminative phenotypic property" (Rossello-Mora and Amann 2001). Numerous studies have revealed clusters of bacterial isolates that share complex phenotypes and these clusters are often designated as species (Shute et al. 1984; Sneath and Stevens 1985; Barrett and Sneath 1994; Mauchline and Keevil 1991; Kirschner et al. 2001). In fact, Cohan uses the mere existence of these clusters as prima facie evidence of the existence of bacterial species. He notes "Bacterial species exist.... bacterial diversity is organized into discrete phenotypic and genotypic clusters, which are separated by large phenotypic and genetic gaps, and these clusters are recognized as species" (Cohan 2002).

The earliest attempts to use molecular data to delineate bacterial species involved the use of DNA–DNA hybridization, in which bacterial species were defined as those isolates sharing at least 70% hybridization under standard conditions (Stackebrandt et al. 2002). Given the enormous range of genetic variation detected in different clearly recognized species, it became obvious that a variability cutoff, such as is imposed with hybridization methods, was not appropriate. Levels of variability will vary over the lifetime of a species and will reflect aspects of its life history, particularly the processes by which it adapts to its habitat.

The use of DNA–DNA hybridization has largely been replaced by the use of 16s rRNA sequences to determine the closest relatives of an isolate, combined with extensive phenotype data. Unfortunately, a disturbingly large number of publications report species diversity based solely upon 16s rRNA sequences (Lindh et al. 2005; Drancourt et al. 2004; Clarridge 2004), an approach which has no sound taxonomic basis.

Recently, investigations into microbial species distinctions have sought to incorporate more robust estimates of molecular diversity into the process of species identification. The assumption is that this molecular diversity will fall into discrete clusters that correspond closely with previously identified phenotype-based clusters. In essence, the question has been raised whether DNA sequence variability can be employed to inform the division of a genus into species, to distinguish among similar species, or to address whether bacterial species exist at all (Godoy et al. 2003; Priest et al. 2004; Baldwin et al. 2005; Hanage et al. 2005; Thompson et al. 2005).

11.3 Whole Genome Comparisons Aid Species Classification

The availability of whole genome sequences for multiple isolates of *Escherichia coli* provided our first glimpse into the dynamic nature of a species genome. Glasner and Perna (2004) and Mau et al. (2006) compared six complete genomes of *E. coli* and revealed a highly conserved genomic backbone of more than 3,000 genes, each with greater than 98% sequence similarity among the isolates. Further, Mau et al. (2006) detected a high level of homologous recombination among these shared genes, confirming earlier studies by Roger Milkman (Milkman et al. 1999; Milkman et al. 2003) that the level of within species recombination for what was then considered a "clonal" species was significant. Even more surprising was the observation that the conserved genomic backbone of *E. coli* was interrupted by hundreds of strain-specific "sequence islands". Edwards et al. (2002) provided a similar comparison of *Salmonella enterica*. As in *E. coli*, a backbone of highly conserved genes was identified, each with an average of greater than 99% sequence similarity and a similar pattern of strain-specific unique sequence islands. This pattern of shared and unique sequences appears to be common among many bacteria (Juhas et al. 2007; Coleman et al. 2006; Waterfield et al. 2003).

Studies with subtractive hybridization and comparative genome hybridization revealed that for *Helicobacter pylori*, *E. coli* and *Staphylococcus aureus*, strains within a species share roughly 75–85% of their genome. A comparison of eight genomes of group B Streptococcus reveals a core of 1,806 genes present in every genome and 907 genes absent in one or more. A similar comparison between five genomes of *Streptococcus pyogenes* revealed a comparable level of genomic diversity and predicted that each new genome added approximately 27 strain-specific genes to the species total (pan) genome. In contrast, eight genomes of *Bacillus anthracis* revealed very few strain-specific genes. In fact, after the addition of four genomes to the comparison, no new unique genes were identified. The general pattern that emerges is that members of a bacterial species share some large fraction of their genomes, but often carry unique, strain-specific sequences. The fraction of the genome shared vs. unique varies greatly from one bacterial species to the next.

11.4 Analyses by Genome-Tree Building

The determination of multiple, complete genome sequences of numerous additional bacteria has enabled a new view of the genomic plasticity of several "well-defined" bacterial species. Such an analysis, referred to as genome-tree building, is based on genome-level phylogenetic analysis. This approach involves scaling-up the traditional tree-building approach and analyzing the phylogenetic trees for multiple gene families (ideally, all families represented in many genomes), in an attempt to derive a consensus, organismal phylogeny (Brown and Volker 2004).

The use of multilocus sequence typing (MLST) is one approach to "genome-tree building" (Maiden et al. 1998; Feil et al. 2003). This method permits the analysis of large numbers of related bacterial isolates, which is essential to the determination of species designations (Feil et al. 2003; Hanage et al. 2005). Such studies have confirmed that species designations based upon phenotypic criteria have a corresponding, underlying MLST-based genotypic clustering (Woodward et al. 2000; Whitaker et al. 2005; Thompson et al. 2005; Godoy et al. 2003).

One of the first multigene-based investigations into the microbial species concept was conducted by Wertz et al. (2003), who sequenced six housekeeping genes from a sample of environmental bacteria representing seven species of Enterobacteriaceae (Gordon and FitzGibbon 1999). Molecular phylogenies for each of the genes were inferred and the branching patterns of the resulting trees compared. In each case, isolates from a species formed a monophyletic group, which corresponded precisely with the clusters identified by phenotypic data, and upon which species distinctions were initially delineated (Holt 1994; Rossello-Mora and Amann 2001).

A molecular-based enteric species phylogeny was inferred from the composite data by concatenating the sequences (Wertz et al. 2003) which contained enough phylogenetic signal to resolve all of the interspecies nodes and thus provided a robust estimate of the enteric phylogeny corresponding with the existing phenotype-based phylogeny. The authors concluded that, at least for this sample of enteric lineages, bacterial species clearly do exist and the same species emerge from phenotypic and genotypic data. More recently, a highly robust phylogenetic tree was constructed for 13 gamma proteobacteria using a concatenated alignment of several hundred conserved orthologous proteins (Lerat et al. 2003). Only two of the proteins had incongruent tree topologies in this analysis.

A similar type of investigation was undertaken with Neisseria, in which a sample of housekeeping genes was obtained from nearly one thousand strains and 11 named species (Hanage et al. 2005). Phylogenetic trees were inferred to investigate whether genotypic clusters can be resolved among these highly recombinogenic bacteria and, if so, the extent to which they correspond to named species. Their analysis suggests that the use of concatenated sequences largely buffered the distorting effect of recombination events and resulted in the resolution of clusters corresponding to the three species most numerous in the sample, *N. meningitidis, N. lactamica* and *N. gonorrhoeae*. Comas et al. (2006) go so far as to propose which genes should be incorporated into these genome-wide surveys for phylogenetic signal. They define "essential" genes, as opposed to "universal" genes, as providing the greatest phylogenetic signal to noise ratio.

These earlier, human-pathogen focused studies have since been expanded to cover a much broader sampling of bacterial diversity. One such investigation involved an Agrobacterium species complex (Popoff et al. 1984; Mougel et al. 2002; Portier et al. 2006). Genome sequence comparisons among members of the genus highlighted a broad range of intra-species divergence within very closely related but distinct species of Agrobacterium. A subsequent study of the same species compared the sensitivity of recombination to DNA sequence divergence across the species complex. Their data supported earlier claims by Majewski (2001) that "bacterial species experience a degree of sexual isolation from genetically divergent organisms since recombination occurs more frequently within species than between species". A genome-based study of the Actinobacteria, a dominant bacterial phylum in the soil, reveals a surprising level of gene syteny and genome conservation. Prior 16S rRNA gene sequence comparisons had recognized 39 families and 130 genera; all of which share a unique molecular synapomorphy: an insertion in the 23S rRNA gene (Roller et al. 1992). How this ancient phylum has retained such a high degree of genome conservation remains a mystery. Similar outcomes have been observed for Haemophilus, Pseudomonas, and Streptococcus, where it was shown that as little as 9% divergence at recombination marker genes results in a drop of three orders of magnitude in recombination efficiency.

Konstantintidis and Tiedje (2004) compared the gene content of 70 closely related and fully sequenced bacterial genomes to identify whether species boundaries exist and to determine the role of the organism's ecology on its shared gene content. They found that levels of sequence similarity on the order of 94% correspond to the traditional 70% DNA–DNA reassociation standard of the current species definition. Notably, a large fraction, e.g., up to 65%, of the differences in gene content within species are associated with bacteriophage and transposase elements, revealing an important role for these elements during bacterial speciation. Their findings are consistent with a species definition that would include a more homogeneous set of strains than provided by the current definition and one that considers the ecology of the strains in addition to their evolutionary distance.

11.5 The Core Genome Hypothesis

Lan and Reeves were the first to recognize the potential link between the observation of shared vs. unique sequences in bacterial genomes and its implication for discriminating bacterial species (Lan and Reeves 1996). They proposed the core genome hypothesis (CGH), which distinguishes between that fraction of the genome (the core) shared by all members of a species and that fraction found in only a subset of the population (the auxiliary). Core genes encode essential metabolic housekeeping and informational processing functions (Feil 2004). They are ubiquitous in a species and define the species-specific characteristics. Auxiliary genes may or may not be present in a strain and are generally genes that encode supplementary biochemical pathways, which are associated with phage or other mobile elements, or encode products that serve to interact with the external environment. Thus, auxiliary genes serve in the adaptation of strains to local competitive or environmental pressures (Cohan 2002). They are likely to encode antibiotic resistance, novel metabolic functions, toxin production, and the like (Dobrindt and Reidl 2000; Karlin 2001; White et al. 2001).

The CGH has dramatically influenced how bacteriologists think about the nature of bacterial species. Prior to the CGH, the strongest argument against the recognition of bacterial "species" was the simple observation of horizontal gene transfer (HGT) between bacterial lineages. The fact that bacterial species gene pools may not be tightly closed was enough reason for many bacteriologists to conclude species cohesion could not survive such exchanges. This contradicts the clearly demonstrated fact that bacteria exist in phenotypic clusters, which many bacteriologists recognize as species. Even more compelling, it is becoming clear that these well-defined phenotypic clusters correspond to underlying genotype clusters (Woodward et al. 2000; Whitaker et al. 2005; Thompson et al. 2005; Godoy et al. 2003).

Some have argued it is futile to expect a bacterial species to ever be characterized fully at the genome level, particularly since as more genome sequences are obtained, the pan genomes (i.e., the sum of all genes identified within a species) of numerous species continues to grow (Medini et al. 2005; Tettelin et al. 2005). In fact, some predict that hundreds of thousands of genome sequences are required to fully define certain bacterial species (Medini et al. 2005). Others suggest that the wide range of intra-species variation observed for bacterial species reflects the lack of a universal and meaningful species definition (Feil 2004).

Many ecological and evolutionary factors will impact how many unique genes a species pan genome may encode and how much genetic variation it harbors. There is no "one size fits all" concept that can, or should, be applied. In fact, no existing species definition requires that either the pan genome or the level of genetic variation be known to delineate members of a species.

One of the more commonly accepted species concepts is that proposed by Ernst Mayr, the biological species concept (BSC) (Mayr 1942). Mayr proposed that a biological species is comprised of groups of actually or potentially interbreeding natural populations, which are reproductively isolated from other such groups (Mayr 1942). Although Mayr developed this definition specifically for eukaryotes, it can easily be modified to apply to bacteria. The CGH provides a perfect backdrop from which to articulate this modification. According to the CGH, a bacterial species is comprised of groups of strains that frequently exchange, or could exchange, core genes, but which are relatively restricted from such exchange with other groups. However, it is important to note that at this juncture, the BSC should not be taken to imply any particular process of speciation, merely that the observation of more gene "sharing" (via recombination and/or lateral gene transfer) is observed within vs. between putative bacterial species.

The CGH predicts that a subset of genes, the core, is present in all, or nearly all, individuals within the species. These are the genes that provide the defining characteristics of the species and are assumed to experience primarily purifying selection, to remove deleterious mutations and maintain existing functions. As a species evolves, its core genome will evolve as a complex of coevolved functions. When transferred between species, such genes will most likely experience a selective disadvantage, as this will disrupt coevolved functions. Such transfer will rarely survive. Thus, core genes will diverge as the species diverge.

In contrast, auxiliary genes will be found in only a subset of individuals within a species. The CGH predicts that these genes experience intermittent positive selection, when their function enhances survival in a varied and ever-changing environment. When such genes are exchanged between species, their functions will often provide a selective advantage to the recipient. Frequent successful transfer between species will serve to limit the divergence of auxiliary genes, relative to the core.

The most specific prediction that emerges from the CGH concerns the rate at which core and auxiliary genes accumulate variability. Core genes will, on average, display a neutral rate of evolution, while auxiliary genes will experience a variety of selective pressures, including diversifying selection (acting to increase levels of variation), directional selection (acting to decrease levels of variation), balancing selection (acting to maintain particular alleles in the species), and purifying selection (acting to weed out deleterious mutations). Thus, the average rate of evolution for auxiliary genes could be just about anything, and the variance around this rate should be extreme. These expectations, based upon the neutral theory (Kimura 1968), are quite useful for testing predictions from the CGH (Fay et al. 2002). However, such tests require population-based samples of multiple genomes per species and, unfortunately, most existing species-based genome samples are chosen to represent the diversity of clinical isolates of human pathogens and thus will often underestimate standing levels of genome diversity. The appropriate data is in the pipeline and should soon be available to permit the sort of population genomics required to address this complex and fascinating matter.

11.6 Conclusion

Although we are on the verge of obtaining the type and amount of genotypic data required to examine bacterial species definitions, it is important to note that there is, in fact, no substantive argument to support the hypothesis that bacterial species do not exist. Hence, the real argument remaining is not do they exist, but rather, how can they exist in the face of potentially high levels of HGT. Our job is to develop an understanding of bacterial evolution rich enough to explain this apparent paradox. The CGH provides a set of testable hypothesis from which to launch this exploration.

References

Baldwin A, Mahenthiralingam E et al (2005) Multilocus sequence typing scheme that provides both species and strain differentiation for the Burkholderia cepacia complex. J Clin Microbiol 43(9):4665–4673

Barrett SJ, Sneath PH (1994) A numerical phenotypic taxonomic study of the genus Neisseria. Microbiology 140(Pt 10):2867–2891

Brown JR, Volker C (2004) Phylogeny of gamma-proteobacteria: resolution of one branch of the universal tree? Bioessays 26(5):463–468

Clarridge JE 3rd (2004) Impact of 16S rRNA gene sequence analysis for identification of bacteria on clinical microbiology and infectious diseases. Clin Microbiol Rev 17(4):840–862, table of contents

Cohan FM (2002) Sexual isolation and speciation in bacteria. Genetica 116(2–3):359–370

Coleman ML, Sullivan MB et al (2006) Genomic islands and the ecology and evolution of Prochlorococcus. Science 311(5768):1768–1770

Comas I, Moya A et al (2006) The evolutionary origin of Xanthomonadales genomes and the nature of the horizontal gene transfer process. Mol Biol Evol 23(11):2049–2057

Dobrindt U, Reidl J (2000) Pathogenicity islands and phage conversion: evolutionary aspects of bacterial pathogenesis. Int J Med Microbiol 290(6):519–527

Drancourt M, Berger P et al (2004) Systematic 16S rRNA gene sequencing of atypical clinical isolates identified 27 new bacterial species associated with humans. J Clin Microbiol 42 (5):2197–2202

Dykhuizen DE, Dean AM (2004) Evolution of specialists in an experimental microcosm. Genetics 167(4):2015–2026

Edwards SV, Fertil B et al (2002) A genomic schism in birds revealed by phylogenetic analysis of DNA strings. Syst Biol 51(4):599–613

Fay JC, Wyckoff GJ et al (2002) Testing the neutral theory of molecular evolution with genomic data from Drosophila. Nature 415(6875):1024–1026

Feil EJ (2004) Small change: keeping pace with microevolution. Nat Rev Microbiol 2(6):483–495

Feil EJ, Cooper JE et al (2003) How clonal is Staphylococcus aureus? J Bacteriol 185 (11):3307–3316

Gans J, Wolinsky M et al (2005) Computational improvements reveal great bacterial diversity and high metal toxicity in soil. Science 309(5739):1387–1390

Glasner JD, Perna NT (2004) Comparative genomics of E. coli. Microbiol Today 31

Godoy D, Randle G et al (2003) Multilocus sequence typing and evolutionary relationships among the causative agents of melioidosis and glanders, Burkholderia pseudomallei and Burkholderia mallei. J Clin Microbiol 41(5):2068–2079

Gordon DM, FitzGibbon F (1999) The distribution of enteric bacteria from Australian mammals: host and geographical effects. Microbiology 145(Pt 10):2663–2671

Hanage WP, Kaijalainen T et al (2005) Using multilocus sequence data to define the pneumococcus. J Bacteriol 187(17):6223–6230

Holt JG, NR K et al (1994) Williams and Wilkins. Baltimore, Bergey's Manual of Determinative Bacteriology: 787

Juhas M, Crook DW et al (2007) Novel type IV secretion system involved in propagation of genomic islands. J Bacteriol 189(3):761–771

Karlin S (2001) Detecting anomalous gene clusters and pathogenicity islands in diverse bacterial genomes. Trends Microbiol 9(7):335–343

Kimura M (1968) Genetic variability maintained in a finite population due to mutational production of neutral and nearly neutral isoalleles. Genet Res 11(3):247–269

Kirschner C, Maquelin K et al (2001) Classification and identification of enterococci: a comparative phenotypic, genotypic, and vibrational spectroscopic study. J Clin Microbiol 39(5):1763–1770

Konstantinidis KT, Tiedje JM (2004) Trends between gene content and genome size in prokaryotic species with larger genomes. Proc Natl Acad Sci USA 101(9):3160–3165

Lan R, Reeves PR (1996) Gene transfer is a major factor in bacterial evolution. Mol Biol Evol 13 (1):47–55

Lerat E, Daubin V et al (2003) From gene trees to organismal phylogeny in prokaryotes: the case of the gamma-Proteobacteria. PLoS Biol 1(1):E19

Lindh JM, Terenius O et al (2005) 16S rRNA gene-based identification of midgut bacteria from field-caught Anopheles gambiae sensu lato and A. funestus mosquitoes reveals new species related to known insect symbionts. Appl Environ Microbiol 71(11):7217–7223

Maiden MC, Bygraves JA et al (1998) Multilocus sequence typing: a portable approach to the identification of clones within populations of pathogenic microorganisms. Proc Natl Acad Sci USA 95(6):3140–3145

Majewski J (2001) Sexual isolation in bacteria. FEMS Microbiol Lett 199(2):161–169

Mau B, Glasner JD et al (2006) Genome-wide detection and analysis of homologous recombination among sequenced strains of Escherichia coli. Genome Biol 7(5):R44

Mauchline WS, Keevil CW (1991) Development of the BIOLOG substrate utilization system for identification of Legionella spp. Appl Environ Microbiol 57(11):3345–3349

Mayr E (1942) Systematics and the origin of species. Columbia University Press, New York

Medini D, Donati C et al (2005) The microbial pan-genome. Curr Opin Genet Dev 15(6):589–594

Milkman R, Jaeger E et al (2003) Molecular evolution of the Escherichia coli chromosome. VI. Two regions of high effective recombination. Genetics 163(2):475–483

Milkman R, Raleigh EA et al (1999) Molecular evolution of the Escherichia coli chromosome. V. Recombination patterns among strains of diverse origin. Genetics 153(2):539–554

Mougel C, Thioulouse J et al (2002) A mathematical method for determining genome divergence and species delineation using AFLP. Int J Syst Evol Microbiol 52(Pt 2):573–586

Popoff MY, Kersters K et al (1984) Taxonomic position of Agrobacterium strains of hospital origin. Ann Microbiol (Paris) 135A(3):427–442

Portier P, Fischer-Le Saux M et al (2006) Identification of genomic species in Agrobacterium biovar 1 by AFLP genomic markers. Appl Environ Microbiol 72(11):7123–7131

Priest FG, Barker M et al (2004) Population structure and evolution of the Bacillus cereus group. J Bacteriol 186(23):7959–7970

Rajilic-Stojanovic M, Smidt H et al (2007) Diversity of the human gastrointestinal tract microbiota revisited. Environ Microbiol 9(9):2125–2136

Roller C, Ludwig W et al (1992) Gram-positive bacteria with a high DNA G+C content are characterized by a common insertion within their 23S rRNA genes. J Gen Microbiol 138 (6):1167–1175

Rossello-Mora R, Amann R (2001) The species concept for prokaryotes. FEMS Microbiol Rev 25 (1):39–67

Sayers EW, Barrett T et al (2010) Database resources of the National Center for Biotechnology Information. Nucleic Acids Res 38:D5–D16

Schloss PD, Handelsman J (2006) Toward a census of bacteria in soil. PLoS Comput Biol 2(7):e92

Shute LA, Gutteridge CS et al (1984) Curie-point pyrolysis mass spectrometry applied to characterization and identification of selected Bacillus species. J Gen Microbiol 130(2):343–355

Sneath PH, Stevens M (1985) A numerical taxonomic study of Actinobacillus, Pasteurella and Yersinia. J Gen Microbiol 131(10):2711–2738

Stackebrandt E, Frederiksen W et al (2002) Report of the ad hoc committee for the re-evaluation of the species definition in bacteriology. Int J Syst Evol Microbiol 52(Pt 3):1043–1047

Tettelin H, Masignani V et al (2005) Genome analysis of multiple pathogenic isolates of Streptococcus agalactiae: implications for the microbial "pan-genome". Proc Natl Acad Sci USA 102 (39):13950–13955

Thompson JR, Pacocha S et al (2005) Genotypic diversity within a natural coastal bacterioplankton population. Science 307(5713):1311–1313

Torsvik V, Goksoyr J et al (1990) High diversity in DNA of soil bacteria. Appl Environ Microbiol 56(3):782–787

Torsvik V, Ovreas L (2002) Microbial diversity and function in soil: from genes to ecosystems. Curr Opin Microbiol 5(3):240–245
Waterfield NR, Daborn PJ et al (2003) The insecticidal toxin makes caterpillars floppy 2 (Mcf2) shows similarity to HrmA, an avirulence protein from a plant pathogen. FEMS Microbiol Lett 229(2):265–270
Wertz JE, Goldstone C et al (2003) A molecular phylogeny of enteric bacteria and implications for a bacterial species concept. J Evol Biol 16(6):1236–1248
Whitaker RJ, Grogan DW et al (2005) Recombination shapes the natural population structure of the hyperthermophilic archaeon Sulfolobus islandicus. Mol Biol Evol 22(12):2354–2361
White PA, McIver CJ et al (2001) Integrons and gene cassettes in the enterobacteriaceae. Antimicrob Agents Chemother 45(9):2658–2661
Woodward MJ, Sojka M et al (2000) The role of SEF14 and SEF17 fimbriae in the adherence of Salmonella enterica serotype Enteritidis to inanimate surfaces. J Med Microbiol 49(5):481–487
Youssef NH, Elshahed MS (2009) Diversity rankings among bacterial lineages in soil. ISME J 3 (3):305–313
Zhang H, Xie X et al (2008) Soil bacteria augment Arabidopsis photosynthesis by decreasing glucose sensing and abscisic acid levels in planta. Plant J 56(2):264–273

Part II
Intercellular and Trans-Kingdom Biocommunication

Chapter 12
Plasmids of the *Rhizobiaceae* and Their Role in Interbacterial and Transkingdom Interactions

Katherine M. Pappas and Miguel A. Cevallos

12.1 Introduction

A primary characteristic of species belonging to the α-proteobacterial family *Rhizobiaceae* (order *Rhizobiales*) is their ability to produce hypertrophies on plants they associate with, whether benevolent or malignant. Using traditional bacterial isolation and characterization methods or molecular typing such as 16S rRNA-based analysis, the genera currently registered as members of the *Rhizobiaceae* are those of the best studied *Agrobacterium*/*Rhizobium* and *Sinorhizobium*/*Ensifer* groups, as also the less known *Amorphomonas*, *Allorhizobium*, *Carbophilus*, *Chelatobacter*, *Liberibacter*, *Kaistia*, and *Shinella* (Young 2008, NCBI registry). Additionally, strains of the *Rhizobiales* such as of *Azorhizobium*, *Bradyrhizobium*, *Mesorhizobium*, *Phyllobacterium* and of the more distant *Bartonella*, *Blastobacter*, and *Brucella* have been reported to belong to this family, although further scrutiny is warranted prior to their final taxonomic placement (van Berkum and Eardly 1998; Young 2008).

Undoubtedly, of all aforementioned genera it is members of the *Agrobacterium*/*Rhizobium* and *Sinorhizobium*/*Ensifer* groups that have been best characterized so far, in fact receiving attention well over a century for their pathogenic or symbiotic properties and decades of study of their plasmids. In recent years, amalgamation of *Agrobacterium* and *Rhizobium* into one genus has been strongly argued in favor (Young 2008), or against (Farrand et al. 2003), and has spawned debate as to whether the basis of genus distinction can mostly owe to plasmid replicons governing bacterial

K.M. Pappas (✉)
Department of Genetics & Biotechnology, University of Athens, Panepistimiopolis, Ilissia, Athens TK 15701, Greece
e-mail: kmpappas@biol.uoa.gr

M.A. Cevallos
Programa de Genómica Evolutiva, Centro de Ciencias Genómicas, Universidad Nacional Autónoma de México, Apartado Postal 565-A, Cuernavaca, Morelos, Mexico

G. Witzany (ed.), *Biocommunication in Soil Microorganisms*, Soil Biology 23,
DOI 10.1007/978-3-642-14512-4_12,

behavior toward the host. Mobility of such replicons leads to further complications, since, for instance, it is well established that agrobacterial plasmids can be hosted in rhizobia and the opposite (Klein and Klein 1953; Hooykaas et al. 1977; Hooykaas et al. 1981; Broothaerts et al. 2005; Rogel et al. 2006). For clarity and in absence of a wider adoption of the genus unification proposition, agrobacterial and rhizobial species will receive their most established denominations in this work. Additionally, nitrogen fixing bacteria outside the boundaries of family or order widely referred to as rhizobia will be also considered in Sect. 12.3.

Regardless of all left to be known and yet discovered, members of these amazing plant-associated pathogens or symbionts have driven knowledge in the fields of host–microbe interactions, transkingdom chemical communication and DNA transfer, modern agronomy and plant biology/biotechnology to unprecedented levels. This chapter focuses on the plasmid replicons of agrobacteria and rhizobia, to which so much of the aforementioned is due. It also discusses the broader genomic organization of species whose genomes have been recently sequenced, which is of high relevance since the genetic content of analyzed representatives is considerably large and split, and the chromosomal or plasmid nature of their largest repicons often in question.

12.2 Pathogenic Members of the *Rhizobiaceae*: Agrobacteria

12.2.1 Agrobacterium and Plant Hypertrophies

Members of the genus *Agrobacterium* are commonly found among soil bacteria (Escobar and Dandekar 2003); they can be also introduced into otherwise virgin or fallow soils via infected plant material (Burr et al. 1987; Otten et al. 2008). In terms of numbers, they range from tens of thousands to tens of millions per gram of infected soil, while their counts may exhibit considerable seasonal fluctuations (Pu and Goodman 1993; Krimi et al. 2002). In natural agrobacterial populations, avirulent and virulent species are often found mixed together; avirulent agrobacteria are classified as *A. radiobacter* and their benign character usually correlates to the absence of a plasmid carrying pathogenic determinants in the cell (which nevertheless does not exclude the possibility of their receiving one; López-López et al. 1999). Virulent species carry pathogenic, disease-causing plasmids that fall into two categories, tumor-inducing (Ti) and root-inducing (Ri). *Agrobacterium* strains causing neoplastic galls, such as *A. tumefaciens* (causative agent of crown gall disease), *A. rubi* (cane gall disease on raspberry), and *A. vitis* (crown gall disease on grapevine), carry Ti plasmids. *A. rhizogenes* strains that cause anomalous rooting (hairy root disease) harbor Ri plasmids.

Host range of pathogens may vary dramatically – for instance *A. tumefaciens* can induce root, stem, and crown galls in hundreds of dicots including stone fruit trees, nut trees, and ornamental plants, being one of the pathogens with the widest host-range known. At the other end, *A. vitis*, *A. rubi*, or a recently described species, *A. larrymoorei* – so far isolated from fig tree tumors (Bouzar and Jones 2001) – are almost host-specific. Irrespective of the disease phenotype against the host, physiological and biochemical criteria have been also used to group agrobacteria into three biovars (Kersters and De Ley 1984), most prominent taxa of each being *A. tumefaciens*, *A. rhizogenes*/*A. radiobacter*, and *A. vitis* for biovars 1, 2, and 3, respectively.

Tumorigenic agrobacteria colonizing the plant host may cause primary tumors at wound sites and also migrate through the plant vasculature and infect sites distal from the origin of entrance (Smith et al. 1912; Cubero et al. 2006). Diseased plants bearing fully developed galls suffer from malnutrition and exhibit less vigor and productivity, susceptibility to further infections or recalcitrance to graft taking, all of which negatively affect horticultural crop yields (Escobar and Dandekar 2003). Interestingly and somewhat oppositely, abnormal root proliferation caused by *A. rhizogenes* has been reported to even benefit plant growth, and this has consequently led to efforts to introduce the trait via genetic engineering in desired hosts (Strobel et al. 1988; Welander et al. 1998; Geier et al. 2008). The last hints to the thin borderline between beneficial or malignant hyperplasias caused by the *Rhizobiaceae*, a motif repeatedly encountered when the behaviors of different members of the family are examined.

The molecular events underlying the *Agrobacterium* pathogenicity cycle have been extensively reviewed throughout the years in numerous publications and dedicated books (i.e., for a most recent edition see Tzfira and Citovsky 2008). In brief, free living agrobacteria seem to be attracted to the plant rhizosphere via sensing plant-released compounds such as phenolics or opines (Ashby et al. 1988; Kim and Farrand 1998). Agrobacteria can colonize intact plant surfaces and form aggregates or biofilms (Fuqua 2008) and often their presence may be asymptomatic (Brencic et al. 2005). However, it is at the highly inducing environments of wound sites that their full-fledged virulence is elicited. Recognition of and attachment to the host are generally regulated by genes located on various replicons of agrobacterial genomes (Setubal et al. 2009), whereas predominant virulence functions are encoded by the *vir* genes that lie on the disease-causing Ti or Ri plasmids. Different *vir* gene subsets mediate in different tasks: in excision and transport of a single-stranded DNA fragment – the T-DNA – to the plant cell nucleus, in providing the proteins that coat and protect the T-DNA or in assembling the type IV secretion apparatus that allows such nucleoprotein export (Atmakuri and Christie 2008; Ream 2008).

In the host plant cell the T-DNA is guided to the nucleus, where it randomly inserts into the plant chromosome (Citovsky et al. 2007). From this position, T-DNA genes driven by eukaryotic promoters are expressed: oncogenic genes that encode for phytohormone production or root differentiation determinants, as also genes that encode for opine production and export. Opines are small condensation products of amino acids, keto acids, and phosphorylated sugars, uptaken and utilized by the infecting bacterium as carbon, nitrogen, and phosphorus sources

(Dessaux et al. 1998). The specificity in the host-pathogen relation derives precisely from the fact that the particular opine molecules encoded for by the T-DNA are consumed by the products of same-opine catabolic genes lying on the T-DNA carrying plasmid. This host-pathogen entanglement creates a microniche that favors the invading pathogen against others and irreversibly acts to enslave the plant host: produced opines may sum up to an impressive 7% of a tumor's dry weight (Dessaux et al. 1993).

Different agrobacterial disease-inducing plasmids encode for the production and utilization of various different opines, which may range from one or two for studied *A. vitis* Ti plasmids to eight or nine for the *A. tumefaciens* so-called octopine- and agropine-type Ti plasmids, respectively (Dessaux et al. 1998). Particular opines from the opine set encoded by certain Ti plasmids also act to induce the conjugal transfer of the respective plasmids and are thus called conjugal opines (i.e., agrocinopines A and B for the nopaline-type Ti or octopine, octopinic acid and lysopine for the octopine-type Ti; Dessaux et al. 1998). The control of the Ti plasmid horizontal transfer by opines is brought about by the activation of the quorum sensing system that is also hosted on the same plasmids, further analyzed in Sect. 12.2.3.

Genes for opine utilization are not restricted to Ti or Ri plasmids. For instance, one of the indigenous plasmids of the avirulent *A. radiobacter* strain K84, pAtK84b, encodes for consumption of nopaline and agrocinopines A and B and can be conjugally induced by them, without engaging in their production (Clare et al. 1990). This is a typical "cheating" behavior, further promoted to full-fledged antagonism, since *A. radiobacter* K84 targets nopaline-type Ti-carrying strains by releasing a bactericidal nucleoside analog, agrocin 84, in addition to two other antagonistic agents (McClure et al. 1998; Penyalver et al. 2001; Kim et al. 2006). Native plasmids of *A. rhizogenes*, other than the Ri plasmids, have been also reported to break-down subsets of opines produced by the hairy roots of transformed hosts (Petit et al. 1983), which, along with the aforementioned, indicates the weight of opine-centered nutrition in agrobacterial physiology.

From the side of the host, plant defense against the invading pathogen involves production of oxidative agents such as hydrogen peroxide (Xu and Pan 2000) or metabolites which have proven to target the *vir* system of *A. tumefaciens*. Salycylic acid or benzoxazinones have been reported to play that role (Zhang et al. 2000; Yuan et al. 2007), but most notably the auxin indole acetic acid (IAA), one of the very compounds whose synthesis is encoded for by the T-DNA, also stands as a powerful *vir* inhibitor (Liu and Nester 2006; see also Sect. 12.2.3). This seems to provide an ultimate barrier to the uncontrollable feedback loop created by the pathogen and its deregulated host, since elevated IAA concentrations are present only after transformation is over and tumorigenesis underway. Interestingly, IAA is known to be produced by *Rhizobium* species during legume nodulation and also appears to be synthesized by mixed bacterial consortia of gall environments (Badenoch-Jones et al. 1982; Prinsen et al. 1991; Vasanthakumar and McManus 2004); in this respect, inhibitors of *vir* released from other bacterial sources are not unlikely to also act in modulating *Agrobacterium* infectivity.

12.2.2 Ti and Ri Plasmids

Several of the nonpathogenic plasmids of agrobacteria such as the tartrate-utilizing plasmids of *A. vitis*, the agrocin-producing plasmids of the biocontrol strain *A. radiobacter* K84, or secondary opine-catabolic plasmids of *A. rhizogenes* have been described in more or less detail (Petit et al. 1983; Szegedi et al. 1992; Farrand et al. 1992; McClure et al. 1998). However, it is the tumor- and root-inducing plasmids that have received years-long attention, not only due to their obvious impact on plants but also because their presence and transmissibility could be monitored in early studies, following their evident traits (van Larebeke et al. 1974; Watson et al. 1975; Moore et al. 1979; White and Nester 1980a). With regard to their ability or disability to coexist in the same host – in other words compatibility or incompatibility – most studied Ti or Ri plasmids have been classified into four incompatibility groups (Otten et al. 2008): IncRh1, comprising of studied octopine- and nopaline-type Ti plasmids of *A. tumefaciens* and *A. vitis* (Hooykaas et al. 1980; Szegedi et al. 1996), IncRh2, with most known representative the succinamopine-type plasmid pTiBo542 (Hood et al. 1986), IncRh3, comprising of agropine-type Ri plasmids (White and Nester 1980b), and IncRh4, comprising of *A. vitis* vitopine-type Ti plasmids (Szegedi et al. 1996). Incompatibility relates to the degree of similarity in replication (*rep*) or partitioning (*par*) machinery components between two replicons (Novick 1987), and in this respect incompatible plasmids are thought to be the most similar in at least their *rep* or *par* elements. However, the marked mosaicism in the genetic organization of Ti and Ri plasmids, along with the substantial variability in the degrees of homology exhibited by syntenic regions, renders difficult any evident conclusion as to their phylogenetic relationships (Moriguchi et al. 2001; Otten et al. 2008).

The sequences of the Ti and Ri plasmid representatives that have been published so far corroborate earlier findings, which were based on genetic and molecular mapping and pointed to their broad similarities (Tanaka and Oka 1994; Farrand 1998). First, they are of same approximate size, ca. 200 kb (Table 12.1); i.e., the *A. tumefaciens* archetypical octopine-type Ti plasmid (pTiR10, pTi15955, pTiB6S3, pTiA6NC, and pTiAch5, considered to be identical; Zhu et al. 2000), pTiC58 and pTiBo542 are 194, 214, and 245 kb respectively, the *A. rhizogenes* plasmids pRi1724 and pRi2659 are 218 kb and 185 kb, while the *A. vitis* pTiS4 is 259 kb. Second, they appear to consist of same functional modules that can be broadly distinguished into those that mediate in replicon housekeeping tasks or those that play role in the host–pathogen relationship. Such modules may vary in gene content and order as also in degree of overall synteny (Fig. 12.1), yet all of them are present in at least the Ti and Ri plasmids that have been analyzed.

A primary housekeeping locus of agrobacterial pathogenic plasmids, extremely widespread in large plasmid replicons and secondary chromosomes of the *Rhizobiales* (Table 12.1), is the *repABC* replication/stabilization unit (Cevallos et al. 2008; Pappas 2008). This is a region consisting of genes and intergenic elements that act in replication, active partitioning and autoregulation, analyzed in more

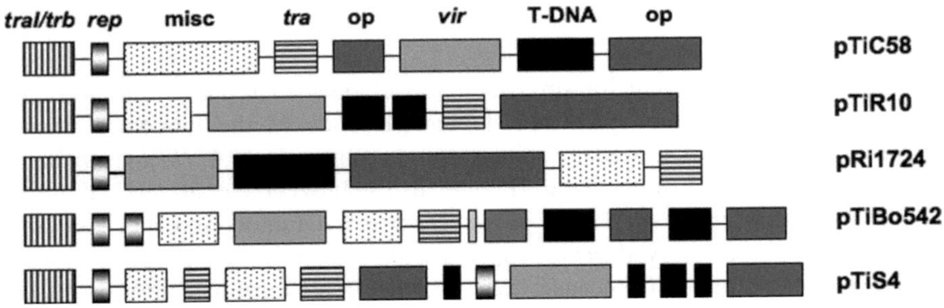

Fig. 12.1 The modular nature of agrobacterial disease-causing plasmids. Depicted are the *A. tumefaciens* tumor-inducing (Ti) plasmids pTiC58, pTiR10 and pTiBo542, the *A. vitis* pTiS4 and the *A. rhizogenes* hairy-root inducing (Ri) plasmid pRi1724, whose sequences are known. Basic modules, indicated for pTiC58 on *top*, are drawn as *gray shaded boxes* for the *repABC* operon (*rep*), *vertically and horizontally lined boxes* for the *traI/trb* and *tra* conjugal transfer operons, respectively, *light gray boxes* for virulence genes and operons (*vir*), *dark gray boxes* for opine uptake and catabolism genes and operons (op), and *black solid boxes* for T-DNA regions. *Black-dotted boxes* on *white* background stand for large blocks of miscellaneous genes, over 20 in number (misc). Genes of lesser numbers and of various functions can be also found in-between assigned modules; module sizes are approximate. Plasmids pTi-SAKURA and pRi2659 that have been also sequenced, were omitted as highly syntenous to pTiC58 and pRi1724, correspondingly

detail in Sect. 12.3.3. In notably two cases – in plasmids pTiBo542 and pTiS4 – two *repABC* copies can be found; in the first case arranged in tandem (Fig. 12.1). This is quite interesting and invites further investigation as to the origins of such duplications, the degree of functionality of either copy or the relations among *cis*- and *trans*-acting elements present in duplicate. The fact that double replicator regions may be encountered on plasmid replicons of this family can plausibly explain certain anomalies in the incompatibility behaviors of particular plasmids, such as the *A. vitis* pTiAT6 and pTiAB3 (Szegedi and Otten 1998).

Two other major housekeeping regions of Ti/Ri plasmids are the highly conserved *tra* and *trb* operons that mediate in the interbacterial conjugal transfer of these plasmids (Farrand 1998). *tra* governs conjugal DNA transfer and replication (Dtr) and comprises of two divergent operons flanking the plasmid conjugal origin of transfer (*oriT*; Cho and Winans 2007), whereas *trb* encodes for the conjugal mating pair formation (Mpf), type IV secretion apparatus (Farrand 1998). *tra* and *trb* are usually separated on plasmid backbones – except for pRi1724, which in every respect seems to be the least rearranged (Fig. 12.1). However and most characteristically, the *trb–rep* operons are syntenic and similarly organized in all Ti/Ri plasmids sequenced so far (Fig. 12.1). First gene of the *trb* locus (and one not participating in mating structures) is *traI*, which codes for the quorum sensing acyl-homoserine lactone (AHL) synthase. In the octopine- and nopaline-type plasmids pTiR10 and pTiC58, it has been shown that the quorum sensing regulator TraR responds to the TraI-generated AHL signal, binds at the *traI/trb–rep* intergenic region and activates both *traI/trb* and *repABC* in a divergent manner (see Sect. 12.2.3). *traI/trb* activation results in autoinduction and Ti conjugal transfer, whereas *repABC* activation results in a pronounced Ti copy number increase. In that respect, *traI/trb–rep* clustering seems crucial in allowing for synchronization of cell-density-dependent responses that end in Ti gene pool amplification, whether in the same or in neighboring cells. TraR also activates the two divergent operons of the octopine-type Ti *tra/oriT* Dtr region that lies afar on the plasmid backbone (Fig. 12.2), thus it is tempting to hypothesize that gene order conservation at *traI/*

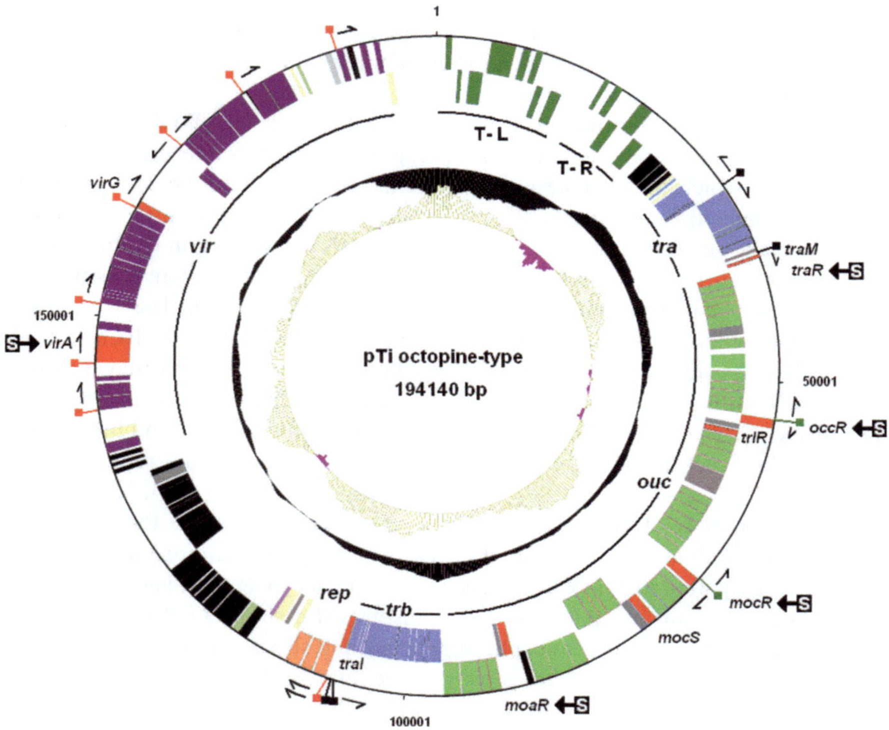

Fig. 12.2 Functional map of the *A. tumefaciens* octopine-type Ti plasmid (Pappas 2008; reproduced with permission from Elsevier). The map is based on GenBank sequence AF2422881. Concentric rings from the outside to the center depict genes on forward strand, genes on reverse strand, GC content, and GC skew (IMG system graphics; Markowitz et al. 2008). Regulatory genes were named and colored *red*; all other genes functionally grouping together were given the following colors: virulence genes (*vir*), *dark purple*; *left* and *right* T-DNA genes (T-L and T-R), *dark green*; conjugal transfer genes (*tra* and *trb*), *light purple*; opine uptake and catabolism genes ("*ouc*"), *light green*; and replication and stabilization genes (*rep*), *orange*. The remaining genes follow the IMG color coding (imgweb.jgi-psf.org), i.e., unassigned genes are *black*. "S" indicates sites of signaling input that affects regulators: OOHL for TraR (also perceived by TrlR, a truncated TraR homolog), octopine and mannopine for OccR and MocR, mannopinic and agropinic acid for MoaR, and plant-signal inducers or inhibitors for VirA. *Pins* and *arrows* point to sites and directions of transcriptional activations: *red* pins correspond to VirG-P activation sites prior to *rep* and *vir* genes or operons; *green* pins, to sites of OccR- and MocR-dependent autoregulation and regulation of mostly opine utilization operons, including the *traR* and *trlR* genes; and *black* pins, to sites of TraR activation at the divergent *tra* operons, prior to *traM*, and at the divergent *traI/trb* and *repABC* operons

trb–rep in particular may involve more than mere operon coregulation and be strongly selected for by yet other factors of functional importance.

Housekeeping *tra/oriT* and *traI/trb–rep* modules homologous to those found on Ti and Ri plasmids can be also met in the *Rhizobium* sp. NGR234 plasmid pNGR234a, arranged in close proximity (Freiberg et al. 1996, 1997). pNGR234a is a nitrogen-fixing, nodule-forming symbiotic plasmid of an extremely wide-host range rhizobial isolate and this backbone conservation, along with a similar to Ti plasmids TraI/TraR/TraM quorum sensing circuitry operating on the plasmid (He et al. 2003; see Sect. 12.2.3), hint to the common evolutionary origins shared by agrobacterial and rhizobial replicons.

Host–pathogen-related modules on Ti/Ri plasmids are the T-DNA, virulence (*vir*), and opine utilization regions. T-DNA regions are bordered by direct sequence repeats upon which the cognate nickases act in order to excise the T-DNA in the form of a single-stranded nucleic acid fragment (Yanofsky et al. 1986; Ream 2008). T-DNAs carry genes that play role in opine synthesis and export, in phytohormone or other tissue differentiation determinant synthesis, as well as in plant sensitivity to the oncogenic or morphogenetic factors. They also carry genes of less well determined function, as well as IS elements (Binns and Costantino 1998; Britton et al. 2008). T-DNAs are thus quite composite and the genes that they carry may be found in various combinations in single or more excisable T-DNA units on the plasmids they reside (Otten et al. 2008). The mode of transfer or integration of multiple small T-DNAs (i.e., such as those of pTiS4; Fig. 12.1) is intriguing and remains to be explored.

T-DNAs in most of the sequenced Ti/Ri representatives can be found in close proximity to virulence (*vir*) regions (Fig. 12.1). *vir* regions comprise of genes and operons participating in T-DNA processing and transfer to the host and stand as the second conjugal transfer system of Ti/Ri plasmids (Lessl and Lanka 1994). Contrary to *tra* and *trb* that are often split, regulatory and structural genes and operons of *vir* regions studied so far are most often colocalized on plasmid backbones and only a few *vir* members (i.e., *virPQH* in pTiBo542 or *virP* in pTiS4) can be circumstantially met at large distances from the respective *vir* clusters. *vir* regions of Ti plasmids that have been studied are not strongly homologous to *tra/trb* regions of same plasmids and, furthermore, gene subsets and intergenic elements within both *tra/trb* and *vir* show mosaicism as to their similarities toward alleles of other broad-host range conjugative plasmids (Zhu et al. 2000).

Coordinate expression of the ~30 *vir* genes of Ti plasmids is succeeded through their activation by the VirA/VirG two-component system, in which VirA is a transmembrane plant signal-receiving histidine kinase and VirG the response regulator that undergoes phosphorylation by VirA upon *vir* induction (Johnson and Das 1998; McCullen and Binns 2006; Fig. 12.2). Indeed, *virA* and *virG* homologs are found in all sequenced Ti/Ri plasmids. Moreover, operons of the *vir* regulon such as *virB*, which forms the type IV T-DNA transport system (Christie et al. 2005), or *virC* and *virD*, which enable T-DNA excision and processing (Yanofsky et al. 1986; Howard et al. 1992; Atmakuri et al. 2007; Ream 2008), are also found in all sequenced plasmids, arranged similarly and in the aforementioned order. In the *A. tumefaciens* Ti plasmids, *virE* operon genes follow; these play crucial role in the protection and nuclear import of the T-DNA (Citovsky et al. 1992; Schrammeijer et al. 2003). However, two *virE* members (*virE1* and *virE2*) are not detected in the *A. rhizogenes* plasmids pRi1724 and pRi2659 (Moriguchi et al. 2001), whereas *virE3*, the third gene of the operon, is present and placed after a *virF*-like gene (VirF, like VirE proteins, is transported to the plant cell and has been recently implicated in targeting the T-DNA protein coat for proteolysis; Tzfira et al. 2004; Lacroix et al. 2008). The *virE* operon in pTiS4 is also found amidst *virA* and the *virB-C-D* complex and is flanked by transposase-like genes; in that respect translocations or deletions of particular *vir* genes are not entirely uncommon in *vir* regions and account for local loss of synteny.

An impressive number of genes on Ti/Ri plasmids are those dedicated to opine uptake and catabolism, in fact numbering more than 40 genes on the octopine-type Ti (Fig. 12.2). Genes implicated in regulation, opine transport (of mostly ABC-type permeases), chemotaxis, and opine degradation, characterized or putative, are found clustered in usually one or more large blocks (Fig. 12.1). From particular studied gene subsets, it has appeared that often, but not always, opine synthesis genes on T-DNAs resemble and may even substitute for same-compound catabolizing genes on the plasmid backbones (Kim and Farrand 1996; Hong et al. 1997). This poses one of the most interesting questions in the plasmid biology of agrobacteria: how did these dual systems that govern opine synthesis/export in the host and opine import/degradation in the pathogen, coordinately evolve – systems that literally split the opine cycle between the eukaryote and the bacterium – without each component losing on spatial and functional autonomy.

12.2.3 Interbacterial and Host-Bacterial Signaling

Studies of past decades on the octopine- and nopaline-type Ti plasmids have led to important insights regarding their role in signaling integration that stands at the center of the *Agrobacterium* pathogenic cycle. They also offer a clear paradigm on the significance of bacterial plasmids in host-pathogen systems in general. As mentioned in Sect. 12.2.2, triggering of *Agrobacterium* virulence proceeds via induction of the Ti *vir* system through the activation of VirA/VirG. The VirA histidine kinase perceives plant wound chemicals such as phenolics, sugars and acidity and in so doing autophosphorylates and passes the phosphor-group to VirG (Stachel and Zambryski 1986; Jin et al. 1990b; Chang and Winans 1992; Winans et al. 1994). VirG-P binds to specific DNA sequence palindromes preceding *vir*-activatable genes, called *vir* boxes (Jin et al. 1990a; Pazour and Das 1990) and activates transcription of all *vir* regulon members that mediate in T-DNA transfer to the plant cell and its nucleus; VirG-P also up-regulates the Ti plasmid *repABC* (Cho and Winans 2005; Fig. 12.2). This results in subtle elevation of the Ti copy number and of the entire Ti transcriptome – an effect highly likely to exacerbate infectivity upon *vir* induction onset.

Following transformation, T-DNA gene expression in the plant cell nucleus leads to tumor formation as well as to opine overproduction and export. Opines, upon import in the agrobacterial cells, activate opine-responsive transcriptional regulators – i.e., agrocinopines A and B or octopine activate the AccR and OccR regulators of nopaline- and octopine-type Ti plasmids, respectively – which in turn activate expression of opine-utilization genes through either derepression or transcriptional activation (Beck von Bodman et al. 1992; Wang et al. 1992). In studied Ti plasmids such as pTiC58, pTiR10, pTiChr5, and pTiBo542, the *traR* gene that belongs to the *luxI*/*luxR* family of proteobacterial quorum sensing regulators is a member of such opine-activatable operons (Farrand 1998; Oger and Farrand 2001, 2002; White and Winans 2007; Fig. 12.2) and thus *traR* expression is coordinately up-regulated by opines, called conjugative after their subsequent effect on Ti transfer.

Synthesis of TraR and its activation by threshold levels of the AHL *N*-3-oxooctanoyl-L-homoserine lactone (OOHL) leads to the subsequent TraR-dependent activation of genes and operons comprising the TraR regulon. Most notable members of the latter are the Ti *tra*, *traI*/*trb*, and *rep* operons, as well as *traM*, which codes for a TraR antagonist (Cho and Winans 2007; Fig. 12.2). Elevated expression of *traI* – the AHL synthase gene – fuels autoinduction, a positive feedback loop where the produced signal promotes its own synthesis (Fuqua et al. 1996). Elevated expression of *tra*, *trb*, and *rep* results in Ti plasmid conjugal transfer (Piper et al. 1993; Fuqua and Winans 1994; Hwang et al. 1994), and maximal – eightfold – Ti copy number increase (Li and Farrand 2000; Pappas and Winans 2003a). Characteristically, TraR, a transcriptional activator structurally similar to prokaryotic and eukaryotic signal-binding regulators (Vannini et al. 2002; Zhang et al. 2002a; Pappas et al. 2004) engulfs OOHL and as a protein-pheromone complex binds to conserved *tra* boxes preceding regulon members. In the case of the *traI*/*trb-rep* region in particular, the TraR-OOHL complex binds to two adjacent boxes and activates no less than five flanking promoters: one facing *traI*/*trb* and four facing the Ti *repABC* (Fuqua and Winans 1994; Pappas and Winans 2003a, b). As would be expected, the increase in the Ti gene pool brought about by *repABC* transcriptional activation and subsequent plasmid amplification is not free of consequence and culminates in a several-fold increase in agrobacterial tumorigenicity (Pappas and Winans 2003a; Fig. 12.3). In this respect, quorum induction in *Agrobacterium* is unequivocally linked to virulence and this complies with what is currently known about most other bacterial pathogens (Whitehead et al. 2001; Williams 2007).

Chemical compounds that inhibit *vir*, proteins that obstruct TraR as well as genes and their products that quench the autoinduction signal, have all proven to play a role in down-regulating agrobacterial virulence or the Ti replicative and conjugative state (White and Winans 2007; Pappas 2008). To start with, plant compounds such as benzoxazinones of graminaceous plant root exudates, indole acetic acid (IAA), salicylic acid (SA), and ethylene have been found to act as potent *vir* inhibitors (Zhang et al. 2000; Liu and Nester 2006; Yuan et al. 2007; Nonaka et al. 2008), while for SA and IAA it has been shown that this inhibition proceeds through direct attenuation of VirA function. Of the aforementioned compounds,

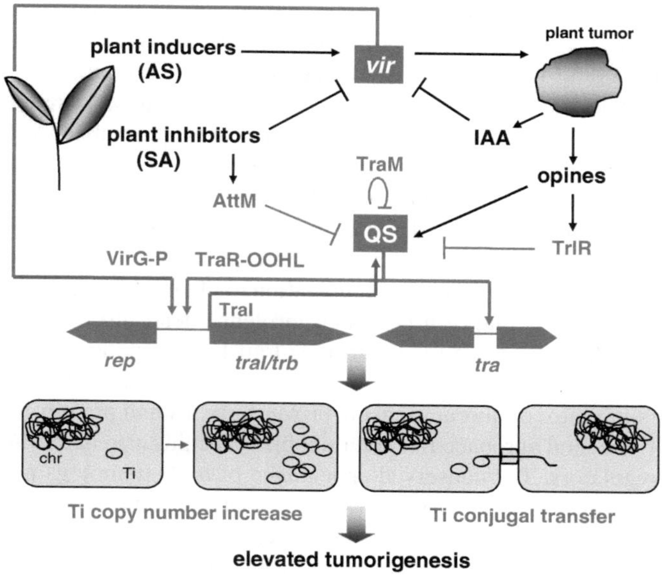

Fig. 12.3 Schematic summary of signaling networks between the plant host and *A. tumefaciens* that involve the agrobacterial Ti plasmid virulence (*vir*) and quorum sensing (QS) systems. The *A. tumefaciens vir* system is activated by plant signals and results in plant transformation and tumorigenesis; *vir*-activation also brings a mild increase in the Ti copy number. QS-activation by tumor-derived opines and autoinduction ends in marked Ti amplification and conjugal spread and reportedly elevates tumorigenesis. Factors that antagonize *Agrobacterium* virulence are plant- and tumor-derived *vir* inhibitors, as also plant and bacterial compounds and proteins that quench the QS system (see text). In the sketch, bacterial networks and functions are depicted in *gray*. OOHL stands for the TraR-inducing AHL. TrlR is a TraR-antagonist met in octopine-type Ti plasmids. For simplicity, only characteristic plant signaling compounds are indicated (*AS* acetosyringone, *SA* salicylic acid, *IAA* indole acetic acid)

IAA stands as the ultimate feedback inhibitor of *vir* since its presence appears to abate its own synthesis (Liu and Nester 2006). SA on the other hand, a known plant defense phenolic (Stacey et al. 2006), further inflicts *Agrobacterium* virulence by down-regulating *vir* regulon members – among which the Ti *repABC* operon – and by independently up-regulating the *attKLM* operon that lies on the 500-kb plasmid of *A. tumefaciens* C58 (Yuan et al. 2007; Anand et al. 2008).

Last gene of the *attKLM* operon, *attM*, encodes for a lactonase whose main function appears to be the hydrolysis of γ-butyrolactone – an intermediate in the γ-aminobutyric acid (GABA) break-down pathway – but which can nevertheless also degrade AHLs (Zhang et al. 2002b; Carlier et al. 2003, 2004; Chai et al. 2007). Indeed, AttM activity was found to relate to AHL decline, inhibition of Ti conjugation and attenuation of agrobacterial virulence (Zhang et al. 2002b; Chevrot et al. 2006), whereas a knock-out mutant of another *A. tumefaciens* AHL lactonase gene, *aiiB*, was recently shown to exhibit the opposite of all the above (Haudecoeur et al. 2009). Like SA, GABA – a major signal of bacteria, plants, and animals – and its degradation pathway intermediates induce expression of the quorum quenching *attKLM* operon (Carlier et al. 2004; Chevrot et al. 2006; Chai et al. 2007). Last but not least and aiding to strengthen previous findings, in genome-wide studies all three plant signals – IAA, SA and GABA – were demonstrated to exert independent yet overlapping effects on agrobacterial virulence and quorum sensing inhibition (Yuan et al. 2008). Taken together, the aforementioned provide remarkable evidence that the plant uses own communication molecules to address the pathogen during the infectious process and, furthermore, that plant defense works through the manipulation of bacterial virulence and interbacterial communication (Zhang 2003). The subjection of quorum communication to plant inhibitors indicates in yet another way its role as that of a bona-fide virulence determinant in agrobacterial biology, rather than of a mere regulator of Ti transfer, as previously thought.

In the agrobacterial cell signaling circuitry, a significant antagonist of TraR and a consequent inhibitor of quorum sensing proves to be a small protein, TraM, which is encoded by several agrobacterial and rhizobial plasmids also harboring genes for TraI/TraR regulators. The conserved appearance of *traM* homologs renders TraM more than a circumstantial oddity – i.e., such as that of a truncated version of TraR, TrlR, encoded by octopine-type plasmids only and known to inhibit TraR by forming inactive dimers (Oger et al. 1998; Zhu and Winans 1998; Fig. 12.2) – and close to an integral component of TraI/TraR networks. In studied octopine- and noplaline-type plasmids, *traM* null mutations profoundly affect Ti conjugal transfer and the Ti copy number in a positive way and also enhance TraR responsiveness to low OOHL concentrations; *traM* overexpression has the opposite effects (Fuqua et al. 1995; Hwang et al. 1995; Li and Farrand 2000). From structural analyses, it appears that the physical interaction between TraR and TraM involves intercalation of TraM in between the amino- and carboxy-terminal domains of TraR and allosteric disruption of TraR binding to target DNA (data provided by the hetero-complex cocrystal of the *Rhizobium* sp. pNGR234 orthologs; Chen et al. 2007). Interestingly and where examined, the *traM* gene is a member of the TraR regulon and, therefore, synthesis of TraM is up-regulated by TraR in what seems to be a quorum-sensing feedback inhibition mechanism (Fuqua et al. 1995; Hwang et al. 1995; Fig. 12.3).

TraI/TraR circuitries encountered in other plasmids of the *Rhizobiaceae*, such as pNGR234a of *Rhizobium* sp. NGR234, p42a of *R. etli* CFN42 and pRL1JI of *R. leguminosarum* bv. *viciae*, act in conjugal transfer activation of the host plasmids (He et al. 2003; Tun-Garrido et al. 2003; McAnulla et al. 2007), while for pRL1JI in specific, the TraR homolog was also shown to activate the plasmid *repABC* operon transcription (McAnulla et al. 2007). In analogy to the agrobacterial quorum sensing system subjection to opine-dependent regulation, in studied rhizobial systems other induction levels also precede final I/R homolog activation, although in this case they appear to involve alternative R-family regulators (see also Sect. 12.3.4). Remarkably, *traI*/*traR* homologs residing on the *Mesorhizobium loti* R7A integrative conjugal element ICE*Ml*Sym(R7A), which carries nitrogen-fixation genes, were recently reported to control the excision and conjugal transfer of the element in a quorum sensing-dependent manner (Ramsay et al. 2009).

Taken together, the aforementioned provide clues that I/R circuitries as those of pTiR10 and pTiC58 are certainly not confined to the Ti plasmid lineage and seem to be of broader presence in replicons and mobile elements of the *Rhizobiaceae*.

traR homologs, devoid of *traI* partners ("orphans," as frequently called), and of no assigned function as yet, are met in the circular chromosome of *A. tumefacines* C58. Likewise, *A. vitis* S4 harbors no less than ten genes encoding for R-type regulators (four residing on chromosome I) and three for I-type AHL synthases (Setubal et al. 2009), which again shows that regulators may exceed numbers of synthases in a given strain. Multiplicity of I/R members and redundancy of R counterparts await further studies to reveal their role; from a novel agrobacterial system that has been recently examined though, the *avsI/avsR* system of *A. vitis* S4 that is located on chromosome I, it appears that new phenotypes may be governed by I/R-mediated quorum sensing in agrobacteria, via also different signaling molecules. In this case, these are the host hypersensitive response and nonhost plant necrosis, via the mediation of long-chain AHLs (Hao and Burr 2006). The traits brought about by *avsI/avsR* regulation are signature pathogenic properties of *A. vitis* besides grape neoplasias, while interestingly, the I/R genes themselves show homology to the *Sinohizobium meliloti sinI/sinR* pair that controls exopolysaccharide production via again long-chain AHL involvement (Marketon et al. 2002). Thus, in every respect, the intricacy in agrobacterial intercellular and host-pathogen signaling networks has only begun to emerge.

12.2.4 Agrobacterial Genomes

Sequenced representatives from all three *Agrobacterium* biovars, namely *A. tumefaciens* C58 (biovar 1), *A. radiobacter* K84 (biovar 2), and *A. vitis* S4 (biovar 3), reveal the degree of division of their genomic material into several replicons per cell: into two to five plasmids ranging in size from tens to hundreds of kilobases and, what is characteristic for agrobacteria compared with the vast majority of rhizobia analyzed so far, into two replicons bearing chromosomal characteristics (Goodner et al. 2001; Wood et al. 2001; Setubal et al. 2009; Slater et al. 2009; Table 12.1). Strains C58 and S4 have two true chromosomes, whereas K84 is under further scrutiny: apart from its 4-Mb circular chromosome and three well-defined plasmids, K84 harbors a 2.65-Mb replicon that bears one gene that is probably essential – an L-seryl-tRNA selenium transferase – in a region that is overall syntenic to regions found in the secondary chromosomes of C58 and S4 (Slater et al. 2009). The 2.65-Mb replicon of K84 is thus considered to be evolving into a chromosome and its case exemplifies an issue regarding mega-replicons of the *Rhizobiaceae* that revolves around the borderline set to discriminate chromosomes from plasmids (see also Sect. 12.3.3). In general, replicons are thought of as essential – and therefore as chromosomes – based on the presence or absence of rRNA genes and genes important for prototrophy. There is not much to convince though that genes conditionally essential are not to be found in megabase-size genetic material and it is ultimately replicon curing, without losses in cell viability or well-being, that may unequivocally answer to this.

The second chromosome of C58 is characteristically linear (Allardet-Servent et al. 1993) and interestingly the gene that safeguards replicon linearity, that of telomerase A (*telA*), seems to be present only in biovar I representatives that, to the extent tested, also bear linear chromosomes (Setubal et al. 2009). This chromosome, along with all secondary chromosomes and large plasmids of sequenced agrobacteria, is a *repABC* replicon (Table 12.1). As has been mentioned, *repABC* replication/stabilization units are found in plasmids and secondary chromosomes of members of the *Rhizobiales*. Examining the chromosomal relations of the latter, it appears that blocks of essential genes and rRNA operons have moved out of what is thought of as an ancestral unit-chromosome and into a *repABC* replicon recipient. This has led to the theory that secondary chromosomes of the *Rhizobiales*, including *Agrobacterium*, have arisen from a plasmid progenitor (Goodner et al. 2001; Wood et al. 2001; Slater et al. 2009).

The first reports on the *A. tumefaciens* C58 genome indicated the degree of similarity of the *A. tumefaciens* primary chromosome to that of the *Sinorhizobium meliloti* 1021 and *Mesorhizobium loti* MAFF303099 chromosomes (Capela et al. 2001; Goodner et al. 2001; Wood et al. 2001). Whole genome comparisons of all sequenced agrobacterial and rhizobial representatives corroborate that gene content and order are far more conserved on primary chromosomes compared with secondary ones. In the latter, as in plasmids, translocations and other rearrangements result in marked mosaicism (Slater et al. 2009). Furthermore, genomic analyses combined to 16S rRNA-based phylogeny indicate that the *A. radiobacter* biovar 2 strain K84 is closer to the fast growing rhizobia, such as *R. etli* and *R. leguminosarum*, than to *Agrobacterium* biovar 1 and 3 members, such as C58 and S4, which also cluster together (Setubal et al. 2009).

Horizontal transfer is a major contributor to lateral gene dissemination and genomic evolution, and it is notable that all plasmids in *A. tumefaciens* C58, *A. radiobacter* K84, and *A. vitis* S4 harbor complete conjugation systems or conjugal DNA metabolism (Dtr) systems, which most likely render them transferable via the type IV systems encoded by other replicons of the same cells (i.e., pAvS4e and pAgK84; Table 12.1). The multiple conjugation systems of the newly sequenced S4 and K84 plasmids relate to agrobacterial orthologs found on the C58 Ti plasmid or on the 500-kb plasmid pAtC58 (called AvhB after "*Agrobacterium* virulence homolog B"), as also to rhizobial orthologs of mostly *Sinorhizobium* species (Setubal et al. 2009). Interestingly, the K84 chromosome also harbors a complete AvhB-like conjugal transfer system, which, if functional, may mean that this 4-Mb replicon is mobile. Apart from true *in-trans* mobilization, plasmid cointegration events may also play role in the shuffling of genetic material and horizontal transfer of nonconjugative replicons; for instance, it has been long known that the *A. rhizogenes* A4 and 8196 Ri plasmids undergo cointegration with a larger coresident replicon in these strains (Petit et al. 1983). The impressive number of insertion elements (i.e., C58 bears 25 putative ISs; Slater et al. 2008) and of genes predicted to encode for recombinases, integrases, and transposases in all recently sequenced agrobacterial replicons is bound to be pivotal in promoting genetic rearrangements that lead to the generation of novel molecules or to the transferability of immobile ones.

Virulence determinants on Ti plasmids, such as the *vir* genes and the T-DNA, enable them to sufficiently transform avirulent bacteria of the *Rhizobiales* into oncogenic derivatives by their mere presence (Hooykaas et al. 1977; López-López et al. 1999; Teyssier-Cuvelle et al. 2004; Broothaerts et al. 2005). However, overall pathogenic determinants that relate to chemosensory mechanisms, attachment to the host, cell wall polymer degradation, biofilm formation, hydrogen peroxide resistance, and other traits crucial to the infectious process are spread to all replicons of *A. tumefaciens*, as judged by C58 at least (Goodner et al. 2001; Wood et al. 2001). Additionally, the 2.65-Mb replicon of K84 harbors a complete type III secretion system that is apparently not related to the *avsI*/*avsR*-regulated system of *A. vitis*, which induces the plant-host hypersensitive response (see above; Setubal et al. 2009). Type III systems form so-called injectisome complexes that mediate in the translocation of effector proteins from bacterial pathogens to plant or animal hosts (Cornelis 2006). They are also encountered in nonpathogenic bacteria such as rhizobia, in strains of which, i.e., *Rhizobium sp*. NGR234, they are involved in host-dependent nodulation (Marie et al. 2001; Schmeisser et al. 2009). It is quite interesting that the type III secretion system of K84 bears homologies to one found in *A. rhizogenes* A4 and differs from other systems such that it might constitute a new class of a type III macromolecular transporter (Setubal et al. 2009). Notably, this locus seems to be another biovar-specific trait, since it has been found present in biovar 2 representatives exclusively, and this inevitably attracts attention as to its role in those members in particular. The wealth of such information derived from already assigned genes and gene systems, as also from the vast number of genes remaining to be assigned – i.e., of the 5,419 originally predicted genes of C58, 756 were hypothetical (Slater et al. 2008) – is expected in future years to shed unprecedented light in *Agrobacterium* biology and its pathogenicity toward the host.

12.3 Symbiotic Members of the *Rhizobiaceae*: Rhizobia

12.3.1 Rhizobia, Legumes, and Nitrogen Fixation

Nitrogen is an essential element for all living organisms and it is highly abundant as molecular nitrogen (N_2) in the atmosphere. However, molecular nitrogen is only useful to organisms that can perform nitrogen fixation, that is, reduce atmospheric nitrogen to ammonia (NH_3) and incorporate it into organic molecules. Plants and algae assimilate ammonia or nitrates to obtain the nitrogen they need to grow and reproduce; animals obtain nitrogenated organic molecules by consumption of plants and algae. Despite the relative small amount of nitrogen fixed by abiotic means, the nitrogen influx into the biosphere relies on biological nitrogen fixation.

Biological nitrogen fixation is carried out by only a numbered species of free-living heterotrophic or autotrophic bacteria that are common in detritus or soil, by bacteria that establish symbioses with vascular plants, and by cyanobacteria that establish symbiotic associations with fungi or plants. The symbiotic associations between bacteria and vascular plants are especially remarkable because they allow very high rates of nitrogen fixation (Boring et al. 1988).

The legume family (*Fabaceae*) has about 650 genera and 18,000 species and is the third largest plant family after the *Asteraceae* and *Orchidaceae*. The *Fabaceae* has the largest number of species that establish symbiotic relationships with nitrogen-fixing bacteria. The interaction between legume roots and nitrogen-fixing bacteria (rhizobia) leads to the development of a new organ, the nodule. The *Fabaceae* are widely distributed; species of this family occur in tropical ecosystems, lowland-wet rainforests, tropical deciduous forests, thorn scrub forests, savannas, and deserts. In temperate ecosystems, some legume trees may be found, but annual or perennial herbaceous legumes or low-stature legume woody shrubs are most common (Crews 1999).

The origin of the legume family has been dated to 60 million years ago (mya). This family has three subfamilies, *Caesalpinioideae*, *Mimosoideae*, and *Papilionoidae*, which are believed to have originated about 50 mya. The capacity to develop nitrogen-fixing nodules is very common in the *Papilionoidae* and *Mimosoidae*, but less frequent in the *Caesalpinioideae* (Lavin et al. 2005; Sprent 2008; Sprent and James 2008). Bacteria capable of inducing nitrogen-fixing nodules are present in members of seven alpha-proteobacterial families, all belonging to the *Rhizobiales*: *Rhizobiaceae*, *Bradyrhizobiaceae*, *Methylobacteriaceae*, *Xanthobacteriaceae*, *Brucellaceae*, *Hiphomicrobiaceae*, and *Phyllobacteriaceae*, and one beta-proteobacterial family, *Burkholderiaceae* (Sprent 2008). The *Rhizobiaceae* has the largest number of legume nodulating genera, including *Rhizobium*, *Allorhizobium*, *Sinorhizobium/Ensifer*, and *Azorhizobium*. The *Phyllobacteriaceae* has two nodulating genera, *Mesorhizobium* and *Phyllobacterium*. The *Bradyrhizobiaceae*, *Methylobacteriaceae*, *Xanthobacteriaceae*, *Brucellaceae*, and *Hiphomicrobiaceae* each have a single nodulating genus: *Bradyrhizobium*, *Methylobacterium*, *Pseudoxanthobacter*, *Ochrobactrum*, and *Devosia*, respectively. The *Burkholderiaceae* has two genera with species that can induce nitrogen-fixing nodules in a restricted number of legume species, *Burkholderia* and *Cupriavidus* (Moulin et al. 2001; Amadou et al. 2008).

The host-range of nodulating bacteria varies widely. Some bacterial species nodulate one or a few legume species, but others can nodulate hundreds of different species. For example, *Azorhizobium caulinodans* can only nodulate *Sesbania rostrata*. In contrast, *Rhizobium* sp. NGR234 and *Rhizobium fredii* USDA257 can nodulate species in 112 and 79 genera, respectively. Legume host specificity also varies over a broad range. *Macroptilium atropurpureum*, *Phaseolus vulgaris*, and *Vigna unguiculata* are recognized as "promiscuous" plants, in that they accept many rhizobial species (Michiels et al. 1998; Broughton and Perret 1999; Pueppke and Broughton 1999). In fact, *M. atropurpureum* can establish symbioses with rhizobia of both alpha- and beta-proteobacteria (alpha- and beta-rhizobia).

In contrast, *Crotalaria podocarpa*, *C. glaucoides*, and *C. perrottetti* are only nodulated by strains of *Methylobacterium nodulans* (Renier et al. 2008). Nodulation promiscuity seems to be more common than nodulation specificity. Promiscuous species occur in primitive and advanced legume tribes, suggesting that this property is a primitive character (Broughton et al. 2000; Perret et al. 2000). Legumes with restricted host ranges may have evolved more recently in response to specific ecological niches.

Research on legume-rhizobia symbiosis has mainly focused on agriculturally important species such as pea (*Pisum sativum*), soybean (*Glycine max*), French bean (*Phaseolus vulgaris*), alfalfa (*Medicago sativa*), and several legume model species (*Lotus japonicus* and *Medicago truncatula*). In contrast, the biology of host-rhizobial symbiosis remains unexplored for the vast majority of tropical legumes. However, recent investigations of noneconomically important legumes and their microsymbionts have led to the discovery of beta-rhizobia and methylotropic rhizobia, as also of novel mechanisms of legume-rhizobial interactions (Moulin et al. 2001; Jourand et al. 2004; Giraud et al. 2007; Amadou et al. 2008).

12.3.2 Nodules and the Nodulation Process

The legume root nodules induced by rhizobia have diverse morphology, physiology, and development. Nevertheless, all nodules allow the microsymbiont to fix atmospheric nitrogen and share it with the plant in exchange for photosynthate from the plant. Usually, nitrogen fixation is carried out within specialized infected plant cells. However, in some legume–rhizobial relationships, this activity occurs between plant cells within the nodule.

The initial interaction between rhizobia and legume roots occurs within the rhizosphere, a highly specialized ecological niche that is within several millimeters of soil around the root system. Plant roots excrete large amounts of low molecular weight compounds, such as sugars, amino acids, phenolic compounds, and organic acids, and high molecular weight compounds, such as polysaccharides and proteins. The amount and diversity of the excreted products depends on the species, its age and physiological condition, and the environment. Root exudates have an important ecological role and the large amount of excreted organic compounds can sustain large bacterial populations and associated bacteriophages. The diversity and quantity of excreted substances profoundly affects the composition of the bacterial communities of the rhizosphere. Legume exudates contain compounds that promote proliferation of specific symbiotic partners. For example, homoserine excreted by *Pisum sativum* (Pea) is specifically catabolized by *Rhizobium leguminosarum*, its compatible microsymbiont (van Egeraat 1975). Legume roots also exude large amounts of flavonoids (2-phenyl-1,4-benzopyrone derivatives) and other compounds that have crucial roles in legume–rhizobial symbiosis. Flavonoids induce rhizobial expression of genes involved in the synthesis of signals that trigger nodule development in plants. In general, mixtures of flavonoids are more efficient than single compounds in promotion of nodule development (Begum et al. 2001).

The initiation of rhizobial invasion requires colonization of the root surface and attachment to root hair tips. In this process, rhizobia sense and respond to flavonoids and other compounds mostly via NodD, a transcriptional regulator of the LysR family that binds and activates promoters, which contain a conserved *nod* box sequence. Rhizobia can harbor several different NodDs, each one recognizing a particular set of flavonoids. NodDs induce the transcription of several sets of genes, including those that encode enzymes involved in the synthesis of lipo-chito-saccharides (Nod factors), which are essential for nodule development. Flavonoid induction of NodD also induces the expression of genes involved in Type III (T3SS) and Type IV (T4SS) secretion systems (Marie et al. 2004; Hubber et al. 2004; Cooper 2007), modification of extracellular polysaccharides (Broughton et al. 2006), and synthesis of indoleacetic acid (Prinsen et al. 1991; Theunis et al. 2004). These all affect the legume–rhizobial relationship. NodD-flavonoid specificity is one of the important factors that modulate legume–rhizobial host-range.

Nod factors consist of three to six b-1,4 linked *N*-acetyl-D-glucosamine residues (chitin oligomers) with an *N*-acyl group attached to the nonreducing end of the oligosaccharide. This *N*-acyl group is usually a fatty acid that is derived from general metabolism, such as vaccenic acid (C18:1) or stearic acid (C18:0) and less frequently an unsaturated C20:1 or C18:4 fatty acid, while other functional groups may be also attached to the nonreducing as well as the reducing end of the sugar residues (D'Haeze and Holsters 2002).

Genes involved in the synthesis of Nod factors can be classified as common nodulation genes (*nodA*, *nodB*, and *nodC*), which are involved in the synthesis of the core Nod factor structure and are conserved among rhizobial species, or those involved in modification of the Nod factor (*nod*, *noe*, *nol*, and others), which are strain- and species-specific. The amount and diversity of Nod factors that are synthesized vary among strains and play a role in determining host-specificity (Spaink 2000).

Detection of Nod factors by legume root hairs triggers multiple responses that are essential for rhizobial invasion and nodule development. A few seconds after the root detects a Nod factor, there is a flux of ions, plasma membrane depolarization, increase of intracellular pH, and an influx of intracellular calcium that enhances the calcium tip gradient of root hairs. Then, there is an oscillation of cytosolic calcium (calcium spiking), cytoskeletal rearrangements of the root hair, expression of early nodulins (ENOD12, RIP1), and inhibition of polar auxin transport in the root cortex (Cullimore et al. 2001). Later, there is a redirection of root hair growth and the roots begin to curl, a process that culminates in the formation of a structure called the "shepherd's crook." Rhizobia become entrapped within this structure, enter the root hair cell and then begin to form the infection thread. The infection thread is a tubular structure surrounded by plant cell walls that conduces rhizobia that divide and grow in its interior, to cells of the inner cortex, susceptible to infection. In addition to Nod factors, development of the infection thread requires certain rhizobial surface polysaccharides, including cyclic beta glucans, exopolysaccharides, lipopolysaccharides, and capsular polysaccharides. The mechanisms of action of these polymers are unknown, but they appear to suppress plant defenses (Broughton et al. 2000). Nodulation outer proteins (Nops), which are secreted by T3SS and T4SS, also have roles in the formation of the infection thread and in the development of the nodule. Some Nops also appear to play roles in the modulation of plant defenses (Fauvart and Michiels 2008; Soto et al. 2009). The infection thread grows into the root hair, penetrates a few layers of cortical cells, and then starts to branch out so that subsequent inner cortical layers are crossed many times.

During this process, Nod factors stimulate inner cortical cells to form nodule meristems. Postmeristematic cells enlarge and the infection threads invade some of these cells. Groups of rhizobial cells are delivered from the infection thread tips as infection droplets, while individual rhizobial cells, surrounded by plant-derived membranes, are released from infection droplets by budding. Bacteria continue to divide for a certain period of time and then undergo a complex differentiation process that culminates in the formation of nitrogen-fixing bacteroids. A differentiated bacteroid that is surrounded by a plant-derived membrane is referred to as the "symbiosome." Nodule vascularization subsequently follows and arises from cells derived from the root pericycle and endoderm (Sprent 2001; Brewin 2004).

In order to fix atmospheric nitrogen, bacteroids must express a new gene set (*nif* and *fix*) including the nitrogenase polypeptides, genes required for the synthesis and incorporation of nitrogenase cofactors and regulatory genes. Nitrogenase is very sensitive to oxidation, but rhizobia are strict aerobes. Thus, it is necessary to protect the nitrogenase complex from oxygen in the nodules and this is carried out by a plant cell layer, without intercellular spaces, which is located below the nodule cortex and acts as an oxygen barrier. In addition, legumes synthesize leghemoglobin, a heme-based O_2-binding protein that provides a low but constant oxygen concentration inside the nodule (Mylona et al. 1995).

12.3.3 Rhizobial Genomes

Complete genome sequences are available for 14 alpha-rhizobia and 1 beta-rhizobium (Table 12.1). These species have large genomes, ranging from 5.36 Mb (*Azorhizobium caulinodans* ORS572) to 9.1 Mb (*Bradyrhizobium japonicum* USDA 110). Such diversity might be expected, since these organisms live in diverse habitats and face different challenges in forming symbioses. Many of the rhizobial genomes are multipartite, in that they are composed of one circular chromosome and several plasmids, some of which have very high molecular weight (more than 1 Mb). Traditionally, chromosomes contain housekeeping genes and plasmids contain genes that provide advantages only in certain circumstances or environments and may be dispensable. Thus, if a replicon contains one or more essential genes, it is considered a chromosome; if a replicon contains dispensable genes, it is considered a plasmid regardless of size. Replicons that contain only one or a few essential genes are sometimes called "minichromosomes" or "secondary chromosomes"; plasmids that are more than 1 Mb in size are sometimes called megaplasmids.

There are large variations in genome size, plasmid number, and gene content of rhizobia, even in strains of the same species. Thus, it is more practical to use the terms "core" genome and "accessory" genome (Tettelin et al. 2008). Core genes are those present in all strains of a species and typically encode genes involved in basic metabolism. Accessory genes are those present only in some strains of a species, are often dispensable, tend to be acquired by horizontal-transfer, and are frequently on plasmids. The "pangenome" of a species is the sum of its core genes and accessory genes.

Some *Bradyrhizobium* strains contain only one very large chromosome (see above), but other species of rhizobia typically contain large pools of plasmids in addition to their chromosome. In fact, plasmid profiles are frequently used as strain markers because the number of plasmids, plasmid size, and plasmid gene content vary widely among isolates, even of the same species (Mozo et al. 1988; Mohmmed et al. 2001; Stiens et al. 2006). The large diversity of plasmid genes is presumably related to the extreme variability of soil habitats occupied by host species.

Only a few attempts have been made to define the core and accessory genes of rhizobia. Based on comparisons between the genome sequences of *Agrobacterium*, *Brucella*, and different sets of alpha-rhizobia, some initial generalizations about core genes and accessory genes can be made. To start with, single chromosomes as well as primary chromosomes of species harboring more than one chromosome (i.e., agrobacteria; see Sect. 12.2.4) tend to have a more conserved gene content and order than plasmids. Second, chromosomes of closely related species contain more syntenic gene blocks than chromosomes of distantly related species; chromosomal syntenic blocks are frequently disrupted by insertion sequences, orphan genes or genes annotated as hypothetical conserved proteins. Third, core genes are more common on single or primary chromosomes, but do not occur there exclusively. Fourth, plasmids are heterogeneous in size and gene content, but some have small blocks of syntenic genes and may have abundant accessory genes. Lastly, genes encoded on the chromosome of some species are on plasmids of other species, suggesting gene shuffling between chromosomes and plasmids or between plasmids (Paulsen et al. 2002; Young et al. 2006; Crossman et al. 2008; Slater et al. 2009, González et al. in preparation).

Most of the genes required for symbiosis (*nol*, *nod*, *nif*, *fix*) are encoded in a single and usually transferable plasmid that is called the symbiotic plasmid. However, in some species (e.g., *Bradyrhizobium japonicum*, *Mesorhizobium loti*), symbiotic genes occur together on "symbiotic islands" on the chromosome. Nonetheless, there are clear differences among symbiotic plasmids with regard to gene content and organization, even of genes directly involved in symbiosis. For example, the symbiotic plasmid pRL10 from *R. leguminosarum* 3841 has 461 protein-encoding genes, whereas p42d from *R. etli* CFN42 has 359. Nodulation and nitrogen fixation genes that are common to other rhizobia are present in p42d and scattered across 125 kb of its genome, but the *nif-nod* region of pRL10 is clustered within a 60-kb region. Plasmid p42d lacks *nodTNMLEF* and *rhiABCR* genes, which are present in pRL10. Contrarily, p42d harbors genes *nifS*, *nifU*, *nifW*, *nifX*, *nifZ*, *nolT*, *nolE*, *noeI*, and *noeJ*, which are absent from pRL10. Lastly, plasmid p42d has three copies of *nifH*, instead of one that pRL10 has (González et al. 2003; Young et al. 2006; Crossman et al. 2008).

Symbiotic genes are an important component of the rhizobial accessory genome and may be dispensable under some circumstances, but are essential for symbiosis. The mobile nature of symbiotic genes is illustrated by the beta-rhizobium *Cupriavidus taiwanensis* LMG 19424. In this species, the nodulation genes are closely related to those present in alpha-rhizobia, indicating horizontal transfer (Amadou et al. 2008).

Insertion Sequences (ISs), archetypical components of the accessory genome, are abundant in rhizobia. These elements are clustered within specific genomic compartments or regions. In *B. japonicum* USDA110, 60% of the ISs are located in the 410-kb symbiotic island. A second cluster of ISs is located in a 206-kb region with a low CG content and an integrase/recombinase gene, suggesting that this region is or was a mobile component (Kaneko et al. 2002). Similarly, of the 580 genes present in the 500-kb symbiotic island of *M. loti* MAFF303099, 111 are related to transposonases, integrases, recombinases, or resolvases (19.6%) and only 0.5% of the genes in one of these four categories are outside the symbiotic island (Kaneko et al. 2000). *Rhizobium etli* CFN42 has 40 complete ISs, with 14 in the symbiotic plasmid p42d (371 kb), 11 in plasmid p42a (194 kb), and 13 in the chromosome (González et al. 2006; Lozano and Gonzáles, manuscript in preparation). In *R. leguminosarum* bv. *viciae* 3841, plasmid pRL7 has the highest density of genes related to transposases (Young et al. 2006). In *Rhizobium* sp. NGR234, there is an uneven distribution of transposable elements, with plasmids pNGR234a, pNGR234b and the chromosome having 16.3, 4.8, and 2.1% of these elements (Schmeisser et al. 2009). The same pattern is present in the genome of *S. meliloti* 1021, in which a large percentage of ISs are located in a single plasmid (pSymA) (Galibert et al. 2001; Barnett et al. 2001). The beta-rhizobium *Cupriavidus taiwanensis* LMG 19424 is an extreme case in this: its 550-kb symbiotic plasmid has 93% of the transposable elements of this genome (Amadou et al. 2008).

An intriguing characteristic of alpha-rhizobial plasmids is that, with a few exceptions (e.g., *Methylobacterium nodulans* ORS2060 plasmids and the rolling circle replication plasmid pRm1132f from *Sinorhizobium meliloti* 1132; Barran et al. 2001), most are under the control of the *repABC* replication/partitioning operon (Table 12.1). The most extreme example of this is represented by the genomes of *Rhizobium etli* CFN42 and *Rhizobium leguminosarum* bv. *viciae* 3841, each of which has 6 *repABC* plasmids (Young et al. 2006; González et al. 2006). This demonstrates that *repABC* plasmids belong to several incompatibility groups.

repABC operons are not exclusive to alpha-rhizobia and are also responsible for the replication and maintenance of plasmids and chromosomes in a wide variety of alpha-proteobacteria that live in very different ecological niches. This suggests that *repABC* plasmids have a wide host-range [for recent reviews: Cevallos et al. (2008)

and Pappas (2008)]. In general, the *repABC* transcriptional unit has three protein-encoding genes (*repA*, *repB*, and *repC*) and a gene that encodes a small antisense or "countertranscript" RNA (ctRNA) located within the *repB–repC* intergenic region. RepA is a Walker-type ATPase, a member of the MinD/ParA superfamily, and can presumably polymerize into filaments. RepB is a DNA-binding protein that can bind to RepA and to the plasmid centromere-like sequence, *parS*, usually located nearby or within the operon (Fig. 12.4). RepA and RepB homologs, largely known as ParA/ParB family proteins, participate in the partitioning of plasmids and chromosomes of many bacteria. RepC, the protein encoded by the last gene of the operon, is present only in alpha-proteobacteria and does not show any similarity with other proteins.

RepA and RepB proteins have dual roles. They form the plasmid segregation machinery in conjunction with the *parS* centromeric site and also act in negative transcriptional regulation of their own expression. RepA represses transcription of the *repABC* operon by directly contacting the operator sequence, while RepB acts as a corepressor. For proper plasmid partitioning, RepB must bind to the *parS* sequence and to RepA (Ramírez-Romero et al. 2001; Pappas and Winans 2003b;

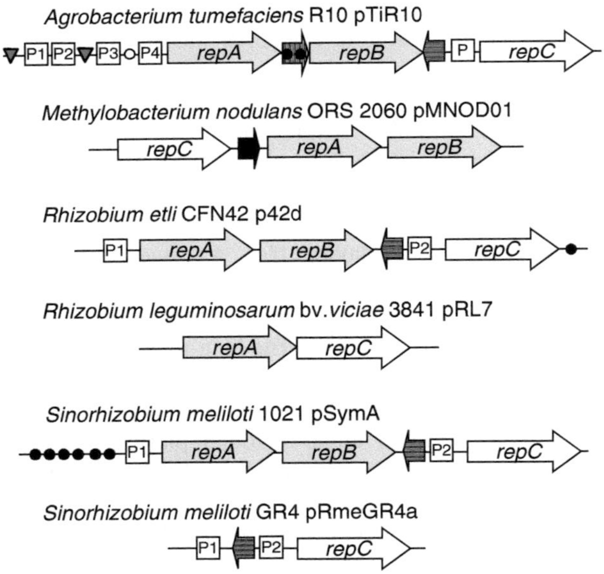

Fig. 12.4 Genetic organization of replication/partitioning regions of *Agrobacterium* and alpha-rhizobial *repABC* replicons, and of close derivatives. *Gray arrows* represent *repA* and *repB* genes, encoding proteins involved in plasmid segregation and in *rep* operon repression. *White arrows* show positions of *repC* genes, encoding replication initiator proteins. *Black circles* indicate partitioning sites (*parS*). *Arrows* filled with horizontal lines indicate the localization of regulatory antisense RNA genes. *Square boxes* indicate promoter sites. In pMNOD01, the *black arrow* represents a hypothetical gene. In pTiR10, the *arrow* filled with *vertical lines* indicates a fourth gene of the *repABC* operon, *repD; triangles* show *tra boxes* and the *open circle*, the location of a *vir* box

Chai and Winans 2005b). The *repC* gene is essential for plasmid replication. It encodes the initiator protein, which functions by binding to the origin of replication that is located within its own coding sequence. The small antisense RNA (ctRNA), located within the *repB–repC* intergenic regions, modulates *repC* expression (Venkova-Canova et al. 2004; Chai and Winans 2005a; MacLellan et al. 2005).

Structurally, *repABC* operons are only superficially similar. In fact, they have highly diverse DNA sequences and some bear specific elements that are only present in a few members of the family. They vary in the number and class of regulatory elements involved in operon transcription, in the number and position of *parS* sequences and in the presence of peptide-encoding minigenes (Fig. 12.4; Cevallos et al. 2008; Pappas 2008).

Some alpha-proteobacterial genomes have *repAB*-like partitioning genes that are not closely associated with a ctRNA/*repC* replication sequence module. Contrarily, in other alpha-proteobacterial plasmids, including some alpha-rhizobia, only the ctRNA/*repC* module is found, organized seperately from the plasmid partitioning genes. This is the case of the pRmeGR4a plasmid of *Sinorhizobium meliloti* GR4 and the pRtc299c plasmid of *R. tropici* (Izquierdo et al. 2005). Thus, fusion of different modules may be the basis for the generation of new *repABC* plasmids and different elements may have different evolutionary histories. Previous events of gene loss and gene shuffling may also play role in the generation of new replication modules. The pRL7 plasmid from *R. leguminosarum* bv. *viciae* 3841 has two *repABC* functional operons and a truncated version that lacks *repB* and the ctRNA gene. This truncated version can confer independent replication of a suicide vector in *Rhizobium* cells (Perez-Segura, unpublished results; Fig. 12.4). The pMNOD01 plasmid from *Methylobacterium nodulans* ORS has a putative replication system with *repA*, *repB*, and *repC* genes all present, although in an unusual order. These three genes are clustered and encoded in the same DNA strand, but *repC* precedes the *repAB* genes. In addition, there is a small hypothetical protein-encoding gene between *repC* and *repAB* (Fig. 12.4).

The balance of all elements that participate in plasmid replication and segregation provides plasmid stability. Perturbation of this balance, which may result from the introduction of a replication or segregation element in excess, could lead to plasmid incompatibility. The *repABC* plasmids have at least three elements involved in plasmid incompatibility: the RepA and RepB proteins, the small antisense RNA, and the *parS* sequences. Phylogenetic analysis of the RepA, RepB, and RepC proteins has shown that different replicons within the same bacterial strain can belong to different clades (Cevallos et al. 2002, 2008; Slater et al. 2009). This suggests that divergent evolution, followed by episodes of horizontal gene transfer, played a central role in the origination of new incompatibility groups. In addition to the different evolutionary histories of RepA, RepB, and RepC, recent analyses have shown that these operons are highly dynamic, despite the conservation of the basic *repABC* structure. This suggests recombinational gene displacement between different *repABCs* within a single bacterial cell (Slater et al. 2009; Castillo-Ramírez et al. 2009).

There is very limited knowledge on the replication/partitioning systems of beta-rhizobia. The genome sequence of *Cupriavidus taiwanensis* LMG19494 indicated that its symbiotic plasmid (pRalta) contains a cluster of genes, *parABrepAB*, with a high degree of similarity to plasmids of *Cupriavidus*, *Ralstonia*, and *Burkholderia*. However, further functional analysis of these genes has not been carried out yet (Amadou et al. 2008). Sheu and coworkers (2007) reported the sequence and the general mechanism of replication of two very small cryptic plasmids (pTJ86-1 and pTJ86-2) in *Cupriavidus taiwanensis* strain TJ86. Both plasmids replicated by a rolling-circle mechanism and both are functional in several species of *Burkholderia*.

12.3.4 Horizontal Transfer and Quorum Sensing

Conjugation is a major mechanism of bacterial evolution, in that it allows for transfer of DNA molecules – plasmids, chromosomes, or integrons – between strains of same or different species. Diverse evidence indicates that alpha-rhizobial plasmids and genomic islands are, or have the potential to be, self-transmissible or mobilizable by other plasmids. To begin with, numerous plasmids are transmissible under laboratory conditions: pRL1JI from *R. leguminosarum* 248, p42a and p42d from *Rhizobium etli* CFN42, pNGR234a from *Rhizobium* sp. NGR234, pRL7 and pRL8 from *R. leguminosarum* bv. *viciae* 3841, and others (Hooykaas et al. 1981; Johnston et al. 1982; Brom et al. 2000; Young et al. 2006; Ding and Hynes 2009). However, their transfer frequencies vary widely, ranging from 10^{-3} to 10^{-9} transconjugants per recipient cells (i.e., Brom et al. 2004; Sepúlveda et al. 2008). Second, there is evidence that a nonmobilizable plasmid can be transferred by conjugation to other strains via cointegrate formation with a self-transmissible plasmid (see also Sect. 12.2.4). This occurs for plasmid p42d of *R. etli* CFN42, whose own transferability is almost undetectable, but increases substantially following cointegration with plasmid p42a (Brom et al. 2004). Third, there is abundant information indicating that mobilization of symbiotic islands and plasmids occurs in the soil (Schofield et al. 1987; Kinkle and Schmidt 1991; Sullivan et al. 1995; Gomes-Barcellos et al. 2007). Fourth, plasmid mobilization has been inferred from phylogenetic and population studies, which have shown that bacterial isolates from diverse phylogenetic lineages have the same or very similar plasmids (i.e., Louvrier et al. 1996; Wernegreen and Riley 1999; Silva et al. 2005). Finally, sequence analysis of alpha-rhizobial genomes has shown that many plasmids encode complete conjugation systems, indicating they are self-transmissible or at least mobilizable (e.g., pSmed01 and pSmed02 of *S. medicae* WSM419, pRetCIAT652b, and pRetCIAT652c of *R. etli* CIAT652 and others; Ding and Hynes 2009).

In a recent review, Ding and Hynes (2009) proposed that the plasmid transfer systems of rhizobia be classified into three groups based on *tra* gene organization, the regulation mechanism of *tra* genes, and the phylogenetic relationships of the TraA protein. Under this scheme, the type I transfer system includes plasmids bearing a genetic organization similar to that of agrobacterial Ti plasmids harboring *tra/trb* gene clusters, which are regulated by quorum sensing (see Sects. 12.2.2, 12.2.3). The type II system includes plasmids with a *tra* gene organization similar to that of the *R. etli* CFN42 plasmid p42d, regulated by the RctA repressor. The putative type III system includes some under-characterized large plasmids of strains 3841, VF39, and WSM1325 of *R. leguminosarum.*

Quorum sensing and its effect on plasmid conjugation have been extensively studied in the octopine- and nopaline-type Ti plasmids of *A. tumefaciens* (see Sect. 12.2.3; Farrand 1998). Compared with agrobacteria, the study of quorum sensing in rhizobia and its role on plasmid transfer is more complicated for many reasons. To start with, members of the alpha-rhizobia can produce multiple *N*-acylated homoserine-lactone (AHL) autoinducer signals. For example, *R. etli* CNPAF512 produces at least seven AHLs, whereas *R. leguminosarum* 8401 (pRL1JI) produces at least six (Daniels et al. 2002; Blosser-Middleton and Gray 2001). Relevant to this is the fact that there are multiple quorum sensing systems in rhizobia. *Bradyrhizobium japonicum* USDA110, in addition to producing multiple AHLs, also produces bradyoxetin, a novel autoinducer molecule (Loh et al. 2002); *S. meliloti* 1021 has at least two quorum-sensing systems, one of which produces several novel AHLs of different size (Marketon et al. 2002). Furthermore, quorum sensing systems have multiple roles in the alpha-rhizobium physiology. Depending on the strain, these systems can modulate symbiosis, motility, growth rate, exopolysaccharide synthesis and additional processes (Marketon et al. 2003; Hoang et al. 2008; Cao et al. 2009; Yang et al. 2009).

Based on the classification system of Ding and Hynes (see above) plasmids pNGR234a from *Rhizobium* sp. NGR234, p42a from *R. etli*, and pRL1JI from *R. leguminosarum* bv. *viciae* 280 contain type I transfer systems. Their gene organization is similar to that of pTiR10 or pTiC58 and they are regulated by quorum sensing. *Rhizobium* sp. NGR234 produces several AHL signals, one of which is *N*-3-oxooctanoyl-L-homoserine lactone (OOHL), synthesized by the *traI* gene on pNGR234a. The other AHL synthases are encoded elsewhere in the plasmid genome. Similar to the regulation system of the transfer genes of Ti plasmids, *traI* is autoregulated by TraR and OOHL, whereas TraR activity is inhibited by TraM (see also Sect. 12.2.3). However, under laboratory conditions, pNGR234a has a low frequency of transfer, suggesting that an unidentified compound in the rhizosphere may enhance plasmid transfer, in a manner analogous to that of opines regulating Ti plasmid transfer in *A. tumefaciens* (He et al. 2003).

The *R. etli* CFN42 plasmid p42a is a highly efficient self-transmissible plasmid that also has *tra* genes similar to those of pTiC58. This plasmid encodes four proteins of the I/R/M circuitry, TraI, TraR, CinR, and TraM, although the last does not seem to be expressed, which highly likely accounts for the constitutive transfer of p42a. TraI controls the synthesis of OOHL, one of the several AHLs produced by *R. etli* CFN42, but the only one that regulates p42a transfer. Expression of *traI* depends on the presence of TraR/OOHL and CinR, while conversely, *cinR* expression is dependent on *traI* (Tun-Garrido et al. 2003).

Three proteins in the LuxR family (TraR, BisR, and CinR) and two AHLs, OOHL synthesized by TraI and *N*-3-hydroxy-7-*cis*-tetradecenoyl-L-homoserine lactone (HTHL) synthesized by CinI, regulate replication and transfer of pRL1JI genes from *Rhizobium leguminosarum* bv. *viciae* 280. TraR and BisR are encoded on the plasmid, while *cinI* and *cinR* on the chromosome. TraR activates the transcription of *repABC* and transfer genes in response to the OOHL signal synthesized by TraI. However, expression of TraR depends on BisR and HTHL. On the other hand, BisR also represses *cinI*. Thus, a strain carrying pRLJ1 synthesizes only small amounts of HTHL and cannot trigger conjugation or, presumably, plasmid copy-number increase. In contrast, "female" strains that lack pRLJ1 and its *bisR*-linked gene synthesize large amounts of HTHL. When a pRL1JI donor strain encounters a recipient strain that produces abundant HTHL, TraR is induced by BisR. In the presence of OOHL, TraR concomitantly induces the conjugative transfer of the plasmid (Danino et al. 2003; McAnulla et al. 2007).

Type II plasmids are exemplified by p42d, the symbiotic plasmid from *R. etli* CFN42. The conjugal auto-transfer of this plasmid is modulated by two genes: *rctA*, a repressor gene adjacent to the *virB* operon, but transcribed in the opposite direction; and *rctB*, a gene located downstream of *traA* that probably acts as an inhibitor of the repressor activity of *rctA*. RctA blocks the transcription of the *virB* operon by binding to a 9-bp motif located within the −10 and −35 hexameric elements of the promoter. The mechanism of action of RctB remains to be elucidated (Sepúlveda et al. 2008). RctA functional homologues are also present in pSymA from *S. meliloti* 1021 and in pAtC58 from *A. tumefaciens* C58. Furthermore, *rctA* homologues can be also met in pSymB from *S. meliloti* 1021, plasmid 1 from *Mesorhizobium* sp. BNC1 and in plasmids pRetCIAT652b and pRetCIAT652c from *R. etli* CIAT652, but their function in these species has not yet been examined (Pérez-Mendoza et al. 2005; Ding and Hynes 2009).

According to Ding and Hynes (2009), a third transfer system that is based on gene homologies of conjugal DNA transfer and replication (Dtr) loci solely seems to be present in plasmids pRL10JI, pRL11JI, and pRL12JI of the *R. leguminosarum* bv. *viciae* strain 3841 and in plasmids pRleVF39d, pRleVF39e, and pRleVF39f of same biovar strain VF39. It may also be encountered in *R. leguminosarum* bv. *trifolii* WSM1325 plasmids, as revealed by the presence of type III-like *traA* alleles. An in-depth study of this novel group of plasmids is currently under way.

Last but not least, the *Mesorhizobium loti* R7A symbiosis island ICE*Ml*Sym^{R7A} was recently shown to undergo excision and conjugal transfer following quorum sensing activation. This island carries genes required for nitrogen fixation and symbiosis with *Lotus* species. It also encodes for two TraI-type and one TraR-type regulator. The quorum sensing circuitry involving one of the I-type counterparts and the TraR homolog uses *N*-3-oxohexanoyl-L-homoserine lactone as signal and controls expression of genes important for the excision and transfer of the element (Ramsay et al. 2009). These last do not appear confined to this strain, which indicates that the repertoire of systems that are mobile and/or induced by inter-bacterial signaling in rhizobia is larger than thought.

12.3.5 Genome Plasticity in Alpha-Rhizobia

The enormous genomic plasticity of alpha-rhizobia depends on the presence of large reiterated DNA sequences dispersed in plasmids and chromosomes of species in this group and on the versatility of the replication/partitioning *repABC* system. Most of the large reiterated DNA sequences in alpha-rhizobial genomes are ISs and duplicated genes. Both of these are good substrates for the homologous recombination machinery and could lead to different genome architectures, depending on position and orientation. Homologous recombination between reiterated sequences in two different replicons can generate a cointegration event. This phenomenon has been studied in detail in two alpha-rhizobia, *S. meliloti* 1021 and *Rhizobim* sp. NGR234. In both strains, it was possible to isolate naturally occurring clones that bore evidence of cointegration events between plasmids or between plasmids and a chromosome that did not reduce or disrupt physiology. Cointegrates can be resolved if a second recombination event uses the same substrates. However, if it occurs in a different pair of duplicated regions, the result will be a pair of replicons each containing a new DNA segment (Mavingui et al. 2002; Guo et al. 2003).

Homologous recombination may also proceed between duplicated regions in the same replicon. Here, the final products depend on the orientation of the target region – on whether it is direct or inverted – and can accordingly lead to deletions or inversions of regions in-between (Romero et al. 1995; Romero and Palacios 1997). Site-specific recombination can also shape genome architecture. The presence of two specific recombination sites and of the cognate recombinase acting thereon can lead to cointegration events and to the subsequent formation of new plasmids if, for instance, the cointegrate is resolved using an alternative duplicated region and a *recA*-dependent recombination process (Brom et al. 2004). Thus, in the case that two replicons share at least two stretches of DNA that can be used as homologous recombination substrates, they can exchange regions. Overall, these properties allow translocation of gene sets among replicons of a single genome.

As discussed before, the replication/partitioning *repABC* system is a conserved component of alpha-rhizobial genomes. This system is widely distributed among alpha-proteobacteria and has diversified into many incompatibility groups, allowing the coexistence of several *repABC* replicons within the same cell. Moreover, genetic recombination and gene shuffling between constituents of different *repABC* operons is an efficient mechanism for the generation of yet more incompatibility groups and new replication and maintenance systems. These processes augment the diversity of plasmids that can be shared in the same strain (Cevallos et al. 2008; Castillo-Ramírez et al. 2009).

Contrarily to previous thinking, latest evidence indicates that *repABC* replicons have a broad host-range. Rogel and coworkers (2006) successfully introduced the symbiotic plasmid of *R. tropici* 299, a member of the *repABC* family, to different strains of *Sinorhizobium*, *Ensifer*, *Ochrobactrum*, *Brucella*, and *Agrobacterium*. This suggests that the genetic information carried by *repABC* plasmids can be shared by very different alpha-proteobacteria. The *repABC* replication unit can replicate and segregate large molecules of DNA, with high stability and at very low copy-numbers. However, natural *Rhizobium* strains that lack the symbiotic plasmid have been recovered from soil, indicating that certain *repABC* plasmids may also be lost. On the other hand, other *repABC* plasmids (e.g., pSymB from *S. meliloti* and p42e from *R. etli* CFN42) are recalcitrant to curing procedures, suggesting that they encode components that are essential for cell viability (Segovia et al. 1991; Brom et al. 2000; Finan et al. 2001). Plasmid acquisition of essential genes is probably a prerequisite for the origination of "secondary" chromosomes. This idea has been indeed developed by Slater and coworkers (2009) to explain the origin of secondary chromosomes whose replication depends on a *repABC* operon in members of the *Rhizobiales* (see also Sect. 12.2.4). According to these authors, secondary chromosomes arose when sets of genes including *minCDE*, *hutIGU*, and *pcaGHID* (at minimum) were transferred by an intragenomic transfer event to an ancestral *repABC* replicon.

With advances in sequencing technologies, we expect that the genomes of more rhizobial species will soon become available. At present, it appears that the alpha-rhizobial pangenome is very large or even completely open. New data will be indispensable in the evaluation of the pangenome size of rhizobia in future years.

12.4 Conclusions

Most prominent members of the *Rhizobiaceae*, agrobacteria and rhizobia, are major inhabitants of soil ecosystems and by far the most studied plant-associated bacteria. The chemical dialogues they engage in with the host, toward pathogenic or mutualistic ends, have been used throughout the years as a fundamental paradigm for transkingdom communication. Moreover, in the applications front, they have provided a wealth of information and a battery of tools that fuel current advancements in plant biotechnology.

Key role in the most examined microbe–plant interactions driven by this group of bacteria is played by plasmid replicons, which receive, integrate and release signals, and engage in functions that determine the outcome of the bacterial-host cohabitation. The plant host, on the other side, has evolved to listen to and differentiate upon receiving beneficial bacterial signals or to defend itself when infected, by intercepting bacterial virulence and interbacterial communication, as recently shown for *Agrobacterium*.

Sequence analysis of representative members of the *Rhizobiaceae* has yielded a massive amount of data to be analyzed in future years. If anything, it has answered as to the size and content of their genome and to its degree of partitioning into well defined plasmid and chromosomal replicons or less well defined mega-replicons that stand in the borderline between the two. It has also provided clues as to the probable paths of evolution of such divided genomes. Lastly, it has illustrated the degree of conservation that characterizes basic chromosomes in species of the family or order and, in sharp contrast to that, the genetic fluidity of the plasmid

Table 12.1 Replicons of sequenced agrobacterial and rhizobial representatives

Species/strain	Replicon	Size (Mb)	Genome size (Mb)	Replication system and properties	Accession number
Agrobacterium					
Agrobacterium radiobacter K84	Chr 1	4.0	7.3	αchr	NC_011985
	Chr 2*	2.65		*repABC* (1) (*min*)	NC_011983
	pAgK84	0.04		NI	NC_011994
	pAgK84b	0.18		*repABC* (1)	NC_011990
	pAgK84c	0.39		*repABC* (1)	NC_011987
Agrobacterium rhizogenes MAFF03-01724	pRi1724	0.22	–	*repABC* (1) (*vir*)	NC_002575
Agrobacterium rhizogenes K599	pRi2659	0.18	–	*repABC* (1) (*vir*)	NC_010841
Agrobacterium tumefaciens	pTi (oct)	0.19	–	*repABC* (1) (*vir*)	NC_002377
Agrobacterium tumefaciens C58	Chr	2.84	5.67	αchr	NC_003062
	circular	2.1		*repABC* (1) (*min*)	NC_003063
	Chr linear	0.54		*repABC* (1)	NC_003064
	pAt58	0.21		*repABC* (1) (*vir*)	NC_003065
	pTiC58				
Agrobacterium vitis S4	Chr 1	3.7	6.31	αchr (*min*)	NC_011989
	Chr 2	1.3		*repABC* (1)	NC_011988
	pAtS4a	0.08		*repABC* (1)	NC_011986
	pAtS4b	0.13		*repABC* (1)	NC_011991
	pAtS4c	0.21		*repABC* (1)	NC_011984
	pTiS4	0.26		*repABC* (2) (*vir*)	NC_011982
	pAtS4e	0.63		*repABC* (1)	NC_011981
alpha-rhizobia					
Azorhizobium caulinodans ORS 571	Chr	5.37	5.37	αchr (*min*)(sym)	NC_009937
Bradyrhizobium japonicum USDA 110	Chr	9.10	9.10	αchr (*min*)(sym)	NC_012226
Bradyrhizobium sp. BTAi1	Chr	8.26	8.49	αchr (*min*)(sym)	NC_009485
	pBBta01	0.23		*repABC* (1)	NC_009475
Bradyrhizobium sp. ORS278	Chr	7.46	7.46	αchr (*min*)(sym)	NC_009445
Mesorhizobium loti MAFF303099	Chr	7.04	7.56	αchr (*min*)(sym*)	NC_002678
	pMLa	0.35		*repABC* (1)	NC_002679
	pMLb	0.21		*repABC* (1)	NC_002682
Mesorhizobium sp. BNC1	Chr	4.41	4.93	αchr (*min*)(sym*)	NC_008254
	plasmid1	0.34		*repABC* (1)	NC_008242
	plasmid2	0.13		*repABC* (1)	NC_008243
	plasmid3	0.04		*repABC* (1)	NC_008244
Methylobacterium nodulans ORS 2060	Chr	7.77	8.84	αchr (sym)	NC_011894
	pMNOD01	0.49		*repCrepAB* (1)	NC_011892
	pMNOD02	0,46		NI	NC_011887
	pMNOD03	0.04		NI	NC_011893
	pMNOD04	0.038		NI	NC_011895
	pMNOD05	0.02		NI	NC_011888
	pMNOD06	0.013		NI	NC_011889
	pMNOD07	0.001		NI	NC_011890
Rhizobium etli CFN42			6.53		

(*continued*)

Table 12.1 (continued)

Species/strain	Replicon	Size (Mb)	Genome size (Mb)	Replication system and properties	Accession number
	Chr	4.38		αchr	NC_007761
	p42a	0.19		*repABC* (2)	NC_007762
	p42b	0.18		*repABC* (1)	NC_007763
	p42c	0.25		*repABC* (1)	NC_007764
	p42d	0.37		*repABC* (1) (sym)	NC_004041
	p42e	0.51		*repABC* (1) (*min*)	NC_007765
	p42f	0.64		*repABC* (2)	NC_007766
Rhizobium etli CIAT652	Chr	4.51	6.45	αchr	NC_010994
	pa	0.41		*repABC* (1) (*min*)	NC_010998
	pb	0.42		*repABC* (1) (sym)	NC_010996
	pc	1.09		*repABC* (2)	NC_010997
Rhizobium leguminosarum bv. *trifolii* WSM1325	Chr	4.78	7.42	αchr	NC_012850
	pR132501	0.83		*repABC* (1) (sym)	NC_012848
	pR132502	0.66		*repABC* (1) (*min*)	NC_012858
	pR132503	0.51		*repABC* (1)	NC_012853
	pR132504	0.35		*repABC* (1)	NC_012852
	pR132505	0.29		*repABC* (1)	NC_012854
Rhizobium leguminosarum bv. *trifolii* WSM2304	Chr	4.54	6.87	αchr	NC_011369
	pRLG01	1.26		*repABC* (2) (sym)	NC_011368
	pRLG02	0.5		*repABC* (1) (*min*)	NC_011366
	pRLG03	0.31		*repABC* (1)	NC_011370
	pRLG04	0.26		*repABC* (1)	NC_011371
Rhizobium leguminosarum bv. *viciae* 3841	Chr	5.06	7.75	αchr	NC_008380
	pRL7	0.15		*repABC* (2) *repAC*	NC_008382
	pRL8	0.15		(1)	NC_008383
	pRL9	0.35		*repABC* (1)	NC_008379
	pRL10	0.49		*repABC* (1)	NC_008381
	pRL11	0.69		*repABC* (1) (sym)	NC_008384
	pRL12	0.87		*repABC* (1) (*min*)	NC_008378
				repABC (1)	
Rhizobium sp. NGR234	Chr	3.92	6.89	αchr	NC_012587
	pNGR234a	0.53		*repABC* (1) (sym)	NC_000914
	pNGR234b	2.43		*repABC* (1) (*min*)	NC_012586
Sinorhizobium medicae WSM419	Chr	3.78	6.82	αchr	NC_009636
	pSMED01	1.57		*repABC* (1) (*min*)	NC_009620
	pSMED02	1.24		*repABC* (1) (sym)	NC_009621
	pSMED03	0.22		*repABC* (1)	NC_009622
Sinorhizobium meliloti 1021	Chr	3.65	6.69	αchr	NC_003047
	pSymA	1.35		*repABC* (1) (sym)	NC_003037
	pSymB	1.68		*repABC* (1) (*min*)	NC_003078
beta-rhizobia					
Cupriavidus taiwanensis LMG 19424	ChrI	3.42	6.48	NI (*min*)	NC_010528
	ChrII	2.5		NI	NC_010530
	pRalta	0.56		NI (sym)	NC_010529

Chr chromosome (*, of chromosomal nature), *NI* not identified, *repABC* (number of *repABC* operons), *αchr* alpha-proteobacterial chromosomal replication system, (*min*) replicon carrying *minCDE* genes, (*vir*) replicon carrying most of virulence determinants, (*oct*) octopine-type, (*sym*) replicon carrying most of the symbiotic genes, (*sym**) symbiotic genes located on a chromosomal symbiotic island, – genome sequence not available

molecules of same species. The latter holds true even for kin plasmids such as the agrobacterial tumor- or root-inducing plasmids, which bear same functional modules in yet a variable, rearranged order.

The majority of agrobacterial and rhizobial plasmids, regardless of their vast differences, share common replication/maintenance control regions of *repABC* type, and many also share broad similarities in their conjugal transfer systems. *repABC* units may be present in multiple replicons per cell; the mechanism underlying this permissiveness opens up a vast area for research on a genetic system that safeguards large plasmid and chromosome maintenance alike, with high accuracy. Furthermore, cell signaling-dependent regulation acting on *repABC* loci of Ti and rhizobial plasmids leads to replicon amplifications thus far registered for Ti representatives – a phenomenon quite unprecedented. The copy number increase of these replicons, in conjunction with their signaling-dependent conjugal dissemination, is an indicator of their impact in the overall biology of their hosts and of soil community organisms in general.

The multitude of pathogenic or symbiotic determinants on agrobacterial and rhizobial plasmids will continue to be analyzed and contribute to our knowledge of what makes these replicons so remarkable. The chromosomal background functions that also shape the associations among members of the *Rhizobiaceae* and their plant hosts are of no less importance. Their elucidation, apart from yielding essential information on the mechanisms underlying transkingdom partnerships, will also clarify significant aspects that pertain to role division in the multipartite genomes of the *Rhizobiaceae*, and to the very specificity of characteristics that drive speciation in the family or order.

Acknowledgements We sincerely wish to thank Ernö Szegedi, Léon Otten, and João Setubal for helpful discussions and information, Michael Hynes for sharing his expertise in rhizobial plasmids and Victor González for unpublished data. This work was supported by grants 70/4/7809 from the NKUA Research Committee to KMP and IN205808 (PAPIIT-UNAM) and 46738-Q (CONACyT) to MAC.

References

Allardet-Servent A, Michaux-Charachon S, Jumas-Bilak E, Karayan L, Ramuz M (1993) Presence of one linear and one circular chromosome in the *Agrobacterium tumefaciens* C58 genome. J Bacteriol 175:7869–7874

Amadou C, Pascal G, Mangenot S, Glew M, Bontemps C, Capela D, Carrère S, Cruveiller S, Dossat C, Lajus A, Marchetti M, Poinsot V, Rouy Z, Servin B, Saad M, Schenowitz C, Barbe V, Batut J, Médigue C, Masson-Boivin C (2008) Genome sequence of the beta-rhizobium *Cupriavidus taiwanensis* and comparative genomics of rhizobia. Genome Res 18:1472–1483

Anand A, Uppalapati SR, Ryu CM, Allen SN, Kang L, Tang Y, Mysore KS (2008) Salicylic acid and systemic acquired resistance play a role in attenuating crown gall disease caused by *Agrobacterium tumefaciens*. Plant Physiol 146:703–715

Ashby AM, Watson MD, Loake GJ, Shaw CH (1988) Ti plasmid-specified chemotaxis of *Agrobacterium tumefaciens* C58C1 toward *vir*-inducing phenolic compounds and soluble factors from monocotyledonous and dicotyledonous plants. J Bacteriol 170:4181–4187

Atmakuri K, Christie P (2008) Translocation of oncogenic T-DNA and effector proteins to plant cells. In: Tzfira T, Citovsky V (eds) *Agrobacterium*: from biology to biotechnology. Springer, New York, pp 315–364

Atmakuri K, Cascales E, Burton OT, Banta LM, Christie PJ (2007) *Agrobacterium* ParA/MinD-like VirC1 spatially coordinates early conjugative DNA transfer reactions. EMBO J 26:2540–2551

Badenoch-Jones J, Summons RE, Djordevic MA, Shine J, Letham DS, Rolfe BG (1982) Mass spectrometric quantification of indole-3-acetic acid in *Rhizobium* culture supernatants – relation to root hair curling and nodule initiation. Appl Environ Microbiol 44:275–280

Barnett MJ, Fisher RF, Jones T, Komp C, Abola AP, Barloy-Hubler F, Bowser L, Capela D, Galibert F, Gouzy J, Gurjal M, Hong A, Huizar L, Hyman RW, Kahn D, Kahn ML, Kalman S, Keating DH, Palm C, Peck MC, Surzycki R, Wells DH, Yeh KC, Davis RW, Federspiel NA, Long SR (2001) Nucleotide sequence and predicted functions of the entire *Sinorhizobium meliloti* pSymA megaplasmid. Proc Natl Acad Sci USA 98:9883–9888

Barran LR, Ritchot N, Bromfield ES (2001) *Sinorhizobium meliloti* plasmid pRm1132f replicates by a rolling-circle mechanism. J Bacteriol 183:2704–2708

Beck von Bodman S, Hayman GT, Farrand SK (1992) Opine catabolism and conjugal transfer of the nopaline Ti plasmid pTiC58 are coordinately regulated by a single repressor. Proc Natl Acad Sci USA 15:643–647

Begum AA, Leibovitch S, Migner P, Zhang F (2001) Specific flavonoids induced *nod* gene expression and pre-activated *nod* genes of *Rhizobium leguminosarum* increased pea (*Pisum sativum* L.) and lentil (*Lens culinaris* L.) nodulation in controlled growth chamber environments. J Exp Bot 52:1537–1543

Binns AN, Costantino P (1998) The *Agrobacterium* oncogenes. In: Spaink HP, Kondorosi A, Hooykaas PJJ (eds) The *Rhizobiaceae*: molecular biology of model plant-associated bacteria. Kluwer Academic Publishers, Dordrecht, The Netherlands, pp 251–266

Blosser-Middleton RS, Gray KM (2001) Multiple N-acyl homoserine lactone signals of *Rhizobium leguminosarum* are synthesized in a distinct temporal pattern. J Bacteriol 183:6771–6777

Boring LR, Swank WT, Waide JB, Henderson GS (1988) Sources, fates and impacts of nitrogen inputs to terrestrial ecosystems: review and synthesis. Biogeochemistry 6:119–159

Bouzar H, Jones JB (2001) *Agrobacterium larrymoorei* sp. nov., a pathogen isolated from aerial tumours of *Ficus benjamina*. Int J Syst Evol Microbiol 51:1023–1026

Brencic A, Angert ER, Winans SC (2005) Unwounded plants elicit *Agrobacterium vir* gene induction and T-DNA transfer: transformed plant cells produce opines yet are tumour free. Mol Microbiol 57:1522–1531

Brewin NJ (2004) Plant cell wall remodeling in the rhizobia legume symbiosis. Crit Rev Plant Sci 23:293–316

Britton MT, Escobar MA, Dandekar AM (2008) The oncogenes of *A. tumefaciens* and *A. rhizogenes*. In: Tzfira T, Citovsky V (eds) *Agrobacterium*: from biology to biotechnology. Springer, New York, pp 523–563

Brom S, García-de los Santos A, Cervantes L, Palacios R, Romero D (2000) In *Rhizobium etli* symbiotic plasmid transfer, nodulation competitivity and cellular growth require interaction among different replicons. Plasmid 44:34–43

Brom S, Girard L, Tun-Garrido C, García-de los Santos A, Bustos P, González V, Romero D (2004) Transfer of the symbiotic plasmid of *Rhizobium etli* CFN42 requires cointegration with p42a, which may be mediated by site-specific recombination. J Bacteriol 186:7538–7548

Broothaerts W, Mitchell HJ, Weir B, Kaines S, Smith LM, Yang W, Mayer JE, Roa-Rodríguez C, Jefferson RA (2005) Gene transfer to plants by diverse species of bacteria. Nature 433:629–633

Broughton WJ, Perret X (1999) Genealogy of legume-*Rhizobium* symbioses. Curr Opin Plant Biol 2:305–311

Broughton WJ, Jabbouri S, Perret X (2000) Keys to symbiotic harmony. J Bacteriol 182:5641–5652
Broughton WJ, Hanin M, Relic B, Kopcińska J, Golinowski W, Simsek S, Ojanen-Reuhs T, Reuhs B, Marie C, Kobayashi H, Bordogna B, Le Quéré A, Jabbouri S, Fellay R, Perret X, Deakin WJ (2006) Flavonoid-inducible modifications to rhamnan O antigens are necessary for *Rhizobium* sp. strain NGR234-legume symbioses. J Bacteriol 188:3654–3663
Burr TJ, Katz BH, Bishop AL (1987) Populations of *Agrobacterium* in vineyard and non vineyard soils and grape roots in vineyards and nurseries. Plant Dis 71:617–620
Cao H, Yang M, Zheng H, Zhang J, Zhong Z, Zhu J (2009) Complex quorum-sensing regulatory systems regulate bacterial growth and symbiotic nodulation in *Mesorhizobium tianshanense*. Arch Microbiol 191:283–289
Capela D, Barloy-Hubler F, Gouzy J, Bothe G, Ampe F, Batut J, Boistard P, Becker A, Boutry M, Cadieu E, Dréano S, Gloux S, Godrie T, Goffeau A, Kahn D, Kiss E, Lelaure V, Masuy D, Pohl T, Portetelle D, Pühler A, Purnelle B, Ramsperger U, Renard C, Thébault P, Vandenbol M, Weidner S, Galibert F (2001) Analysis of the chromosome sequence of the legume symbiont *Sinorhizobium meliloti* strain 1021. Proc Natl Acad Sci USA 98:9877–9882
Carlier A, Uroz S, Smadja B, Fray R, Latour X, Dessaux Y, Faure D (2003) The Ti plasmid of *Agrobacterium tumefaciens* harbors an *attM*-paralogous gene, *aiiB*, also encoding N-acyl homoserine lactonase activity. Appl Environ Microbiol 69:4989–4993
Carlier A, Chevrot R, Dessaux Y, Faure D (2004) The assimilation of gamma-butyrolactone in *Agrobacterium tumefaciens* C58 interferes with the accumulation of the N-acyl-homoserine lactone signal. Mol Plant Microbe Interact 17:951–957
Castillo-Ramírez S, Vázquez-Castellanos JF, González V, Cevallos MA (2009) Horizontal gene transfer and diverse functional constrains within the commonest plasmid replication-partitioning system in *Alphaproteobacteria*: the *repABC* operon. BMC Genomics 10:536
Cevallos MA, Porta H, Izquierdo J, Tun-Garrido C, García-de-los-Santos A, Dávila G, Brom S (2002) *Rhizobium etli* CFN42 contains at least three plasmids of the *repABC* family: a structural and evolutionary analysis. Plasmid 48:104–116
Cevallos MA, Cervantes-Rivera R, Gutiérrez-Ríos RM (2008) The *repABC* plasmid family. Plasmid 60:19–37
Chai Y, Winans SC (2005a) A small antisense RNA downregulates expression of an essential replicase protein of an *Agrobacterium tumefaciens* Ti plasmid. Mol Microbiol 56:1574–1585
Chai Y, Winans SC (2005b) RepB protein of an *Agrobacterium tumefaciens* Ti plasmid binds to two adjacent sites between *repA* and *repB* for plasmid partitioning and autorepression. Mol Microbiol 58:1114–1129
Chai Y, Tsai CS, Cho H, Winans SC (2007) Reconstitution of the biochemical activities of the AttJ repressor and the AttK, AttL, and AttM catabolic enzymes of *Agrobacterium tumefaciens*. J Bacteriol 189:3674–3679
Chang CH, Winans SC (1992) Functional roles assigned to the periplasmic, linker and receiver domains of the *Agrobacterium tumefaciens* VirA protein. J Bacteriol 174:7033–7039
Chen G, Jeffrey PD, Fuqua C, Shi Y, Chen L (2007) Structural basis for antiactivation in bacterial quorum sensing. Proc Natl Acad Sci USA 104:16474–16479
Chevrot R, Rosen R, Haudecoeur E, Cirou A, Shelp BJ, Ron E, Faure D (2006) GABA controls the level of quorum-sensing signal in *Agrobacterium tumefaciens*. Proc Natl Acad Sci USA 103:7460–7464
Cho H, Winans SC (2005) VirA and VirG activate the Ti plasmid *repABC* operon, elevating plasmid copy number in response to wound-released chemical signals. Proc Natl Acad Sci USA 102:14843–14848
Cho H, Winans SC (2007) TraA, TraC and TraD autorepress two divergent quorum-regulated promoters near the transfer origin of the Ti plasmid of *Agrobacterium tumefaciens*. Mol Microbiol 63:1769–1782
Christie PJ, Atmakuri K, Krishnamoorthy V, Jakubowski S, Cascales E (2005) Biogenesis, architecture, and function of bacterial type IV secretion systems. Annu Rev Microbiol 59:451–485

Citovsky V, Zupan J, Warnick D, Zambryski P (1992) Nuclear localization of *Agrobacterium* VirE2 protein in plant cells. Science 256:1802–1805

Citovsky V, Kozlovsky SV, Lacroix B, Zaltsman A, Dafny-Yelin M, Vyas S, Tovkach A, Tzfira T (2007) Biological systems of the host cell involved in *Agrobacterium* infection. Cell Microbiol 9:9–20

Clare BG, Kerr A, Jones DA (1990) Characteristics of the nopaline catabolic plasmid in *Agrobacterium* strains K84 and K1026 used for biological control of crown gall disease. Plasmid 23:126–137

Cooper JE (2007) Early interactions between legumes and rhizobia: disclosing complexity in a molecular dialogue. J Appl Microbiol 103:1355–1365

Cornelis GR (2006) The type III secretion injectisome. Nat Rev Microbiol 4:811–825

Crews TE (1999) The presence of nitrogen fixing legumes in terrestrial communities: evolutionary vs. ecological considerations. Biogeochemistry 46:233–246

Crossman LC, Castillo-Ramírez S, McAnnula C, Lozano L, Vernikos GS, Acosta JL, Ghazoui ZF, Hernández-González I, Meakin G, Walker AW, Hynes MF, Young JP, Downie JA, Romero D, Johnston AW, Dávila G, Parkhill J, González V (2008) A common genomic framework for a diverse assembly of plasmids in the symbiotic nitrogen fixing bacteria. PLoS ONE 3:e2567

Cubero J, Lastra B, Salcedo CI, Piquer J, López MM (2006) Systemic movement of *Agrobacterium tumefaciens* in several plant species. J Appl Microbiol 101:412–421

Cullimore JV, Ranjeva R, Bono JJ (2001) Perception of lipo-chitooligosaccharidic Nod factors in legumes. Trends Plant Sci 6:24–30

Daniels R, De Vos DE, Desair J, Raedschelders G, Luyten E, Rosemeyer V, Verreth C, Schoeters E, Vanderleyden J, Michiels J (2002) The *cin* quorum sensing locus of *Rhizobium etli* CNPAF512 affects growth and symbiotic nitrogen fixation. J Biol Chem 277:462–468

Danino VE, Wilkinson A, Edwards A, Downie JA (2003) Recipient-induced transfer of the symbiotic plasmid pRL1JI in *Rhizobium leguminosarum* bv. *viciae* is regulated by a quorum-sensing relay. Mol Microbiol 50:511–525

Dessaux Y, Petit A, Tempe J (1993) Chemistry and biochemistry of opines, chemical mediators of parasitism. Phytochemistry 34:31–38

Dessaux Y, Petit A, Farrand SK, Murphy PJ (1998) Opines and opine-like molecules involved in plant-*Rhizobiaceae* interactions. In: Spaink HP, Kondorosi A, Hooykaas PJJ (eds) The *Rhizobiaceae*: molecular biology of model plant-associated bacteria. Kluwer Academic Publishers, Dordrecht, The Netherlands, pp 173–197

D'Haeze W, Holsters M (2002) Nod factor structures, responses and perception during initiation of nodule development. Glycobiology 12:79–105

Ding H, Hynes MF (2009) Plasmid transfer systems in the rhizobia. Can J Microbiol 55:1–11

Escobar MA, Dandekar AM (2003) *Agrobacterium tumefaciens* as an agent of disease. Trends Plant Sci 8:380–386

Farrand SK (1998) Conjugal plasmids and their transfer. In: Spaink HP, Kondorosi A, Hooykaas PJJ (eds) The *Rhizobiaceae*: molecular biology of model plant-associated bacteria. Kluwer Academic Publishers, Dordrecht, The Netherlands, pp 199–233

Farrand SK, Wang CL, Hong SB, O'Morchoe SB, Slota JE (1992) Deletion derivatives of pAgK84 and their use in the analysis of *Agrobacterium* plasmid functions. Plasmid 28:201–212

Farrand SK, Van Berkum PB, Oger P (2003) *Agrobacterium* is a definable genus of the family *Rhizobiaceae*. Int J Syst Evol Microbiol 53:1681–1687

Fauvart M, Michiels J (2008) Rhizobial secreted proteins as determinants of host specificity in the rhizobium-legume symbiosis. FEMS Microbiol Lett 285:1–9

Freiberg C, Perret X, Broughton WJ, Rosenthal A (1996) Sequencing the 500-kb GC-rich symbiotic replicon of *Rhizobium* sp NGR234 using dye terminators and a thermostable "sequenase": a beginning. Genome Res 6:590–600

Freiberg C, Fellay R, Bairoch A, Broughton WJ, Rosenthal A, Perret X (1997) Molecular basis of symbiosis between *Rhizobium* and legumes. Nature 387:394–401

Finan TM, Weidner S, Wong K, Buhrmester J, Chain P, Vorhölter FJ, Hernandez-Lucas I, Becker A, Cowie A, Gouzy J, Golding B, Pühler A (2001) The complete sequence of the 1,683-kb pSymB megaplasmid from the N2-fixing endosymbiont *Sinorhizobium meliloti*. Proc Natl Acad Sci USA 98:9889–9894

Fuqua C (2008) *Agrobacterium*-host attachment and biofilm formation. In: Tzfira T, Citovsky V (eds) *Agrobacterium*: from biology to biotechnology. Springer, New York, pp 243–277

Fuqua WC, Winans SC (1994) A LuxR-LuxI type regulatory system activates *Agrobacterium* Ti plasmid conjugal transfer in the presence of a plant tumor metabolite. J Bacteriol 176:2796–2806

Fuqua C, Burbea M, Winans SC (1995) Activity of the *Agrobacterium* Ti plasmid conjugal transfer regulator TraR is inhibited by the product of the *traM* gene. J Bacteriol 177:1367–1373

Fuqua C, Winans SC, Greenberg EP (1996) Census and consensus in bacterial ecosystems: the LuxR-LuxI family of quorum-sensing transcriptional regulators. Annu Rev Microbiol 50:727–751

Galibert F, Finan TM, Long SR, Puhler A, Abola P, Ampe F, Barloy-Hubler F, Barnett MJ, Becker A, Boistard P, Bothe G, Boutry M, Bowser L, Buhrmester J, Cadieu E, Capela D, Chain P, Cowie A, Davis RW, Dreano S, Federspiel NA, Fisher RF, Gloux S, Godrie T, Goffeau A, Golding B, Gouzy J, Gurjal M, Hernandez-Lucas I, Hong A, Huizar L, Hyman RW, Jones T, Kahn D, Kahn ML, Kalman S, Keating DH, Kiss E, Komp C, Lelaure V, Masuy D, Palm C, Peck MC, Pohl TM, Portetelle D, Purnelle B, Ramsperger U, Surzycki R, Thebault P, Vandenbol M, Vorholter FJ, Weidner S, Wells DH, Wong K, Yeh KC, Batut J (2001) The composite genome of the legume symbiont *Sinorhizobium meliloti*. Science 293:668–672

Geier T, Eimert K, Scherer R, Nickel C (2008) Production and rooting behaviour of *rol* B-transgenic plants of grape rootstock 'Richter 110' (*Vitis berlandieri* × *V rupestris*). Plant Cell Tissue Organ Cult 94:269–280

Giraud E, Moulin L, Vallenet D, Barbe V, Cytryn E, Avarre JC, Jaubert M, Simon D, Cartieaux F, Prin Y, Bena G, Hannibal L, Fardoux J, Kojadinovic M, Vuillet L, Lajus A, Cruveiller S, Rouy Z, Mangenot S, Segurens B, Dossat C, Franck WL, Chang WS, Saunders E, Bruce D, Richardson P, Normand P, Dreyfus B, Pignol D, Stacey G, Emerich D, Verméglio A, Médigue C, Sadowsky M (2007) Legumes symbioses: absence of Nod genes in photosynthetic bradyrhizobia. Science 316:1307–1312

Gomes-Barcellos F, Menna P, da Silva Batista JS, Hungria M (2007) Evidence of horizontal transfer of symbiotic genes from a *Bradyrhizobium japonicum* inoculant strain to indigenous diazotrophs *Sinorhizobium* (*Ensifer*) *fredii* and *Bradyrhizobium elkanii* in a Brazilian savannah soil. Appl Environ Microbiol 73:2635–2643

González V, Bustos P, Ramírez-Romero MA, Medrano-Soto A, Salgado H, Hernández-González I, Hernández-Celis JC, Quintero V, Moreno-Hagelsieb G, Girard L, Rodríguez O, Flores M, Cevallos MA, Collado-Vides J, Romero D, Dávila G (2003) The mosaic structure of the symbiotic plasmid of *Rhizobium etli* CFN42 and its relation to other symbiotic genome compartments. Genome Biol 4:R36

González V, Santamaría RI, Bustos P, Hernández-González I, Medrano-Soto A, Moreno-Hagelsieb G, Janga SC, Ramírez MA, Jiménez-Jacinto V, Collado-Vides J, Dávila G (2006) The partitioned *Rhizobium etli* genome: genetic and metabolic redundancy in seven interacting replicons. Proc Natl Acad Sci USA 103:3834–3839

Goodner B, Hinkle G, Gattung S, Miller N, Blanchard M, Qurollo B, Goldman BS, Cao Y, Askenazi M, Halling C, Mullin L, Houmiel K, Gordon J, Vaudin M, Iartchouk O, Epp A, Liu F, Wollam C, Allinger M, Doughty D, Scott C, Lappas C, Markelz B, Flanagan C, Crowell C, Gurson J, Lomo C, Sear C, Strub G, Cielo C, Slater S (2001) Genome sequence of the plant pathogen and biotechnology agent *Agrobacterium tumefaciens* C58. Science 294:2323–2328

Guo X, Flores M, Mavingui P, Fuentes SI, Hernández G, Dávila G, Palacios R (2003) Natural genomic design in *Sinorhizobium meliloti*: novel genomic architectures. Genome Res 13:1810–1817

Haudecoeur E, Tannières M, Cirou A, Raffoux A, Dessaux Y, Faure D (2009) Different regulation and roles of lactonases AiiB and AttM in *Agrobacterium tumefaciens* C58. Mol Plant Microbe Interact 22:529–537

Hao G, Burr TJ (2006) Regulation of long-chain N-acyl-homoserine lactones in *Agrobacterium vitis*. J Bacteriol 188:2173–2183

He X, Chang W, Pierce DL, Seib LO, Wagner J, Fuqua C (2003) Quorum sensing in *Rhizobium sp.* strain NGR234 regulates conjugal transfer (*tra*) gene expression and influences growth rate. J Bacteriol 185:809–822

Hoang HH, Gurich N, González JE (2008) Regulation of motility by the ExpR/Sin quorum-sensing system in *Sinorhizobium meliloti*. J Bacteriol 190:861–871

Hong SB, Hwang I, Dessaux Y, Guyon P, Kim KS, Farrand SK (1997) A T-DNA gene required for agropine biosynthesis by transformed plants is functionally and evolutionarily related to a Ti plasmid gene required for catabolism of agropine by *Agrobacterium* strains. J Bacteriol 179:4831–4840

Hood EE, Helmer GL, Fraley RT, Chilton MD (1986) The hypervirulence of *Agrobacterium tumefaciens* A281 is encoded in a region of pTiBo542 outside of T-DNA. J Bacteriol 168:1291–1301

Hooykaas PJ, Klapwijk PM, Nuti MP, Schilperoort RA, Rorsch A (1977) Transfer of the *Agrobacterium tumefaciens*-Ti plasmid to avirulent agrobacteria and to *Rhizobium explanta*. J Gen Microbiol 98:477–484

Hooykaas PJ, den Dulk-Ras H, Ooms G, Schilperoort RA (1980) Interactions between octopine and nopaline plasmids in *Agrobacterium tumefaciens*. J Bacteriol 143:1295–1306

Hooykaas PJJ, van Brussel AAN, den Dulk-Ras H, van Slogteren GM, Schilperoort RA (1981) Sym plasmid of *Rhizobium trifolii* expressed in different rhizobial species and *Agrobacterium tumefaciens*. Nature 291:351–353

Howard EA, Zupan JR, Citovsky V, Zambryski PC (1992) The VirD2 protein of *A. tumefaciens* contains a C-terminal bipartite nuclear localization signal: implications for nuclear uptake of DNA in plant cells. Cell 68:109–118

Hubber A, Vergunst AC, Sullivan JT, Hooykaas PJ, Ronson CW (2004) Symbiotic phenotypes and translocated effector proteins of the *Mesorhizobium loti* strain R7A VirB/D4 type IV secretion system. Mol Microbiol 54:561–574

Hwang I, Li PL, Zhang L, Piper KR, Cook DM, Tate ME, Farrand SK (1994) TraI, a LuxI homologue, is responsible for production of conjugation factor, the Ti plasmid N-acylhomoserine lactone autoinducer. Proc Natl Acad Sci USA 91:4639–4643

Hwang I, Cook DM, Farrand SK (1995) A new regulatory element modulates homoserine lactone-mediated autoinduction of Ti plasmid conjugal transfer. J Bacteriol 177:449–458

Izquierdo J, Venkova-Canova T, Ramírez-Romero MA, Téllez-Sosa J, Hernández-Lucas I, Sanjuan J, Cevallos MA (2005) An antisense RNA plays a central role in the replication control of a *repC* plasmid. Plasmid 54:259–277

Jin SG, Roitsch T, Christie PJ, Nester EW (1990a) The regulatory VirG protein specifically binds to a cis-acting regulatory sequence involved in transcriptional activation of *Agrobacterium tumefaciens* virulence genes. J Bacteriol 172:531–537

Jin SG, Prusti RK, Roitsch T, Ankenbauer RG, Nester EW (1990b) Phosphorylation of the VirG protein of *Agrobacterium tumefaciens* by the autophosphorylated VirA protein: essential role in biological activity of VirG. J Bacteriol 172:4945–4950

Johnson TM, Das A (1998) Organization and regulation of expression of the *Agrobacterium* virulence genes. In: Spaink HP, Kondorosi A, Hooykaas PJJ (eds) The *Rhizobiaceae*: molecular biology of model plant-associated bacteria. Kluwer Academic Publishers, Dordrecht, The Netherlands, pp 267–279

Johnston AWB, Hombrecher G, Brewin NJ, Cooper MC (1982) Two transmissible plasmids in *Rhizobium leguminosarum* strain 300. J Gen Microbiol 128:85–93

Jourand P, Giraud E, Béna G, Sy A, Willems A, Gillis M, Dreyfus B, de Lajudie P (2004) *Methylobacterium nodulans* sp. nov., for a group of aerobic, facultatively methylotrophic, legume root-nodule-forming and nitrogen-fixing bacteria. Int J Syst Evol Microbiol 54:2269–2273

Kaneko T, Nakamura Y, Sato S, Asamizu E, Kato T, Sasamoto S, Watanabe A, Idesawa K, Ishikawa A, Kawashima K, Kimura T, Kishida Y, Kiyokawa C, Kohara M, Matsumoto M, Matsuno A, Mochizuki Y, Nakayama S, Nakazaki N, Shimpo S, Sugimoto M, Takeuchi C, Yamada M, Tabata S (2000) Complete genome structure of the nitrogen-fixing symbiotic bacterium *Mesorhizobium loti*. DNA Res 7:331–338

Kaneko T, Nakamura Y, Sato S, Minamisawa K, Uchiumi T, Sasamoto S, Watanabe A, Idesawa K, Iriguchi M, Kawashima K, Kohara M, Matsumoto M, Shimpo S, Tsuruoka H, Wada T, Yamada M, Tabata S (2002) Complete genomic sequence of nitrogen-fixing symbiotic bacterium *Bradyrhizobium japonicum* USDA110. DNA Res 9:189–197

Kersters K, De Ley J (1984) Genus III *Agrobacterium* Conn. In: Krieg NR, Holt JG (eds) Bergey's manual of systematic bacteriology, vol 1, 8th edn. The Williams and Wilkins Co, Baltimore-London, pp 244–254

Kim KS, Farrand SK (1996) Ti plasmid-encoded genes responsible for catabolism of the crown gall opine mannopine by *Agrobacterium tumefaciens* are homologs of the T-region genes responsible for synthesis of this opine by the plant tumor. J Bacteriol 178:3275–3284

Kim H, Farrand SK (1998) Opine catabolic loci from *Agrobacterium* plasmids confer chemotaxis to their cognate substrates. Mol Plant Microbe Interact 11:131–143

Kim JG, Park BK, Kim SU, Choi D, Nahm BH, Moon JS, Reader JS, Farrand SK, Hwang I (2006) Bases of biocontrol: sequence predicts synthesis and mode of action of agrocin 84, the Trojan horse antibiotic that controls crown gall. Proc Natl Acad Sci USA 103:8846–8851

Kinkle BK, Schmidt EL (1991) Transfer of the pea symbiotic plasmid pJB5JI in nonsterile soil. Appl Environ Microbiol 57:3264–3269

Klein DT, Klein RM (1953) Transmittance of tumor-inducing ability to avirulent crown-gall and related bacteria. J Bacteriol 66:220–228

Krimi Z, Petit A, Mougel C, Dessaux Y, Nesme X (2002) Seasonal fluctuations and long-term persistence of pathogenic populations of *Agrobacterium* spp. in soils. Appl Environ Microbiol 68:3358–3365

Lacroix B, Loyter A, Citovsky V (2008) Association of the *Agrobacterium* T-DNA-protein complex with plant nucleosomes. Proc Natl Acad Sci USA 105:15429–15434

Lavin M, Herendeen PS, Wojciechowski MF (2005) Evolutionary rates analysis of *Leguminosae* implicates a rapid diversification of lineages during the tertiary. Syst Biol 54:575–594

Lessl M, Lanka E (1994) Common mechanisms in bacterial conjugation and Ti-mediated T-DNA transfer to plant cells. Cell 77:321–324

Li PL, Farrand SK (2000) The replicator of the nopaline-type Ti plasmid pTiC58 is a member of the *repABC* family and is influenced by the TraR dependent quorum-sensing regulatory system. J Bacteriol 182:179–188

Liu P, Nester EW (2006) Indoleacetic acid, a product of transferred DNA, inhibits *vir* gene expression and growth of *Agrobacterium tumefaciens* C58. Proc Natl Acad Sci USA 103:4658–4662

Loh J, Carlson RW, York WS, Stacey G (2002) Bradyoxetin, a unique chemical signal involved in symbiotic gene regulation. Proc Natl Acad Sci USA 99:14446–14451

López-López MJ, Vicedo B, Orellana N, Piquer J, López MM (1999) Behavior of a virulent strain derived from *Agrobacterium radiobacter* strain K84 after spontaneous Ti plasmid acquisition. Phytopathology 89:286–292

Louvrier P, Laguerre G, Amarger N (1996) Distribution of symbiotic genotypes in *Rhizobium leguminosarum* biovar *viciae* populations isolated directly from soils. Appl Environ Microbiol 62:4202–4205

MacLellan SR, Smallbone LA, Sibley CD, Finan TM (2005) The expression of a novel antisense gene mediates incompatibility within the large *repABC* family of alpha-proteobacterial plasmids. Mol Microbiol 55:611–623

Marie C, Broughton WJ, Deakin WJ (2001) *Rhizobium* type III secretion systems: legume charmers or alarmers? Curr Opin Plant Biol 4:336–342

Marie C, Deakin WJ, Ojanen-Reuhs T, Diallo E, Reuhs B, Broughton WJ, Perret X (2004) TtsI, a key regulator of *Rhizobium* species NGR234 is required for type III-dependent protein secretion and synthesis of rhamnose-rich polysaccharides. Mol Plant Microbe Interact 17:958–966

Marketon MM, Gronquist MR, Eberhard A, González JE (2002) Characterization of the *Sinorhizobium meliloti sinR/sinI* locus and the production of novel N-acyl homoserine lactones. J Bacteriol 184:5686–5695

Marketon MM, Glenn SA, Eberhard A, González JE (2003) Quorum sensing controls exopolysaccharide production in *Sinorhizobium meliloti*. J Bacteriol 185:325–331

Markowitz VM, Szeto E, Palaniappan K, Grechkin Y, Chu K, Chen IM, Dubchak I, Anderson I, Lykidis A, Mavromatis K, Ivanova NN, Kyrpides NC (2008) The integrated microbial genomes (IMG) system in 2007: data content and analysis tool extensions. Nucleic Acids Res (Database issue) 36:D528–D533

Mavingui P, Flores M, Guo X, Dávila G, Perret X, Broughton WJ, Palacios R (2002) Dynamics of genome architecture in *Rhizobium* sp. strain NGR234. J Bacteriol 184:171–176

McAnulla C, Edwards A, Sanchez-Contreras M, Sawers RG, Downie JA (2007) Quorum-sensing-regulated transcriptional initiation of plasmid transfer and replication genes in *Rhizobium leguminosarum* biovar *viciae*. Microbiology 153:2074–2082

McClure NC, Ahmadi AR, Clare BG (1998) Construction of a range of derivatives of the biological control strain *Agrobacterium rhizogenes* K84: a study of factors involved in biological control of crown gall disease. Appl Environ Microbiol 64:3977–3982

McCullen CA, Binns AN (2006) *Agrobacterium tumefaciens* and plant cell interactions and activities required for interkingdom macromolecular transfer. Annu Rev Cell Dev Biol 22:101–127

Michiels J, Van Soom T, D'hooghe I, Dombrecht B, Benhassine T, de Wilde P, Vanderleyden J (1998) The *Rhizobium etli rpoN* locus: DNA sequence analysis and phenotypical characterization of *rpoN*, *ptsN*, and *ptsA* mutants. J Bacteriol 180:1729–1740

Mohmmed A, Sharma RS, Ali S, Babu CR (2001) Molecular diversity of the plasmid genotypes among *Rhizobium* gene pools of Sesbanias from different habitats of a semi-arid region (Delhi). FEMS Microbiol Lett 205:171–178

Moore L, Warren G, Strobel G (1979) Involvement of a plasmid in the hairy root disease of plants caused by *Agrobacterium rhizogenes*. Plasmid 2:617–626

Moriguchi K, Maeda Y, Satou M, Hardayani NS, Kataoka M, Tanaka N, Yoshida K (2001) The complete nucleotide sequence of a plant root-inducing (Ri) plasmid indicates its chimeric structure and evolutionary relationship between tumor-inducing (Ti) and symbiotic (Sym) plasmids in *Rhizobiaceae*. J Mol Biol 307:771–784

Moulin L, Munive A, Dreyfus B, Boivin-Masson C (2001) Nodulation of legumes by members of the beta-subclass of proteobacteria. Nature 411:948–950

Mozo T, Cabrera E, Ruiz-Argüeso T (1988) Diversity of plasmid profiles and conservation of symbiotic nitrogen fixation genes in newly isolated *Rhizobium* strains nodulating sulla (*Hedysarum coronarium* L.). Appl Environ Microbiol 54:1262–1267

Mylona P, Pawlowski K, Bisseling T (1995) Symbiotic nitrogen fixation. Plant Cell 7:869–885

Nonaka S, Yuhashi K, Takada K, Sugaware M, Minamisawa K, Ezura H (2008) Ethylene production in plants during transformation suppresses *vir* gene expression in *Agrobacterium tumefaciens*. New Phytol 178:647–656

Novick RP (1987) Plasmid incompatibility. Microbiol Rev 51:381–395

Oger P, Kim KS, Sackett RL, Piper KR, Farrand SK (1998) Octopine-type Ti plasmids code for a mannopine-inducible dominant-negative allele of *traR*, the quorum-sensing activator that regulates Ti plasmid conjugal transfer. Mol Microbiol 27:277–288

Oger P, Farrand SK (2001) Co-evolution of the agrocinopine opines and the agrocinopine-mediated control of TraR, the quorum-sensing activator of the Ti plasmid conjugation system. Mol Microbiol 41:1173–1185

Oger P, Farrand SK (2002) Two opines control conjugal transfer of an *Agrobacterium* plasmid by regulating expression of separate copies of the quorum-sensing activator gene *traR*. J Bacteriol 184:1121–1131

Otten L, Burr T, Szegedi E (2008) *Agrobacterium*: a disease-causing bacterium. In: Tzfira T, Citovsky V (eds) *Agrobacterium*: from biology to biotechnology. Springer, New York, pp 1–46

Pappas KM (2008) Cell-cell signaling and the *Agrobacterium tumefaciens* Ti plasmid copy number fluctuations. Plasmid 60:89–107

Pappas KM, Winans SC (2003a) A LuxR-type regulator from *Agrobacterium tumefaciens* elevates Ti plasmid copy number by activating transcription of plasmid replication genes. Mol Microbiol 48:1059–1073

Pappas KM, Winans SC (2003b) The RepA and RepB autorepressors and TraR play opposing roles in the regulation of a Ti plasmid repABC operon. Mol Microbiol 49:441–455

Pappas KM, Weingart CL, Winans SC (2004) Chemical communication in proteobacteria: biochemical and structural studies of signal synthases and receptors required for intercellular signalling. Mol Microbiol 53:755–769

Paulsen IT, Seshadri R, Nelson KE, Eisen JA, Heidelberg JF, Read TD, Dodson RJ, Umayam L, Brinkac LM, Beanan MJ, Daugherty SC, Deboy RT, Durkin AS, Kolonay JF, Madupu R, Nelson WC, Ayodeji B, Kraul M, Shetty J, Malek J, Van Aken SE, Riedmuller S, Tettelin H, Gill SR, White O, Salzberg SL, Hoover DL, Lindler LE, Halling SM, Boyle SM, Fraser CM (2002) The *Brucella suis* genome reveals fundamental similarities between animal and plant pathogens and symbionts. Proc Natl Acad Sci USA 99:13148–13153

Pazour GJ, Das A (1990) Characterization of the VirG binding site of *Agrobacterium tumefaciens*. Nucleic Acids Res 18:6909–6913

Penyalver R, Oger P, López MM, Farrand SK (2001) Iron-binding compounds from *Agrobacterium* spp. biological control strain *Agrobacterium rhizogenes* K84 produces a hydroxamate siderophore. Appl Environ Microbiol 67:654–664

Pérez-Mendoza D, Sepúlveda E, Pando V, Muñoz S, Nogales J, Olivares J, Soto MJ, Herrera-Cervera JA, Romero D, Brom S, Sanjuán J (2005) Identification of the *rctA* gene, which is required for repression of conjugative transfer of rhizobial symbiotic megaplasmids. J Bacteriol 187:7341–7350

Perret X, Staehelin C, Broughton WJ (2000) Molecular basis of symbiotic promiscuity. Microbiol Mol Biol Rev 64:180–201

Petit A, David C, Dahl GA, Ellis JG, Guyon P, Casse-Delbart F, Tempé J (1983) Further extension of the opine concept: plasmids in *Agrobacterium rhizogenes* cooperate for opine degradation. Mol Gen Genet 190:204–214

Piper KR, Beck von Bodman S, Farrand SK (1993) Conjugation factor of *Agrobacterium tumefaciens* regulates Ti plasmid transfer by autoinduction. Nature 362:448–450

Prinsen E, Chauvaux N, Schmidt J, John M, Wieneke U, De Greef J, Schell J, Van Onckelen H (1991) Stimulation of indole-3-acetic acid production in *Rhizobium* by flavonoids. FEBS Lett 282:53–55

Pu XA, Goodman RN (1993) Effects of fumigation and biological control on infection of indexed crown gall free grape plants. Am J Enol Vitic 44:241–248

Pueppke SG, Broughton WJ (1999) *Rhizobium* sp. strain NGR234 and *R. fredii* USDA257 share exceptionally broad, nested host ranges. Mol Plant Microbe Interact 12:293–318

Ramírez-Romero MA, Téllez-Sosa J, Barrios H, Pérez-Oseguera A, Rosas V, Cevallos MA (2001) RepA negatively autoregulates the transcription of the *repABC* operon of the *Rhizobium etli* symbiotic plasmid basic replicon. Mol Microbiol 42:195–204

Ramsay JP, Sullivan JT, Jambari N, Ortori CA, Heeb S, Williams P, Barrett DA, Lamont IL, Ronson CW (2009) A LuxRI-family regulatory system controls excision and transfer of the *Mesorhizobium loti* strain R7A symbiosis island by activating expression of two conserved hypothetical genes. Mol Microbiol 73:1141–1155

Ream W (2008) Production of a mobile T-DNA by *Agrobacterium tumefaciens*. In: Tzfira T, Citovsky V (eds) *Agrobacterium*: from biology to biotechnology. Springer, New York, pp 279–313

Renier A, Jourand P, Rapior S, Poinsot V, Sy A, Dreyfus B, Moulin L (2008) Symbiotic properties of *Methylobacterium nodulans* ORS 2060T: a classic process for an atypical symbiont. Soil Biol Biochem 40:1404–1412

Rogel MA, Torres C, Lloret L, Rosenblueth M, Hernández-Lucas I, Martínez L, Martínez J, Martínez-Romero E (2006) Lateral transfer of *Rhizobium* symbiotic plasmids leading to genomic innovation. In: Sánchez F, Quinto C, López-Lara IM, Geiger O (eds) Biology of plant microbe interactions, vol 5. International society for molecular plant–microbe interactions, St. Paul, Minnesota, pp 310–318

Romero D, Palacios R (1997) Gene amplification and genomic plasticity in prokaryotes. Annu Rev Genet 31:91–111

Romero D, Martínez-Salazar J, Girard L, Brom S, Dávilla G, Palacios R, Flores M, Rodríguez C (1995) Discrete amplifiable regions (amplicons) in the symbiotic plasmid of *Rhizobium etli* CFN42. J Bacteriol 177:973–980

Schofield PR, Gibson AH, Dudman WF, Watson JM (1987) Evidence for genetic exchange and recombination of *Rhizobium* symbiotic plasmids in a soil population. Appl Environ Microbiol 53:2942–2947

Schmeisser C, Liesegang H, Krysciak D, Bakkou N, Le Quéré A, Wollherr A, Heinemeyer I, Morgenstern B, Pommerening-Röser A, Flores M, Palacios R, Brenner S, Gottschalk G, Schmitz RA, Broughton WJ, Perret X, Strittmatter AW, Streit WR (2009) *Rhizobium* sp. strain NGR234 possesses a remarkable number of secretion systems. Appl Environ Microbiol 75:4035–4045

Schrammeijer B, den Dulk-Ras A, Vergunst AC, Jurado Jácome E, Hooykaas PJ (2003) Analysis of Vir protein translocation from *Agrobacterium tumefaciens* using *Saccharomyces cerevisiae* as a model: evidence for transport of a novel effector protein VirE3. Nucleic Acids Res 31:860–868

Segovia L, Piñero D, Palacios R, Martínez-Romero E (1991) Genetic structure of a soil population of nonsymbiotic *Rhizobium leguminosarum*. Appl Environ Microbiol 57:426–433

Sepúlveda E, Pérez-Mendoza D, Ramírez-Romero MA, Soto MJ, López-Lara IM, Geiger O, Sanjuán J, Brom S, Romero D (2008) Transcriptional interference and repression modulate the conjugative ability of the symbiotic plasmid of *Rhizobium etli*. J Bacteriol 190:4189–4197

Setubal JC, Wood D, Burr T, Farrand S, Goldman B, Goodner B, Otten L, Slater S (2009) The genomics of *Agrobacterium*: insights into its pathogenicity, biocontrol and evolution. In: Jackson R (ed) Plant pathogenic bacteria: genomics and molecular biology. Caister Academic Press, Norfolk, UK, pp 91–112

Sheu SY, Chen WM, Lin GH (2007) Characterization and application of a rolling-circle-type plasmid from *Cupriavidus taiwanensis*. Plasmid 57:275–285

Silva C, Vinuesa P, Eguiarte LE, Souza V, Martínez-Romero E (2005) Evolutionary genetics and biogeographic structure of *Rhizobium gallicum* sensu lato, a widely distributed bacterial symbiont of diverse legumes. Mol Ecol 14:4033–4050

Slater SC, Goodner BW, Setubal JC, Goldman BS, Wood DW, Nester EW (2008) The *Agrobacterium tumefaciens* C58 genome. In: Tzfira T, Citovsky V (eds) *Agrobacterium*: from biology to biotechnology. Springer, New York, pp 149–181

Slater SC, Goldman BS, Goodner B, Setubal JC, Farrand SK, Nester EW, Burr TJ, Banta L, Dickerman AW, Paulsen I, Otten L, Suen G, Welch R, Almeida NF, Arnold F, Burton OT, Du Z, Ewing A, Godsy E, Heisel S, Houmiel KL, Jhaveri J, Lu J, Miller NM, Norton S, Chen Q, Phoolcharoen W, Ohlin V, Ondrusek D, Pride N, Stricklin SL, Sun J, Wheeler C, Wilson L, Zhu H, Wood DW (2009) Genome sequences of three *Agrobacterium* biovars help elucidate the evolution of multichromosome genomes in bacteria. J Bacteriol 191:2501–2511

Smith EF, Brown NA, Townsend CO (1912) The structure and development of crown gall: a plant cancer. US Dept Agric Bur Plant Ind Bull 255:1–61

Soto MJ, Domínguez-Ferreras A, Pérez-Mendoza D, Sanjuán J, Olivares J (2009) Mutualism versus pathogenesis: the give-and-take in plant-bacteria interactions. Cell Microbiol 11:381–388

Spaink HP (2000) Root nodulation and infection factors produced by rhizobial bacteria. Annu Rev Microbiol 54:257–288

Sprent JI (2001) Nodulation of legumes. The Cromwell press Ltd, Royal Botanical Gardens, Kew, London

Sprent JI (2008) 60 Ma of legume nodulation. What's new? What's changing? J Exp Bot 59:1081–1084

Sprent JI, James EK (2008) Legume-rhizobial symbiosis: an anorexic model? New Phytol 179:3–5

Stacey G, McAlvin CB, Kim SY, Olivares J, Soto MJ (2006) Effects of endogenous salicylic acid on nodulation in the model legumes *Lotus japonicus* and *Medicago truncatula*. Plant Physiol 141:1473–1481

Stachel SE, Zambryski PC (1986) *virA* and *virG* control the plant-induced activation of the T-DNA transfer process of *A. tumefaciens*. Cell 46:325–333

Stiens M, Schneiker S, Keller M, Kuhn S, Pühler A, Schlüter A (2006) Sequence analysis of the 144-kilobase accessory plasmid pSmeSM11a, isolated from a dominant *Sinorhizobium meliloti* strain identified during a long-term field release experiment. Appl Environ Microbiol 72:3662–3672

Strobel GA, Nachimias A, Hess WM (1988) Improvements in the growth and yield of olive trees by transformation with the Ri plasmid of *Agrobacterium rhizogenes*. Can J Bot 66:2581–2585

Sullivan JT, Patrick HN, Lowther WL, Scott DB, Ronson CW (1995) Nodulating strains of Rhizobium loti arise through chromosomal symbiotic gene transfer in the environment. Proc Natl Acad Sci USA 92:8985–8989

Szegedi E, Otten L (1998) Incompatibility properties of tartrate utilization plasmids derived from *Agrobacterium vitis* strains. Plasmid 39:35–40

Szegedi E, Otten L, Czakó M (1992) Diverse types of tartrate plasmids in *Agrobacterium tumefaciens*. Mol Plant Microbe Interact 5:435–438

Szegedi E, Czako M, Otten L (1996) Further evidence that the vitopine-type pTi's of *Agrobacterium vitis* represent a novel group of Ti plasmids. Mol Plant Microbe Interact 9:139–143

Tanaka N, Oka A (1994) Restriction endonuclease map of the root-inducing plasmid (pRi1724) of *Agrobacterium rhizogenes* strain MAFF03-01724. Biosci Biotech Biochem 58:297–299

Tettelin H, Riley D, Cattuto C, Medini D (2008) Comparative genomics: the bacterial pan-genome. Curr Opin Microbiol 11:472–477

Teyssier-Cuvelle S, Oger P, Mougel C, Groud K, Farrand SK, Nesme X (2004) A highly selectable and highly transferable Ti plasmid to study conjugal host range and Ti plasmid dissemination in complex ecosystems. Microb Ecol 48:10–18

Theunis M, Kobayashi H, Broughton WJ, Prinsen E (2004) Flavonoids, NodD1, NodD2, and *nod*-box NB15 modulate expression of the y4wEFG locus that is required for indole-3-acetic acid synthesis in *Rhizobium* sp. strain NGR234. Mol Plant Microbe Interact 17:1153–1161

Tun-Garrido C, Bustos P, González V, Brom S (2003) Conjugative transfer of p42a from *Rhizobium etli* CFN42, which is required for mobilization of the symbiotic plasmid, is regulated by quorum sensing. J Bacteriol 185:1681–1692

Tzfira T, Vaidya M, Citovsky V (2004) Involvement of targeted proteolysis in plant genetic transformation by *Agrobacterium*. Nature 431:87–92

Tzfira T, Citovsky V (2008) *Agrobacterium*: from biology to biotechnology. Springer, New York

van Berkum P, Eardly BD (1998) Molecular evolutionary systematics of the *Rhizobiaceae*. In: Spaink HP, Kondorosi A, Hooykaas PJJ (eds) The *Rhizobiaceae*: molecular biology of model plant-associated bacteria. Kluwer Academic Publishers, Dordrecht, The Netherlands, pp 1–24

van Egeraat AWSM (1975) The possible role of homoserine in the development of *Rhizobium leguminosarum* in the rhizosphere of pea seedlings. Plant Soil 42:381–386

van Larebeke N, Engler G, Holsters M, Van den Elsacker S, Zaenen I, Schilperoort RA, Schell J (1974) Large plasmid in *Agrobacterium tumefaciens* essential for crown gall-inducing ability. Nature 252:169–170

Vannini A, Volpari C, Gargioli C, Muraglia E, Cortese R, De Francesco R, Neddermann P, Marco SD (2002) The crystal structure of the quorum sensing protein TraR bound to its autoinducer and target DNA. EMBO J 21:4393–4401

Vasanthakumar A, McManus PS (2004) Indole-3-acetic acid-producing bacteria are associated with cranberry stem gall. Phytopathology 94:1164–1171

Venkova-Canova T, Soberón NE, Ramírez-Romero MA, Cevallos MA (2004) Two discrete elements are required for the replication of a *repABC* plasmid: an antisense RNA and a stem-loop structure. Mol Microbiol 54:1431–1444

Wang L, Helmann JD, Winans SC (1992) The *A tumefaciens* transcriptional activator OccR causes a bend at a target promoter, which is partially relaxed by a plant tumor metabolite. Cell 69:659–667

Watson B, Currier TC, Gordon MP, Chilton MD, Nester EW (1975) Plasmid required for virulence of *Agrobacterium tumefaciens*. J Bacteriol 123:255–264

Welander M, Pawlicki N, Holefors A, Wilson F (1998) Genetic transformation of the apple rootstock M26 with the *rolB* gene and its influence on rooting. J Plant Physiol 153:371–380

Wernegreen JJ, Riley MA (1999) Comparison of the evolutionary dynamics of symbiotic and housekeeping loci: a case for the genetic coherence of rhizobial lineages. Mol Biol Evol 16:98–113

White FF, Nester EW (1980a) Hairy root: plasmid encodes virulence traits in *Agrobacterium rhizogenes*. J Bacteriol 141:1134–1141

White FF, Nester EW (1980b) Relationship of plasmids responsible for hairy root and crown gall tumorigenicity. J Bacteriol 144:710–720

White CE, Winans SC (2007) Cell-cell communication in the plant pathogen *Agrobacterium tumefaciens*. Philos Trans R Soc Lond B Biol Sci 362:1135–1148

Whitehead NA, Barnard AM, Slater H, Simpson NJ, Salmond GP (2001) Quorum-sensing in gram-negative bacteria. FEMS Microbiol Rev 25:365–404

Williams P (2007) Quorum sensing, communication and cross-kingdom signalling in the bacterial world. Microbiology 153:3923–3938

Winans SC, Mantis NJ, Chen CY, Chang CH, Han DC (1994) Host recognition by the VirA, VirG two-component regulatory proteins of *Agrobacterium tumefaciens*. Res Microbiol 145:461–473

Wood DW, Setubal JC, Kaul R, Monks DE, Kitajima JP, Okura VK, Zhou Y, Chen L, Wood GE, Almeida NF Jr, Woo L, Chen Y, Paulsen IT, Eisen JA, Karp PD, Bovee D Sr, Chapman P, Clendenning J, Deatherage G, Gillet W, Grant C, Kutyavin T, Levy R, Li MJ, McClelland E, Palmieri A, Raymond C, Rouse G, Saenphimmachak C, Wu Z, Romero P, Gordon D, Zhang S, Yoo H, Tao Y, Biddle P, Jung M, Krespan W, Perry M, Gordon-Kamm B, Liao L, Kim S, Hendrick C, Zhao ZY, Dolan M, Chumley F, Tingey SV, Tomb JF, Gordon MP, Olson MV, Nester EW (2001) The genome of the natural genetic engineer *Agrobacterium tumefaciens* C58. Science 294:2317–2323

Xu XQ, Pan SQ (2000) An *Agrobacterium* catalase is a virulence factor involved in tumorigenesis. Mol Microbiol 35:407–414

Yang M, Sun K, Zhou L, Yang R, Zhong Z, Zhu J (2009) Functional analysis of three AHL autoinducer synthase genes in *Mesorhizobium loti* reveals the important role of quorum sensing in symbiotic nodulation. Can J Microbiol 55:210–214

Yanofsky MF, Porter SG, Young C, Albright LM, Gordon MP, Nester EW (1986) The *virD* operon of *Agrobacterium tumefaciens* encodes a site-specific endonuclease. Cell 47:471–477

Young JP, Crossman LC, Johnston AW, Thomson NR, Ghazoui ZF, Hull KH, Wexler M, Curson AR, Todd JD, Poole PS, Mauchline TH, East AK, Quail MA, Churcher C, Arrowsmith C, Cherevach I, Chillingworth T, Clarke K, Cronin A, Davis P, Fraser A, Hance Z, Hauser H, Jagels K, Moule S, Mungall K, Norbertczak H, Rabbinowitsch E, Sanders M, Simmonds M, Whitehead S, Parkhill J (2006) The genome of *Rhizobium leguminosarum* has recognizable core and accessory components. Genome Biol 7:R34

Young JM (2008) *Agrobacterium* – taxonomy of plant pathogenic *Rhizobium* species. In: Tzfira T, Citovsky V (eds) *Agrobacterium*: from biology to biotechnology. Springer, New York, pp 183–220

Yuan ZC, Edlind MP, Liu P, Saenkham P, Banta LM, Wise AA, Ronzone E, Binns AN, Kerr K, Nester EW (2007) The plant signal salicylic acid shuts down expression of the *vir* regulon and activates quormone-quenching genes in *Agrobacterium*. Proc Natl Acad Sci USA 104:11790–11795

Yuan ZC, Haudecoeur E, Faure D, Kerr KF, Nester EW (2008) Comparative transcriptome analysis of *Agrobacterium tumefaciens* in response to plant signal salicylic acid, indole-3-acetic acid and gamma-amino butyric acid reveals signalling cross-talk and *Agrobacterium*-plant co-evolution. Cell Microbiol 10:2339–2354

Zhang J, Boone L, Kocz R, Zhang C, Binns AN, Lynn DG (2000) At the maize/*Agrobacterium* interface: natural factors limiting host transformation. Chem Biol 7:611–621

Zhang RG, Pappas T, Brace JL, Miller PC, Oulmassov T, Molyneaux JM, Anderson JC, Bashkin JK, Winans SC, Joachimiak A (2002a) Structure of a bacterial quorum-sensing transcription factor complexed with pheromone and DNA. Nature 417:971–974

Zhang HB, Wang LH, Zhang LH (2002b) Genetic control of quorum-sensing signal turnover in *Agrobacterium tumefaciens*. Proc Natl Acad Sci USA 99:4638–4643

Zhang LH (2003) Quorum quenching and proactive host defense. Trends Plant Sci 8:238–244

Zhu J, Winans SC (1998) Activity of the quorum-sensing regulator TraR of *Agrobacterium tumefaciens* is inhibited by a truncated, dominant defective TraR-like protein. Mol Microbiol 27:289–297

Zhu J, Oger PM, Schrammeijer B, Hooykaas PJ, Farrand SK, Winans SC (2000) The bases of crown gall tumorigenesis. J Bacteriol 182:3885–3895

Chapter 13
Quorum Sensing and Quorum Quenching in Soil Ecosystems

Yves Dessaux, Emilie Chapelle, and Denis Faure

Abbreviations

3 OH-PAME	3-Hydroxy-palmitic acid methyl ester
AHL	*N*-acyl homoserine lactone
AI-2	Autoinducer 2
DKP	Diketo-peptides
DSF	Diffusible signal factor or cis-11-methyl-2-dodecenoic acid
GABA	Gamma-aminobutyrate
GBL	Gamma-butyrolactone
GCL	Gamma-caprolactone (or gamma-hexanolactone)
GHB	Gamma-hydroxybutyrate
GHL	Gamma-heptanolactone
PQS	Pseudomonas quinolone signal
QQ	Quorum quenching
QS	Quorum sensing
QSI	QS inhibitors
SSA	Succinic semialdehyde

Y. Dessaux (✉), E. Chapelle, and D. Faure
Institut des sciences du végétal, UPR 2355, CNRS, avenue de la terrasse, 91198 Gif-sur-Yvette CEDEX, France
e-mail: dessaux@isv.cnrs-gif.fr

G. Witzany (ed.), *Biocommunication in Soil Microorganisms*, Soil Biology 23,
DOI 10.1007/978-3-642-14512-4_13, © Springer-Verlag Berlin Heidelberg 2011

AHL abbreviations

BHL	*N*-butyroyl-homoserine lactone
dDHL	*N*-dodecanoyl-homoserine lactone
DHL	*N*-decanoyl-homoserine lactone
HHL	*N*-hexanoyl-homoserine lactone
OdDHL	3-Oxo-*N*-dodecanoyl-homoserine lactone
OHΔtDeHL	3-Hydroxy-*N*-7-cis-tetradecenoyl-homoserine lactone
OHHL	3-Oxo-*N*-hexanoyl-homoserine lactone
OHL	*N*-octanoyl-homoserine lactone
OOHL	3-Oxo-*N*-octanoyl-homoserine lactone
OtDHL	3-Oxo-*N*-tetradecanoyl-homoserine lactone

13.1 Introduction

13.1.1 Quorum Sensing

13.1.1.1 A Brief History

Quorum sensing (QS; Fuqua et al. 1994) is a bacterial regulatory process that couples gene expression to cell density. This process is mediated by low molecular weight signal molecules that are synthesized by a bacterial population and accumulate in the environment. Beyond a threshold concentration, i.e., a threshold cell density, the presence of these molecules is sensed by the bacteria and induces either the expression or the repression of QS-regulated gene(s) (for recent reviews: Turovskiy et al. 2007; Hooshangi and Bentley 2008; Jayaraman and Wood 2008; Popat et al. 2008).

The prototypic and first investigated QS system was light emission by the fish and calmar symbiont *Photobacterium fischeri* (e.g., *Vibrio fischeri*). Light is emitted by bacteria living at high cell density in dedicated animal organs, as the result from a chemical reaction involving flavine mononucleotide and aldehydes and the products of the *lux* genes (Rogers and McElroy 1955; Engebrecht et al. 1983; Engebrecht and Silverman 1984; for review: Pappas et al. 2004). This phenomenon can be reproduced in vitro and depends on accumulation in the environment (both in vivo and in vitro) of the signal 3-oxo-*N*-hexanoyl-homoserine lactone (OHHL) (Eberhard et al. 1981). In *P. fisheri*, two genetic determinants are crucial for QS regulation: the *luxI* and *luxR* genes (Engebrecht and Silverman 1984, 1987). They encode the eponym proteins LuxI, responsible for the production of the signal molecule, and LuxR, the sensor protein that perceives and binds the signal OHHL and acts as a transcriptional activator of the expression of the *lux* genes, including *luxI* (reviews Fuqua et al. 1996; Pappas et al. 2004). This last feature generates a positive feedback loop that boosts the production of the *P. fischeri* signal and strongly contributes to the synchronization of light emission within the bacterial population.

Light emission is a "costly" process for the bacteria in metabolic terms. However, it has been selected for and maintained during evolution likely for two possible reasons. First, both partners may indeed benefit from the association. In the dedicated light emitting organs, and as the result of a complex colonization process, the animal provides the bacteria with an exclusive carbon-rich niche. The bacterium emits photons, a function that may both attract prey and/or deters predators in the light-deprived environment of the ocean or at night and increase the fitness of the animal in this environment. Second, light emission depends upon QS regulation. As a consequence *P. fischeri* only produces light in the fish organ where it lives at very high density and in a carbon-rich environment, and not in the open sea where bacteria are very diluted and face an oligotrophic milieu (for reviews on the ecology of the symbiosis, see Nealson and Hastings 1979; Nyholm and McFall-Ngai 2004; Visick and Ruby 2006).

Since the elucidation of the regulatory process controlling light emission in *P. fischeri*, several QS regulatory systems have been found in other bacterial systems, including proteobacteria, firmicutes, and actinobacteria. QS has therefore been the focus of multiple reviews (for recent examples: Barnard et al. 2007; Van Houdt et al. 2007; White and Winans 2007; He and Zhang 2008; Jayaraman and Wood 2008; Novick and Geisinger 2008; Popat et al. 2008; Willcox et al. 2008). Additional information on this regulatory mode can also be found in this opus (see the companion Chap. 14). Some of these systems are also described below in more details.

13.1.1.2 Multiple Systems, Multiple Signals

QS regulation was observed in several soil-borne bacteria, including plant pathogens and plant growth promoting rhizobacteria (Von Bodman et al. 2003; Barnard et al. 2007; Sanchez-Contreras et al. 2007; Van Houdt et al. 2007; White and Winans 2007). Table 13.1 presents some of these bacteria as well as their respective signals and the regulated functions. As it can be deduced from this table, firmicutes rely upon peptide signals (Lyon and Novick 2004), actinobacteria on gamma-butyrolactones (Chater 1993), and proteobacteria mostly on low molecular weight lipid or cyclic compounds (Barnard et al. 2007; Van Houdt et al. 2007; White and Winans 2007; He and Zhang 2008; Jayaraman and Wood 2008; Novick and Geisinger 2008; Popat et al. 2008; Willcox et al. 2008). Among these, the plant pathogen *Ralstonia solanacearum* produces 3-hydroxy-palmitic acid methyl ester (3 OH-PAME; Flavier et al. 1997; review Von Bodman et al. 2003) and members of the *Xanthomonas* genus a diffusible signal factor (DSF or cis-11-methyl-2-dodecenoic acid; Wang et al. 2004; review Fouhy et al. 2006). The most common signals, however, belong to the *N*-acyl-homoserine lactone (AHL) class. Specificity for a given AHL signal depends upon substitution at carbon 3 (unsubstituted, oxo, or hydroxy) and length (from 4 to 18 carbon atoms) and saturation of the acyl side

Table 13.1 QS signals and regulated functions in several soil-borne bacteria

Taxonomic class	Representative genus or species	QS signals produced	Known regulated functions	References
Actinobacteria	*Streptomyces* sp.	Gamma-butyrolactones[a]	Antibiotic compound synthesis, differentiation	Chater (1993)
Alpha-proteobacteria	*Agrobacterium tumefaciens*	OOHL[b]	Ti plasmid transfer, virulence	White and Winans (2007)
	Bradyrhizobium sp.	AHLs	nd[c]	Pongsilp et al. (2005)
		Bradyoxetin	Expression of *nod* genes	Loh et al. (2002)
		pCHL	nd	Schaefer et al. (2008)
	Rhizobium leguminosarum bv. viciae	OHtDeHL, HHL, HHL, OHL, OHOHL, OOHL	Growth inhibition, induction of *rhi* genes, plasmid transfer, nodulation	González and Marketon (2003)
	Sinorhizobium meliloti	OHL, OHOHL, OOHL, DDHL, OtdHL, HDHL, OhDHL, . . .	EPSII synthesis, motility, nodulation kinetics, plasmid transfer	González and Marketon (2003)
	Mesorhizobium loti	OHHL, OHL, DHL, dDHL	Nodulation	Yang et al. (2009)
	Rhodopseudomonas palustris	pCHL	Chemotaxis	Schaefer et al. (2008), (2008)
Beta-proteobacteria	*Burkholderia glumae*	OHL	Toxoflavin synthesis	Ferluga et al. (2008)
	Burkholderia kururiensis	OHHL, OHL, DHL, dDHL	nd	Ferluga et al. (2008)
	Ralstonia solanacearum	OHL	nd	Ferluga et al. (2008)
		3-OH-PAME	Virulence	Flavier et al. (1997)
Firmicutes	*Staphylococcus aureus*	AIPs (thiolactone peptides)	Virulence, antimicrobial peptides synthesis, genetic competence	Sturme et al. (2007), Lyon and Novick (2004)
Gamma-proteobacteria	*Erwinia amylovora*	OHHL, OHHHL	Virulence, EPS synthesis, hydrogen peroxyde tolerance	Ferluga et al. (2008)
	Pectobacterium carotovorum	OHHL	Varbapenem production, exoenzymes, hrp secretion system	Braeken et al. (2008)
	Dickeya dadantii	HHL, OHHL, DHL	nd	Ferluga et al. (2008)

Pantoea stewartii	OHHL, OOHL	EPS synthesis, biofilm, adhesion, xylem dissemination, pathogenicity	Braeken et al. (2008)
Serratia liquefaciens	BHL	Swarming, biofilm, protease	Ferluga et al. (2008)
Pseudomonas aeruginosa	BHL, OdDHL	Biofilm, elastase, lipase, alkaline protease, HCN, pyocyanin, exotoxin A, swarming, lectins, rhamnolipids, virulence...	Braeken et al. (2008), Ferluga et al. (2008)
	PQS	Elastase, pyocyanin synthesis, LecA lectin, biofilm, AHL signaling, motility + intrinsec functions (antibiosis, iron chelation)	Dubern and Diggle (2008)
	DKPs (e.g., cyclo(Δ-Ala-$_{L}$-Val))	Unclear, crosslinked to AHL signaling	Holden et al. (1999)
Pseudomonas syringae pv. syringae	OHHL	Cell aggregation, epiphytic fitness, disease development, hydrogen peroxyde tolerance, motility	Elasri et al. (2001), Quiñones et al. (2005)
Pseudomonas aureofaciens	HHL	Phenazine production, rhizosphere colonization, protease activty	Braeken et al. (2008), Ferluga et al. (2008)
Pseudomonas fluorescens 2P24	OHHL, OHL	Biofilm formation, wheat rhizosphere colonization, biocontrol ability	Wei and Zhang (2006)
Xanthomonas campestris	DSF (cis-11-methyl-2-dodecenoic acid)	Extracellular enzymes, LPS and EPS synthesis, multidrug resistance and detoxification, motility and chemotaxis, HR and pathogenicity (hrp), iron uptake, primary metabolism	He and Zhang (2008)
	DF (butyrolactones)	Xanthomonadin production (light protection), colonization	He and Zhang (2008)

[a] In some species or genera (e.g., pseudomonads), QS signals may be strain specific
[b] Abbreviations are as in text and described under abbreviations above
[c] nd, not determined

chain. Though these decorations ensure some strong specificity, they do not prevent some cross-talk to occur (Dulla and Lindow 2009). This trait permitted the development of several signal biosensors (for examples: McClean et al. 1997; Shaw et al. 1997). These are engineered bacteria that do not produce their cognate signal molecule(s) but still respond to it or closely related ones. Biosensors are of particular interest considering that QS signals are generally accumulated and sensed in the environment at pM to μM concentrations, preventing easy routine analysis of bacterial supernatants.

Aside from the above mentioned signals, at least four other signal molecules (or class of molecules) have been reported. Diketo-peptides (DKPs) such as cyclo (delta-Ala-L-Val) and cyclo(L-Pro-L-Tyr) were first identified in *Pseudomonas aeruginosa* and later in the soil and water bacteria *Proteus mirabilis*, *Citrobacter freundii*, and *Enterobacter agglomerans*. Another DKP, cyclo(L-Phe-L-Pro), was found in *Pseudomonas* spp. All DKPs can activate AHL-responding biosensors (Holden et al. 1999). Quinolones (PQSs) are specific for pseudomonads (review: Dubern and Diggle 2008). A third signal type, p-coumaroyl-homoserine lactone, is partly related to AHLs. It was detected in culture supernatant of strains belonging to the genera *Rhodopseudomonas palustris*, *Bradyrhizobium* sp., and *Silicibacter pomeroyi* (Schaefer et al. 2008). The synthesis of this molecule depends upon the plant-produced p-coumaric acid molecule, one of the main constituent of lignin from which it derives, for instance upon wounding. Interestingly, *Bradyrhizobium* sp. strains are nitrogen fixing bacteria forming nodules on roots and stems of various water plants. That QS regulation in these bacteria is placed under partial control of a plant signal constitutes a novel instance of plant bacteria communication. Last, autoinducer 2 (AI-2), a furanosyl borate diester QS signal detected in spent culture supernatant of several bacteria (Bassler et al. 1994; Surette et al. 1999), has been described as a putative universal QS signal (Miller and Bassler 2001). This view, however, has been challenged as AI-2 can also be described as a byproduct of a metabolic signaling system (Vendeville et al. 2005; Krin et al. 2006; Rezzonico and Duffy 2007).

An intriguing question deals with the diversity of signals observed in nature. At least two reasons may explain this diversity. The first one lies in the possible evolutionary selection of different signals as a way to ensure sufficient communication "privacy" within a given bacterial population among the microflora. This explanation is debatable, however, since several proteobacteria relying upon related – or even identical – AHL signals have been found in the same plant root environment (D'Angelo-Picard et al. 2005). A second explanation accounting for signal diversity may be related to the various physico-chemical constraints faced by the bacteria. AHLs, for instance, are sensitive to alkaline pH and elevated temperature (Byers et al. 2002; Yates et al. 2002; Delalande et al. 2005). *Ralstonia* and *Xanthomonas*, two phytopathogenic proteobacteria living in warm regions of Earth, do not produce AHL as QS signals (Cf. above). It is tempting to speculate that the synthesis, by these bacteria, of the more heat-stable 3 OH-PAME and DSF molecules has been selected by evolution as a way to ensure efficient communication under those specific climate conditions.

13.1.2 Quorum Quenching

Quorum quenching (QQ) designates all natural and engineered phenomena that interfere with QS regulation. Two major interference processes are known. The first one leads to a limited accumulation of QS signal in the environment, the second to the preclusion of the sensing of the QS signal by its cognate protein receptor (e.g., the LuxR-type sensors for AHL molecules) (review: Zhang 2003).

Regarding the last process, prevention of QS signal recognition by its receptor has been found to result often from the presence in the bacterial environment of specific molecules termed QS inhibitors (QSI). Typical examples of inhibitors are halogenated furanones. First identified in the red algae *Delisea pulchra* (Givskov et al. 1996), furanones interfere with AHL-based QS regulation (as well as with AI-2 signalization) and weaken the AHL-mediated stability of bacterial biofilms (Rasmussen et al. 2000). This finding has led to the formulation of boat paintings preventing biofouling (Armstrong et al. 2000; review: Dobretsov et al. 2009). Similarly, multiple compounds that interfere with AHL sensing in bacteria have been detected in the unicellular green algae *Chlamydomonas reinhardtii* (Teplitski et al. 2004; González and Keshavan 2006; see also the Chap. 16 in this book). Aside from algal extracts, plant extracts from *Allium sativum*, *Fragaria vesca*, *Glycine max*, *Lycopersicum esculentum*, *Medicago truncatula*, *Oryza sativa*, *Pisum sativum*, and vanilla have been found to prevent QS signal recognition (Teplitski et al. 2000; Gao et al. 2003; Rasmussen et al. 2005). The responsible QSI molecules have not been always identified. Some legumes, however, appear to produce large amounts of L-canavanine, a QSI that interferes with QS in *Sinorrhizobium meliloti* (Keshavan et al. 2005; reviews Rasmussen and Givskov 2006; Dobretsov et al. 2009).

Microbial cyclopeptides (e.g., cyclo-Ala-Val), penicillic acid or patulin also exert a QSI effect (Holden et al. 1999; Rasmussen and Givskov 2006). These two last molecules are produced by fungi. Significantly, production of very active antifungal molecules such as phenazine and diphloroglucinol by several soil-borne bacteria depends upon AHL-mediated, QS regulation (Chin-A-Woeng et al. 2001; Zhang and Pierson 2001). Production of patulin and penicillic acid may therefore be a trait evolved by some fungi to efficiently "jam foe bacterial communication".

Last, natural inhibition of AHL signal sensing may involve AHLs themselves. Thus *N*-decanoyl-homoserine lactone (DHL) and 3-oxo-*N*-tetradecanoyl-homoserine lactone (OtDHL), two long-chain AHLs, inhibit the production of the antibiotic pigment violacein by *Chromobacterium violaceum*, a process regulated by QS via the short chain AHL *N*-hexanoyl-homoserine lactone (HHL) (McClean et al. 1997).

Similarly, in some *Rhizobiaceae*, the "small bacteriocin" known to inhibit the growth of several strains of *Rhizobium* and related genera, was later identified as 3-hydroxy-*N*-7-cis-tetradecenoyl-homoserine lactone (OHΔtDeHL), the first described unsaturated AHL (Schripsema et al. 1996). This signal is produced by the AHL-synthase CinI expressed in several *R. leguminosarum* bv. viciae strains devoid of the symbiotic plasmid pRL1JI. OHΔtDeHL activates in part the conjugation of this plasmid from a donor strain and at the same time prevent the growth of the donor (via the bacteriocin effect; Blosser-Middleton and Gray 2001). Interestingly, the CinI-synthesized AHL regulates bacterial growth and some bean nodulation traits in *Rhizobium etli* (Daniels et al. 2002). All these observation provide additional clues to address the above mentioned question of signal diversity.

The other major QQ process involves the degradation (or the modification) of the QS molecules by enzymatic activities. It leads to a limited accumulation of QS signal in the environment. The QQ enzymes mechanisms involved are diverse and the biological significance and potential applications of these processes quite substantial. Several of these data are therefore reviewed below in more detail.

13.2 Biological Significance of QQ

13.2.1 Known Organisms and Activities Involved in AHL Signal Degradation

The best documented QS signal degradation phenomena deal with AHL molecules. First, AHLs are sensitive to alkaline pH and temperature (Byers et al. 2002; Delalande et al. 2005; Yates et al. 2002). The mechanism involved is lactonolysis, i.e., the opening of the lactone ring, which leads to the conversion of AHLs to their QS-inactive, acyl homoserine derivatives. The shorter is the acyl side chain, the faster the degradation occurs (Yates et al. 2002). This phenomenon may play a role in the modulation of their accumulation in microbial or plant environment. For instance, in *Pectobacterium carotovorum* pv. *carotovorum* (*Pcc*), the expression of the bacterial pathogenicity functions are regulated by QS, via the signal OHHL. Degradation of this signal is documented in the culture supernatant of the plant pathogen during the stationary phase (Byers et al. 2002). It correlates with medium alkalinization as the result of microbial metabolism. Remarkably, the activation of a proton pump is one of the plant defense mechanisms induced upon *Pcc* infection; it leads to a strong medium alkalinization at the infection site (Nachin and Barras 2000). pH-mediated degradation of AHLs also occur in another complex environment, namely the biofilm covering marine stromatolites (Decho et al. 2009). This exopolysaccharide-rich biofilm is composed mostly of four major functional groups (cyanobacteria, sulfur reducing, sulfur oxidizing, and fermentating bacteria) that form cyclic series of three communities responsible for the deposition of calcium carbonate at the surface of the stromatolite. The photosynthetic activity of the cyanobacteria occurs only during daylight, and induces an elevation of the environmental pH values. This change along with that of oxygen concentration imposes modifications of the levels of short chain AHLs accumulated in this environment, a feature that most likely plays a role in the control of several bacterial metabolic activities –and possibly biofilm-related traits in a day/night fashion.

Aside from these abiotic parameters, biotic factors are certainly critical to AHL stability in nature. Thus, AHL molecules may be degraded by a broad range of organisms (Table 13.2). First, degradation of AHLs was reported in bacteria. This ability is known to occur within members of some 20 genera and also within bacterial consortia (Cirou et al. 2007). Second, fungi from both the Ascomycota and Basidiomycota divisions also degrade AHLs (Uroz and Heinonsalo 2008), a feature shared with some more complex eukaryotic organisms. Some plants from the legume clade (e.g., *Lotus corniculatus*, *Pachyrhizus erosus*, *Trifolium pratense*) efficiently degrade AHLs (Delalande et al. 2005; Götz et al. 2007), as do mammalian cells such as those from human lung epithelia or porcine kidney (Xu et al. 2003; Yang et al. 2005). Remarkably, organisms that degrade AHLs are potentially in contact with large AHL-producing bacterial communities, a situation first demonstrated in a biofilm isolated from a water reclamation system (Hu et al. 2003). It is therefore tempting to speculate that degradation of AHLs may result from a coevolution of degradative organisms and AHL-producing bacteria. In agreement, AHL degradation activity is stronger at the root system of legume plants than at the aerial parts (Chapelle, unpublished). Indeed, the aerial parts host a bacterial population one thousand time less dense than that hosted at the root system. AHL degradation in plants has so far only been detected in legumes, a clade known to interact specifically and very intimately with several nitrogen-fixing *Rhizobiaceae*, which rely upon AHL-based QS signaling for the regulation of various rhizosphere and symbiosis-related functions (reviews: González and Marketon 2003; Sanchez-Contreras et al. 2007).

Enzymatic activities involved in AHL degradation, collectively termed AHLases, fall essentially into three classes: lactone hydrolases, acylases (or amidohydrolases), and oxidase or reductase.

13.2.1.1 Oxidase and Reductase

This class of enzymes comprises two members. The first is a cytochrome P450 monooxygenase from *Bacillus megaterium* (Chowdhary et al. 2007). This enzyme preferentially oxidizes long-chain AHLs, e.g., *N*-dodecanoyl-homoserine lactone (dDHL) and 3-oxo-*N*-dodecanoyl-homoserine lactone (OdDHL) and was previously described for its ability to oxidize fatty acids and *N*-fatty acyl amino acids. The second activity is a reductase. First detected in a strain of *Rhodococcus erythropolis*, it catalyzes the conversion of a broad range of 3-oxo-AHLs to 3-OH-AHLs (Uroz et al. 2005). This activity therefore does not lead to the degradation of the AHL molecules into QS-inactive molecules but rather affect the specificity and recognition of the AHL signal. In other words, this reductase translates one bacterial language to another.

Table 13.2 Organisms with AHL degradative ability

Domain	Class	Genus and species	AHLase detected	Genetic determinant	References
Prokaryotes					
	Acidobacteria	nd[a]	Lactonase	*qlcA*	Riaz et al. (2008)
	Actinobacteria	*Rhodococcus erythropolis*	Acylase	nd	Uroz et al. (2005)
			Reductase	nd	Uroz et al. (2005)
		Rhodococcus spp.	Lactonase[b]	*qsdA*	Park et al. (2006), Uroz et al. (2008)
		Streptomyces sp.	Acylase	*ahlM*	Park et al. (2006)
	Alpha-proteobacteria	*Agrobacterium tumefaciens*	Lactonase	*attM, aiiB*	Carlier et al. (2003), Zhang et al. (2002)
		Agrobacterium radiobacter	Lactonase	nd	Uroz, personal communication
		Bosea sp.	nd	nd	D'Angelo-Picard et al. (2005)
		Ochrobactrum sp.	nd	nd	Jafra et al. (2006)
		Sphingomonas sp.	nd	nd	D'Angelo-Picard et al. (2005)
		Sphingopyxis sp.	nd	nd	D'Angelo-Picard et al. (2005)
	Beta-proteobacteria	*Comamonas* sp.	Acylase	nd	Uroz et al. (2007)
		Delftia acidovorans	nd	nd	Jafra et al. (2006)
		Ralstonia sp.	Acylase	*aiiD*	Lin et al. (2003)
		Variovorax pardoxus	Acylase	nd	Leadbetter and Greenberg (2000)
	Cyanobacteria	*Anabaena* sp.	Acylase	*aiiC*	Romero et al. (2008)
	Firmicute	*Arthrobacter* sp.	Lactonase	*ahlD*	Park et al. (2003)
		Bacillus spp.	Lactonase	*aiiA*	Dong et al. (2002)
		Bacillus megaterium	Oxidase	nd	Chowdhary et al. (2007)
	Gamma-proteobacteria	*Acinetobacter*	nd	nd	Kang et al. (2004)
		Klebsiella pneumoniae	Lactonase	*ahlK*	Park et al. (2003)
		Pseudomonas aeruginosa	Acylase	*quiP*	Huang et al. (2006)
				pvdQ	Huang et al. (2003)
		Pseudomonas sp.	Acylase	nd	Huang et al. (2003)
		Shewanella sp.	Acylase	*aac*	Morohoshi et al. (2008)
		Unknown (soil metagenome)	Lactonase	nd	Schipper et al. (2009)
Eukaryotes					
	Leotiomycetes (fungi)	*Meliniomyces variabilis*	Lactonase	nd	Uroz and Heinonsalo (2008)
	Liliopsida (plant)	*Hordeum vulgare* (barley)	nd	nd	Götz et al. (2007)

(*continued*)

Table 13.2 (continued)

Domain	Class	Genus and species	AHLase detected	Genetic determinant	References
	Magnoliopsida (plant)	*Lotus corniculatus* (lotus)	Lactonase-like	nd	Delalande et al. (2005), Chapelle Personal communication
		Pachyrhizus erosus (yam bean)	nd	nd	Götz et al. (2007)
	Mammalia (animal)	*Equus caballus* (horse)	Lactonase[c]	nd	Yang et al. (2005)
		Homo sapiens (human)	Lactonase[c]	*pon1,2,3*	Draganov et al. (2005), Yang et al. (2005)
		Mus musculus (mouse)	Lactonase[c]	*pon2*	Stoltz et al. (2007), Yang et al. (2005)
		Oryctolagus cuniculus (rabbit)	Lactonase[c]	nd	Yang et al. (2005)
		Sus scrofa domesticus (pork)	Acylase	nd	Xu et al. (2003)
	ND (fungi phylum Basidiomycota)		nd	Lactonase	Uroz and Heinonsalo (2008)
	ND (fungi phylum Ascomycota)		nd	nd	Uroz and Heinonsalo (2008)

[a]nd, not determined
[b]All these enzymes are paroxonases with lactonase activity
[c]This enzyme is related to phosphotriesterases

13.2.1.2 Amidohydrolases

AHL amidohydrolases (also termed acylases) catalyze the hydrolysis of the amide bond of AHLs to release homoserine lactone and the corresponding fatty acid. The hydrolysis of AHLs by acylases leads to products that have no QS signaling activity. The acylases therefore prevent AHL-mediated, QS communication to occur.

AHL acylase activity has been mostly found in bacteria, for instance in *Variovorax paradoxus* (Leadbetter and Greenberg 2000), and later in several other bacterial genera such as *Anabaena*, *Comamonas*, *Nostoc*, *Pseudomonas*, *Ralstonia*, *Rhodococcus*, *Shewanella*, or *Streptomyces* (review: Uroz et al. 2009). The substrate specificity varies as a function of the acylase considered, but as a rule, AHL acylases generally degrade long-chain AHLs more efficiently than short-chain ones. Several genetic determinants have been identified: they fall within the family of penicillin acylases. An AHL acylase activity has also been reported in a mammal organ, i.e., pork kidney (Xu et al. 2003). The AHL substrate specificity of the pork acylase is broader than that of the microbial ones. The mammal enzyme was also characterized earlier as an *N*-acyl-L-amino acid hydrolase.

13.2.1.3 Lactonases

AHL lactonases are hydrolases that cleave the internal ester bond of these QS signal, permitting the release of the corresponding, QS-inactive, acyl homoserines. The catalyzed reaction is therefore similar to the chemical lactonolysis described above (see Sect. 13.2.1). Numerous AHL lactonases have been described, first in bacteria belonging to several *Bacillus* species (Dong et al. 2000, 2002) within the Firmicute clad. AHL lactonases have also been described in another Firmicute genus *Arthrobacter*, in Actinobacteria (*R. erythropolis*), and in several proteobacteria (e.g., *Klebsiella pneumoniae*, *Agrobacterium* sp.) (review: Uroz et al. 2009). These lactonases fall into two groups: the largest one gathers zinc-metalo hydrolases of the beta-lactamase family (Dong et al. 2001). The second type of AHL-lactonase is represented by the QsdA lactonase from *R. erythropolis* that belong to the phosphotriesterase (PTE) family (Uroz et al. 2008). Remarkably, PTEs were first described as enzymes capable to cleave the P–O bond in phosphotriesters, but some indeed harbor lactonase or amidohydrolase activities (Roodveldt and Tawfik 2005; Elias et al. 2008). Furthermore in mammal sera or in murine tracheal epithelia, the detected AHL lactonase activity is that of enzymes known as paraoxonases (Draganov et al. 2005; Stoltz et al. 2007). The lactonases generally exhibit a very broad AHL substrate range, and as the acylases, they may be used in biocontrol or ecological engineering approaches to perturb QS communication in bacteria (Uroz et al. 2003).

13.2.2 The Biology of QS Signal Degradation

The above data clearly establish that QS signal could be degraded. The key question, however, remains the biological significance of the phenomenon. Several elements suggest that this process is biologically significant in nature. The first one deals with long-chain 3-oxo-AHL signals which spontaneously form rearrangement compounds that are toxic to several *Bacillus* species (Kaufmann et al. 2005). Interestingly, numerous *Bacillus* species harbor strong lactonase activities that efficiently degrade the 3-oxo-AHL molecules (Dong et al. 2002), possibly acting as detoxifying enzymes. The second argument deals with fungi. As stated above, production of very active antifungal molecules such as phenazine and diphloroglucinol by several soil-borne bacteria depends upon AHL-mediated, QS regulation

(Chin-A-Woeng et al. 2001; Zhang and Pierson 2001). The production of AHL hydrolyzing enzymes may therefore be interpreted as a defense mechanisms evolved by fungi to prevent the deleterious production of antifungal molecules by bacteria. One of the most investigated QQ system, however, occurs in *Agrobacterium tumefaciens*, a bacteria that has both the ability to produce its own signal (3-oxo-octanoyl-homoserine lactone or OOHL) and to degrade it. A review of this important model is therefore presented below.

13.2.2.1 The *Agrobacterium* Paradigm

A. tumefaciens can infect a broad range of dicotyledonous plants and transfer a specific DNA fragment, called the T-DNA, from its tumor-inducing (Ti) plasmid to the nuclear genome of the host plants (reviews: White and Winans 2007; Pappas 2008). In transformed plant tissues, expression of T-DNA genes leads to the synthesis of plant growth factors, auxin and cytokinin, resulting in the formation of a tumor (crown gall disease). T-DNA also codes for the synthesis of opines, which are specific growth substrates and signals used by the Agrobacteria that colonize tumoral tissues (Dessaux et al. 1998). Some opines are required for the synthesis of QS signal OOHL. QS-pathway controls the amplification of Ti plasmid (i.e., copy-number) and its conjugative activity (Li and Farrand 2000; Piper et al. 1993; Zhang et al. 1993). Accumulation of QS signal at a high level also increases the number of the emerging tumors, suggesting a link between QS and magnitude of the symptoms on host plants (Pappas and Winans 2003).

In *A. tumefaciens* C58, two AHL lactonase-encoding genes, *attM* (=atu5139) and *aiiB* (=atu6071), are carried on At and Ti plasmids, respectively (Zhang et al. 2002; Carlier et al. 2003). They belong to two subclusters of a large family of Zn-hydrolases (Riaz et al. 2008) as confirmed by the structure analysis of AiiB (Liu et al. 2007). The expression of lactonases, AiiB and AttM, is controlled by different environmental signals, including plant compounds that may stimulate the degradation of QS-signals at different steps in the course of the *A. tumefaciens*–plant host interaction. The transcription of *aiiB* is increased in the presence of agrocinopines A and B, which are also required for synthesis of the QS signal OOHL in *A. tumefaciens* C58 (Haudecoeur et al. 2009a). The expression of *attM*, which belongs to the *attKLM* operon, is stimulated in the presence of gamma-hydroxybutyrate (GHB), succinic semialdehyde (SSA), and gamma-aminobutyrate (GABA), as well as salicylic acid (Carlier et al. 2004; Chevrot et al. 2006; Yuan et al. 2007). These molecules are produced by plants in response to biotic and abiotic stresses. Bacterial growth conditions, such as starvation, or mutations, may provoke accumulation of one of these molecules and stimulate the expression of the *attKLM* operon (Wang et al. 2006). In vitro, SSA and GHB, but neither GBL nor GABA, are able to inhibit the DNA-binding ability of the transcriptional repressor AttJ at the promoter region of attKLM (Wang et al. 2006; Chai et al. 2007). In cell cultures, the expression of *attKLM* observed in the presence of GABA and GBL should therefore require their conversion into SSA and GHB by GABA-transaminase (unidentified enzyme) and lactonase (AttM), respectively.

The role of the AHL lactonases AttM and AiiB in the control of QS-regulated functions, especially conjugation of plasmid Ti was evaluated. Conjugation of plasmid Ti was compared using donor strains in which lactonase-encoding genes were mutated or not. These experiments were performed in planta and in minimal medium in the presence of agrocinopines or in *traM* and/or *accR* genetic backgrounds. AiiB mutant and its corresponding wild-type strain were used as donors in conjugation experiments: Ti plasmid transconjugants emerged earlier in crosses involving the lactonase AiiB-defective strain. Moreover, the higher conjugation efficiency of *aiiB* mutant was correlated both with an early appearance and a higher level of OOHL in culture medium, confirming the implication of AiiB in QS and Ti plasmid conjugation (Haudecoeur et al. 2009). Regarding AttM, the other AHL lactonase, an *attM* mutant, and an *attJ* mutant (this latter constitutively express the *AttM* lactonase) were used as donors in conjugation experiments. The *attJ* mutant was affected for transfer of plasmid Ti in minimal medium (Zhang et al. 2002) and in planta (Khan and Farrand 2009). However, an *attM* mutant was a slightly more efficient a donor than its wild type counterpart for conjugation of Ti plasmid in planta (Khan and Farrand 2009). More precisely, transconjugants appeared earliest with *attM* mutant than with wild type as the donor, suggesting that *attM* exerts a transitory effect on conjugation. Two main nonexclusive hypotheses may be proposed to explain this transitory effect: (1) the AttM-mediated inactivation of OOHL could be balanced by a stronger signal synthesis in plant tumor; (2) the GABA-controlled expression of *attKLM* in planta could be attenuated by another plant signal. The hypothesis of a mechanism that down-regulates the GABA-mediated expression of *attKLM* operon in plant tumor is supported by the following observation: expression of *attKLM* was lower in the presence of tumor than stem tissues, while the level of GABA was higher in tumor than stem tissues (Haudecoeur et al. 2009a). Furthermore, a recent paper demonstrated that free proline, which accumulates in the tumor tissues, antagonizes the transport of GABA into *A. tumefaciens*, therefore GABA-induced degradation of OOHL (Haudecoeur et al. 2009b).

The implication of QS in the modulation of virulence (number of emerging tumors) was initially reported in *A. tumefaciens* strain R10 overexpressing *traR* (Pappas and Winans 2003) and then observed in other *A. tumefaciens* isolates (Molina et al. 2003; Chevrot et al. 2006). It was hypothesized that overaccumulation of QS signal could increase Ti plasmid copy number, thereby virulence of *Agrobacterium* (Pappas and Winans 2003). This amplification loop that increases plant symptoms required an excess of QS signal. Mutations affecting lactonase-mediated degradation of QS signals are predicted to increase accumulation of QS signal, thereby seriousness of plant symptoms. Virulence assays performed on tomato and tobacco plants confirmed this hypothesis (Haudecoeur et al. 2009). Noticeably, the *aiiB* mutant was more virulent than the *attM* mutant, revealing the stronger effect of AHL lactonase AiiB on QS.

13.2.2.2 QS and QQ Among Natural Microbial Communities in Soils

To address the biological significance of QS and QQ in nature, several approaches could also be taken. The first one is to investigate how "frequent" QS and QQ phenomena are. This approach has been used for the AHL-mediated QS regulation. However, other QS signals exist and investigation on the stability of these signals may provide some additional clues on the biological significance of the quenching phenomenona.

AHL-Based QS and QQ in the Soil Environment

As indicated earlier, AHL QS signals are produced and sensed at low concentration, within the pM to the μM range in vitro, in the test tube confined environment. In the "open" soil environment, a critical question is whether and "how far" AHL molecules could be sensed. Two studies revealed that AHL-based QS communication occurs in soils (Steidle et al. 2001; Burmølle et al. 2005). AHL concentrations in soil lie within the nM to μM range, barely less than what is observed in vitro (Gantner et al. 2006). The same authors suggested a possible diffusion zone of 4–80 μm (depending of the location) allowing sensing by other bacteria. AHLs appear to diffuse freely and become component of soils (Scott et al. 2006), a feature possibly related to their amphiphilic nature (at least for those with short and medium size acyl side chain) that permits them to cross the bacterial cell membranes and walls.

The biological significance of QS and QQ may also be inferred from a series of observations that aimed at comparing the occurrence of AHL-producing bacteria in bare soil and the plant root environment (also termed rhizosphere). Overall, some 8% of the cultivable bacteria isolated from bare soil produce AHLs while this ratio reaches 16–19% in the tobacco root environment (D'Angelo-Picard et al. 2005). This result is in agreement with an earlier observation on AHL-communicating pseudomonads that are more frequent within the rhizosphere than within bare soil (Elasri et al. 2001). The ratio of AHL-degrading strains is however not higher in the plant rhizosphere than in bare soil (ca. 3%) (D'Angelo-Picard et al. 2005). Considering that bacterial density is an average 1,000 times higher in the rhizosphere than in bare soil, it is tempting to speculate that the higher ratios of both AHL-producing and AHL-degrading bacteria in the rhizosphere reflect the selective value of both functions in a dense microbial environment, hence their significant biological role.

For obvious technical reasons, most investigations on AHL-producing and AHL-degrading bacteria have been performed on members of the culturable bacterioflora. In the soil and in the rhizosphere, these members represent only a part of the total bacterioflora (ca. 5% as judged from enumeration on classical growth media vs. microscopic examination) though modifications of cultivation conditions could in theory greatly increase this ratio (Janssen et al. 2002). Nevertheless, most of QS and QQ determinants may have been "missed" by previous studies – unless unculturable bacteria do not harbor such functions. To address this important ecological question, metagenomic (reviews: Riesenfeld et al. 2004; Green and Keller 2006) studies were undertaken. These studies have led to the identification of genetic determinants of novel AHL synthase and lactonases (Williamson et al. 2005; Riaz et al. 2008; Schipper et al. 2009). Among these, the QlcA lactonase, though a Zn-metallo hydrolase is only distantly related to the prototypic AiiA and AttM lactonases. Interestingly, the analysis of *qlcA* and neighboring genes permitted the authors to propose that *qlcA* originated most likely from a horizontal transfer from an alpha proteobacteria to an acidobacterial host (Riaz et al. 2008). In published metagenomic studies, the number of screened clones was relatively limited. However, the frequency of identification of QS and QQ determinants among the metagenomic clones suggests that the corresponding functions may be neither much more frequent nor much rarer in this community than in culturable bacteria. All these data first confirm the value of metagenomic approaches to investigate functions of unculturable microorganisms. They also suggest that QS and QQ functions are widespread among soil bacteria, irrespectively of their culturability, a feature that strengthens the view that both functions should play important ecological role in the soil environment.

Degradation of Non-AHL, QS Signals

Most of investigations on QQ have dealt with the AHL class of molecules, likely because they are frequent QS signals in proteobaceria, and especially those produced by several pathogens (mostly plant pathogens) to control the expression of virulence or virulence-related functions. Recently, however, similar investigations have targeted non-AHL signals produced by other phytopathogenic agents. A *Pseudomonas* sp. strain degrading the DSF signals produced by members of the *Xanthomonas* genus (see Sect. 13.1.1.2) has been isolated. The genes encoding this function indeed encode a carbamoyl phosphate synthetase. Coinoculation of this strain with the pathogen *X. campestris* pv. campestris into mustard and cabbage leaves reduced the severity of black rot symptoms compared with plants inoculated with the pathogen alone (Newman et al. 2008). Similarly, strains degrading the *Ralstonia* QS signal 3 OH-PAME were obtained. The key gene encoding the degradative function of one such strain from the *Ideonella* genus was isolated and found to be an acyl methyl ester methylase (Shinohara et al. 2007), but no demonstration of the quenching activity of this isolate was reported in planta. In another study, several 3 OH-PAME degrading strains were identified and tested for their potential value to set up a biocontrol strategy aimed at *Ralstonia*. Some of these strains indeed exhibited biocontrol ability directed at the pathogen, but neither the mode of action nor the genetic determinants were identified (Riaz, personal communication).

The two above cases demonstrate first that quenching of QS communication process that control virulence of various pathogens may be a general strategy, independent of the QS signal type (i.e., it is "language" independent). They also demonstrate that this approach indeed constitutes a very powerful way to develop innovative and environmentally friendly strategies aimed at these pathogens. These strategies are developed below.

13.3 Applied Outcomes: Ecological Engineering of QQ-Bacteria in the Rhizosphere

QQ-microorganisms are natural inhabitants of soils and should therefore contribute to AHL-degrading capacity of soils (Wang and Leadbetter 2005). QQ-bacteria represent up to 10% of the total culturable bacteria recovered from several soils and rhizospheres (Dong et al. 2000; D'Angelo-Picard et al. 2005). They belong to different phyla (Table 13.2), including Proteobacteria, Firmicutes, Actinobacteria, and probably unculturated Acidobacteria (Uroz et al. 2003; D'Angelo-Picard et al. 2004; Riaz et al. 2008). In vitro experiments showed that some QQ-bacteria may act as efficient biocontrol strains to protect the plants against some pathogens in which QS control the expression of major virulence factors (Uroz et al. 2003; Dong et al. 2004; Jafra et al. 2006). Noteworthy, some biocontrol strains rely upon QS to control the virulence of some pathogens. Disrupting QS regulation in the rhizosphere may therefore affect the biocontrol ability of plant beneficial bacteria, as reported for biocontrol *Pseudomonas* strains (Molina et al. 2003; Kang et al. 2004). The potential antagonism between QS-biocontrol strains and QQ-biocontrol strains therefore remains to be deeply evaluated before extended use of these biocontrol agents in the field.

In situ, different strategies may be proposed to increase the percentage of QQ-bacteria in the soils to develop crop protection (Faure and Dessaux 2007). These are: (1) introduction of selected signal degrading bacteria in soil (see above); (2) introduction of biodegradable compounds that stimulate the growth of the QQ-bacteria community naturally inhabiting soils; (3) a combination of both. The use of QQ-microorganisms in crop protection remains an emerging scientific domain; and no paper has reported field experiments yet. Remarkably, a work describes the application of biodegradable compounds for stimulating the growth of native QQ-bacteria in the rhizosphere of *Solanum tuberosum* grown under hydroponic conditions (Cirou et al. 2007). The two investigated compounds, gamma-caprolactone (GCL or gamma-hexanolactone) and gamma-heptanolactone (GHL), exhibit a gamma-butyrolactone ring and a short aliphatic carbon chain. When applied to hydroponic culture of *S. tuberosum*, they stimulated the growth of QQ-bacteria, mainly *Rhodococcus* and *Delftia* populations that can use GCL and GHL as a carbon source. Both introduced compounds were undetectable 4-week after their application. Consequently, application of GCL and GHL appears as environmentally friendly since they are rapidly and fully biodegraded by the bacterial community they stimulate. Furthermore, these two molecules have been used for years as flavoring agents by the food industry, and generally regarded as very low or nontoxic compounds. These features and their moderate cost make them suitable compounds to develop ecological, sustainable, disease control procedures in the field. However, many efforts are still required to evaluate the performance and environmental impact of QQ-strategies under those conditions.

13.4 QS: A Broad Communication System

13.4.1 QS Regulation Is Not Restricted to Bacteria

The discovery of coordinated gene transcription systems (operon structure, positive, and negative regulation) constituted a first evidence that bacterial populations were capable to partly coordinate some biological functions. Another proof, however, came in the 1980s with the identification of QS regulation of light emission in *P. fischeri* (see above). This finding was confirmed later by the identification of similar mechanisms regulating traits as diverse as motility modes, production of siderophores, plasmid transfer, antibiotic production, exopolysaccharide production, biofilm stability, etc. (a non limitative list, see for instance von Bodman et al. 2003). Interestingly, some sort of QS regulation also occurs in other microbes, such as fungi and amoebae. In the *Candida albicans* dimorphic fungus, the growth mode (budding yeast vs. filamentous growth) is cell density dependent (Hazen and Cutler 1979); it relies upon QS and the signal farnesol (Hornby et al. 2001; Oh et al. 2001). Other QS or QS-like systems have been reported in fungi. For instance, growing cells of the Dutch elm disease agent *Ceratocystis ulmi* excreted a QS factor responsible for the morphological shift from mycelia to budding yeast. This factor is specific for the species and is not farnesol (Hornby et al. 2004). In *Glomerella cingulata* cultures, a diffusible factor of unknown structure, termed auto-inhibitor, decreases mycelia formation and concomitantly increases conidia formation at cell densities greater than 10^6 per ml (Lingappa and Lingappa 1969; reviews: Hogan 2006; Kruppa 2009). In the amoebae *Dictostelium discoideum*, the formation of a multicellular reproductive structure depends upon multiple signals, including starvation and QS-like signals that differ from those found in bacteria (Clarke and Gomer 1995). Sensing the population density therefore appears as a trait that extends beyond the only bacterial world.

13.4.2 Beyond Sensing a Quorum

QS permits bacteria to couple gene expression to cell density and to synchronize – at the population level – one or more biological process(es). Though these traits are straightforward in confined environments (such as the light organs of marine organisms or in the plant tumor), the classical scheme of QS regulation predicts that this mode of regulation should be affected by diffusion of the signals in "open environments" such as river streams, sewer reclamation systems, or watered soils. A theory therefore proposes that QS systems may rather be a diffusion sensing system (Redfield 2002), i.e., a mechanism evolved by bacteria to sense how confined the environment is. This view, however, is challenged by the frequent occurrence of QS and QQ functions in the typical open environment that the soil is. Another theory, termed efficiency sensing (Hense et al. 2007), reunites QS and diffusion sensing. This theory states that QS permits bacteria to perceive at the same time their own cell density, and the diffusion constraints (including those related to signal degradation) in relation with ever changing environmental conditions. Though it is a somewhat "anthropomorphic" view, efficiency sensing predicts that QS is a global strategy by which bacteria determine whether the amount of energy required to turn on or off a biological phenomenon, such as antifungal or antibiotics synthesis, plasmid transfer machinery, or exoenzyme production is "worthwhile" in a particular environmental situation.

13.4.3 QS Signals May Be Involved in Interkingdom Communications

Bacterial QS signals are not solely recognized by prokaryotes. The DSF from *Xanthomonas campestris*, for instance, is perceived by the yeast species *C. albicans*. DSF mimics the yeast farnesol signal (see above) that inhibits filamentation (Wang et al. 2004). A DSF-related molecule, cis-2-dodecenoic acid, produced by strains of *Burkholderia cenocepacia* complement *X. campestris* mutants affected in DSF production and also affects *C. albicans* filamentation (Boon et al. 2008).

Plants also respond to AHL production by bacteria in different ways. First, short chain AHLs such as butyroyl-*N*-homoserine lactone (BHL), HHL, and *N*-octanoyl homoserine lactone (OHL) diffuse from root tissues to the aerial parts as seen in barley (*Hordeum vulgare*), the tropical legume *Pachyrhizus erosus* L., and in *Arabidopsis thaliana* (Götz et al. 2007; von Rad et al. 2008). In this species, exposure of the root system to HHL increases root elongation and affects the ratio of the plant hormones auxin and cytokinin in root cells. In leaves, the transcription of several genes is also affected. Some 40% of these genes encode proteins of unknown functions, but several code for peptides involved in cell wall organization and general metabolism.

More remarkably, the legume *M. truncatula* responds to the exogenous presence of the signals OdDHL and OHΔtDeHL signals, brought to the plant at physiological concentration (nM range; Mathesius et al. 2003). These two molecules are precisely those produced and sensed by the Medicago microbial symbiont *Sinorhizobium meliloti* (Sanchez-Contreras et al. 2007) that form nodules at the root system of the plant. In the presence of OdDHL and OHΔtDeHL, the concentration of ca. 8% of the plant proteins is affected, including some proteins involved in auxin response, plant defense reaction, and general regulatory and metabolic functions.

Stimulation of plant defenses by bacterial AHLs, namely BHL and HHL, is also documented. Strains of proteobacteria *Serratia liquefaciens* and *Pseudomonas putida* that produce these AHLs appear to be able to elicit (i.e., to stimulate) both the ethylene- and salicylic acid-mediated plant defense pathways of the plants (Schuhegger et al. 2006). In agreement, bacterial mutants affected in AHL production do not elicit defense reaction, while isolated compounds trigger them. This phenomenon is thus largely responsible for the biocontrol activity of both bacterial strains.

There are many other examples of recognition of bacterial QS signals by eukaryotes (reviews Hughes and Sperandio 2008; Lowery et al. 2008). These include (a non limitative list): the sensing of OdDHL produced by *P. aeruginosa* by lung epithelial cell in which this signal stimulates the synthesis of interleukin-8 and favors the inflammatory response (Smith et al. 2001); the inhibition of porcine arterial smooth muscle contraction by the same above AHL (Lawrence et al. 1999); the acceleration of apoptosis in macrophages and neutrophils by OdDHL (Telford et al. 2003); the sensing by olfaction of the OdDHL by the free-living terrestrial nematode *Caenorhabditis elegans* (Beale et al. 2006); and the attraction of green alga *Ulva* zoospores to bacterial biofilms in response to the production of AHLs of various types (Tait et al. 2005). This wide occurrence of such interkingdom communications suggest that cell–cell signaling (and communication) in bacteria is not simply a matter of quorum (Boyer and Wisniewski-Dyé 2009).

13.5 Concluding Remarks

For a long time, bacteria have been described as fast multiplying unicellular organisms, unable to "communicate". Twenty years of investigations on QS have drastically changed this view that biologist had at bacteria. Indeed bacteria not only communicate, but they did so in multiple languages, using for the only QS system multiple signals. Some of these communications are quite important in the "war" that opposes bacterial groups, or bacteria and fungi in the very competitive environment that the soil is. As a consequence, it is from an evolutionary point of view explicable that both classes of microbes have evolved some counter-measures to jam the adversary communication systems. Considering that several plant pathogenic bacteria also rely upon QS molecules to regulate virulence or virulence-related functions, the same evolutionary reading provides potential explanation for the plant capacity to detect the presence of the bacterial signal molecules.

Regarding QS regulation, two major counter measures have been described and collectively termed QQ: the production of QSI molecules that prevent sensing of the signal by the receptor protein and that of enzymes degrading the signals. Though these traits become largely documented, the biological significance of the QQ phenomenon remains intriguing. The general occurrence of functions capable to induce QS signal degradation in a range of fungi and bacteria, including non cultivable ones, strongly suggests that these functions might play a significant biological role. This role is indeed understood in several species of the

Gram-positive *Bacillus* genus and in the Gram-negative *Agrobacterium* genus. It could also be inferred from partial clues, for example, the degradation ability of mammalian cells potentially in contact with high microbial densities.

Whatever this role, QQ strategies have been developed and present a multifaceted value. They may be developed to prevent or limit biofilm formation on several structures, or the impact bacterial diseases in plants as seen from the *Pectobacterium* example. One of the most interesting aspects of QQ lies in the fact that it is a nonlethal approach that limits the selective pressure for the survival of pathogenic bacteria. The second most interesting aspect is the possibility to develop ecological engineering approaches that are environmentally friendly as they may use non toxic molecules.

Last, the above data indicate that the quenching phenomenon must be reintroduced in the global QS regulation scheme, in between synthesis and sensing, especially in complex environments. One reason to do so lies in this answer to the question: what is a signal? By definition, a signal is produced at certain times while a constant production of a signal (whatever the biological of physical model) eventually generates a background noise blurring communications. As a consequence, to be perceived as such, QS molecules must appear and disappear. The various quenching processes limit the accumulation of QS signals, or their sensing, and possibly contribute to their necessary fading. In other words, QS and QQ are complementary phenomena that ensure efficient bacterial communication in nature.

References

Armstrong E, Boyd KG, Burgess JG (2000) Prevention of marine biofouling using natural compounds from marine organisms. Biotechnol Annu Rev 6:221–241

Barnard AM, Bowden SD, Burr T, Coulthurst SJ, Monson RE, Salmond GP (2007) Quorum sensing, virulence and secondary metabolite production in plant soft-rotting bacteria. Philos Trans R Soc Lond B Biol Sci 362:1165–1183

Bassler BL, Wright M, Silverman MR (1994) Multiple signalling systems controlling expression of luminescence in Vibrio harveyi: sequence and function of genes encoding a second sensory pathway. Mol Microbiol 13:273–278

Beale E, Li G, Tan MW, Rumbaugh KP (2006) Caenorhabditis elegans senses bacterial autoinducers. Appl Environ Microbiol 72:5135–5137

Blosser-Middleton RS, Gray KM (2001) Multiple N-acyl homoserine lactone signals of *Rhizobium leguminosarum* are synthesized in a distinct temporal pattern. J Bacteriol 183:6771–6777

Boon C, Deng Y, Wang LH, He Y, Xu JL, Fan Y, Pan SQ, Zhang LH (2008) A novel DSF-like signal from *Burkholderia cenocepacia* interferes with *Candida albicans* morphological transition. ISME J 2:27–36

Boyer M, Wisniewski-Dyé F (2009) Cell-cell signalling in bacteria: not simply a matter of quorum. FEMS Microbiol Ecol 70(1):1–19. doi:10.1111/j.1574-6941.2009.00745

Burmølle M, Hansen LH, Sørensen SJ (2005) Use of a whole-cell biosensor and flow cytometry to detect AHL production by an indigenous soil community during decomposition of litter. Microb Ecol 50:221–229

Braeken K, Daniels R, Ndayizeye M, Vanderleyden J, Michiels J (2008) Quorum sensing in bacteria-plant interactions. In: Nautiyal CS, Dion P (eds) Molecular mechanisms of plant and microbe coexistence, vol 15, Soil biology. Springer, Berlin, Heidelberg, pp 265–290

Byers JT, Lucas C, Salmond GP, Welch M (2002) Nonenzymatic turnover of an *Erwinia carotovora* quorum-sensing signaling molecule. J Bacteriol 184:1163–1171

Carlier A, Uroz S, Smadja B, Latour X, Fray R, Dessaux Y, Faure D (2003) The Ti plasmid of *Agrobacterium tumefaciens* harbors an attM-paralogous gene, aiiB, also encoding N-acyl homoserine lactonase activity. Appl Environ Microbiol 69:4989–4993

Carlier A, Chevrot R, Dessaux Y, Faure D (2004) The assimilation of gamma-butyrolactone in *Agrobacterium tumefaciens* C58 interferes with the accumulation of the N-acyl-homoserine lactone signal. Mol Plant Microbe Interact 17:951–957

Chai Y, Tsai CC, Cho H, Winans SC (2007) Reconstitution of the biochemical activities of the AttJ repressor and the AttK, AttL and AttM catbolic enzymes of *Agrobacterium tumefaciens.* J Bacteriol 189:3674–3679

Chater KF (1993) Genetics of differentiation in *Streptomyces.* Annu Rev Microbiol 47:685–713

Chevrot R, Rosen R, Haudecoeur E, Cirou A, Shelp BJ, Ron E, Faure D (2006) GABA controls the level of quorum-sensing signal in *Agrobacterium tumefaciens.* Proc Natl Acad Sci USA 103:7201–7202

Chin-A-Woeng TF, van den Broek D, de Voer G, van der Drift KM, Tuinman S, Thomas-Oates JE, Lugtenberg BJ, Bloemberg GV (2001) Phenazine-1-carboxamide production in the biocontrol strain *Pseudomonas chlororaphis* PCL1391 is regulated by multiple factors secreted into the growth medium. Mol Plant Microbe Interact 14:969–979

Chowdhary PK, Keshavan N, Nguyen HQ, Peterson JA, González JE, Haines DC (2007) *Bacillus megaterium* CYP102A1 oxidation of acyl homoserine lactones and acyl homoserines. Biochemistry 46:14429–14437

Clarke M, Gomer RH (1995) PSF and CMF, autocrine factors that regulate gene expression during growth and early development of *Dictyostelium.* Experientia 51:1124–1134

Cirou A, Diallo S, Kurt C, Latour X, Faure D (2007) Growth promotion of quorum-quenching bacteria in the rhizosphere of *Solanum tuberosum.* Environ Microbiol 9:1511–1522

D'Angelo-Picard C, Faure D, Carlier A, Uroz S, Raffoux A, Fray R, Dessaux Y (2004) Dynamics of bacterial populations in the rhizosphere of tobacco plants producing – or not – the quorum sensing signals hexanoyl- and 3-oxohexanoyl-homoserine lactone. FEMS Microbiol Ecol 51:19–29

D'Angelo-Picard C, Faure D, Penot I, Dessaux Y (2005) Diversity of N-acylhomoserine lactone-producing and -degrading bacteria in soil and tobacco rhizosphere. Environ Microbiol 7:1796–1808

Daniels R, De Vos DE, Desair J, Raedschelders G, Luyten E, Rosemeyer V, Verreth C, Schoeters E, Vanderleyden J, Michiels J (2002) The *cin* quorum sensing locus of *Rhizobium etli* CNPAF512 affects growth and symbiotic nitrogen fixation. J Biol Chem 277:462–468

Decho AW, Visshel PT, Ferry J, Kawaguchi T, He L, Przekop KM, Norman RS, Reid RP (2009) Autoinducers extracted from microbial mats reveal a surprising diversity of N-acylhomoserine lactones (AHLs) and abundance changes that may relate to diel pH. Environ Microbiol 11:409–420

Delalande L, Faure D, Raffoux A, Uroz S, D'Angelo-Picard C, Elasri M, Carlier A, Berruyer R, Petit A, Williams P, Dessaux Y (2005) N-hexanoyl-L-homoserine lactone, a mediator of bacterial quorum-sensing regulation, exhibits plant-dependent stability and may be inactivated by germinating *Lotus corniculatus* seedlings. FEMS Microbiol Ecol 52:13–20

Dessaux Y, Petit A, Farrand SK, Murphy PJ (1998) Opines and opine-like molecules involved in plant–Rhizobiaceae interactions. In: Spaink HP, Kondorosi A, Hooykaas PJJ (eds) The Rhizobiaceae: molecular biology of model plant-associated bacteria. Kluwer Academic Publishers, Dordrecht, pp 173–197

Dobretsov S, Teplitski M, Paul V (2009) Mini-review: quorum sensing in the marine environment and its relationship to biofouling. Biofouling 25:413–427

Dong YH, Xu JL, Li XZ, Zhang LH (2000) AiiA, an enzyme that inactivates the acylhomoserine lactone quorum-sensing signal and attenuates the virulence of *Erwinia carotovora*. Proc Natl Acad Sci USA 97:3526–3531

Dong YH, Wang LH, Xu JL, Zhang HB, Zhang XF, Zhang LH (2001) Quenching quorum-sensing-dependent bacterial infection by an N-acyl homoserine lactonase. Nature 411:813–817

Dong YH, Gusti AR, Zhang Q, Xu JL, Zhang LH (2002) Identification of quorum-quenching N-acyl homoserine lactonases from *Bacillus* species. Appl Environ Microbiol 68:1754–1759

Dong YH, Zhang XF, Xu JL, Zhang LH (2004) Insecticidal Bacillus thuringiensis silences *Erwinia carotovora* virulence by a new form of microbial antagonism, signal interference. Appl Environ Microbiol 70:954–960

Draganov DI, Teiber JF, Speelman A, Osawa Y, Sunahara R, La Du BN (2005) Human paraoxonases (PON1, PON2, and PON3) are lactonases with overlapping and distinct substrate specificities. J Lipid Res 46:1239–1247

Dubern JF, Diggle SP (2008) Quorum sensing by 2-alkyl-4-quinolones in *Pseudomonas aeruginosa* and other bacterial species. Mol Biosyst 4:882–888

Dulla GF, Lindow SE (2009) Acyl-homoserine lactone-mediated cross talk among epiphytic bacteria modulates behavior of *Pseudomonas syringae* on leaves. ISME J 3:825–834

Eberhard A, Burlingame AL, Eberhard C, Kenyon GL, Nealson KH, Oppenheimer NJ (1981) Structural identification of autoinducer of *Photobacterium fischeri* luciferase. Biochemistry 20:2444–2449

Elasri M, Delorme S, Lemanceau P, Stewart G, Laue B, Glickmann E, Oger PM, Dessaux Y (2001) Acyl-homoserine lactone production is more common among plant-associated *Pseudomonas* spp. than among soilborne *Pseudomonas* spp. Appl Environ Microbiol 67:1198–1209

Elias M, Dupuy J, Merone L, Mandrich L, Porzio E, Moniot S, Rochu D, Lecomte C, Rossi M, Masson P, Manco G, Chabriere E (2008) Structural basis for natural lactonase and promiscuous phosphotriesterase activities. J Mol Biol 379:1017–1028

Engebrecht J, Silverman M (1984) Identification of genes and gene products necessary for bacterial bioluminescence. Proc Natl Acad Sci USA 81:4154–4158

Engebrecht J, Silverman M (1987) Nucleotide sequence of the regulatory locus controlling expression of bacterial genes for bioluminescence. Nucleic Acids Res 15:10455–10467

Engebrecht J, Nealson K, Silverman M (1983) Bacterial bioluminescence: isolation and genetic analysis of functions from *Vibrio fischeri*. Cell 32:773–781

Faure D, Dessaux Y (2007) Quorum sensing as a target for developing biocontrol strategies towards the plant pathogen Pectobacterium. Eur J Plant Pathol 119:353–365

Ferluga S, Staindler L, Ventiru V (2008) N-acyl homoserine lactone quorum sensing in Gram negative rhizobacteria. In: Karlovsky P (ed) Secondary metabolites in soil ecology, vol 14, Soil biology. Springer, Berlin, Heidelberg, pp 69–92

Flavier AB, Clough SJ, Schell MA, Denny TP (1997) Identification of 3-hydroxypalmitic acid methyl ester as a novel autoregulator controlling virulence in *Ralstonia solanacearum*. Mol Microbiol 26:251–259

Fouhy Y, Lucey JF, Ryan RP, Dow JM (2006) Cell-cell signaling, cyclic di-GMP turnover and regulation of virulence in *Xanthomonas campestris*. Res Microbiol 157:899–904

Fuqua WC, Winans SC, Greenberg EP (1994) Quorum sensing in bacteria: the LuxR-LuxI family of cell density-responsive transcriptional regulators. J Bacteriol 176:269–275

Fuqua C, Winans SC, Greenberg EP (1996) Census and consensus in bacterial ecosystems: the LuxR-LuxI family of quorum-sensing transcriptional regulators. Annu Rev Microbiol 50:727–751

Gantner S, Schmid M, Dürr C, Schuhegger R, Steidle A, Hutzler P, Langebartels C, Eberl L, Hartmann A, Dazzo FB (2006) In situ quantitation of the spatial scale of calling distances and population density-independent N-acylhomoserine lactone-mediated communication by rhizobacteria colonized on plant roots. FEMS Microbiol Ecol 56:188–194

Gao M, Teplitski M, Robinson JB, Bauer WD (2003) Production of substances by *Medicago truncatula* that affect bacterial quorum sensing. Mol Plant Microbe Interact 16:827–834

Givskov M, de Nys R, Manefield M, Gram L, Maximilien R, Eberl L, Molin S, Steinberg PD, Kjelleberg S (1996) Eukaryotic interference with homoserine lactone-mediated prokaryotic signalling. J Bacteriol 178:6618–6622

Green BD, Keller M (2006) Capturing the uncultivated majority. Curr Opin Biotechnol 17:236–240

González JE, Marketon MM (2003) Quorum sensing in nitrogen-fixing rhizobia. Microbiol Mol Biol Rev 67:574–592

González JE, Keshavan ND (2006) Messing with bacterial quorum sensing. Microbiol Mol Biol Rev 70:859–875

Götz C, Fekete A, Gebefuegi I, Forczek ST, Fuksová K, Li X, Englmann M, Gryndler M, Hartmann A, Matucha M, Schmitt-Kopplin P, Schröder P (2007) Uptake, degradation and chiral discrimination of N-acyl-D/L-homoserine lactones by barley (*Hordeum vulgare*) and yam bean (*Pachyrhizus erosus*) plants. Anal Bioanal Chem 389:1447–1457

Haudecoeur E, Tannières M, Cirou A, Raffoux A, Dessaux Y, Faure D (2009a) Different regulation and roles of lactonases AiiB and AttM in *Agrobacterium tumefaciens* C58. Mol Plant Microbe Interact 22:529–537

Haudecoeur E, Planamente S, Cirou A, Tannières M, Shelp BJ, Moréra S, Faure D (2009b) Proline antagonizes GABA-induced quenching of quorum-sensing in *Agrobacterium tumefaciens*. Proc Natl Acad Sci USA 106:14587–14592

Hazen KC, Cutler JE (1979) Autoregulation of germ tube formation by *Candida albicans*. Infect Immun 24:661–666

He YW, Zhang LH (2008) Quorum sensing and virulence regulation in *Xanthomonas campestris*. FEMS Microbiol Rev 32:842–857

Hense BA, Kuttler C, Müller J, Rothballer M, Hartmann A, Kreft J-U (2007) Does efficiency sensing unify diffusion and quorum sensing? Nat Rev Microbiol 5:230–239

Hogan DA (2006) Talking to themselves: autoregulation and quorum sensing in fungi. Eukaryot Cell 5:613–619

Holden MT, Ram Chhabra S, de Nys R, Stead P, Bainton NJ, Hill PJ, Manefield M, Kumar N, Labatte M, England D, Rice S, Givskov M, Salmond GP, Stewart GS, Bycroft BW, Kjelleberg S, Williams P (1999) Quorum-sensing cross talk: isolation and chemical characterization of cyclic dipeptides from *Pseudomonas aeruginosa* and other gram-negative bacteria. Mol Microbiol 33:1254–1266

Hornby JM, Jensen EC, Lisec AD, Tasto JJ, Jahnke B, Shoemaker R, Dussault P, Nickerson KW (2001) Quorum sensing in the dimorphic fungus *Candida albicans* is mediated by farnesol. Appl Environ Microbiol 67:2982–2992

Hornby JM, Jacobitz-Kizzier SM, McNeel DJ, Jensen EC, Treves DS, Nickerson KW (2004) Inoculum size effect in dimorphic fungi: extracellular control of yeast-mycelium dimorphism in *Ceratocystis ulmi*. Appl Environ Microbiol 70:1356–1359

Hooshangi S, Bentley WE (2008) From unicellular properties to multicellular behavior: bacteria quorum sensing circuitry and applications. Curr Opin Biotechnol 19:550–555

Hu JY, Fan Y, Lin YH, Zhang HB, Ong SL, Dong N, Xu JL, Ng WJ, Zhang LH (2003) Microbial diversity and prevalence of virulent pathogens in biofilms developed in a water reclamation system. Res Microbiol 154:623–629

Huang JJ, Han JI, Zhang LH, Leadbetter JR (2003) Utilization of acyl-homoserine lactone quorum signals for growth by a soil pseudomonad and *Pseudomonas aeruginosa* PAO1. Appl Environ Microbiol 69:5941–5949

Huang JJ, Petersen A, Whiteley M, Leadbetter JR (2006) Identification of QuiP, the product of gene PA1032, as the second acyl-homoserine lactone acylase of *Pseudomonas aeruginosa* PAO1. Appl Environ Microbiol 72:1190–1197

Hughes DT, Sperandio V (2008) Inter-kingdom signalling: communication between bacteria and their hosts. Nat Rev Microbiol 6:111–120

Jafra S, Przysowa J, Czajkowski R, Michta A, Garbeva P, van der Wolf JM (2006) Detection and characterization of bacteria from the potato rhizosphere degrading N-acyl-homoserine lactone. Can J Microbiol 52:1006–1015

Janssen PH, Yates PS, Grinton BE, Taylor PM, Sait M (2002) Improved culturability of soil bacteria and isolation in pure culture of novel members of the divisions Acidobacteria, Actinobacteria, Proteobacteria, and Verrucomicrobia. Appl Environ Microbiol 68:2391–2396

Jayaraman A, Wood TK (2008) Bacterial quorum sensing: signals, circuits, and implications for biofilms and disease. Annu Rev Biomed Eng 10:145–167

Kang BR, Lee JH, Ko SJ, Lee YH, Cha JS, Cho BH, Kim YC (2004) Degradation of acyl-homoserine lactone molecules by *Acinetobacter* sp. strain C1010. Can J Microbiol 50:935–941

Kaufmann GF, Sartorio R, Lee SH, Rogers CJ, Meijler MM, Moss JA, Clapham B, Brogan AP, Dickerson TJ, Janda KD (2005) Revisiting quorum sensing: discovery of additional chemical and biological functions for 3-oxo-N-acylhomoserine lactones. Proc Natl Acad Sci USA 102:309–314

Keshavan ND, Chowdhary PK, Haines DC, González JE (2005) L-Canavanine made by *Medicago sativa* interferes with quorum sensing in *Sinorhizobium meliloti*. J Bacteriol 187:8427–8436

Khan SR, Farrand SK (2009) The BlcC (AttM) Lactonase of *Agrobacterium tumefaciens* does not quench the quorum-sensing system that regulates Ti plasmid conjugative transfer. J Bacteriol 191:1320–1329

Krin E, Chakroun N, Turlin E, Givaudan A, Gaboriau F, Bonne I, Rousselle JC, Frangeul L, Lacroix C, Hullo MF, Marisa L, Danchin A, Derzelle S (2006) Pleiotropic role of quorum-sensing autoinducer 2 in *Photorhabdus luminescens*. Appl Environ Microbiol 72:6439–6451

Kruppa M (2009) Quorum sensing and *Candida albicans*. Mycoses 52:1–10

Lawrence RN, Dunn WR, Bycroft B, Camara M, Chhabra SR, Williams P, Wilson VG (1999) The *Pseudomonas aeruginosa* quorum-sensing signal molecule, N-(3-oxododecanoyl)-l-homoserine lactone, inhibits porcine arterial smooth muscle contraction. Br J Pharmacol 128:845–848

Leadbetter JR, Greenberg EP (2000) Metabolism of acyl-homoserine lactone quorum-sensing signals by *Variovorax paradoxus*. J Bacteriol 182:6921–6926

Li PL, Farrand SK (2000) The replicator of the nopaline-type Ti plasmid is a member of the repABC family and is influenced by the TraR-dependent quorum-sensing regulatory system. J Bacteriol 182:179–188

Lin YH, Xu JL, Hu J, Wang LH, Ong SL, Leadbetter JR, Zhang LH (2003) Acyl-homoserine lactone acylase from *Ralstonia* strain XJ12B represents a novel and potent class of quorum-quenching enzymes. Mol Microbiol 47:849–860

Lingappa BT, Lingappa Y (1969) Role of auto-inhibitors on mycelial growth and dimorphism of *Glomerella cingulata*. J Gen Microbiol 56:35–45

Liu D, Thomas PW, Momb J, Hoang QQ, Petsko GA, Ringe D, Fast W (2007) Structure and specificity of a quorum-quenching lactonase (AiiB) from *Agrobacterium tumefaciens*. Biochemistry 46:11789–11799

Loh J, Carlson RW, York WS, Stacey G (2002) Bradyoxetin, a unique chemical signal involved in symbiotic gene regulation. Proc Natl Acad Sci USA 99:14446–14451

Lowery CA, Dickerson TJ, Janda KD (2008) Interspecies and interkingdom communication mediated by bacterial quorum sensing. Chem Soc Rev 37:1337–1346

Lyon GJ, Novick RP (2004) Peptide signaling in *Staphylococcus aureus* and other Gram-positive bacteria. Peptides 25:1389–1403

Mathesius U, Mulders S, Gao M, Teplitski M, Caetano-Anolles G, Rolfe BG, Bauer WD (2003) Extensive and specific responses of a eukaryote to bacterial quorum-sensing signals. Proc Natl Acad Sci USA 100:1444–1449

McClean KH, Winson MK, Fish L, Taylor A, Chhabra SR, Camara M, Daykin M, Lamb JH, Swift S, Bycroft BW, Stewart GS, Williams P (1997) Quorum sensing and *Chromobacterium violaceum*: exploitation of violacein production and inhibition for the detection of N-acylhomoserine lactones. Microbiology 143:3703–3711

Miller MB, Bassler BL (2001) Quorum sensing in bacteria. Annu Rev Microbiol 55:165–199

Molina L, Constantinescu F, Michel L, Reimmann C, Duffy B, Défago G (2003) Degradation of pathogen quorum-sensing molecules by soil bacteria: a preventive and curative biological control mechanism. FEMS Microbiol Ecol 45:71–81

Morohoshi T, Nakazawa S, Ebata A, Kato N, Ikeda T (2008) Identification and characterization of N-acylhomoserine lactone-acylase from the fish intestinal *Shewanella* sp. strain MIB015. Biosci Biotechnol Biochem 72:1887–1893

Nachin L, Barras F (2000) External pH: an environmental signal that helps to rationalize pel gene duplication in *Erwinia chrysanthemi*. Mol Plant Microbe Interact 13:882–886

Nealson KH, Hastings JW (1979) Bacterial bioluminescence: its control and ecological significance. Microbiol Rev 43:496–518

Newman KL, Chatterjee S, Ho KA, Lindow SE (2008) Virulence of plant pathogenic bacteria attenuated by degradation of fatty acid cell-to-cell signaling factors. Mol Plant Microbe Interact 21:326–334

Novick RP, Geisinger E (2008) Quorum sensing in *Staphylococci*. Annu Rev Genet 42:541–564

Nyholm SV, McFall-Ngai M (2004) The winnowing: establishing the squid–Vibrio symbiosis. Nat Rev Microbiol 2:632–642

Oh KB, Miyazawa H, Naito T, Matsuoka H (2001) Purification and characterization of an autoregulatory substance capable of regulating the morphological transition in *Candida albicans*. Proc Natl Acad Sci USA 98:4664–4668

Pappas KM, Winans SC (2003) A LuxR-type regulator from *Agrobacterium tumefaciens* elevates Ti plasmid copy number by activating transcription of plasmid replication genes. Mol Microbiol 48:1059–1073

Pappas KM (2008) Cell-cell signaling and the *Agrobacterium tumefaciens* Ti plasmid copy number fluctuations. Plasmid 60:89–107

Pappas KM, Weingart CL, Winans SC (2004) Chemical communication in proteobacteria: biochemical and structural studies of signal synthases and receptors required for intercellular signalling. Mol Microbiol 53:755–769

Park SY, Lee SJ, Oh TK, Oh JW, Koo BT, Yum DY, Lee JK (2003) AhlD, an N-acylhomoserine lactonase in *Arthrobacter* sp., and predicted homologues in other bacteria. Microbiology 149:1541–1550

Park SY, Hwang BJ, Shin MH, Kim JA, Kim HK, Lee JK (2006) N-acylhomoserine lactonase producing *Rhodococcus* spp. with different AHL-degrading activities. FEMS Microbiol Lett 261:102–108

Piper KR, Beck von Bodman S, Farrand SK (1993) Conjugation factor of *Agrobacterium tumefaciens* regulates Ti plasmid transfer by autoinduction. Nature 362:448–450

Pongsilp N, Triplett EW, Sadowsky MJ (2005) Detection of homoserine lactone-like quorum sensing molecules in *Bradyrhizobium* strains. Curr Microbiol 51:250–254

Popat R, Crusz SA, Diggle SP (2008) The social behaviours of bacterial pathogens. Br Med Bull 87:63–75

Quiñones B, Dulla G, Lindow SE (2005) Quorum sensing regulates exopolysaccharide production, motility, and virulence in *Pseudomonas syringae*. Mol Plant Microbe Interact 18:682–693

Rasmussen TB, Givskov M (2006) Quorum sensing inhibitors: a bargain of effects. Microbiology 152:895–904

Rasmussen TB, Manefield M, Andersen JB, Eberl L, Anthoni U, Christophersen C, Steinberg P, Kjelleberg S, Givskov M (2000) How *Delisea pulchra* furanones affect quorum sensing and swarming motility in *Serratia liquefaciens* MG1. Microbiology 146:3237–3244

Rasmussen TB, Bjarnsholt T, Skindersoe ME, Hentzer M, Kristoffersen P, Kote M, Nielsen J, Eberl L, Givskov M (2005) Screening for quorum-sensing inhibitors (QSI) by use of a novel genetic system, the QSI selector. J Bacteriol 187:1799–1814

Redfield RJ (2002) Is quorum sensing a side effect of diffusion sensing? Trends Microbiol 10:365–370

Rezzonico F, Duffy B (2007) The role of *luxS* in the fire blight pathogen *Erwinia amylovora* is limited to metabolism and does not involve quorum sensing. Mol Plant Microbe Interact 20:1284–1297

Riaz K, Elmerich C, Moreira D, Raffoux A, Dessaux Y, Faure D (2008) A metagenomic analysis of rhizospheric bacteria extends the diversity of quorum-quenching lactonases. Environ Microbiol 10:560–570

Riesenfeld CS, Schloss PD, Handelsman J (2004) Metagenomics: genomic analysis of microbial communities. Annu Rev Genet 38:525–552

Rogers P, McElroy WD (1955) Biochemical characteristics of aldehyde and luciferase mutants of luminous bacteria. Proc Natl Acad Sci USA 41:67–70

Romero M, Diggle SP, Heeb S, Cámara M, Otero A (2008) Quorum quenching activity in *Anabaena* sp. PCC 7120: identification of AiiC, a novel AHL-acylase. FEMS Microbiol Lett 280:73–80

Roodveldt C, Tawfik DS (2005) Shared promiscuous activities and evolutionary features in various members of the amidohydrolase superfamily. Biochemistry 44:12728–12736

Sanchez-Contreras M, Bauer WD, Gao M, Robinson JB, Allan Downie J (2007) Quorum-sensing regulation in rhizobia and its role in symbiotic interactions with legumes. Philos Trans R Soc Lond B Biol Sci 362:1149–1163

Schaefer AL, Greenberg EP, Oliver CM, Oda Y, Huang JJ, Bittan-Banin G, Peres CM, Schmidt S, Juhaszova K, Sufrin JR, Harwood CS (2008) A new class of homoserine lactone quorum-sensing signals. Nature 454:595–599

Schipper C, Hornung C, Bijtenhoorn P, Quitschau M, Grond S, Streit WR (2009) Metagenome-derived clones encoding two novel lactonase family proteins involved in biofilm inhibition in *Pseudomonas aeruginosa*. Appl Environ Microbiol 75:224–233

Schripsema J, de Rudder KEE, van Vliet TB, Lankhorst PP, de Vroom E, Kijne JW, van Brussel AAN (1996) Bacteriocin small of *Rhizobium leguminosarum* belongs to the class of N-acyl-L-homoserine lactone molecules known as autoinducers and as quorum sensing co-transcription factors. J Bacteriol 178:366–371

Scott RA, Weil J, Le PT, Williams P, Fray RG, von Bodman SB, Savka MA (2006) Long- and short-chain plant-produced bacterial N-acyl-homoserine lactones become components of phyllosphere, rhizosphere, and soil. Mol Plant Microbe Interact 19:227–239

Schuhegger R, Ihring A, Gantner S, Bahnweg G, Knappe C, Vogg G, Hutzler P, Schmid M, Van Breusegem F, Eberl L, Hartmann A, Langebartels C (2006) Induction of systemic resistance in tomato by N-acyl-L-homoserine lactone-producing rhizosphere bacteria. Plant Cell Environ 29:909–918

Shaw PD, Ping G, Daly SL, Cha C, Cronan JE Jr, Rinehart KL, Farrand SK (1997) Detecting and characterizing N-acyl-homoserine lactone signal molecules by thin-layer chromatography. Proc Natl Acad Sci USA 94:6036–6041

Shinohara M, Nakajima N, Uehara Y (2007) Purification and characterization of a novel esterase (beta-hydroxypalmitate methyl ester hydrolase) and prevention of the expression of virulence by *Ralstonia solanacearum*. J Appl Microbiol 103:152–162

Smith RS, Fedyk ER, Springer TA, Mukaida N, Iglewski BH, Phipps RP (2001) IL-8 production in human lung fibroblasts and epithelial cells activated by the *Pseudomonas* autoinducer N-3-oxododecanoyl homoserine lactone is transcriptionally regulated by NF-kappa B and activator protein-2. J Immunol 167:366–374

Steidle A, Sigl K, Schuhegger R et al (2001) Visualization of N-acylhomoserine lactone (AHL)-mediated cell-cell communication between bacteria colonizing the tomato rhizosphere. Appl Environ Microbiol 67:5761–5770

Stoltz DA, Ozer EA, Ng CJ, Yu JM, Reddy ST, Lusis AJ, Bourquard N, Parsek MR, Zabner J, Shih DM (2007) Paraoxonase-2 deficiency enhances *Pseudomonas aeruginosa* quorum sensing in murine tracheal epithelia. Am J Physiol Lung Cell Mol Physiol 292:L852–L860

Sturme MH, Francke C, Siezen RJ, de Vos WM, Kleerebezem M (2007) Making sense of quorum sensing in lactobacilli: a special focus on *Lactobacillus plantarum* WCFS1. Microbiology 153:3939–3947

Surette MG, Miller MB, Bassler BL (1999) Quorum sensing in *Escherichia coli*, *Salmonella typhimurium*, and *Vibrio harveyi*: a new family of genes responsible for autoinducer production. Proc Natl Acad Sci USA 96:1639–1644

Tait K, Joint I, Daykin M, Milton DL, Williams P, Cámara M (2005) Disruption of quorum sensing in seawater abolishes attraction of zoospores of the green alga *Ulva* to bacterial biofilms. Environ Microbiol 7:229–240

Telford G, Wheeler D, Williams P, Tomkins PT, Appleby P, Sewell H, Stewart GS, Bycroft BW, Pritchard DI (2003) The *Pseudomonas aeruginosa* autoinducer N-3-oxododecanoyl homoserine lactone accelerates apoptosis in macrophages and neutrophils. Infect Immun 71:5785–5793

Teplitski M, Robinson JB, Bauer WD (2000) Plants secrete substances that mimic bacterial N-acyl homoserine lactone signal activities and affect population density-dependent behaviors in associated bacteria. Mol Plant Microbe Interact 13:637–664

Teplitski M, Chen H, Rajamani S, Gao M, Merighi M, Sayre RT, Robinson JB, Rolfe BG, Bauer WD (2004) *Chlamydomonas reinhardtii* secretes compounds that mimic bacterial signals and interfere with quorum sensing regulation in bacteria. Plant Physiol 134:137–146

Turovskiy Y, Kashtanov D, Paskhover B, Chikindas ML (2007) Quorum sensing: fact, fiction, and everything in between. Adv Appl Microbiol 62:191–234

Uroz S, Heinonsalo J (2008) Degradation of N-acyl homoserine lactone quorum sensing signal molecules by forest root-associated fungi. FEMS Microbiol Ecol 65:271–278

Uroz S, D'Angelo C, Carlier A, Faure D, Petit A, Oger P, Sicot C, Dessaux Y (2003) Novel bacteria degrading N-acyl homoserine lactones and their use as quenchers of quorum-sensing regulated functions of plant pathogenic bacteria. Microbiology 149:1981–1989

Uroz S, Chhabra SR, Cámara M, Williams P, Oger P, Dessaux Y (2005) N-Acylhomoserine lactone quorum-sensing molecules are modified and degraded by *Rhodococcus erythropolis* W2 by both amidolytic and novel oxidoreductase activities. Microbiology 151:3313–3322

Uroz S, Oger P, Chhabra SR, Cámara M, Williams P, Dessaux Y (2007) N-acyl homoserine lactones are degraded via an amidolytic activity in *Comamonas* sp. strain D1. Arch Microbiol 187:249–256

Uroz S, Oger PM, Chapelle E, Adeline MT, Faure D, Dessaux Y (2008) A *Rhodococcus* qsdA-encoded enzyme defines a novel class of large-spectrum quorum-quenching lactonases. Appl Environ Microbiol 74:1357–1366

Uroz S, Dessaux Y, Oger P (2009) Quorum sensing and quorum quenching: the yin and yang of bacterial communication. Chembiochem 10:205–216

Van Houdt R, Givskov M, Michiels CW (2007) Quorum sensing in *Serratia*. FEMS Microbiol Rev 31:407–424

Vendeville A, Winzer K, Heurlier K, Tang CM, Hardie KR (2005) Making "sense" of metabolism: autoinducer-2, LuxS and pathogenic bacteria. Nat Rev Microbiol 3:383–396

Visick KL, Ruby EG (2006) *Vibrio fischeri* and its host: it takes two to tango. Curr Opin Microbiol 9:632–638

Von Bodman SB, Bauer WD, Coplin DL (2003) Quorum sensing in plant-pathogenic bacteria. Annu Rev Phytopathol 41:455–482

von Rad V, Klein I, Dobrev PI, Kottova J, Zazimalova E, Fekete A, Hartmann A, Schmitt-Kopplin P, Durner J (2008) Response of *Arabidopsis thaliana* to N-hexanoyl-DL-homoserine-lactone, a bacterial quorum sensing molecule produced in the rhizosphere. Planta 229:73–85

Wang C, Zhang HB, Wang LH, Zhang LH (2006) Succinic semialdehyde couples stress responses to quorum-sensing signal decay in *Agrobacterium tumefaciens*. Mol Microbiol 62:45–56

Wang LH, He Y, Gao Y, Wu JE, Dong YH, He C, Wang SX, Weng LX, Xu JL, Tay L, Fang RX, Zhang LH (2004) A bacterial cell-cell communication signal with cross-kingdom structural analogues. Mol Microbiol 51:903–912

Wang YJ, Leadbetter JR (2005) Rapid acyl-homoserine lactone quorum signal biodegradation in diverse soils. Appl Environ Microbiol 71:1291–1299

Wei HL, Zhang LQ (2006) Quorum-sensing system influences root colonization and biological control ability in *Pseudomonas fluorescens* 2P24. Antonie Van Leeuwenhoek 89:267–280

White CE, Winans SC (2007) Cell-cell communication in the plant pathogen *Agrobacterium tumefaciens*. Philos Trans R Soc Lond B Biol Sci 362:1135–1148

Willcox MD, Zhu H, Conibear TC, Hume EB, Givskov M, Kjelleberg S, Rice SA (2008) Role of quorum sensing by *Pseudomonas aeruginosa* in microbial keratitis and cystic fibrosis. Microbiology 154:2184–2194

Williamson LL, Borlee BR, Schloss PD, Guan C, Allen HK, Handelsman J (2005) Intracellular screen to identify metagenomic clones that induce or inhibit a quorum-sensing biosensor. Appl Environ Microbiol 71:6335–6344

Xu F, Byun T, Deussen HJ, Duke KR (2003) Degradation of N-acylhomoserine lactones, the bacterial quorum-sensing molecules, by acylase. J Biotechnol 101:89–96

Yang F, Wang LH, Wang J, Dong YH, Hu JY, Zhang LH (2005) Quorum quenching enzyme activity is widely conserved in the sera of mammalian species. FEBS Lett 579:3713–3717

Yang M, Sun K, Zhou L, Yang R, Zhong Z, Zhu J (2009) Functional analysis of three AHL autoinducer synthase genes in *Mesorhizobium loti* reveals the important role of quorum sensing in symbiotic nodulation. Can J Microbiol 55:210–214

Yates EA, Philipp B, Buckley C, Atkinson S, Chhabra SR, Sockett RE, Goldner M, Dessaux Y, Cámara M, Smith H, Williams P (2002) N-acylhomoserine lactones undergo lactonolysis in a pH-, temperature-, and acyl chain length-dependent manner during growth of *Yersinia pseudotuberculosis* and *Pseudomonas aeruginosa*. Infect Immun 70:5635–5646

Yuan ZC, Edlind MP, Liu P, Saenkham P, Banta LM, Wise AA, Ronzone E, Binns AN, Kerr K, Nester EW (2007) The plant signal salicylic acid shuts down expression of the vir regulon and activates quormone-quenching genes in *Agrobacterium*. Proc Natl Acad Sci USA 104: 11790–11795

Zhang HB, Wang LH, Zhang LH (2002) Genetic control of quorum-sensing signal turnover in *Agrobacterium tumefaciens*. Proc Natl Acad Sci USA 99:4638–4643

Zhang L, Murphy PJ, Kerr A, Tate ME (1993) *Agrobacterium* conjugation and gene regulation by N-acyl homoserine lactones. Nature 362:446–448

Zhang LH (2003) Quorum quenching and proactive host defense. Trends Plant Sci 8:238–244

Zhang Z, Pierson LS 3rd (2001) A second quorum-sensing system regulates cell surface properties but not phenazine antibiotic production in *Pseudomonas aureofaciens*. Appl Environ Microbiol 67:4305–4315

Chapter 14
Integration of Cell-to-Cell Signals in Soil Bacterial Communities

Max Teplitski, Massimo Merighi, Mengsheng Gao, and Jayne Robinson

14.1 Introduction

In complex and uncertain environments, such as soils, unicellular organisms depend on various cues to coordinately regulate genes and behaviors important for survival of their populations. How microbes perceive environmental and metabolic signals and how they integrate this information to modulate gene expression is a central question in microbial ecology. Unlike obligate parasites and symbionts of eukaryotes, genomes of which have undergone considerable reductions, soil bacteria, as well as opportunistic pathogens contain a large number of genes involved in environmental sensing. For example, of the approximately half of the *Pseudomonas aeruginosa* genes (for which functions could be assigned at least tentatively), 2% are two-component regulatory systems and 7.2% are transcriptional regulators (Stover et al. 2000). Regulatory proteins make up 8.7% of the genome of *S. meliloti* 1021 (including 36 pairs of sensor kinases and response regulators) (Galibert et al. 2001). These genome-wide comparisons indicate that the ability to accurately sense the environment and to change gene expression accordingly is crucial to the bacteria that spend at least a part of their life cycle in soils.

Quorum sensing (QS) is one of the mechanisms by which microbes change global patterns of gene expression in response to increases in their population densities within a diffusion-limited environment [rev. Waters and Bassler (2005) and Hense et al. (2007)]. At least a dozen kinds of small molecules and short

M. Teplitski (✉) and M. Gao
Soil and Water Sciences Department, Genetics Institute, University of Florida – Institute of Food and Agricultural Sciences (IFAS), 1376 Mowry Rd, Gainesville, FL, USA
e-mail: maxtep@ufl.edu

M. Merighi
Glycosyn Inc, Boston, MA, USA

J. Robinson
Department of Biology, University of Dayton, Dayton, OH, USA

G. Witzany (ed.), *Biocommunication in Soil Microorganisms*, Soil Biology 23,
DOI 10.1007/978-3-642-14512-4_14,

peptides have been designated as signals for QS-like regulatory cascades in proteobacteria, firmicutes, and yeast. Even though the ability to sense the presence of other bacteria (including siblings) appears to be evolutionarily conserved in different clades of unicellular organisms, the signals used in QS and mechanisms by which they are perceived are distinct [rev. Waters and Bassler (2005), Diggle et al. (2007a), and Dobretsov et al. (2009)]. Since the original discoveries of QS in *Streptomyces* and *Vibrio fischeri* (Khokhlov et al. 1967; Eberhard et al. 1981), QS research has exploded with over 2,400 peer-reviewed publications on the subject cataloged in Pubmed (as of October 2009). Therefore, instead of surveying behaviors controlled by QS in various soil bacteria, this chapter will focus on the mechanisms by which various regulatory inputs (including QS signals) are integrated in well-characterized soil and plant-associated bacteria.

Our goals for this Chapter are to highlight signal integration mechanisms that are common to many proteobacteria and then to identify regulatory mechanisms that are unique to specific systems. Due to space limitations, this review will not be comprehensive. Rather, we will briefly survey what is known about QS in soil microbial communities, then focus on QS in two model bacteria (*Sinorhizobium meliloti* and *Pseudomonas aeruginosa*) and then review the role of orthologs of GacS/GacA in the integration of environmental inputs and QS in γ-proteobacteria.

14.2 Examples of QS in Soil Bacterial Communities

Proteobacteria rely on several types of QS systems to regulate the expression of genes important for structuring of microbial communities, for their interactions with other organisms and eukaryotic hosts, and for production of secondary metabolites. The best-characterized QS signals in Gram-negative bacteria are *N*-acyl-L-homoserine lactones (AHLs). Soil and plant-associated rhizobia, pseudomonads, serratia, chromobacteria, pectobacteria, pantoeae are among those that produce AHL QS signals (Elasri et al. 2001; Hirsch et al. 2003; Von Bodman et al. 2003; Sanchez-Contreras et al. 2007; Van Houdt et al. 2007). AHLs produced by different bacterial species differ in the length of their side chains, degree of saturation, and the decorations at the third carbon of the side chain.

Synthesis of AHLs from acyl-acyl carrier proteins (acyl-ACP) and S-adenosyl methionine (SAM) is typically catalyzed by the homologs of LuxI (Fuqua and Eberhard 1999). Once produced, AHLs with shorter acyl side chains diffuse freely across cell membranes. However, molecules with longer acyl side chains (e.g., 3-oxo-C_{12}-HSL of *P. aeruginosa*) can be actively transported (Pearson et al. 1999). Inside bacterial cells, AHLs are typically detected by the homologs of LuxR proteins (Zhang et al. 2002; Koch et al. 2005). Homologs of LuxR have an AHL-binding N-terminal domain and a DNA-binding C-terminal binding domain. Binding of an AHL to the cognate receptor protein typically leads to the oligomerization of the protein and allows the LuxR-AHL complex to bind its target DNA sequences, "*lux* boxes" (Stevens et al. 1994; Zhang et al. 2002; Bartels et al. 2007). This is the most common mechanism of AHL signal perception in bacteria, although some variations on this common theme have been reported.

Mechanisms of AHL synthesis and perception differ in some bacteria. For example, some species of *Vibrio* can synthesize AHLs via AinS and LuxM enzymes (Gilson et al. 1995; Ng and Bassler 2009). In *V. harveyi* and *V. cholerae*, AHLs are perceived by a hybrid histidine kinase LuxN, which upon perception of an AHL signal contributes to phosphorylation of a phosphorelay protein LuxU, and this eventually sets off the QS regulatory cascade in *V. harveyi* and *V. cholerae* (Ng and Bassler 2009). The AHL receptor, ExpR in *Pantoeae stewartii* functions as a repressor, which upon binding of a cognate AHL is released from DNA sequences within regulated promoters and this allows transcription of the downstream genes (Von Bodman et al. 2003). *Salmonella*, *Klebsiella* and *E. coli* do not synthesize their own AHL signals, but have a functional AHL receptor, SdiA (Ahmer 2004; Kim and Surette 2006; Smith et al. 2008). Even though the function of SdiA during interactions of *Salmonella* with its animal hosts is not yet completely understood, it is clear that the ability to perceive AHLs is dispensable during plant-associated growth (Noel et al. 2010). Collectively, these results indicate that even though the mechanisms of AHL synthesis and perception differ in some bacteria, AHL-mediated QS is widespread in Gram-negative bacteria.

Not all Gram-negative bacteria have AHL-mediated QS, and some use other signals as well as AHLs for QS regulation. Even members of the same family or closely related bacteria may completely differ in regard to their QS machinery (Cha et al. 1998; Elasri et al. 2001; Charkowski 2009). Interestingly, the ability to synthesize AHLs is more common in plant-associated bacteria recovered from the rhizosphere than among bacteria isolated from bulk soil (Elasri et al. 2001). Such heterogeneity in the distribution of QS genes within the genome in closely related bacteria could be explained – at least in part – by the observation that AHL synthases and AHL receptors are often carried on self-transmissible plasmids or integrative and conjugative elements (Marketon and Gonzalez 2002; Cho et al. 2009; Ramsay et al. 2009), or are sometimes found flanked by phage or transposon sequences (Williamson et al. 2005; Wei et al. 2006; Hao et al. 2010), suggestive of horizontal transfer. Curiously, genomes of many bacteria (including *S. meliloti* and *P. aeruginosa*, discussed below in detail) contain "orphan" AHL receptors, which are not encoded next to an AHL synthase (Stover et al. 2000; Pellock et al. 2002; Bartels et al. 2007). Unlike synthase-receptor QS gene clusters that are often carried on plasmids or contained within variable regions of genomes of soil bacteria, "orphan" AHL receptors are typically orthologous in closely related bacteria. These observations raise interesting questions regarding the evolutionary origin and ecological functions of AHL-mediated QS (Lerat and Moran 2004; Diggle et al. 2007a; Hense et al. 2007), questions for which no satisfactory explanations currently exist.

Screens of metagenomic libraries using AHL reporter bioassays have identified AHL synthases encoded next to *luxR* AHL receptors (Williamson et al. 2005; Hao et al. 2010). In addition to a canonical *luxI* AHL synthase, Williamson et al. (2005) also identified a clone homologous to either 2-isopropylmalate synthase or a homocitrate synthase, encoded next to a transcriptional activator. Overexpression of this synthase gene in *E. coli* cells carrying an AHL-responsive *gfp* reporter triggered activation of the AHL GFP reporter, indicating that it likely encoded a novel non-*luxI* AHL synthase or a synthase of an AHL-mimic compound (Williamson et al. 2005). A functional screen for QS inhibitors also identified a clone encoding a cytochrome-like molecule, capable of inhibiting responses of an AHL reporter, although the mechanism of this inhibition is currently unknown (Williamson et al. 2005).

The role of QS regulation in natural environments, where there are no sharp limits to diffusion of QS signals, is much less well characterized. AHL synthase and receptor mutants generally have reduced virulence in both plant and animal pathogens, indicating that such regulation has important functions in the natural environments of these hosts. Likewise there is evidence that AHL-mediated signaling takes places in situ in natural aquatic biofilms and soil microcosms (Gantner et al. 2006; Gao and Teplitski 2008). In the root zone (rhizosphere) of tomato and wheat, *Pseudomonas putida* produced AHLs (as measured by activation of an AHL-responsive GFP reporter). AHLs produced by *P. putida* formed a quorum gradient with highest responses within 5 μm but extending as far as 40–80 μm over the root surface (Gantner et al. 2006). Recombinant-based in vivo expression technology (RIVET) demonstrated that the AHL synthase gene *sinI* in *S. meliloti* was expressed during the colonization of rhizosphere of *M. truncatula* (Gao and Teplitski 2008). Expression of *sinI* was variable at early stages of the rhizosphere colonization, but stabilized once bacteria reached the stage where they are predicted to form microcolonies on the root surfaces (Gao and Teplitski 2008). Collectively, these observations indicate that AHL-mediated QS is common in such natural environments.

14.2.1 The Role of Sinorhizobium meliloti *Quorum Sensing in the Interaction with Its Plant Hosts*

14.2.1.1 Signal Exchange Leading to the Establishment of the *S. meliloti–Medicago* Symbiosis

Most commonly, *S. meliloti* is found in the symbiosis with alfalfa and related legumes from genera *Medicago* and *Trigonella* (Jones et al. 2007). It is also a common soil bacterium, capable of living saprophytically outside of plant hosts. The establishment and maintenance of the symbiosis between *S. meliloti* and its legume hosts is complex and requires precisely timed signal exchange between the partners (Gage 2004; Mitra et al. 2004; Jones et al. 2007). The symbiosis begins with initial contacts via extracellular signals produced by each partner (specific flavonoids from the plant hosts and chitin lipo-oligosaccharide Nod factors from the bacterium) (Mulligan and Long 1985; Peters et al. 1986; Caetano-Anolles et al. 1992). This is followed by the initiation of new root nodule meristems and the formation of infection threads (Long et al. 1982; Jones et al. 2007). Infection threads are tubular cell wall invaginations of host root hairs that carry multiplying *Sinorhizobium* cells into the differentiating nodule tissue. Bacterial exopolysaccharides (EPS) and host lectins play important signaling roles in the infection process (Bauer 1977; Gage 2004; Mitra et al. 2004; Skorupska et al. 2006). Infecting bacteria are released from the threads into envelopes of host plasma membrane, where *Sinorhizobium* differentiates into "bacteroids" that multiply to fill most of the infected host cell. Bacteroid formation and the differentiation of tissues within the nodule generate a zone of symbiotic nitrogen fixation (Timmers et al. 2000; Barnett and Fisher 2006; Capela et al. 2006). *Sinorhizobium* provides the host with fixed nitrogen in the form of NH_4, in exchange for the photosynthetically fixed C, mostly in the form of C4-dicarboxylic acids (Poole and Allaway 2000; Djordjevic 2004). Many of the N-fixing host and bacterial cells subsequently senesce, with renewed multiplication of bacteria in older nodules (Timmers et al. 2000). The establishment and maintenance of this symbiosis provides many opportunities for QS regulation and for host manipulation of QS.

QS regulation is common in rhizobia (Gonzalez and Marketon 2003; Sanchez-Contreras et al. 2007). Mutants in various QS-regulated genes have proven to be defective in nodule initiation (Gao et al. 2005). In addition, cell division, symbiotic plasmid transfer, gene expression in the rhizosphere, symbiosome development and N-fixation, and nodule number are QS-regulated in *Rhizobium* species [rev. Gonzalez and Marketon (2003) and Sanchez-Contreras et al. (2007)]. In general, QS mutants of rhizobia can still infect and form nodules on their legume hosts. This suggests that the primary role of QS in rhizobia is to optimize many aspects of host interaction rather than determine any single essential step.

14.2.1.2 QS Signal Generation During *S. meliloti–Medicago* Interactions

QS in *Sinorhizobium* is complex and consists of multiple AHL signals and receptors. In laboratory shake cultures, the genomically sequenced strain 1021 and its derivative 8530 produce a diversity of AHLs with long acyl chains: C_{12}-HSL, C_{14}-HSL, 3-oxo-C_{14}-HSL, C_{16}-HSL, $C_{16:1}$-HSL, 3-oxo-$C_{16:1}$-HSL, and C_{18}-HSL (Marketon et al. 2002; Teplitski et al. 2003a; Gao et al. 2005). *S. meliloti* strains also produce QS-active substances that are chemically distinct from AHLs in that they are more hydrophyllic than known AHLs and are resistant to hydrolysis by AiiA, an AHL lactonase (Gao et al. 2007). In strains 1021 and 8530, all AHLs are produced in vitro by the sole AHL synthase, SinI, which is encoded on the chromosome (Gao et al. 2005), although the presence of a second cryptic AHL synthase has not been completely ruled out (Marketon et al. 2002). In addition to *sinI*, *S. meliloti* AK631 also contains another AHL synthase encoded on a plasmid (Marketon and Gonzalez 2002), which directs synthesis of AHLs with short side chains (Marketon and Gonzalez 2002; Teplitski et al. 2003a).

To find out whether QS signal synthesis is activated in *S. meliloti* during symbiosis with *M. truncatula*, regulation of merodiploid reporters (*sinI*+ *sinI::gus* and *sinI*+ *sinI::tnpR-gus*) was tested during different stages of the symbiosis (Gao and Teplitski 2008). When the expression of *sinI* was quantified using the RIVET reporter, it was determined that the expression of *sinI* was variable during the early stages of rhizosphere colonization, then reached a plateau within a week (Gao and Teplitski 2008), a time period sufficient to form steady-state population densities in the rhizosphere (Caetano-Anolles et al. 1992). The *sinI* promoter was induced strongly in older parts of 3-week-old nodules, indicating that SinI AHLs are actively synthesized during symbiosis, but not at detectable levels in the tip region where the nodule meristem is located (Gao and Teplitski 2008). Thus, the synthesis of AHL QS signals may be limited in younger nodule tissues and in the rhizosphere near the growing tip where infections take place. This could be biologically relevant. The observation that *sinI* is expressed in *M. truncatula* nodules is consistent with the earlier report that QS signals are produced by *R. etli* bacteroids in bean nodules (Daniels et al. 2002). Transcriptomic analyses indicate that *sinI* was expressed in the bacteroids formed by the *expR*- strain *S. meliloti* 1021 (Barnett et al. 2004). Interestingly, despite the fact that the *sinI* transcriptional fusions were strongly expressed in the nodules formed by the wild type *S. meliloti* (Gao and Teplitski 2008), qualitative RT-PCR assays indicate that the accumulation of the *sinI* mRNA was downregulated inside the nodules (Gurich and Gonzalez 2009). Technical limitations of the two assays aside, collectively these results could suggest posttranscriptional regulation of the *sinI* mRNA stability in *expR*-dependent manner. Evidence that levels of *sinI* mRNA (as indicated by both microarray and quantitative RT-PCR analyses) decrease in the *hfq* mutant in vitro (M. Gao and M. Barnett, manuscript in preparation) supports the hypothesis that posttranscriptional regulation of *sinI* by RNA binding proteins like Hfq may affect QS during normal symbiotic interactions between the bacterium and its plant hosts, adding another level of regulatory complexity to QS regulation in the host.

14.2.1.3 Responses of *S. meliloti* to AHL QS Signals

In *S. meliloti* strain 8530, there are two known AHL receptors (SinR and ExpR) and perhaps several additional, but untested AHL receptors capable of recognizing AHLs (Pellock et al. 2002; Bartels et al. 2007). The SinR receptor modulates expression of ~30 genes, including *sinI*, the AHL synthase (Marketon et al. 2002; Hoang et al. 2004; Gao et al. 2005). The ExpR receptor is a functional "orphan" AHL receptor (Bartels et al. 2007). Orthologs of *expR* are present in the genomes of *S. medicae*, *Rhizobium leguminosarum*, and *Agrobacterium tumefaciens*. In *S. meliloti* 1021, *expR* is interrupted by a native insertion sequence, but this sequence excises spontaneously to provide the functional AHL receptor in the strain 8530 (Pellock et al. 2002). ExpR regulates surface swarming and activates synthesis of a symbiotically important EPSII (Becker et al. 1997; Pellock et al. 2002; Gao et al. 2005). Transcriptome studies indicate that ExpR is involved in regulating the expression of about 140 genes under laboratory shake culture conditions (Hoang et al. 2004; Gurich and Gonzalez 2009). Parallel proteome studies identified additional 50 proteins subject to ExpR-dependent QS regulation (Gao et al. 2005). In addition to the ~200 genes/proteins that are QS-regulated in laboratory cultures via the ExpR receptor, another 200 proteins have been identified as QS-regulated by AHL receptors other than ExpR (Chen et al. 2003; Hoang et al. 2004; Teplitski et al. 2004; Gao et al. 2005).

One of the most surprising results from the global studies of QS in *S. meliloti* is the lack of overlap between the sets of QS-regulated genes/proteins. Only about 5% of the genes/proteins identified as QS-regulated were seen in more than one study. Different culture methods, different receptor backgrounds, and different means of distinguishing QS-regulated genes/proteins probably account much of this lack of overlap. But the lack of overlap between these several studies teaches us two important things: first, that the we are far from identifying the full range of QS-regulated functions in *Sinorhizobium* and the contribution of posttranscriptional regulatory systems (like those mediated by an RNA binding protein Hfq); and second that further analysis of QS in laboratory cultures would not be very useful in helping us understand the roles of QS in the real biology of the bacterium.

14.2.1.4 Plant Hosts Detect Rhizobial AHLs and Manipulate Bacterial Signaling

Since bacterial pathogens and symbionts rely on QS regulation to optimize their infection of host organisms, it is not surprising that eukaryotic hosts have evolved the means to take advantage of this dependence. Eukaryotes have different mechanisms to deal with QS in the bacteria they encounter (Teplitski et al. 2010). Both plants and animals can inactivate AHLs (please see a chapter by Dessaux et al. in this book). Plant hosts, including *M. truncatula*, can detect bacterial AHLs and activate potent defense responses (Mathesius et al. 2003; Schuhegger et al. 2006). These extensive and specific responses of the host to the AHL signals from bacteria are likely an integral part of the QS-related interplay between *Sinorhizobium* and *M. truncatula*.

Plants, algae, and fungi also produce a variety of compounds that disrupt or manipulate bacterial QS circuitry and affect QS-regulated behaviors in situ. In pioneering studies, the marine red seaweed *Delisea pulchra* was found to produce a set of 20–30 different halogenated furanones. These algal furanones inhibit QS in Gram-negative bacteria (Givskov et al. 1996). Seedlings of many plants, including pea, rice, tomato, soybean, and *M. truncatula*, secrete compounds that stimulate or inhibit AHL receptor-dependent responses in QS reporter bacteria (Teplitski et al. 2000; Gao et al. 2003, 2007; Mathesius et al. 2003; Karamanoli and Lindow 2006). The plant compounds are therefore able to mimic bacterial AHL QS signals, and they appear to be secreted at levels that would affect QS regulation in associated bacteria during natural encounters (Teplitski et al. 2000), although not all bacterial AHL receptors are susceptible to manipulation by plant AHL signal-mimics in situ (Gantner et al. 2006). Pea and *M. truncatula* were found to produce 10–20 chromatographically separable AHL mimic compounds. Most of these AHL mimic compounds were preferentially extracted into methanol rather than ethyl acetate, indicating that they are chemically different from bacterial AHLs (Teplitski et al. 2000; Gao et al. 2003).

In addition to AHL-mimics, L-canavanine, a compound commonly found in exudates of alfalfa, was shown to inhibit QS-mediated behaviors in reporter strains and ExpR-responsive genes in *S. meliloti* (Keshavan et al. 2005). L-canavanine is an analog of arginine, and a strong inhibitor of bacterial arginine deaminase (Lu et al. 2005; Li et al. 2008). It is likely that down-regulation of QS responses may result from a general toxicity or could be an indirect consequence of the production of O-ureido-L-homoserine, a known intermediary produced from L-canavanine via a reactive thiouronium intermediate (Li et al. 2008). Because the addition of L-arginine at least partially reversed QS inhibitory effects of L-canavanine (Keshavan et al. 2005), the indirect effect of L-canavanine on metabolism and/or protein synthesis (Keshavan et al. 2005) is one explanation of its biological activity in QS assays.

The discoveries that rhizobia produce AHLs during normal interactions with the legume hosts, and that the plant hosts detect bacterial AHLs, respond to them and also produce compounds capable of modulating bacterial QS behaviors indicate that the bi-directional exchange of AHL signals may function as another regulatory layer in the establishment of this tightly coevolved symbiosis. In contrast to the red algae, which only produce inhibitory halogenated furanones, the ability of legumes to produce stimulatory AHL mimics, AHL-degrading enzymes, and compounds that directly or indirectly inhibit QS suggests that these plants may use their battery of QS-related signals and proteins to manipulate QS in the bacteria they encounter rather than just broadly suppressing QS.

14.2.1.5 *S. meliloti* Responds to Non-AHL QS Signals from Other Microbes

In soils, sinorhizobia are exposed to signals from other bacteria and eukaryotes. Recent evidence suggests that even though *S. meliloti* does not produce some of the known chemical classes of QS signals, it recognizes and responds to non-AHL QS signals from other soil microorganisms. For example, in addition to AHL QS signals, most Gram-negative bacteria and firmicutes produce furanone signals collectively known as autoinducer-2 (AI-2) [rev. Waters and Bassler (2005)]. Because chemical structures of AI-2 produced by different bacteria are closely related and because these signals appear to be wide spread in different clades of bacteria, their function in interspecies bacterial communication has been intensively studied [e.g., rev. Waters and Bassler (2005)].

Even though *S. meliloti* does not produce AI-2 signals, nor do published genomes of sinorhizobia contain homologues of the AI-2 synthase *luxS*, it detects and responds to AI-2 signals from other bacteria (Pereira et al. 2008). The symbiotic megaplasmid B (pSymB) of *S. meliloti* 1021 contains an *ait* operon (*autoinducer-2 transporter*, Smb21023-Smb21016), which is organizationally similar and functionally homologous to the AI-2 transporter operon *lsr* from *Salmonella enterica* (Pereira et al. 2008). The protein encoded by Smb21016 binds to a synthetic AI-2

((*2R*, *4S*)-2-methyl-2,3,3,4-tetrahydroxytetrahydrofuran), as demonstrated by crystallography and by the *aitK*- and *aitA*-dependent removal of exogenously supplied AI-2 to the growth medium with the associated changes in the transcription of the wild-type *ait* operon (Pereira et al. 2008). Consistent with the lack of AI-2 synthesis by the wild-type *S. meliloti*, coinoculation of the wild type and AI-2-insensitive strain onto *Medicago* seedlings demonstrated no effect of the *ait* mutations on seedling colonization (Pereira et al. 2008). In its ability to detect and respond to AI-2 from other bacteria, *S. meliloti* is similar to *P. aeruginosa*, which also does not produce but responds to AI-2 signals from other bacteria (Duan et al. 2003). Now that it is clear that *S. meliloti* responds to the synthetic and bacterially produced AI-2 signals (Pereira et al. 2008), it will be interesting to delineate the role of this behavior during the interactions of *S. meliloti* with AI-2-producing soil microbiota.

Within soil microbial communities, *S. meliloti* likely encounters QS-related signals produced by eukaryotic microbes. Proteomic and gene expression studies using synthetic signal analogs and HPLC-purified fractions containing eukaryotic signals, preliminary screens using 10 mM farnesol, a compound known to effect QS-like responses in dimorphic yeast (Hornby et al. 2001), identified promoters for *ndvA* and *bhbA* as those up-regulated by farnesol in the *sinI*-strain of *S. meliloti* (M. Gao and K. Jogelekar, unpublished). In another study, treatment of *S. meliloti* 1021 cultures with an HPLC-purified fraction containing an active compound produced by the soil alga *Chlamydomonas reinhardtii* affected accumulation of 34 polypeptides. This fraction contained an activity capable of stimulating the LasR AHL receptor (Teplitski et al. 2004). Many of the polypeptides accumulated in response to an activity from *C. reinhardtii* were also accumulated in response to AHL signals produced by the wild-type cultures of *S. meliloti*, indicating that the purified AHL mimic compound produced by *C. reinhardtii* affects QS in the wild-type soil bacteria (Teplitski et al. 2004). The chemical structure of this algal signal is currently unknown (Teplitski et al. 2004). However, similar studies identified lumichrome as a stimulatory signal produced by *C. reinhardtii* (Rajamani et al. 2008). The role of lumichrome in the interactions between soil algae and bacteria is discussed in the accompanying Chap. 16.

14.2.2 QS in **Pseudomonas aeruginosa,** *a Model Environmental Bacterium and Opportunistic Pathogen*

P. aeruginosa is a model environmental bacterium and an important opportunistic pathogen of animals, humans, and plants. The adaptation of this microorganism to such variety of environmental niches is reflected in the large size of its genome and by the large proportion of genes devoted to transcriptional regulation and signal transduction (Stover et al. 2000). These include functions encoding complex QS multisignaling systems. The current understanding is that in this microorganism QS is mediated by AHLs synthesized by LasI (a 3-oxo-C_{12}-HSL synthase) and RhlI (a C_4-HSL synthase) and by 2-alkyl-4-quinolone (AQ) signal molecules like the *Pseudomonas* quinolone signal, PQS (2-hepty-2-hydroxy-4-quinolone), and HHQ (2-heptyl-4-quinolone) (Xiao et al. 2006a, b; Williams and Camara 2009). This last class of signals includes more than 50 AQS produced by *P. aeruginosa* upon expression of the *pqsABCDE operon*. Figure 14.1 summarizes the complex epistatic interactions between these QS systems. AHL signaling depends on the *lasRI* and *rhlRI* genes, where LasR and RhlR are LuxR-type transcriptional regulators binding 3-oxo-C_{12}-HSL and C_4-HSL, respectively. The *lasIR* and *rhlIR* systems are hierarchically arranged, and *lasI* is autoregulated by LasR, and LasR activates *rhlRI*, which is in turn autoregulated. A negative autoregulatory loop is also triggered by LasR binding of the RsaL repressor (Rampioni et al. 2007).

In *P. aeruginosa*, 10% of the genome transcriptionally responds to AHL signals (Schuster et al. 2003; Wagner et al. 2003). Transcriptome and genetic studies have identified many direct and indirect targets for the LasR and RhlR regulators (Whiteley and Greenberg 2001). LasR activates virulence factors such as elastase (LasB), proteases (LasA), exotoxin A (toxA), and type II secretion (XcpR, XcpP). RhlR activates the production of rhamnolipids, but it is also necessary for the optimal production of LasB elastase, LasA protease, pyocyanin, and cyanide in vitro [reviewed in Williams and Camara (2009)]. The differential regulation of QS target genes depends on the recognition of conserved *las*, *rhl* boxes acting as binding sites for either or both regulators. The LasIR system has been initially shown to be involved in the differentiation of *P. aeruginosa* biofilms. For instance, a mutant defective in 3-oxo-C_{12}-HSL production forms an abnormal, flat, and undeveloped biofilm that is sensitive to low concentrations of the detergent sodium dodecyl sulfate (SDS) (Davies et al. 1998). The Las system, and to some extent the Rhl system, was later shown to contribute to the expression of the *pel* polysaccharide, one of the components of the *P. aeruginosa* pellicle in model biofilms (Sakuragi and Kolter 2007). More controversial is the role of *rhlIR* in model biofilm production, as rhamnolipids trigger Pseudomonas detachment from biofilms (Boles et al. 2005), but rhamnolipids have been shown to play a role in various aspects of biofilms development, including microcolony formation, maintenance of open channels, mushroom cap formation and dispersion from the biofilm (Davey et al. 2003; Espinosa-Urgel 2003).

The AHL-based QS is further complicated by the presence of two orphan LuxR-type regulators, QscR and VqsR. QscR has its own regulon of more than 400 genes. QcsR can function as either a repressor or an activator, but it has not been shown to bind AHLs (Dong et al. 2005; Juhas et al. 2005). VsqR is required for AHL production, it has been shown to bind 3-oxo-C_{12}-HSL and to regulate *lasI* (Fig. 14.1).

The other branch of QS in Pseudomonas involves AQS molecules produced by the *pqsABCDE* operon under the positive control of PqsR (MvfR), a LysR regulator activated by LasR. PSQ and HHQ (a precursor of PQS) signals both act as coinducers of PqsR (Diggle et al. 2006, 2007b; Xiao et al. 2006a; Coleman et al. 2008; Farrow et al. 2008, and Fig. 14.1). Exported HHQ is taken up by adjacent bacterial cells and converted into PQS by PqsH, a putative mono-oxygenase (Diggle et al. 2006;

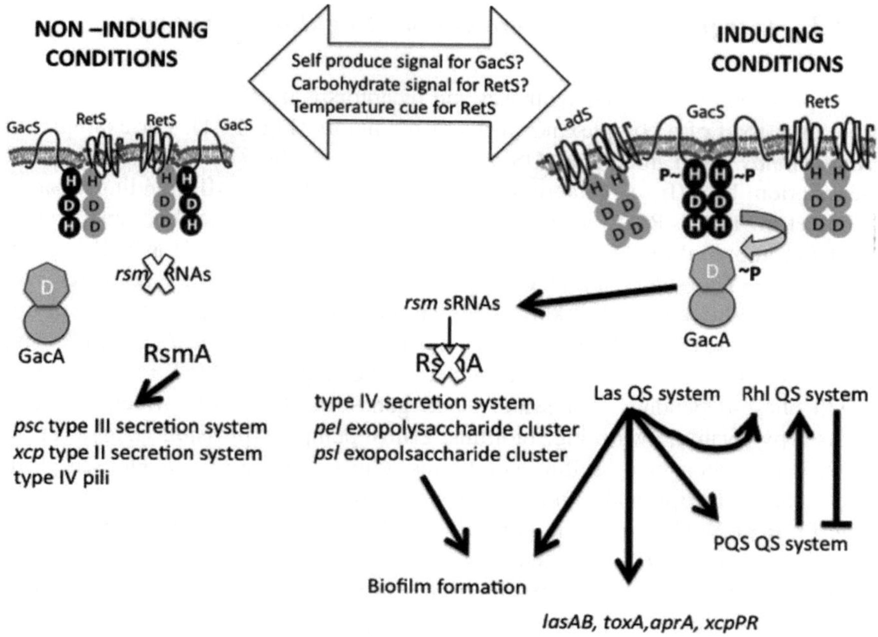

Fig. 14.1 Signal integration in pseudomonads. Under noninducing conditions, the GacS–GacA regulatory cascade is blocked by binding of RetS to GacS. Under these conditions, *rsm* sRNAs are not produced. Under inducing conditions, GacS autophosphorylates and then transphosphorylates GacA (Asp and His residues involved in phosphotransfer are indicated as D and H, respectively; ~P indicates phosphate group). Phosphorylated GacA induces transcription of *rsm sRNAs*, which titrate RsmA. Under these conditions, quorum sensing genes are induced and production of biofilm is favored. Las QS system is mediated by 3-oxo-C_{12}-HSL, Rhl QS system is mediated by C_4-HSL and the PQS system requires quinolone signals. In addition to Gac-Rsm system (and not shown on the diagram), Las QS system is positively regulated by an orphan AHL receptor VqsR, cAMP-binding protein Vfr, polyphosphate kinase Ppk and DksA, a transcriptional factor that interacts with RNA polymerase through a secondary channel. Negative regulation of Las QS system is effected by a global transcriptional regulator AlgR2, HN-S like regulator MvaT, and an orphan AHL receptor QscR

Xiao et al. 2006a; Bera et al. 2009). PqsR positively feeds back on *rhlR* expression. A direct negative regulatory feedback involves RhlR, which represses both *pqsR* and *pqsABCDE*. Mutations in either QS systems reciprocally reduce expression and therefore reduce virulence, as PQS signaling also affects elastase, *lecA*, pyocianin, biofilm formation. Transcriptome studies have revealed that PQS induces an iron starvation response upregulating genes involved in iron scavenging and siderophore biosynthesis; PQS can directly chelate ferric iron at high affinity, probably trapping the iron complexes outside the cell envelope (Bredenbruch et al. 2006; Fletcher et al. 2007). PQS also interacts with *P. aeruginosa* envelope causing the release of membrane vesicles, which plays a role in cell autolysis in biofilms (Baysse et al. 2005; Diggle et al. 2007b).

14.2.3 Integration of QS into Global Regulatory Networks

The expression of QS signal synthases and receptors is itself tightly controlled and at the crossroad of a complex networks of regulators (Fig. 14.1). For example, in *Pseudomonas*, AHL-mediated QS circuits are controlled at the level of *lasR* transcription by Vfr, a catabolite regulator and cyclic-AMP binding protein (Albus et al. 1997). RsaL, a regulatory gene divergently transcribed upstream of the *lasI* promoter, directly represses *lasI* and is itself controlled by LasR (Rampioni et al. 2007). The stationary phase sigma factor RpoS negatively regulates *rhlI* expression (Whiteley et al. 2000), while an alternate sigma factor RpoN can up- or down-regulate *rhlI* and functions as a negative regulatory of *lasI* (Heurlier et al. 2003; Thompson et al. 2003). The nucleoid-like regulator MvaT is involved in growth phase-dependent expression of QS genes and behaviors affecting LecA, elastase, pyocyanin, swarming, biofilms, efflux pumps expression (Diggle et al. 2003). Under phosphate-limiting conditions, PhoB stimulates *rhlR* (Jensen et al. 2006). The transcriptional regulators DksA, AlgR, PtxR negatively controls *rhlI*, while PtxR or VqsM positively control *lasI* (Dong et al. 2005). Finally, positive regulation of *lasR*, *rhlR* is exerted by GacS/GacA, a two component signal transduction system acting as global regulator in γ-proteobacteria. Among other effects, mutations in this two-component system reduce production of the AHL signals and accumulation of the mRNAs encoding the AHL synthase and AHL receptor (Parkins et al. 2001; Quinones et al. 2004). The contribution of the GacS/GacA to the regulation of QS appears to be evolutionarily conserved in γ-proteobacteria and is further discussed below.

14.3 GacS/GacA Is a Two-Component System Controlling Environmental Adaptation, Biofilm Formation, and Motility in γ-Proteobacteria

14.3.1 Discovery of GacA and GacS in γ-Proteobacteria

gacA (*global antibiotic and cyanide control*) was first discovered in biocontrol pseudomonads in screens for mutants defective in the production of fungicidal secondary metabolites (Laville et al. 1992). The regulatory role for *gacA* orthologs in controlling production of toxins and antibiotic compounds is conserved in many soil and host-associated bacteria from the genus *Pseudomonas* [rev. Haas et al. (2002) and Teplitski and Ahmer (2004)]. In *P. aeruginosa*, *gacA* mutants were identified as less virulent in nematodes, plants, and mice (Tan et al. 1999). In the tomato pathogen *Pseudomonas syringae* pv *tomato* DC3000 (*Pst* DC3000), disruption of *gacA* reduced the virulence of the bacterium and reduced its ability to elicit a hypersensitive (apoptosis-like) response in plants. In several strains of *P. syringae*, *gacA* orthologs control the production of *N*-acyl homoserine lactone signals and other traits associated with fitness (Chatterjee et al. 2003; Quinones et al. 2004). Therefore, in pseudomonads *gacA* orthologs have a global regulatory role in controlling behaviors related to environmental fitness and adaption.

In various pathogenic bacteria, *gacA* orthologs were identified in screens of mutants defective in aspects of virulence. *Salmonella sirA*, for example, was identified in a screen for mutants with reduced expression of genes encoded within the *Salmonella* Pathogenicity Island 1 (SPI-1) (Johnston et al. 1996). Orthologs of *gacA* are also present in symbiotic and saprophytic γ-proteobacteria. In the soil bacterium, *Azotobacter vinelandii gacA* controls production of alginate (Castaneda et al. 2001). In *Vibrio fischeri*, *gacA* is required for colonization of the squid host (*Euprymna scolopes*) and regulates genes involved with chemotaxis, substratum-dependent motility, and production of *N*-acyl homoserine lactone signals (Whistler et al. 1998; Whistler et al. 2007). Of all the sequenced genomes of γ-proteobacteria (as of October 2009), *gacA* orthologues are only absent from genomes of some insect endosymbionts: genomes of these bacteria have undergone considerable reduction (Gil et al. 2004). This suggests the central role for GacA in all γ-proteobacteria in controlling behaviors that are possibly related to environmental sensing and adaptation.

GacA orthologs are "orphan" response regulators, that is, unlike many two-component response regulators, they are not encoded near a sensor kinase. Of the published genome sequences, the *gacA* stop codon either overlaps the *uvrC* start codon or lies just within the *uvrC* coding sequence in *Salmonella*, *Photobacterium*, *Vibrio*, *Yersinia*, *Serratia*, and *Pectobacterium*. The *gacA-uvrC* tandem gene arrangement is conserved throughout the γ-proteobacteria, with the exception of *Xanthomonas* spp. and *Xylella* spp. These genetic arrangements suggest that *uvrC* and *gacA* are often translationally coupled. Indeed, *E. coli gacA* ortholog (*uvrY*) was identified as an ORF containing regulatory elements for *uvrC*, a gene involved in nucleotide excision repair (Moolenaar et al. 1987). Similarly, in *P. fluorescens*, *P. aeruginosa*, *E. coli*, and *Pectobacterium carotovorum*, polar mutations in *gacA* result in altered UV tolerance, because a portion of the *gacA* transcript extends through *uvrC* (Moolenaar et al. 1987; Laville et al. 1992; Reimmann et al. 1997; Eriksson et al. 1998). However, *gacA* orthologs themselves do not function in UV resistance. The genomic regions upstream of *gacA* orthologs are much more variable than the downstream regions [rev. Teplitski and Ahmer (2004)]. Among the genera *Salmonella*, *Klebsiella*, and *Escherichia*, the *gacA* orthologs are located downstream of *sdiA* (but within a separate transcriptional unit). SdiA is a LuxR-type AHL receptor used by these bacteria to detect the quorum sensing signals of other bacterial species (Ahmer 2004). However, this arrangement is not conserved in other genera.

Because *gacA* is not encoded next to a sensor kinase, genetic (and later, biochemical) evidence was used to establish a link between *gacA* and a sensor kinase encoded by *gacS*. The first identified *gacS* ortholog ("*lemA*" for "non lesion forming") was detected in *P. syringae* pv *syringae* based on the inability of the mutant to cause disease lesions on bean leaves (Hrabak and Willis 1992;

Kitten et al. 1998). The *gacS* ortholog of a free-living soil nitrogen fixing bacterium *A. vinelandii* was identified in a screen of clones that complemented a nonmucoid (alginate-deficient) phenotype (Castaneda et al. 2000). Consistent with the postulated GacS–GacA interaction, various mutant screens identified *gacS* mutants as having phenotypes similar to those of the *gacA* mutants (Chancey et al. 1999).

14.3.2 GacS–GacA Signal Transduction

14.3.2.1 Structure/Function Analysis of GacS Orthologs

GacS orthologs are highly conserved among the γ-proteobacteria [Fig. 14.2 and Heeb and Haas (2001)]. The *gacS* gene encodes an unorthodox sensor kinase with a HAMP phosphatase domain (residues 188–240), histidine kinase A (dimerization/phosphoacceptor) domain (residues 281–344), HATPase_c domain (Histidine kinase-like ATPase, residues 400–507); REC (signal receiver domain, residues 668–784) and a histidine phosphotransfer (HPT) domain (residues 860–910). Similar to other two-component sensor kinases, these three domains have conserved H295, D718, and H861 residues, predicted to have a role in autophosphorylation (H295) and phosphotransfer to GacA (Zuber et al. 2003). Corresponding single amino acid replacements in *Pseudomonas* GacS (H294, D717, H863) resulted in a *gacS* null phenotype, suggesting that these residues play an important role in affecting downstream gene regulation (Zuber et al. 2003). GacS orthologs have a predicted periplasmic loop (M33 to L162 of GacS) and the deletion of the predicted periplasmic domain did not affect the in vivo function of *gacS* (Zuber et al. 2003). Similarly, the ability to auto- and trans-phosphorylate in vitro was not affected by a deletion of the first 198 residues of BarA (an ortholog of GacS in *Salmonella* and *E. coli*) (Pernestig et al. 2001; Teplitski et al. 2003b, 2006).

In *P. fluorescens*, the histidine kinase (HisKA), the ATPase, and the receiver REC domains of GacS were all needed for the interactions with GacA, based on the results of a bacterial two-hybrid assay (Workentine et al. 2009). The 60 amino acid residues immediately upstream of the REC domain were also required for the interactions of GacS with GacA (Workentine et al. 2009). The HPT domain was dispensable in the interactions of GacS with GacA (Workentine et al. 2009). Only HAMP domain was required for the interactions of two GacS proteins (Workentine et al. 2009). Deletion of the M190-E270 residues of HAMP domain and their replacement with Glu-Ala-Phe resulted in a signal-independent ON *gacS* (Zuber et al. 2003).

Important functional differences exist between GacS orthologs of pseudomonads and enteric bacteria. In pseudomonads, GacS interacts with two other sensor kinases, LadS and RetS (see below). This interaction leads to either inhibition or activation of GacS autokinase activity, presumably upon perception of unknown signals. Direct interaction between GacS and LadS or RetS has been shown (Goodman et al. 2009; Workentine et al. 2009). Enteric bacteria, such as *E. coli*,

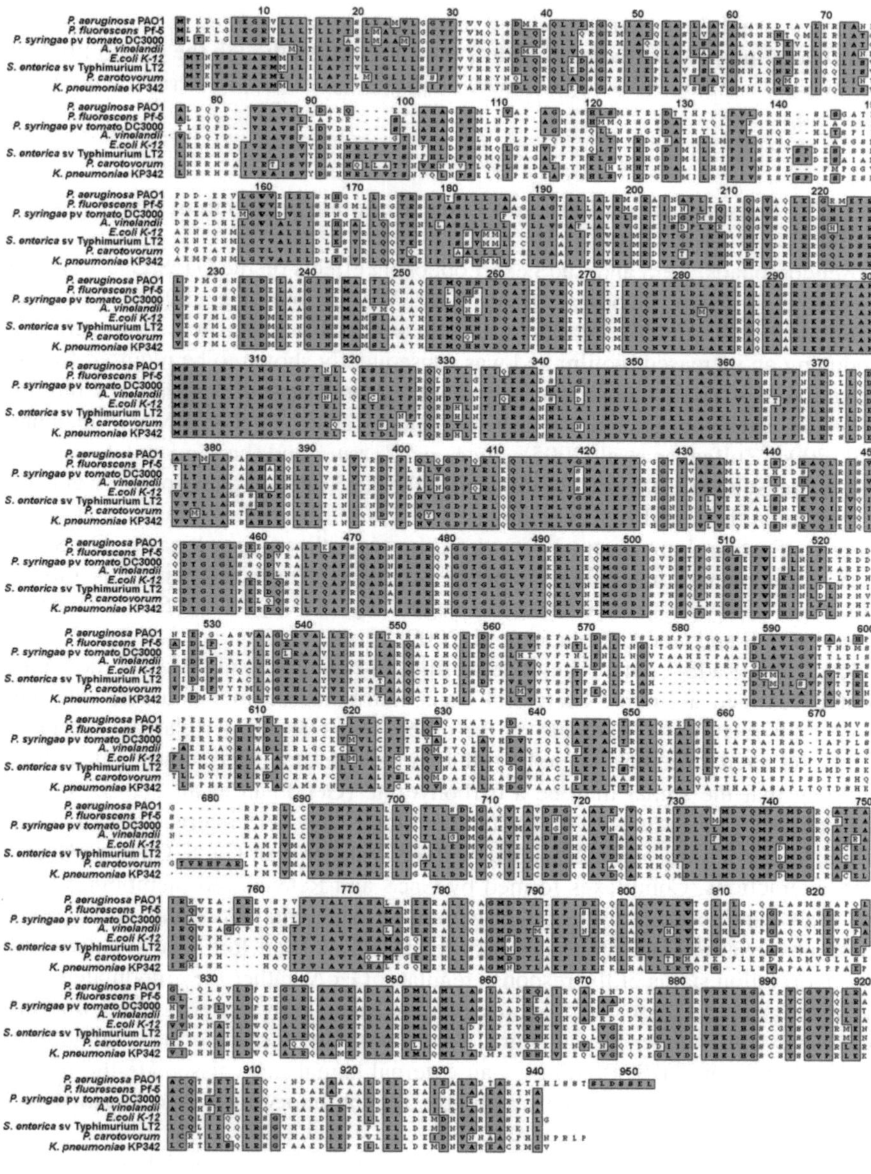

Fig. 14.2 CLUSTALW alignment of GacS orthologs. GacS is an unorthodox sensor kinase with a HAMP phosphatase domain (residues 188–240), histidine kinase A (dimerization/phosphoacceptor) domain (residues 281–344, with a conserved H295), HATPase_c domain (Histidine kinase-like ATPase, residues 400–507); REC (signal receiver domain, residues 668–784, with a conserved D718) and a histidine phosphotransfer (HPT) domain (residues 860–910, with a conserved H861). GacS orthologs from *P. aeruginosa* PAO1 (GenBank accession # NP_249619), *P. fluorescens* Pf-5 (#YP_261539), *P. syringae* pv *tomato* DC3000 (#NP_791516), *A. vinelandii* (#AAF04854), *E. coli* K12 (#YP_003035194), *S. enterica* sv Typhimurium LT2 (#NP_461879), *Pectobacterium carotovorum* (#CAA74020), and *K. pneumoniae* KP342 (YP_002236863) are shown

Salmonella, and their plant-associated relatives (*Klebsiella*, *Erwinia*, *Pantoea*, and *Pectobacterium*) lack orthologs of RetS and LadS. Even though the possibility that functionally similar sensor kinases may be present in enteric bacteria has not been ruled out.

14.3.2.2 Sensor Kinases RetS and LadS Modulate Function of GacS

In pseudomonads, RetS and LadS exert opposing effects on the GacS/GacA system (Ventre et al. 2006; Goodman et al. 2009; Workentine et al. 2009). The gene encoding the unorthodox hybrid sensor kinase RetS was isolated from a mutant producing an increased biofilm and was subsequently shown to be required for the activation of a type 3 secretion system, concomitant with the repression of biofilm formation, and colonization/dissemination in murine acute infection models (Goodman et al. 2004; Zolfaghar et al. 2005). *gacA*, *gacS*, *rsmA* and the diguanylate cyclase *sadC* were isolated as suppressor mutants of the RetS phenotype, implying a genetic link between them and *retS* (Goodman et al. 2004). It was later shown that genes under RetS control are inversely regulated by two other sensor kinases, GacS and LadS (Goodman et al. 2004; Ventre et al. 2006). Consistent with this model, microarray analyses revealed that the regulatory activity of RetS is channeled through GacS and GacA (Brencic et al. 2009).

RetS and LadS contain putative periplasm-facing signal receiver domains, located between a single N-terminal trans-membrane segment on one side and seven trans-membrane segments on the other. This class of periplasmic receptors is hypothesized to assume a jelly roll folding conformation as seen in lectins, implying that the signal may be a carbohydrate (Goodman et al. 2009). Neither LadS nor RetS interacts with GacA, but both LadS and RetS interacted with GacS, although interactions of GacS with LadS were weaker than GacS–GacS or GacS–RetS interactions. Complexes formed by GacS and RetS in cell membranes were very stable and resistant to treatment with mild detergents (Goodman et al. 2009). These observations indicate that sensor kinases LadS and RetS do not affect the phosphorylation status of the response regulator GacA, rather they modulate activity of the sensor kinase GacS in response to an unknown signal. One working hypothesis is that under ground signal conditions, RetS could form a homodimer, sequestered from GacS. Binding of an external signal may dissociate the homodimer and make RetS monomers available for inhibitory interactions with GacS (Fig. 14.1 and (Goodman et al. 2009).

The RetS/GacS system constitutes a novel paradigm for bacterial signal transduction systems. Contrary to what other known sensor kinase do, RetS directly regulates the activity of GacS via formation of heteromeric membrane complexes (Goodman et al. 2009). This activity is consistent with the absolute dispensability of its conserved phosphorelay residues of RetS (Goodman et al. 2009). Sensor kinases involved in multistep phosphorelay reactions, such as GacS, share a conserved modular architecture and a common mechanism of phosphate transfer. They function as homodimers in which the g-phosphate from ATP bound by one monomer is transferred intermolecularly to the conserved histidine on the histidine kinase domain of the opposing monomer. The experiments described by Goodman et al. (2009) are consistent with a model in which RetS blocks GacS phosphorylation at this very early step of the signal transduction pathway, by disrupting the formation of GacS homodimers and instead forming a nonproductive GacS:RetS heterodimer or a higher-order structure of GacS and RetS homodimer (Fig. 14.1 and Goodman et al. 2009).

RetS and GacS are constitutively coexpressed under most conditions [Goodman et al. (2009) and references therein]. Therefore, the effect of RetS on GacS must come at the posttranscriptional or posttranslation level, likely in response to a yet unknown signal. For example, in the pathogenesis model for *P. aeruginosa* PAO1, RetS is hypothesized to perceive a cue important for the switch between acute and chronic infections (Goodman et al. 2009). In fluorescent soil pseudomonads, RetS was responsible for the temperature-dependent regulation of sRNAs and the downstream genes involved in the production of secondary products, while effects of LadS were independent of the temperature (Humair et al. 2009).

It appears, therefore, that the Gac signaling network in pseudomonads consists of RetS, LadS, and GacS/GacA and controls the expression of a significant number of genes at the level of mRNA translation and/or stability. Taken together, these genetic approaches identified multiple components of a molecular switch controlling adaptation to various niches relevant to its pathogenesis and adaptation to different environmental niches.

14.3.3 Structure/Function Analysis of GacA Orthologs

Similarly to other FixJ/LuxR-type transcriptional regulators, GacA functions as a dimer (as demonstrated by the bacterial two-hybrid experiments (Workentine et al. 2009). Unlike LuxR AHL receptor proteins (multimerization of which is facilitated by binding of a ligand directly to the N-terminus of the protein), orthologs of GacA become active once phosphorylated by a sensor kinase GacS (as discussed above). The crystal structure for NarL, a member of the FixJ family of regulators, has been determined (Baikalov et al. 1996). By comparison with NarL, GacA orthologs are predicted to have an N-terminal receiver domain and a C-terminal DNA binding domain (Baikalov et al. 1996). The two domains are connected by an a-helix (a6, residues S127–S136). The N-terminal receiver domain (M1-S127) is comprised of five parallel b-sheets surrounded by five a-helices (Baikalov et al. 1996). D54 is the predicted phosphorylation site. In some two-component response regulators a D54E mutation leads to a constitutive ON (kinase-independent) phenotype by mimicking the phosphorylated state of the aspartate residue. However, in most cases D54E mutants are unable to accept the phosphate group and have a null phenotype.

Several amino acid residues are conserved in the N-terminus of all FixJ-type regulators. The following regions of GacA are highly conserved and are predicted to interact with the phosphorylation site: D8-D9, S34-C35, P58-I61, T82-E86, S103-A107 (Fig. 14.3). Consistent with the prediction that these conserved loops interact with the phosphorylation site, C86T, C84T, and G105R yielded a *gacS*-independent, constitutively ON, GacA in *P. fluorescens* (Gaffney and Lam 1999). In several systems, mutations in the residues corresponding to D8-D9 and V83 alter the function of the protein and result in constitutively active response regulators (Smith et al. 2004). A V88A mutation in NarL (and also in the corresponding residue of CheY) results in a ligand or cofactor-independent, but phosphorylation-dependent regulator (Egan and Stewart 1991). Curiously, while V83 (corresponding to V88 of NarL) is conserved in the GacA orthologs of *E. coli*, *Salmonella*, *P. aeruginosa*, *Xylella fastidiosa*, and *Azotobacter*, natural V83A variations are found in GacA orthologs of *Yersinia pestis*, *P. syringae* DC3000, and *L. pneumophila* (isoleucine is found at position 83 of *P. carotovorum*, *Photobacteirum profundum*, and *V. fischeri*). The consequences, if any, of the V83A and V83I variations are unknown. It will be of interest to explore whether, like in CheY and NarL, V83 functions as a site for interaction of GacA orthologs with a ligand, a cofactor or docking with another protein to provide a hydrophobic pocket for V83, and whether A83 or I83 found in some bacteria may result in a different set of regulatory requirements for a particular GacA ortholog. Additionally, point mutations in *P. fluorescens gacA*, yielding M20I (in α1), and Q132R (in α6) resulted in kinase-independent regulators (Gaffney and Lam 1999). M20 is conserved only in GacA orthologs of *Pseudomonas* spp., *Azotobacter*, and *Yersinia*; isoleucine (or leucine) residues in the corresponding position are found in the GacA orthologs of *E. coli*, *Salmonella*, *Legionella*, *Xylella*, *Photobacterium*, *Pectobacterium*, and *Vibrio* and also in the related response regulators NarL, UhpA, FimZ, and RcsB. Q132 is present in most orthologs of GacA, but not in the GacA ortholog of *Xylella* or the related two-component response regulators, NarL, UhpA, FimZ, and RcsB. Because these two residues (M20 and Q132) are found in a limited number of proteins, it is difficult to speculate about the biochemical basis of the constitutive ON phenotype observed in the corresponding *P. fluorescens gacA* mutants.

The C-terminal DNA binding domain of the GacA protein consists of four α-helices (α7-10). The central helices α8 and α9 are predicted to form a helix-turn-helix motif, supported by the flanking hydrophobic core of helices α7 and α10. Based on the comparison with the solved crystal structure for NarL, residues I157, I170, and V181 are predicted to build a hydrophobic cluster that fixes α7–α9 in their positions. L175 is predicted to anchor the α8–α9 loop to the cluster, while G164 assures a proper angle between α7 and α8 (Baikalov et al. 1996). Spontaneous mutations predicted to affect the structure of the DNA-binding helices in *P. fluorescens* GacA resulted in a null phenotype (L173P in α8 of the HTH and A203V in α10 of the HTH) (Whistler et al. 1998; Bull et al. 2001). This is consistent with the reduction in activity of the corresponding mutants in the HTH-domain of LuxR from *V. fischeri* (Egland and Greenberg 2001; Trott and Stevens 2001). Though not yet demonstrated biochemically, these mutants are likely to be defective in their ability to interact with target DNA sequences. As shown in Fig. 14.3, these residues are highly conserved in the GacA orthologs of different γ-proteobacteria.

In addition to the transphosphorylation by GacS, GacA orthologs appear to be capable of being phosphorylated from the cellular pool of acetyl phosphate (Ac-P)

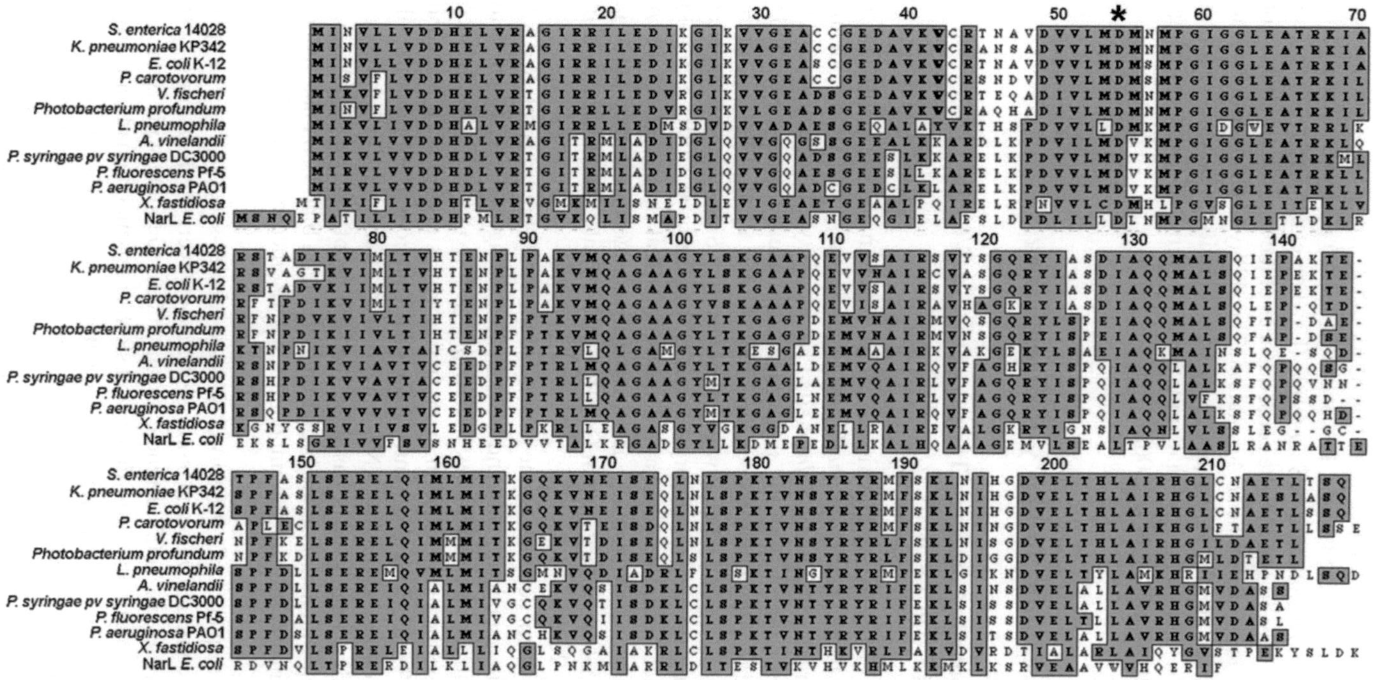

Fig. 14.3 CLUSTALW alignment of GacA orthologs and NarL. The predicted protein sequences of GacA orthologs show a high degree of conservation. For example, *S.* Typhimurium SirA and *E. coli* UvrY share 96% amino acid identity, well above the average of 90% for these two genera. GacA of *P. aeruginosa* and UvrY are 60% identical. Conserved amino acid residues are shaded. The N-terminus (residues 5–119) contains a signal receiver domain, the C-terminus (149–205) contains DNA-binding domain typical of all LuxR-type transcriptional regulators. GacA orthologs from *S. enterica* (GenBank Accession # AAC08300), plant-associated *K. pneumoniae* 342 (#ACI09250), *E. coli* K-12 (#AP_002529), soft rot pathogen *P. carotovorum* (#YP_003018235), *V. fischeri* (#AAQ83589), *Photobacterium profundum* (YP_130426), *L. pneumophila* (#AAL79360), *A. vinelandii* (#AAK97431), *P. syringae pv. syringae* DC3000 (#NP_792819), *P. fluorescens* Pf-5 (#YP_260665), *P. aeruginosa* PAO1 (#AAG05974), *X. fastidiosa* (#NP_780164), and NarL from *E. coli* (#CAA48935) are shown. Conserved Asp54 residue, the predicted site of phosphorylation, is indicated by an *asterisk*

independently of the sensor kinases (Lawhon et al. 2002). At pH 6.7, acetate (or Acetyl-CoA) is phosphorylated to Ac-P via *ackA* (or *pta*) and at least partially compensates for the *barA* mutation in *S. enterica* (Lawhon et al. 2002). This effect of Ac-P may explain why, in general, the consequences of the *barA* mutation are not as severe as the consequences of the *sirA* mutation. The ability of GacA orthologs to be phosphorylated from alternate P-donors under some conditions may also explain why the phenotypes and transcriptional units under control of *gacS* orthologs overlap only partially (Lawhon et al. 2002).

14.3.4 GacA Regulons in Soil γ-Proteobacteria

14.3.4.1 Evolutionarily Conserved Targets of the GacS/GacA Orthologs: The *csr* RNA

Since their original discovery, systematic efforts have been made to identify GacA regulons in various bacteria. In *Salmonella*, for example, the initial studies identified only the horizontal acquisitions *Salmonella* Pathogenicity Island (SPI) 1, SPI4, and SPI5, which are controlled by the *sirA* (*gacA*)-dependent expression of SPI1 regulators *hilA* and/or *hilC* (Johnston et al. 1996; Ahmer et al. 1999). A more ancient function was identified when it was determined that SirA represses flagellar genes via controlling the Csr regulatory system (Pernestig et al. 2001; Teplitski et al. 2003b). In fact, recent studies have demonstrated that the effects of GacA orthologs on motility and horizontally acquired virulence genes are mediated via the posttranscriptional regulatory system Csr (Rsm) (Kay et al. 2005; Lapouge et al. 2008; Brencic et al. 2009).

The Csr (Rsm) system depends on the activity of the RNA-binding protein CsrA (known as RsmA in pseudomonads and erwinias). Binding of CsrA (RsmA) to mRNA either stabilizes or destabilizes transcripts (Romeo et al. 1993; Romeo 1998b). Published genomes of pseudomonads and erwinias also contain a second CsrA homologue, RsmE, with apparently redundant functions (Reimmann et al. 2005). Small noncoding regulatory RNA antagonizes effects of CsrA proteins. Mechanistically, this antagonism is based on the folding of *csr*/*rsm* sRNAs into secondary structures which expose loops and bulges, which bind CsrA and reduce the intracellular concentration of the unbound CsrA (Romeo et al. 1993; Romeo 1998b).

The number and sequence of *csr* (*rsm*) sRNAs are species-specific. For example, all pseudomonads contain *rsmY* and *rsmZ* sRNA; however, *P. fluorescens* also contains *rsmX* (Kay et al. 2005; Brencic et al. 2009). Genome-wide searches identified at least one additional *gacA*-dependent sRNA in pseudomonads, although its role in the GacS/GacA-Rsm signal transduction is not yet fully elucidated (González et al. 2008). Functional homologs of *rsm* sRNAs in *E. coli* and *S. enterica* are *csrB* and *csrC*, and *rsmB* and *rsmC* in erwinias. There is little sequence conservation among these sRNAs; however, they all form stem-loop secondary structures with exposed GGA bulges (Romeo 1998a, b; Valverde et al. 2004; Kay et al. 2005; Lapouge et al. 2008). These features appear to be sufficient for their function, as demonstrated by the ability of over-expressed *rsmY* or *rsmZ* from *P. fluorescens* CHA0 to mimic a *csrA* mutation in *E. coli* (Valverde et al. 2004) and further supported by the experiments in which synthetic sRNA containing exposed GGA bulges functionally complemented the triple *rsmX rsmY rsmZ* mutant of *P. fluorescens* (Valverde 2009).

The link between orthologs of *gacA* and the Csr system was first demonstrated by showing that the effects of *Pseudomonas gacA* on the downstream genes were posttranscriptional and required the Rsm system (Blumer et al. 1999; Kay 2005 #3014). Consistent with the predicted role of the Csr system in the GacA regulatory cascade, the overexpression of RsmA mimics a *gacA* mutation in *Pseudomonas* (Blumer et al. 1999), and a double *rsmA rsmE* mutant overcame the *gacS* mutation (Reimmann et al. 2005). Conversely, overexpression of the genes encoding regulatory sRNAs can complement mutations in *gacS* or *gacA* orthologs and a deletion of all *csr* RNAs mimics a *gacA* mutation in different bacterial species (Aarons et al. 2000; Altier et al. 2000a, b; Heeb et al. 2002; Valverde et al. 2003; Weilbacher et al. 2003). A more rigorous support for this hypothesis was provided by proteomic and transcriptomic profiling of *P. aeruginosa gacA* and *rsmY rsmZ* mutants (Kay et al. 2006; Brencic et al. 2009), which demonstrated that in *P. aeruginosa gacA*-regulated proteins and transcripts are controlled via the *rsmY* and/or *rsmZ*. It is of note, that while the effects of *gacA* in *P. aeruginosa* are due to the regulatory effects of GacA on the Csr system, CsrA has additional targets: CsrA regulates expression of the ribosomal protein S1 independently of the *csr* sRNAs (Kay et al. 2006).

The effects of GacA on the Csr (Rsm) posttranscriptional regulatory system are due to the control of the expression of *csr*/*rsm* sRNA by GacA orthologs. Phosphorylated GacA of *P. aeruginosa* most likely binds to the conserved sequences (TGTAAGCATTAACTTACA or TGTAAGCCAAGGCTTACA) within the promoter regions of the target sRNA genes (Heeb et al. 2002; Kay et al. 2006). A similar sequence (TGTAAGACAAGGTGAAAC) was identified within the predicted promoter region for the *S. enterica csrB*, which was shifted by phosphorylated SirA (Teplitski et al. 2003b). These observations suggest that the binding site for GacA orthologs may be conserved in γ-proteobacteria.

Interestingly, in vitro gel mobility shift assays demonstrated that a fragment upstream of *hilA* (the horizontally acquired regulator of SPI1) was shifted by phosphorylated SirA (Teplitski et al. 2003b). This shifted fragment contained the sequence TTAAAGCACAGGCATAAG, which is similar to the hypothetical conserved GacA DNA binding site. Furthermore, this shifted fragment is distinct from the *hilA* promoter controlled by other regulators (Ellermeier et al. 2005). It remains to be established whether/how this and other horizontally acquired genes became integrated into the SirA (GacA) regulon of *Salmonella* both at the level of direct transcriptional regulation by SirA and by posttranscriptional regulation via the Csr system.

14.3.4.2 Orthologs of *gacS/gacA* Are Central to Biofilm Formation in γ-Proteobacteria

Almost invariably, mutants in *gacS*, *gacA* are defective in biofilm formation. Biofilms are multicellular highly structured, sessile microbial communities that form on submerged or wet surfaces, including surfaces of plant leaves, roots, and plant-associated fungi [rev. Davey and O'toole (2000), Hall-Stoodley et al. (2004), and Danhorn and Fuqua (2007)]. Formation of a biofilm is a complex, multistep process that requires surface conditioning, reversible attachment, adhesion, cell aggregation, biofilm maturation, and eventual dispersal (Hall-Stoodley et al. 2004). It has been hypothesized that one of the early steps preceding biofilm formation requires that the bacterial cells "make a decision" as to whether to continue active colonization of surfaces via surfactant-mediated flagella-dependent swarming motility or to form sessile biofilms (Verstraeten et al. 2008). Consistent with this hypothesis, swarming motility and biofilm formation are often divergently regulated [rev. Verstraeten et al. (2008)]. Furthermore, consistent with this hypothesis, nonmotile nonflagellated *flhDC S. enterica* mutants formed more biofilm compared with the wild type (Teplitski et al. 2006). Mutant analysis revealed that flagellated but chemotaxis-defective mutants or those with paralyzed flagella formed reduced biofilms, while mutants in structural flagellar genes increased biofilm formation (Teplitski et al. 2006). It should be noted, however, that the effects of *flhDC* on biofilm formation even by the same bacteria are not always the same in different studies, especially when different media or surfaces are used [rev. Davey and O'Toole (2000)].

Orthologs of *gacA* downregulate swarming or swimming motility in pseudomonads and vibrios (Goodier and Ahmer 2001) and also positively regulate biofilm formation in the same bacteria (Parkins et al. 2001). Both of these effects are due to the effects of the GacS/GacA cascade on the Csr (Rsm) posttranscriptional regulatory system. As described above, GacA up-regulates transcription of the *csr* (*rsm*) regulatory sRNAs, which antagonize activity of the CsrA (RsmA) RNA-binding protein and this promotes synthesis of polymers required for biofilm formation (Suzuki et al. 2002; Wang et al. 2005; Brencic et al. 2009). It is not yet known whether the GacS/GacA-Csr regulatory cascade contributes to this swarming vs. biofilm "decision making" or whether it is one of the regulatory cascades, which sets off as the consequence of this decision.

14.4 The Elusive GacS Signal

Currently, there is little information regarding the signal(s) detected by GacS. Gene activation mediated by GacS/GacA orthologs has been observed during growth on common laboratory media and within plant and animal tissues. It is not yet clear whether the signal that ultimately triggers the function of GacA is the same in all bacteria. As discussed above, in pseudomonads (but not in *Enterobacteriacea*), phosphorylation status of GacS depends on the input of two other sensor kinases RetS and LadS. Even though homologues of *retS*, *ladS* are absent from the genomes of *Enterobacteriacea*, it may be premature to rule out the possibility that RetS or LadS-like sensor kinases alter the function of GacS in enterobacteria. On another extreme, in *Xyllela* a functional GacA is present, while a GacS ortholog is missing (Shi et al. 2009). As shown in Fig. 14.3, the N-terminus of $GacA_{X.f.}$ contains the conserved amino acid residues involved in phosphorylation. Therefore, it is likely that $GacA_{X.f.}$ does not directly bind a ligand, rather is phosphorylated by a yet-unknown sensor kinase. As described below, several research groups are focusing their efforts on identifying a signal for the GacS/GacA regulatory system.

14.4.1 Evidence for the Self-Produced GacS Signal

During batch culture, GacS of *P. fluorescens* responds to a compound in its own culture supernatant by increasing the expression of *rsmZ* fivefold as the cultures entered stationary phase (Heeb et al. 2002). Follow-up studies demonstrated that *hcnA* (another gene under *gacA* control) was similarly activated by the culture supernatant extract. The ability of the signal to affect expression of *hcnA::lacZ* depended on the presence of the sensor kinase *gacS*. Production of the signal is controlled by the GacS/GacA-Rsm system (Kay et al. 2005). The signal perceived by GacS was extractable from the stationary phase cultures with dichloromethane (Heeb et al. 2002; Zuber et al. 2003). Even though the GacS signal is present in stationary phase cultures and is hydrophobic, it is not an AHL (Heeb et al. 2002; Heurlier et al. 2004).

14.4.2 Evidence for the Eukaryotic Contribution to the GacS/GacA-Mediated Signaling

A compound from seed exudates was reported to stimulate the expression of a *luxAB* fusion in a gene under GacS/GacA control in a plant-associated *Pseudomonas* spp. DSS73 (Koch et al. 2002). The effect of the plant compound required functional *gacS*, suggesting that the signal (or signal-mimic) for GacS may be produced by eukaryotes as well. Expression of *rsmB* from *Dickeya dadantii* was modestly (~1.5-fold) up-regulated in the presence of 100 mM *o*-coumaric acid, a common plant phenolic acid (Yang et al. 2008). Expression of *gacA* was not affected by the compound; however, the effect of *o*-coumaric acid on *rsmB* required a functional *gacS*, suggesting the *o*-coumaric acid may be affecting the cascade at the level of signal perception or transduction (Yang et al. 2008).

14.5 Conclusions and Future Directions

One could argue that soils in temperate regions truly represent the most extreme environments on Earth. After all, environmental conditions in hot springs or permafrost are reasonably predictable. In temperate soils, on the other hand, microorganisms deal with cycles of flooding and drought, oscillations of temperature, predation, intra- and interspecies competition, as well as uncertain availability of nutrients. To sense environmental conditions and to change patterns of gene regulation accordingly, soil bacteria have evolved sophisticated signal transduction mechanisms.

In γ-proteobacteria, the GacS/GacA two-component regulatory system is one of the mechanisms for posttranscriptional regulation of gene expression in response to cues that reflect environmental conditions. Orthologs of *gacA* are found in the genomes of all γ-proteobacteria, with the exception of those that have undergone extreme reductions. The structure and function of GacA orthologs is highly conserved in γ-proteobacteria: upon phosphorylation of a conserved aspartate residue, these FixJ-type regulators bind within promoters of sRNA genes and thus set off the regulatory cascade, which is mediated by a posttranscriptional regulatory protein CsrA (RsmA). The regulatory inputs leading to the phosphorylation of GacA are variable. GacA orthologs of *Pseudomonas* spp., *S. enterica*, and *E. coli* are phosphorylated by GacS orthologs. The same mechanism of GacA activation is highly likely in other γ-proteobacteria based on the wide distribution of *gacS* orthologs in genomes of most other γ-proteobacteria. Interestingly, genomes of *Xylella* and *Xanthomonas* lack identifiable orthologues of *gacS*. However, $gacA_{X.f.}$ was shown to be functional and complemented *gacA* mutations in heterologous hosts. This discovery may provide important clues for the understanding of the function and the evolution of the GacS–GacA signal transduction. In pseudomonads, the phosphorylation status of GacS depends on two other sensor kinases, RetS and LadS; orthologs of *retS* and *ladS* are not present in the genomes of enteric bacteria. Signals required for the autophosphorylation of GacS and activation of RetS are not known. However, the available evidence indicates that GacS responds to a self-produced hydrophobic molecule, while activity of RetS is dependent on the temperature. Several research groups are working on identifying the signal(s) that set off the GacS/GacA regulatory cascade as well as inhibitors that would block the function of this regulatory system.

While the GacS/GacA system is present only in γ-proteobacteria, quorum sensing mediated by *N*-acyl homoserine lactone (AHL) signals has been characterized in all subgroups of proteobacteria. Even though AHL-mediated signaling is wide spread, it is not universal: closely related bacteria may differ substantially in the repertoire of the AHLs they produce and perceive. At least in part, this is due to the fact that the genes responsible for the synthesis of AHLs and their perception are borne on plasmids, transposons, or other mobile elements. Interestingly, those AHL receptor genes that are evolutionarily conserved among distantly related bacteria are often encoded on chromosome and are unlinked to AHL synthase genes in most soil bacteria. In light of these observations, one may wonder whether the

AHL-mediated QS is a "social behavior" or whether it is more akin to a "social disease," which is transmitted by mobile genetic elements. Anthropomorphic comparisons aside, these observations highlight the need for an unbiased evaluation of the evolutionary origins of AHL-mediated quorum sensing and its ecological consequences.

Acknowledgments Preparation of this manuscript was supported by USDA NRI 2007-35319-18158 (M.T. and J.B.R.), and CRIS project FLA-SWS-004984 of Florida Agricultural Experimental Station.

References

Aarons S, Abbas A, Adams C, Fenton A, O'Gara F (2000) A regulatory RNA (PrrB RNA) modulates expression of secondary metabolite genes in *Pseudomonas fluorescens* F113. J Bacteriol 182:3913–3919

Ahmer BM (2004) Cell-to-cell signalling in *Escherichia coli* and *Salmonella enterica*. Mol Microbiol 52:933–945

Ahmer BM, van Reeuwijk J, Watson PR, Wallis TS, Heffron F (1999) *Salmonella* SirA is a global regulator of genes mediating enteropathogenesis. Mol Microbiol 31:971–982

Albus AM, Pesci EC, Runyen-Janecky LJ, West SE, Iglewski BH (1997) Vfr controls quorum sensing in *Pseudomonas aeruginosa*. J Bacteriol 179:3928–3935

Altier C, Suyemoto M, Lawhon SD (2000a) Regulation of *Salmonella enterica* serovar Typhimurium invasion genes by *csrA*. Infect Immun 68:6790–6797

Altier C, Suyemoto M, Ruiz AI, Burnham KD, Maurer R (2000b) Characterization of two novel regulatory genes affecting *Salmonella* invasion gene expression. Mol Microbiol 35:635–646

Baikalov I, Schroder I, Kaczor-Grzeskowiak M, Grzeskowiak K, Gunsalus RP, Dickerson RE (1996) Structure of the *Escherichia coli* response regulator NarL. Biochemistry 35:11053–11061

Barnett MJ, Fisher RF (2006) Global gene expression in the rhizobial-legume symbiosis. Symbiosis 42:1–24

Barnett MJ, Toman CJ, Fisher RF, Long SR (2004) A dual-genome symbiosis chip for coordinate study of signal exchange and development in a prokaryote–host interaction. Proc Natl Acad Sci USA 101:16636–16641

Bartels FW et al (2007) Effector-stimulated single molecule protein-DNA interactions of a quorum-sensing system in *Sinorhizobium meliloti*. Biophys J 92:4391–4400

Bauer WD (1977) Lectins as determinants of specificity in legume-*Rhizobium* symbiosis. Basic Life Sci 9:283–297

Baysse C et al (2005) Modulation of quorum sensing in *Pseudomonas aeruginosa* through alteration of membrane properties. Microbiology 151:2529–2542

Becker A et al (1997) The 32-kilobase *exp* gene cluster of *Rhizobium meliloti* directing the biosynthesis of galactoglucan: genetic organization and properties of the encoded gene products. J Bacteriol 179:1375–1384

Bera AK et al (2009) Structure of PqsD, a *Pseudomonas* quinolone signal biosynthetic enzyme, in complex with anthranilate. Biochemistry 48:8644–8655

Blumer C, Heeb S, Pessi G, Haas D (1999) Global GacA-steered control of cyanide and exoprotease production in *Pseudomonas fluorescens* involves specific ribosome binding sites. Proc Natl Acad Sci USA 96:14073–14078

Boles BR, Thoendel M, Singh PK (2005) Rhamnolipids mediate detachment of *Pseudomonas aeruginosa* from biofilms. Mol Microbiol 57:1210–1223

Bredenbruch F, Geffers R, Nimtz M, Buer J, Haussler S (2006) The *Pseudomonas aeruginosa* quinolone signal (PQS) has an iron-chelating activity. Environ Microbiol 8:1318–1329

Brencic A et al (2009) The GacS/GacA signal transduction system of *Pseudomonas aeruginosa* acts exclusively through its control over the transcription of the RsmY and RsmZ regulatory small RNAs. Mol Microbiol 73:434–445

Bull CT, Duffy B, Voisard C, Defago G, Keel C, Haas D (2001) Characterization of spontaneous *gacS* and *gacA* regulatory mutants of *Pseudomonas fluorescens* biocontrol strain CHA0. Antonie Van Leeuwenhoek 79:327–336

Caetano-Anolles G, Wrobel B, Bauer WD (1992) Growth and movement of spot inoculated *Rhizobium meliloti* on the root surface of alfalfa. Plant Physiol 98(3):1181–1189

Capela D, Filipe C, Bobilk C, Batut J, Bruand C (2006) *Sinorhizobium meliloti* differentiation during symbiosis with alfalfa: a transcriptomic dissection. Mol Plant Microbe Interact 19:363–372

Castaneda M, Guzman J, Moreno S, Espin G (2000) The GacS sensor kinase regulates alginate and poly-beta-hydroxybutyrate production in *Azotobacter vinelandii*. J Bacteriol 182:2624–2628

Castaneda M, Sanchez J, Moreno S, Nunez C, Espin G (2001) The global regulators GacA and sigma(S) form part of a cascade that controls alginate production in *Azotobacter vinelandii*. J Bacteriol 183:6787–6793

Cha C, Gao P, Chen YC, Shaw PD, Farrand SK (1998) Production of acyl-homoserine lactone quorum-sensing signals by gram-negative plant-associated bacteria. Mol Plant Microbe Interact 11:1119–1129

Chancey ST, Wood DW, Pierson LS 3rd (1999) Two-component transcriptional regulation of N-acyl-homoserine lactone production in *Pseudomonas aureofaciens*. Appl Environ Microbiol 65:2294–2299

Charkowski AO (2009) Decaying signals: will understanding bacterial-plant communications lead to control of soft rot? Curr Opin Biotechnol 20:178–184

Chatterjee A, Cui Y, Yang H, Collmer A, Alfano JR, Chatterjee AK (2003) GacA, the response regulator of a two-component system, acts as a master regulator in *Pseudomonas syringae* pv. *tomato* DC3000 by controlling regulatory RNA, transcriptional activators, and alternate sigma factors. Mol Plant Microbe Interact 16:1106–1117

Chen H, Teplitski M, Robinson JB, Rolfe BG, Bauer WD (2003) Proteomic analysis of wild-type *Sinorhizobium meliloti* responses to *N*-acyl homoserine lactone quorum sensing signals and the transition to stationary phase. J Bacteriol 185:5029–5036

Cho H, Pinto UM, Winans SC (2009) Transsexuality in the rhizosphere: Quorum Sensing reversibly converts *Agrobacterium tumefaciens* from phenotypically female to male. J Bacteriol 191:3375–3383

Coleman JP et al (2008) *Pseudomonas aeruginosa* PqsA is an anthranilate-coenzyme A ligase. J Bacteriol 190:1247–1255

Danhorn T, Fuqua C (2007) Biofilm formation by plant-associated bacteria. Annu Rev Microbiol 61:401–422

Daniels R et al (2002) The *cin* quorum sensing locus of *Rhizobium etli* CNPAF512 affects growth and symbiotic nitrogen fixation. J Biol Chem 277:462–468

Davey ME, Caiazza NC, O'Toole GA (2003) Rhamnolipid surfactant production affects biofilm architecture in *Pseudomonas aeruginosa* PAO1. J Bacteriol 185:1027–1036

Davey ME, O'Toole GA (2000) Microbial biofilms: from ecology to molecular genetics. Microbiol Mol Biol Rev 64:847–867

Davies DG, Parsek MR, Pearson JP, Iglewski BH, Costerton JW, Greenberg EP (1998) The involvement of cell-to-cell signals in the development of a bacterial biofilm. Science 280:295–298

Diggle SP, Cornelis P, Williams P, Camara M (2006) 4-quinolone signalling in *Pseudomonas aeruginosa*: old molecules, new perspectives. Int J Med Microbiol 296:83–91

Diggle SP, Gardner A, West SA, Griffin AS (2007a) Evolutionary theory of bacterial quorum sensing: when is a signal not a signal? Philos Trans R Soc Lond B Biol Sci 362:1241–1249

Diggle SP et al (2007b) The *Pseudomonas aeruginosa* 4-quinolone signal molecules HHQ and PQS play multifunctional roles in quorum sensing and iron entrapment. Chem Biol 14:87–96

Diggle SP, Winzer K, Chhabra SR, Worrall KE, Camara M, Williams P (2003) The *Pseudomonas aeruginosa* quinolone signal molecule overcomes the cell density-dependency of the quorum sensing hierarchy, regulates *rhl*-dependent genes at the onset of stationary phase and can be produced in the absence of LasR. Mol Microbiol 50:29–43

Djordjevic MA (2004) *Sinorhizobium meliloti* metabolism in the root nodule: a proteomic perspective. Proteomics 4:1859–1872

Dobretsov S, Teplitski M, Paul V (2009) Mini-review: quorum sensing in the marine environment and its relationship to biofouling. Biofouling 25:413–427

Dong YH, Zhang XF, Xu JL, Tan AT, Zhang LH (2005) VqsM, a novel AraC-type global regulator of quorum-sensing signalling and virulence in *Pseudomonas aeruginosa*. Mol Microbiol 58:552–564

Duan KM, Dammel C, Stein J, Rabin H, Surette MG (2003) Modulation of *Pseudomonas aeruginosa* gene expression by host microflora through interspecies communication. Mol Microbiol 50:1477–1491

Eberhard A, Burlingame AL, Eberhard C, Kenyon GL, Nealson KH, Oppenheimer NJ (1981) Structural identification of autoinducer of *Photobacterium fischeri* luciferase. Biochemistry 20:2444–2449

Egan SM, Stewart V (1991) Mutational analysis of nitrate regulatory gene *narL* in *Escherichia coli* K-12. J Bacteriol 173:4424–4432

Egland KA, Greenberg EP (2001) Quorum sensing in *Vibrio fischeri*: analysis of the LuxR DNA binding region by alanine-scanning mutagenesis. J Bacteriol 183:382–386

Elasri M et al (2001) Acyl-homoserine lactone production is more common among plant-associated *Pseudomonas* spp. than among soilborne *Pseudomonas* spp. Appl Environ Microbiol 67:1198–1209

Ellermeier CD, Ellermeier JR, Slauch JM (2005) HilD, HilC and RtsA constitute a feed forward loop that controls expression of the SPI1 type three secretion system regulator *hilA* in *Salmonella enterica* serovar Typhimurium. Mol Microbiol 57:691–705

Eriksson AR, Andersson RA, Pirhonen M, Palva ET (1998) Two-component regulators involved in the global control of virulence in *Erwinia carotovora* subsp. *carotovora*. Mol Plant Microbe Interact 11:743–752

Espinosa-Urgel M (2003) Resident parking only: rhamnolipids maintain fluid channels in biofilms. J Bacteriol 185:699–700

Farrow JM 3rd, Sund ZM, Ellison ML, Wade DS, Coleman JP, Pesci EC (2008) PqsE functions independently of PqsR-*Pseudomonas* quinolone signal and enhances the *rhl* quorum-sensing system. J Bacteriol 190:7043–7051

Fletcher MP, Diggle SP, Crusz SA, Chhabra SR, Camara M, Williams P (2007) A dual biosensor for 2-alkyl-4-quinolone quorum-sensing signal molecules. Environ Microbiol 9:2683–2693

Fuqua C, Eberhard A (1999) Signal generation in autoinduction systems: synthesis of acylated homoserine lactones by LuxI-type proteins. In: Dunny GM, Winans SC (eds) Cell-cell signalling in bacteria. American Society for Microbiology, Washington, DC, pp 211–242

Gaffney TD, Lam ST (1999) DNA encoding LemA-independent GacA and its use in activating gene expression. In: Official gazette of the United States Patent and Trademark Office Patents. Novartis Finance Corporation, USA

Gage DJ (2004) Infection and invasion of roots by symbiotic, nitrogen-fixing rhizobia during nodulation of temperate legumes. Microbiol Mol Biol Rev 68:280–300

Galibert F et al (2001) The composite genome of the legume symbiont *Sinorhizobium meliloti*. Science 293:668–672

Gantner S et al (2006) *In situ* quantitation of the spatial scale of calling distances and population density-independent N-acylhomoserine lactone-mediated communication by rhizobacteria colonized on plant roots. FEMS Microbiol Ecol 56:188–194

Gao M et al (2007) Effects of AiiA-mediated quorum quenching in *Sinorhizobium meliloti* on quorum-sensing signals, proteome patterns, and symbiotic interactions. Mol Plant Microbe Interact 20:843–856

Gao M et al (2005) *sinI*- and *expR*-dependent quorum sensing in *Sinorhizobium meliloti*. J Bacteriol 187:7931–7944

Gao M, Teplitski M (2008) RIVET-a tool for *in vivo* analysis of symbiotically relevant gene expression in *Sinorhizobium meliloti*. Mol Plant Microbe Interact 21:162–170

Gao M, Teplitski M, Robinson JB, Bauer WD (2003) Production of substances by *Medicago truncatula* that affect bacterial quorum sensing. Mol Plant Microbe Interact 16:827–834

Gil R, Latorre A, Moya A (2004) Bacterial endosymbionts of insects: insights from comparative genomics. Environ Microbiol 6:1109–1122

Gilson L, Kuo A, Dunlap PV (1995) AinS and a new family of autoinducer synthesis proteins. J Bacteriol 177:6946–6951

Givskov M et al (1996) Eukaryotic interference with homoserine lactone-mediated prokaryotic signalling. J Bacteriol 178:6618–6622

Gonzalez JE, Marketon MM (2003) Quorum sensing in nitrogen-fixing rhizobia. Microbiol Mol Biol Rev 67:574–592

González N, Heeb S, Valverde C, Kay E, Reimmann C, Junier T, Haas D (2008) Genome-wide search reveals a novel GacA-regulated small RNA in *Pseudomonas* species. BMC Genomics 9:167–181

Goodier RI, Ahmer BM (2001) SirA orthologs affect both motility and virulence. J Bacteriol 183:2249–2258

Goodman AL, Kulasekara B, Rietsch A, Boyd D, Smith RS, Lory S (2004) A signaling network reciprocally regulates genes associated with acute infection and chronic persistence in *Pseudomonas aeruginosa*. Dev Cell 7:745–754

Goodman AL, Merighi M, Hyodo M, Ventre I, Filloux A, Lory S (2009) Direct interaction between sensor kinase proteins mediates acute and chronic disease phenotypes in a bacterial pathogen. Genes Dev 23:249–259

Gurich N, Gonzalez JE (2009) Role of quorum sensing in *Sinorhizobium meliloti*-alfalfa symbiosis. J Bacteriol 191:4372–4382

Haas D, Keel C, Reimmann C (2002) Signal transduction in plant-beneficial rhizobacteria with biocontrol properties. Antonie Van Leeuwenhoek 81:385–395

Hall-Stoodley L, Costerton JW, Stoodley P (2004) Bacterial biofilms: from the natural environment to infectious diseases. Nat Rev Microbiol 2:95–108

Hao Y, Winans SC, Glick BR, Charles TC (2010) Identification and characterization of new LuxR/LuxI-type quorum sensing systems from metagenomic libraries. Environ Microbiol 12 (1):105–117

Heeb S, Blumer C, Haas D (2002) Regulatory RNA as mediator in GacA/RsmA-dependent global control of exoproduct formation in *Pseudomonas fluorescens* CHA0. J Bacteriol 184: 1046–1056

Heeb S, Haas D (2001) Regulatory roles of the GacS/GacA two-component system in plant-associated and other gram-negative bacteria. Mol Plant Microbe Interact 14:1351–1363

Hense BA, Kuttler C, Muller J, Rothballer M, Hartmann A, Kreft JU (2007) Does efficiency sensing unify diffusion and quorum sensing? Nat Rev Microbiol 5:230–239

Heurlier K, Denervaud V, Pessi G, Reimmann C, Haas D (2003) Negative control of quorum sensing by RpoN (sigma54) in *Pseudomonas aeruginosa* PAO1. J Bacteriol 185:2227–2235

Heurlier K et al (2004) Positive control of swarming, rhamnolipid synthesis, and lipase production by the posttranscriptional RsmA/RsmZ system in *Pseudomonas aeruginosa* PAO1. J Bacteriol 186:2936–2945

Hirsch AM, Bauer WD, Bird DM, Cullimore J, Tyler B, Yoder JI (2003) Molecular signals and receptors: controlling rhizosphere interactions between plants and other organisms. Ecology 84:858–868

Hoang HH, Becker A, Gonzalez JE (2004) The LuxR homolog ExpR, in combination with the Sin quorum sensing system, plays a central role in *Sinorhizobium meliloti* gene expression. J Bacteriol 186:5460–5472

Hornby JM et al (2001) Quorum sensing in the dimorphic fungus *Candida albicans* is mediated by farnesol. Appl Environ Microbiol 67:2982–2992

Hrabak EM, Willis DK (1992) The *lemA* gene required for pathogenicity of *Pseudomonas syringae* pv. *syringae* on bean is a member of a family of two-component regulators. J Bacteriol 174:3011–3020

Humair B, Gonzalez N, Mossialos D, Reimmann C, Haas D (2009) Temperature-responsive sensing regulates biocontrol factor expression in *Pseudomonas fluorescens* CHA0. ISME J 3:955–965

Jensen V et al (2006) RhlR expression in *Pseudomonas aeruginosa* is modulated by the Pseudomonas quinolone signal via PhoB-dependent and -independent pathways. J Bacteriol 188:8601–8606

Johnston C, Pegues DA, Hueck CJ, Lee A, Miller SI (1996) Transcriptional activation of *Salmonella typhimurium* invasion genes by a member of the phosphorylated response-regulator superfamily. Mol Microbiol 22:715–727

Jones KM, Kobayashi H, Davies BW, Taga ME, Walker GC (2007) How rhizobial symbionts invade plants: the *Sinorhizobium-Medicago* model. Nat Rev Microbiol 5:619–633

Juhas M, Wiehlmann L, Salunkhe P, Lauber J, Buer J, Tummler B (2005) GeneChip expression analysis of the VqsR regulon of *Pseudomonas aeruginosa* TB. FEMS Microbiol Lett 242:287–295

Karamanoli K, Lindow SE (2006) Disruption of N-acyl homoserine lactone-mediated cell signaling and iron acquisition in epiphytic bacteria by leaf surface compounds. Appl Environ Microbiol 72:7678–7686

Kay E, Dubuis C, Haas D (2005) Three small RNAs jointly ensure secondary metabolism and biocontrol in *Pseudomonas fluorescens* CHA0. Proc Natl Acad Sci USA 102:17136–17141

Kay E et al (2006) Two GacA-dependent small RNAs modulate the quorum-sensing response in *Pseudomonas aeruginosa*. J Bacteriol 188:6026–6033

Keshavan ND, Chowdhary PK, Haines DC, Gonzalez JE (2005) L-Canavanine made by *Medicago sativa* interferes with quorum sensing in *Sinorhizobium meliloti*. J Bacteriol 187:8427–8436

Khokhlov AS et al (1967) A-factor responsible for biosynthesis of streptomycin in mutant strains of *Actinomyces streptomycini*. Dokl Akad Nauk SSSR (Proc Natl Acad Sci USSSR) 177:232

Kim W, Surette MG (2006) Coordinated regulation of two independent cell-cell signaling systems and swarmer differentiation in *Salmonella enterica* serovar Typhimurium. J Bacteriol 188:431–440

Kitten T, Kinscherf TG, McEvoy JL, Willis DK (1998) A newly identified regulator is required for virulence and toxin production in *Pseudomonas syringae*. Mol Microbiol 28:917–929

Koch B, Liljefors T, Persson T, Nielsen J, Kjelleberg S, Givskov M (2005) The LuxR receptor: the sites of interaction with quorum-sensing signals and inhibitors. Microbiology 151:3589–3602

Koch B et al (2002) Lipopeptide production in *Pseudomonas* sp. strain DSS73 is regulated by components of sugar beet seed exudate via the Gac two-component regulatory system. Appl Environ Microbiol 68:4509–4516

Lapouge K, Schubert M, Allain FH, Haas D (2008) Gac/Rsm signal transduction pathway of gamma-proteobacteria: from RNA recognition to regulation of social behaviour. Mol Microbiol 67:241–253

Laville J, Voisard C, Keel C, Maurhofer M, Defago G, Haas D (1992) Global control in *Pseudomonas fluorescens* mediating antibiotic synthesis and suppression of black root rot of tobacco. Proc Natl Acad Sci USA 89:1562–1566

Lawhon SD, Maurer R, Suyemoto M, Altier C (2002) Intestinal short-chain fatty acids alter *Salmonella typhimurium* invasion gene expression and virulence through BarA/SirA. Mol Microbiol 46:1451–1464

Lerat E, Moran NA (2004) The evolutionary history of quorum-sensing systems in bacteria. Mol Biol Evol 21:903–913

Li L et al (2008) Inactivation of microbial arginine deiminases by L-canavanine. J Am Chem Soc 130:1918–1931

Long SR, Buikema WJ, Ausubel FM (1982) Cloning *Rhizobium meliloti* nodulation genes by direct complementation of Nod-mutants. Nature 298:485–488

Lu X et al (2005) L-canavanine is a time-controlled mechanism-based inhibitor of *Pseudomonas aeruginosa* arginine deiminase. J Am Chem Soc 127:16412–16413

Marketon MM, Gonzalez JE (2002) Identification of two quorum-sensing systems in *Sinorhizobium meliloti*. J Bacteriol 184:3466–3475

Marketon MM, Gronquist MR, Eberhard A, Gonzalez JE (2002) Characterization of the *Sinorhizobium meliloti sinR/sinI* locus and the production of novel N-acyl homoserine lactones. J Bacteriol 184:5686–5695

Mathesius U et al (2003) Extensive and specific responses of a eukaryote to bacterial quorum-sensing signals. Proc Natl Acad Sci USA 100:1444–1449

Mitra RM, Shaw SL, Long SR (2004) Six nonnodulating plant mutants defective for Nod factor-induced transcriptional changes associated with the legume-rhizobia symbiosis. Proc Natl Acad Sci USA 101:10217–10222

Moolenaar GF, van Sluis CA, Backendorf C, van de Putte P (1987) Regulation of the *Escherichia coli* excision repair gene *uvrC*. Overlap between the *uvrC* structural gene and the region coding for a 24 kD protein. Nucleic Acids Res 15:4273–4289

Mulligan JT, Long SR (1985) Induction of *Rhizobium meliloti nodC* expression by plant exudate requires *nodD*. Proc Natl Acad Sci USA 82:6609–6613

Ng W-L, Bassler BL (2009) Bacterial quorum-sensing network architectures. Annu Rev Genet 43:197–222

Noel JT et al (2010) *Salmonella* SdiA recognizes N-acyl homoserine lactones from *Pectobacterium carotovorum in vitro* but not in a bacterial soft rot. Mol Plant Microbe Interact 23(3):273–282

Parkins MD, Ceri H, Storey DG (2001) *Pseudomonas aeruginosa* GacA, a factor in multihost virulence, is also essential for biofilm formation. Mol Microbiol 40:1215–1226

Pearson JP, Van Delden C, Iglewski BH (1999) Active efflux and diffusion are involved in transport of *Pseudomonas aeruginosa* cell-to-cell signals. J Bacteriol 181:1203–1210

Pellock BJ, Teplitski M, Boinay RP, Bauer WD, Walker GC (2002) A LuxR homolog controls production of symbiotically active extracellular polysaccharide II by *Sinorhizobium meliloti*. J Bacteriol 184:5067–5076

Pereira CS, McAuley JR, Taga ME, Xavier KB, Miller ST (2008) *Sinorhizobium meliloti*, a bacterium lacking the autoinducer-2 (AI-2) synthase, responds to AI-2 supplied by other bacteria. Mol Microbiol 70:1223–1235

Pernestig AK, Melefors O, Georgellis D (2001) Identification of UvrY as the cognate response regulator for the BarA sensor kinase in *Escherichia coli*. J Biol Chem 276:225–231

Peters NK, Frost JW, Long SR (1986) A plant flavone, luteolin, induces expression of *Rhizobium meliloti* nodulation genes. Science 233:977–980

Poole P, Allaway D (2000) Carbon and nitrogen metabolism in *Rhizobium*. Adv Microb Physiol 43:117–163

Quinones B, Pujol CJ, Lindow SE (2004) Regulation of AHL production and its contribution to epiphytic fitness in *Pseudomonas syringae*. Mol Plant Microbe Interact 17:521–531

Rajamani S et al (2008) The vitamin riboflavin and its derivative lumichrome activate the LasR bacterial quorum-sensing receptor. Mol Plant Microbe Interact 21:1184–1192

Rampioni G et al (2007) RsaL provides quorum sensing homeostasis and functions as a global regulator of gene expression in *Pseudomonas aeruginosa*. Mol Microbiol 66:1557–1565

Ramsay JP et al (2009) A LuxRI-family regulatory system controls excision and transfer of the *Mesorhizobium loti* strain R7A symbiosis island by activating expression of two conserved hypothetical genes. Mol Microbiol 73:1141–1155

Reimmann C et al (1997) The global activator GacA of *Pseudomonas aeruginosa* PAO positively controls the production of the autoinducer N-butyryl-homoserine lactone and the formation of the virulence factors pyocyanin, cyanide, and lipase. Mol Microbiol 24:309–319

Reimmann C, Valverde C, Kay E, Haas D (2005) Posttranscriptional repression of GacS/GacA-controlled genes by the RNA-binding protein RsmE acting together with RsmA in the biocontrol strain *Pseudomonas fluorescens* CHA0. J Bacteriol 187:276–285

Romeo T (1998a) *Escherichia coli csrB* gene and RNA encoded thereby. In: United States Patent and Trademark Office. University of North Texas, Health Science Center at Fort Worth, Fort Worth, TX

Romeo T (1998b) Global regulation by the small RNA-binding protein CsrA and the non-coding RNA molecule *csrB*. Mol Microbiol 29:1321–1330

Romeo T, Gong M, Liu MY, Brun-Zinkernagel AM (1993) Identification and molecular characterization of *csrA*, a pleiotropic gene from *Escherichia coli* that affects glycogen biosynthesis, gluconeogenesis, cell size, and surface properties. J Bacteriol 175:4744–4755

Sakuragi Y, Kolter R (2007) Quorum-sensing regulation of the biofilm matrix genes (*pel*) of *Pseudomonas aeruginosa*. J Bacteriol 189:5383–5386

Sanchez-Contreras M, Bauer WD, Gao M, Robinson JB, Downie AJ (2007) Quorum-sensing regulation in rhizobia and its role in symbiotic interactions with legumes. Philos Trans R Soc Lond B Biol Sci 362:1149–1163

Schuhegger R et al (2006) Induction of systemic resistance in tomato by N-acyl-L-homoserine lactone-producing rhizosphere bacteria. Plant Cell Environ 29:909–918

Schuster M, Lostroh CP, Ogi T, Greenberg EP (2003) Identification, timing, and signal specificity of *Pseudomonas aeruginosa* quorum-controlled genes: a transcriptome analysis. J Bacteriol 185:2066–2079

Shi XY, Dumenyo CK, Hernandez-Martinez R, Azad H, Cooksey DA (2009) Characterization of regulatory pathways in *Xylella fastidiosa*: genes and phenotypes controlled by *gacA*. Appl Environ Microbiol 75:2275–2283

Skorupska A, Janczarek M, Marczak M, Mazur A, Krol J (2006) Rhizobial exopolysaccharides: genetic control and symbiotic functions. Microb Cell Fact 5:7

Smith JG, Latiolais JA, Guanga GP, Pennington JD, Silversmith RE, Bourret RB (2004) A search for amino acid substitutions that universally activate response regulators. Mol Microbiol 51:887–901

Smith JN et al (2008) SdiA, an N-acylhomoserine lactone receptor, becomes active during the transit of *Salmonella enterica* through the gastrointestinal tract of turtles. PLoS ONE 3:e2826

Stevens AM, Dolan KM, Greenberg EP (1994) Synergistic binding of the *Vibrio fischeri* LuxR transcriptional activator domain and RNA polymerase to the *lux* promoter region. Proc Natl Acad Sci USA 91:12619–12623

Stover CK et al (2000) Complete genome sequence of *Pseudomonas aeruginosa* PAO1, an opportunistic pathogen. Nature 406:959–964

Suzuki K et al (2002) Regulatory circuitry of the CsrA/*csrB* and BarA/UvrY systems of *Escherichia coli*. J Bacteriol 184:5130–5140

Tan MW, Rahme LG, Sternberg JA, Tompkins RG, Ausubel FM (1999) *Pseudomonas aeruginosa* killing of *Caenorhabditis elegans* used to identify *P. aeruginosa* virulence factors. Proc Natl Acad Sci USA 96:2408–2413

Teplitski M, Ahmer BM (2004) The control of secondary metabolism, motility, and virulence by the two-component regulatory system BarA/SirA of *Salmonella* and other γ-proteobacteria. In: Pruss BM (ed) Global regulatory networks in enteric bacteria. Research Signpost, Kerala, India

Teplitski M, Al-Agely A, Ahmer BM (2006) Contribution of the SirA regulon to biofilm formation in *Salmonella enterica* serovar Typhimurium. Microbiology 152:3411–3424

Teplitski M et al (2004) *Chlamydomonas reinhardtii* secretes compounds that mimic bacterial signals and interfere with quorum sensing regulation in bacteria. Plant Physiol 134:1–10

Teplitski M, Eberhard A, Gronquist MR, Gao M, Robinson JB, Bauer WD (2003a) Chemical identification of N-acyl homoserine lactone quorum-sensing signals produced by *Sinorhizobium meliloti* strains in defined medium. Arch Microbiol 180:494–497

Teplitski M, Goodier RI, Ahmer BM (2003b) Pathways leading from BarA/SirA to motility and virulence gene expression in *Salmonella*. J Bacteriol 185:7257–7265

Teplitski M, Robinson JB, Bauer WD (2000) Plants secrete substances that mimic bacterial N-acyl homoserine lactone signal activities and affect population density-dependent behaviors in associated bacteria. Mol Plant Microbe Interact 13:637–648

Teplitski M, Mathesius U, Rumbaugh KP (2010) Perception and Degradation of N-Acyl Homoserine Lactone Quorum Sensing Signals by Mammalian and Plant Cells. Chem Rev. doi:10.1021/cr100045m E-Publication Date (Web): June 10, 2010

Thompson LS, Webb JS, Rice SA, Kjelleberg S (2003) The alternative sigma factor RpoN regulates the quorum sensing gene *rhlI* in *Pseudomonas aeruginosa*. FEMS Microbiol Lett 220:187–195

Timmers AC et al (2000) Saprophytic intracellular rhizobia in alfalfa nodules. Mol Plant Microbe Interact 13:1204–1213

Trott AE, Stevens AM (2001) Amino acid residues in LuxR critical for its mechanism of transcriptional activation during quorum sensing in *Vibrio fischeri*. J Bacteriol 183:387–392

Valverde C (2009) Artificial sRNAs activating the Gac/Rsm signal transduction pathway in *Pseudomonas fluorescens*. Arch Microbiol 191:349–359

Valverde C, Heeb S, Keel C, Haas D (2003) RsmY, a small regulatory RNA, is required in concert with RsmZ for GacA-dependent expression of biocontrol traits in *Pseudomonas fluorescens* CHA0. Mol Microbiol 50:1361–1379

Valverde C, Lindell M, Wagner EGH, Haas D (2004) A repeated GGA motif is critical for the activity and stability of the riboregulator RsmY of *Pseudomonas fluorescens*. J Biol Chem 279:25066–25074

Van Houdt R, Givskov M, Michiels CW (2007) Quorum sensing in *Serratia*. FEMS Microbiol Rev 31:407–424

Ventre I et al (2006) Multiple sensors control reciprocal expression of *Pseudomonas aeruginosa* regulatory RNA and virulence genes. Proc Natl Acad Sci USA 103:171–176

Verstraeten N et al (2008) Living on a surface: swarming and biofilm formation. Trends Microbiol 16:496–506

Von Bodman SB, Bauer WD, Coplin DL (2003) Quorum sensing in plant-pathogenic bacteria. Annu Rev Phytopathol 41:455–482

Wagner VE, Bushnell D, Passador L, Brooks AI, Iglewski BH (2003) Microarray analysis of *Pseudomonas aeruginosa* quorum-sensing regulons: effects of growth phase and environment. J Bacteriol 185:2080–2095

Wang X, Dubey AK, Suzuki K, Baker CS, Babitzke P, Romeo T (2005) CsrA post-transcriptionally represses *pgaABCD*, responsible for synthesis of a biofilm polysaccharide adhesin of *Escherichia coli*. Mol Microbiol 56:1648–1663

Waters CM, Bassler BL (2005) Quorum sensing: cell-to-cell communication in bacteria. Annu Rev Cell Dev Biol 21:319–346

Wei JR et al (2006) Regulatory roles of *spnT*, a novel gene located within transposon TnTIR. Biochem Biophys Res Commun 348:1038–1046

Weilbacher T et al (2003) A novel sRNA component of the carbon storage regulatory system of *Escherichia coli*. Mol Microbiol 48:657–670

Whistler CA, Corbell NA, Sarniguet A, Ream W, Loper JE (1998) The two-component regulators GacS and GacA influence accumulation of the stationary-phase sigma factor sigmaS and the stress response in *Pseudomonas fluorescens* Pf-5. J Bacteriol 180:6635–6641

Whistler CA, Koropatnick TA, Pollack A, McFall-Ngai MJ, Ruby EG (2007) The GacA global regulator of *Vibrio fischeri* is required for normal host tissue responses that limit subsequent bacterial colonization. Cell Microbiol 9:766–778

Whiteley M, Greenberg EP (2001) Promoter specificity elements in *Pseudomonas aeruginosa* quorum-sensing-controlled genes. J Bacteriol 183:5529–5534

Whiteley M, Parsek MR, Greenberg EP (2000) Regulation of quorum sensing by RpoS in *Pseudomonas aeruginosa*. J Bacteriol 182:4356–4360

Williams P, Camara M (2009) Quorum sensing and environmental adaptation in *Pseudomonas aeruginosa*: a tale of regulatory networks and multifunctional signal molecules. Curr Opin Microbiol 12:182–191

Williamson LL, Borlee BR, Schloss PD, Guan C, Allen HK, Handelsman J (2005) Intracellular screen to identify metagenomic clones that induce or inhibit a quorum-sensing biosensor. Appl Environ Microbiol 71:6335–6344

Workentine ML, Chang LM, Ceri H, Turner RJ (2009) The GacS-GacA two-component regulatory system of *Pseudomonas fluorescens*: a bacterial two-hybrid analysis. FEMS Microbiol Lett 292:50–56

Xiao G et al (2006a) MvfR, a key *Pseudomonas aeruginosa* pathogenicity LTTR-class regulatory protein, has dual ligands. Mol Microbiol 62:1689–1699

Xiao G, He J, Rahme LG (2006b) Mutation analysis of the *Pseudomonas aeruginosa mvfR* and *pqsABCDE* gene promoters demonstrates complex quorum-sensing circuitry. Microbiology 152:1679–1686

Yang SH, Peng Q, San Fracisco M, Wang Y, Zeng Q, Yang CH (2008) Type III secretion system genes of *Dickeya dadantii* 3937 are induced by plant phenolic acids. PLoS ONE e2973:2971–2979

Zhang RG et al (2002) Structure of a bacterial quorum-sensing transcription factor complexed with pheromone and DNA. Nature 417:971–974

Zolfaghar I, Angus AA, Kang PJ, To A, Evans DJ, Fleiszig SM (2005) Mutation of *retS*, encoding a putative hybrid two-component regulatory protein in *Pseudomonas aeruginosa*, attenuates multiple virulence mechanisms. Microbes Infect 7:1305–1316

Zuber S et al (2003) GacS sensor domains pertinent to the regulation of exoproduct formation and to the biocontrol potential of *Pseudomonas fluorescens* CHA0. Mol Plant Microbe Interact 16:634–644

Chapter 15
Beneficial Rhizobacteria Induce Plant Growth: Mapping Signaling Networks in *Arabidopsis*

Paul W. Paré, Huiming Zhang, Mina Aziz, Xitao Xie, Mi-Seong Kim, Xin Shen, and Jinlin Zhang

15.1 Agricultural Impact of Plant Growth-Promoting Rhizobacteria

Plant growth-promoting rhizobacteria (PGPR) are naturally occurring soil microorganisms that colonize roots and stimulate shoot and root growth. Such bacteria have been applied to a wide range of agricultural crops for the purpose of growth enhancement, including increased seed emergence, plant weight, and crop yields (Kloepper et al. 1991). For example, emergence increases of 10–40% result for canola when seeds are coated with PGPR before planting, and plant weight of tuber-treated potatoes increases by 80% on average by midseason (Kloepper et al. 1980). Yield increases between 10 and 20% with PGPR applications have also been documented for several other agricultural crops (Kloepper et al. 1991).

Plant-PGPR associations are mediated through an exchange of chemical metabolites. Root exudates provide energy-rich organic acids that are metabolized within hours by soil microorganisms (Jones et al. 2003) while specialized microbes generate an array of biologically active compounds that elicit plant growth promotion (Kloepper et al. 1999; Ryu et al. 2003; Paré et al. 2005). Widely accepted mechanisms for plant growth promotion by PGPR include bacterial synthesis of plant hormones (Loper and Schroth 1986; Timmusk et al. 1999; MacDonald et al. 1986), breakdown of plant produced ethylene (Glick et al. 1999), and increased mineral availability in the soil by bacterial produced siderophores (Lin et al. 1983).

P.W. Paré (✉), H. Zhang, M. Aziz, X. Xie, M.-S. Kim, X. Shen, and J. Zhang
Department of Chemistry and Biochemistry, Texas Tech University, Lubbock, TX 79409, USA
e-mail: Paul.Pare@ttu.edu

G. Witzany (ed.), *Biocommunication in Soil Microorganisms*, Soil Biology 23,
DOI 10.1007/978-3-642-14512-4_15,

15.2 Low-Molecular Weight Bacterial Signals

To differentiate between bacterial-defined signaling and bacterial mimicry of plant signaling, often times by nonvolatile components such as hormone production and siderophores, blends of volatile organic chemicals (VOCs) emitted from specific strains of PGPR, in the absence of physical contact with plant roots, have been assayed for growth induction in the model plant *Arabidopsis* (Ryu et al. 2003). Of the several PGPR strains assayed, two species *Bacillus subtilis* GB03 and *B. amyloliquefaciens* IN937a elicited plant growth via volatile emissions. To provide a comprehensive chemical profile of bacterial volatiles emitted from these biologically active strains, headspace solid phase micro extraction (SPME) coupled with software extraction of overlapping GC-separated components was employed (Farag et al. 2006). Thirty eight volatile metabolites from GB03 and IN937a were characterized. Most of the identified compounds were branched-chain alcohols released from IN937a, at much higher levels than in GB03. Principal component analysis clearly separated GB03 from IN937a, with GB03 producing higher amounts of 3-methyl-1-butanol, 2-methyl-1-butanol, and butane-1-methoxy-3-methyl. The branched-chain alcohols share a similar functional motif to that of 2,3-butanediol and may afford alternative structural patterns for elicitors from bacterial sources.

15.3 Probing Bacterial-Mediated Plant Growth Signaling Pathways

To elucidate signaling networks involved in growth promotion via PGPR VOCs, a series of mutant lines were tested (Ryu et al. 2003). Enhanced total leaf surface area was observed from exposure to GB03 VOCs for mutant lines ethylene insensitive (*etr*1), auxin-transporter-deficient and ethylene insensitive (*eir*1), gibberellic acid-insensitive (*gai*2), and brassinosteroid-insensitive (*cbb*1) mutants, thereby negating the essential involvement of brassinosteroid-, gibberellic acid-, or ethylene-signaling in the activation of growth promotion by VOCs. Considering that the auxin efflux carrier EIR1 (also known as AGR and AtPIN2) is a root-specific protein (Luschnig et al. 1998; Sieberer et al. 2000), a mutation in this auxin transporter (*eirl*) does not necessarily affect basipetal auxin transport or auxin action in leaves. Therefore, participation of auxin in VOC-induced growth promotion cannot be excluded. The cytokinin receptor-deficient (*cre*1) and cytokinin- and ethylene-insensitive (*ein*2) mutants were also tested and exhibited no growth promotion with exposure to GB03 VOCs. Although the function of EIN2 has not been clearly resolved, *ein2* mutants have been recovered in screens for *Arabidopsis* mutants resistant to auxin transport inhibitors, cytokinins, or abscisic acid and in screens for delayed senescence (Alonso et al. 1999), implicating auxin, cytokinin, and/or abscisic acid as possible mediators of growth promotion by GB03 VOCs.

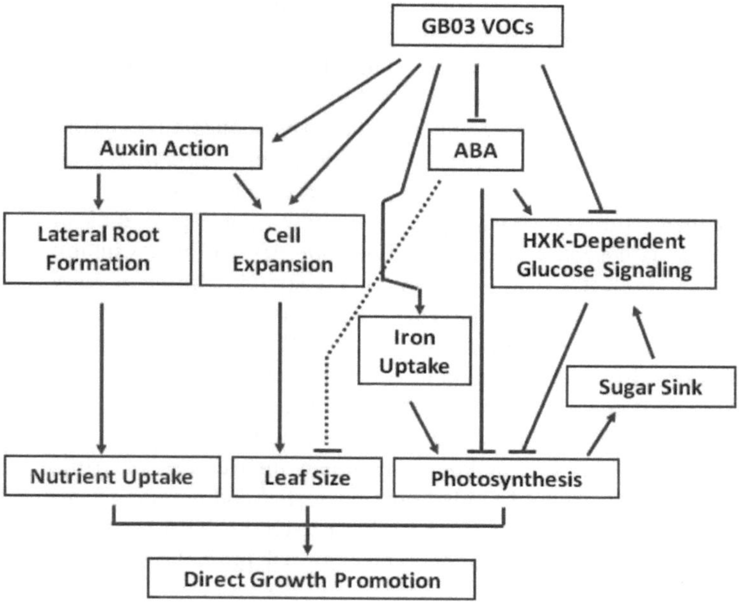

Fig. 15.1 Integrated mechanistic links for growth promotion in *Arabidopsis* by *Bacillus subtilis* (GB03) volatile emissions. *Arrows* and *blocked lines* indicate up- and down-regulated pathways, respectively; *unbroken lines* represent direct experimental evidence while *dotted lines* are proposed responses. Abbreviations: *VOCs* volatile organic chemicals, *ABA* abscisic acid, *HXK* hexokinase. See text for appropriate references

15.3.1 Bacterial Regulation of Auxin Synthesis, Transport, and Distribution in Planta

To directly probe how PGPR VOCs trigger growth in plants, RNA transcript levels of *Arabidopsis* seedlings exposed to *B. subtilus* (GB03) were examined using oligonucleotide microarrays (Zhang et al. 2007). In screening over 26,000 protein-coded transcripts, a group of approximately 600 differentially expressed genes related to cell wall modifications, primary and secondary metabolism, stress responses, hormone regulation, and other expressed proteins were identified. Transcriptional and histochemical data indicate that VOCs from the PGPR strain GB03 trigger growth promotion in *Arabidopsis* by regulating auxin homeostasis. Specifically, gene expression for auxin synthesis was up-regulated in aerial regions of GB03-exposed plants; auxin accumulation decreased in leaves and increased in roots with GB03 exposure as revealed in a transgenic DR5::GUS marker line for auxin accumulation (Ulmasov et al. 1997), suggesting activation of basipetal auxin transport. Application of the auxin transport inhibitor 1-naphthylphthalamic acid (NPA) restricted auxin accumulation to sites of synthesis (Keller et al. 2004), thereby preventing GB03-mediated decreases in shoot auxin levels as well as thwarting GB03-mediated growth promotion.

Although the role auxin plays in controlling leaf expansion remains unclear, increased auxin levels have been reported to have negative effects on leaf expansion. *Arabidopsis* mutants *sur1* and *sur2* that overproduce auxin have reduced leaf expansion (Boerjan et al. 1995). Transgenic petunia (*Petunia hybrida*), which overproduces auxin, develops leaves that are epinastic as well as smaller and narrower than wild-type plants (Klee et al. 1987). And application of exogenous auxin to bean (*Phaseolus vulgaris*) and *Arabidopsis* leaf blades inhibits long-term blade elongation (Keller et al. 2004). Auxin transport inhibition by NPA application to petioles traps auxin within the leaf and selectively increases leaf auxin content thereby inhibited leaf growth (Keller et al. 2004). On the other hand, low auxin concentrations drive cell elongation, cell enlargement, and cell differentiation in tobacco cell culture (Winicur et al. 1998; Zazimalova et al. 1995). Although NPA-induced reduction of plant growth due to a nonspecific response such as diminished plant perception of VOCs and/or downstream interference of critical nonauxin signaling pathways cannot be ruled out, the spatial and temporal redistribution of auxin observed with GB03 VOCs as well as exogenous NPA treatment demonstratively implicates auxin participation in plant growth promotion (Zhang et al. 2007).

15.3.2 Transcriptional Regulation of Cell Wall Rigidity by GB03

In addition, microarray data revealed coordinated regulation of cell wall loosening enzymes that implicated cell expansion with GB03 exposure, which was confirmed by comparative cytological measurements. Genes differentially regulated with exposure to GB03 emissions and associated with cell wall modifications were examined for possible regulatory control of cell enlargement. The induction of a group of pectin-related genes, including pectin methylesterase inhibitor, pectinase, and pectate lyase (Zhang et al. 2007) suggested that cell-wall loosening contributes to cell expansion. Pectins, a mixture of heterogeneous, branched, and highly hydrated polysaccharides rich in D-galacturonic acid, are thought to perform several cell-wall functions including determination of wall porosity, provide charged surfaces that can modulate wall pH and ion balance, and serve as recognition molecules that alert plant cells to the presence of symbiotic organisms, pathogens, and insects (Buchanan et al. 2000). Pectin methylesterases hydrolyze methyl ester of pectins and thus create free carboxyl groups that can link with Ca^{2+} ions and stabilize two polygalacturonan (PGA) chains, thereby making the pectic network more rigid. Moreover, binding of GalA units in PGA to Ca^{2+} results in calcium pectate, which is the cementing substance for cell adhesion. Besides the action of cell wall reassembly enzymes, enzymes involved in the biosynthesis of cell walls also seemed to be altered. UDP-glucose 4-epimerase is involved in channeling UDP-D-galactose into cell wall polymers and was down-regulated with plant exposure to GB03 VOCs. Mutation in *UGE4* has been reported to weaken cell structure as observed with a dramatic bulging of root epidermal cells (Schiefelbein and Somerville 1990; Baskin et al. 1992).

Traditional models of rhizobacterial-stimulated plant growth promotion suggest that soil microbes produce auxin and/or other phytohormones that drive plant growth promotion (Loper and Schroth 1986; Timmusk et al. 1999; MacDonald et al. 1986; Glick et al. 1999; Lin et al. 1983); however, in the case of biologically active bacterial VOCs, blends of volatile chemicals devoid of traditional auxins can trigger plant growth promotion mediated by auxin synthesis and transport *in planta* (Zhang et al. 2007). This observation that bacterial VOCs devoid of auxin or other known plant hormones regulate auxin homeostasis, and cell expansion provides a new paradigm as to how rhizobacteria promote plant growth.

15.3.3 GB03 Volatile Organic Compounds Elevate Plant Energy Acquisition

Although cell wall loosening in combination with increase water uptake can explain short-term leaf expansion, an increase in energy uptake seems necessary for sustained growth promotion. In measuring photosynthetic efficiency with plant exposure to GB03 volatiles, it was observed that photosynthetic capacity is augmented in *Arabidopsis* by increasing photosynthetic efficiency and chlorophyll content (Zhang et al. 2008b). Mechanistic studies revealed an elevation of sugar accumulation as well as the suppression of classic glucose signaling responses, including hypocotyl elongation and seed germination, with exposure to GB03. Compared with wild-type plants, two *Arabidopsis* mutants defective in hexokinase-dependent sugar signaling exhibited increased photosynthetic capacity, which is not further enhanced with GB03 exposure. Overlap in sugar/ABA sensing is observed in GB03-exposed plants, with a reduction of ABA biosynthetic transcripts as well as downstream metabolite levels in leaves. Moreover, exogenous ABA abrogates GB03-triggered increases in photosynthetic efficiency and chlorophyll content. These results demonstrate that certain rhizobacteria elevate photosynthesis through the modulation of endogenous sugar/ABA signaling and establish a regulatory role for soil microbes in plant acquisition of energy.

GB03 enhances *Arabidopsis* photosynthetic capacity by improving the efficiency of the conversion of light energy, as well as by enhancing the photosynthetic apparatus, as evidenced by increases in photosynthetic efficiency and chlorophyll content, respectively (Zhang et al. 2008b). Consistent with elevated photosynthesis, transcripts of genes encoding chloroplast proteins known to be associated with photosynthesis are up-regulated with GB03 exposure (Zhang et al. 2007). A plastic partition separating bacteria from the plant excludes the possibility that GB03 represses the glucose inhibition of hypocotyl elongation or of germination through competition for plant sugar uptake. Moreover, the higher hexose level in GB03-exposed plants establishes that GB03 attenuates glucose inhibitory effects through the repression of sugar signaling, rather than by lowering sugar accumulation.

Furthermore, studies using *Arabidopsis* mutants defective in hexokinase-dependent sugar signaling (Zhang et al. 2008b) indicate that GB03 augments photosynthesis through repressing hexokinase-dependent, rather than hexokinase-independent, sugar signaling. Glucose signaling largely overlaps with ABA signal transduction, as revealed by the fact that *Arabidopsis* ABA synthesis (aba) and ABA insensitive (abi) mutants are, to varying degrees, sugar-sensing mutants (Rolland et al. 2006; Smeekens 2000). Therefore, the reduction of ABA levels explains the repressed glucose signaling in GB03-exposed plants. Although ABA initiates various protective responses in plants upon biotic or abiotic stress (Finkelstein et al. 2002), ABA can also suspend the postgerminative growth of seedlings, representing an early developmental checkpoint (Lopez-Molina et al. 2001). Genetic and biochemical studies have provided enzymatic information for ABA biosynthesis, which is subject to complex regulation during plant development and in response to abiotic stresses (Finkelstein et al. 2002; Nambara and Marion-Poll 2005; Xiong and Zhu 2002). Results of transcriptional studies show that GB03 reduces plant ABA production (Zhang et al. 2008b). The future identification of cis-elements for ABA synthesis genes, and their respective transcriptional factors, will assist in the elucidation of how GB03 regulates ABA levels in plants.

An additional observation in employing the split-plate Petri dish assay is that bacterial emissions are sufficient to regulate observed photosynthetic changes (Zhang et al. 2008b). Volatiles from a nongrowth-promoting Escherichia coli strain DH5a failed to augment photosynthetic efficiency via a nonspecific microbial signal, such as the generation of elevated CO_2.

15.3.4 GB03 Regulates Iron Assimilation Independent of Metal Chelation

With the photosynthetic complex being one of the most iron-enriched cellular systems in nature with over 20 atoms per complex, photosynthetic activity is highly sensitive to the availability of iron (Spiller and Terry 1980; Sandmann 1985; Behrenfeld et al. 1996). With mounting evidence that GB03 plays a signaling role in activating growth responses in plants, the question arises as to whether such soil microbes regulate iron assimilation. Indeed GB03 activates the plant's own iron acquisition machinery to increase assimilation of metal ions in *Arabidopsis* (Zhang et al. 2009). Mechanistic studies reveal that GB03 transcriptionally up-regulates the Fe-deficiency-induced transcription factor 1 (FIT1), which is necessary for GB03-induction of ferric reductase FRO2 and the iron transporter IRT1. In addition, GB03 causes acidification of the rhizosphere by enhancing root proton release and by direct bacterial acidification, thereby facilitating iron mobility. As a result, GB03-exposed plants have elevated endogenous iron levels as well as increased photosynthetic capacity compared with water-treated controls. In contrast, loss-of-function fit1-2 mutants are compromised in terms of enhanced iron assimilation and photosynthetic efficiency triggered by GB03.

Given the importance of iron homeostasis in plant growth and development, a highly orchestrated series of root localized and shoot-borne signals have been characterized that tightly control the production of proteins involved in Fe^{3+}-chelate reductase activity and Fe^{2+} uptake, although initiation of these signals remains elusive (Romera et al. 1992; Grusak 1995; Schmidt and Schuck 1996; Curie and Briat 2003; Vert et al. 2003). GB03-triggered induction of IRT1 and FRO2 gene is not observable in the fit1 mutant, suggesting that FIT1 mediates regulation of IRT1 and FRO2 by GB03. Since FIT1 itself is iron-regulated (Colangelo and Guerinot 2004), an iron sensor upstream of FIT1 would be predicted. Characterization of activating signals in GB03 VOCs that stimulate acquisition of iron by plants may well reveal insights into early regulatory steps in plant iron uptake and homeostasis.

15.4 GB03 Volatile Organic Compounds Augment *Arabidopsis* Reproductive Success

While utilizing Petri-dish grown *Arabidopsis* seedlings has proven to be an effective model system to mechanistically probe early growth responses activated by volatiles from beneficial bacteria (Fig. 15.1) (Zhang et al. 2007, 2008a), an examination of long-term growth promotion by PGPR volatiles requires expanded growth conditions. Closed Magenta boxes (75 × 75 × 100 mm) coupled together as pairs by a plastic collar and containing solid growth medium allow for long-term plant exposure to GB03 VOCs that resulted in sustained growth promotion compared with water controls in *Arabidopsis* (Xie et al. 2009). Persistent GB03 signaling appears to be necessary and sufficient as indicated by the loss of enhanced growth when GB03 is withdrawn early in plant development. In addition to larger root mass (Zhang et al. 2007), greater rosette number likely contributes to the significantly higher fresh and dry weight numbers (Xie et al. 2009). Although flowering time was delayed with sustained GB03 exposure, seed set as a metric of reproductive success was significantly increased compared with water controls at a 10-week harvest period.

15.5 Induced Growth Promotion Beyond the Model Plant

Bacterial volatiles emitted by GB03 significantly increase biomass and essential oil production in laboratory-grown sweet basil (*Ocimum basilicum*) (Banchio et al. 2009). Because plant volatiles are utilized for flavor enhancement in foods as well as for natural defense against herbivore pests, how bacterial-induced plant growth promotion influences essential oil production on a fresh weight basis is of particular interest in the growing of commercial spices.

For field applications, GB03 is thought to survive on seeds until planted and then uses seed exudates during seed germination, directionally multiplying to reach young roots, and maintaining a robust population in the presence of field crops via plant–microbial interactions (Kloepper et al. 2004). The minimum bacterial density in the soil for triggering observable plant responses is ca. 10^4 colony forming units (cfu)/root. GB03 can maintain soil populations of 10^5 cfu/root for over 60 days after planting; negative plant development effects are not observed with bacterial populations as high as 10^{10} cfu/root (Kokalis-Burelle et al. 2006). The inoculum in Petri dish bioassays for VOC-mediated growth promotion contains ca. 10^7 cfu.

Acknowledgments This research was partially supported by the Robert Welch Foundation (D-1478) and the ACS/Frasch Foundation for Chemical Research.

References

Alonso JM, Hirayama T, Roman G, Nourizadeh S, Ecker JR (1999) EIN2, a bifunctional transducer of ethylene and stress responses in *Arabidopsis*. Science 284:2148–2152

Banchio E, Xie X, Zhang H, Paré PW (2009) Soil bacteria elevate essential oil accumulation and emissions in sweet basil. J Agric Food Chem 57:653–657

Baskin TI, Betzner AS, Hoggart R, Cork A, Williamson RE (1992) Root morphology mutants in *Arabidopsis thaliana*. Aust J Plant Physiol 19:427–437

Behrenfeld MJ, Bale AJ, Kolber ZS, Aiken J, Falkowski PG (1996) Confirmation of iron limitation of phytoplankton photosynthesis in the equatorial Pacific Ocean. Nature 383:508–511

Boerjan W, Cervera MT, Delarue M, Beeckman T, Dewitte W, Bellini C, Caboche M, Van Onckelen H, Van Montagu M, Inze D (1995) Superroot, a recessive mutation in *Arabidopsis*, confers auxin over production. Plant Cell 7:1405–1419

Buchanan BB, Gruissem W, Jones RL (2000) In: Biochemistry and molecular biology of plants. American Society of Plant Physiologists, Rockville, MD, pp 52–89

Colangelo EP, Guerinot ML (2004) The essential basic helix-loop-helix protein FIT1 is required for the iron deficiency response. Plant Cell 16:3400–3412

Curie C, Briat JF (2003) Iron transport and signaling in plants. Annu Rev Plant Biol 54:183–206

Farag MA, Ryu CM, Sumner LW, Paré PW (2006) GC-MS SPME profiling of rhizobacterial volatiles reveals prospective inducers of growth promotion and induced systemic resistance in plants. Phytochemistry 67:2262–2268

Finkelstein RR, Gampala SSL, Rock CD (2002) Abscisic acid signaling in seeds and seedlings. Plant Cell 14:S15–S45

Glick BR, Patten CN, Holguin G, Penrose DM (1999) Biochemical and genetic mechanisms used by plant growth promotion bacteria. Imperial College Press, London, pp 1–13

Grusak MA (1995) Whole-root iron (III)-reductase activity throughout the life cycle of iron grown *Pisum sativum* L. (Fabaceae): relevance to the iron nutrition of developing seeds. Planta 197:111–117

Jones DL, Dennis PG, Owen AG, van Hees PAW (2003) Organic acid behavior in soils-misconceptions and knowledge gaps. Plant Soil 248:31–41

Keller CP, Stahlberg R, Barkawi LS, Cohen JD (2004) Long-term inhibition by auxin of leaf blade expansion in bean and *Arabidopsis*. Plant Physiol 134:1217–1226

Klee HJ, Horsch RB, Hinchee MA, Hein MB, Hoffman NL (1987) The effects of overproduction of two *Agrobacterium tumefaciens* T-DNA auxin biosynthetic gene products in transgenic petunia plants. Genes Dev 1:86–96

Kloepper JW, Leong J, Teintze M, Schroth MN (1980) Enhanced plant growth by siderophores produced by plant growth-promoting rhizobacteria. Nature 286:885–886

Kloepper JW, Zablotowicz RM, Tipping EM, Lifshitz R (1991). Plant growth promotion mediated by bacterial rhizosphere colonizers. In: Keister KL, Cregan PB (eds) The rhizosphere and plant growth. Kluwer Academic Publishers, Dordecht, pp 315–326

Kloepper JW, Rodriguez-Kabana R, Zehnder GW, Murphy J, Sikora E, Fernandez C (1999) Plant root–bacterial interactions in biological control of soil borne diseases and potential extension to systemic and foliar diseases. Austr Plant Pathol 28:27–33

Kloepper JW, Ryu CM, Zhang S (2004) Induced systemic resistance and promotion of plant growth by Bacillus spp. Phytopathology 94:1259–1266

Kokalis-Burelle N, Kloepper JW, Reddy MS (2006) Plant growth-promotion rhizobacteria as transplant amendments and their effects on indigenous rhizosphere microorganisms. Appl Soil Ecol 31:91–100

Lin W, Okon Y, Hardy RWF (1983) Enhanced mineral uptake by *Zea mays* and sorghum bicolor roots inoculated with *Azospirillum brasilense*. Appl Environ Microbiol 45:1775–1779

Loper JE, Schroth MN (1986) Influence of bacterial sources of indole-3-acetic acid on root elongation of sugar beet. Phytopathology 76:386–389

Lopez-Molina L, Mongrand S, Chua NH (2001) A post-germination developmental arrest check-point is mediated by abscisic acid and requires the ABI5 transcription factor in *Arabidopsis*. Proc Natl Acad Sci USA 98:4782–4787

Luschnig C, Gaxiola RA, GrisaW P, Fink GR (1998) EIR1, a root-specific protein involved in auxin transport, is required for gravitropism in *Arabidopsis thaliana*. Genes Dev 12:2175–2187

MacDonald EMS, Powell GK, Regier DA, Glass NL, Roberto F, Kosuge T, Morris RO (1986) Secretion of zwatin, Ribosylzeatin, and ribosyl-1″-methylzeatin by *Pseudomonas savastanoi* plasmidcoded cytokinin biosynthesis. Plant Physiol 82:742–747

Nambara E, Marion-Poll A (2005) Abscisic acid biosynthesis and catabolism. Annu Rev Plant Biol 56:165–185

Paré PW, Farag MA, Krishnamachari V, Zhang H, Ryu CM, Kloepper JW (2005) Elicitors and priming agents initiate plant defense responses. Photosynth Res 85:149–159

Rolland F, Baena-Gonzalez E, Sheen J (2006) Sugar sensing and signaling in plants: conserved and novel mechanisms. Annu Rev Plant Biol 57:675–709

Romera FJ, Alcantara E, de Diaz LA, Guardia M (1992) Role of roots and shoots in the regulation of the Fe efficiency responses in sunflower and cucumber. Physiol Plant 85:141–146

Ryu CM, Farag MA, Hu CH, Reddy MS, Wei HX, Paré PW, Kloepper JW (2003) Bacterial volatiles promote growth in *Arabidopsis*. Proc Natl Acad Sci USA 100:4927–4932

Sandmann G (1985) Consequences of iron-deficiency on photosynthetic and respiratory electron-transport in blue-green-algae. Photosynth Res 6:261–271

Schiefelbein JW, Somerville C (1990) Genetic control of root hair development in *Arabidopsis thaliana*. Plant Cell 2:235–243

Schmidt W, Schuck C (1996) Pyridine nucleotide pool size changes in iron-deficient *Plantago lanceolata* roots during reduction of external oxidants. Physiol Plant 98:215–221

Sieberer T, Seifert GJ, Hauser MT, Grisafi P, Fink GR, Luschnig C (2000) Post-transcriptional control of the *Arabidopsis* auxin eflux carrier EIR1 requires AXR1. Curr Biol 10:1595–1598

Smeekens S (2000) Sugar-induced signal transduction in plants. Annu Rev Plant Physiol Plant Mol Biol 51:49–81

Spiller S, Terry N (1980) Limiting factors in photosynthesis II. Iron stress diminishes photochemical capacity by reducing the number of photosynthetic units. Plant Physiol 65:121–125

Timmusk S, Nicander B, Granhall U, Tillberg E (1999) Cytokinin production by *Paenibacillus polymyxa*. Soil Biol Biochem 31:1847–1852

Ulmasov T, Murfett J, Hagen G, Guilfoyle TJ (1997) Aux/IAA proteins repress expression of reporter genes containing natural and highly active synthetic auxin response elements. Plant Cell 9:1963–1971

Vert GA, Briat JF, Curie C (2003) Dual regulation of the *Arabidopsis* high-affinity root iron uptake system by local and long-distance signals. Plant Physiol 132:796–804

Winicur ZM, Zhang GF, Staehelin LA (1998) Auxin deprivation induces synchronous golgi differentiation in suspension-cultured tobacco BY-2 cells. Plant Physiol 117:501–513

Xie X, Zhang H, Paré PW (2009) Sustained growth promotion in *Arabidopsis* with long-term exposure to the beneficial soil bacterium *Bacillus subtilis* (GB03). Plant Signal Behav 4:1–6

Xiong L, Zhu JK (2002) Regulation of abscisic acid biosynthesis. Plant Physiol 133:29–36

Zazimalova E, Opatrny Z, Brezinova A (1995) The effect of auxin starvation on the growth of auxin-dependent tobacco cell culture dynamics of auxin binding activity and endogenous free IAA content. J Exp Bot 46:1205–1213

Zhang H, Kim MS, Krishnamachari V, Payton P, Sun Y, Grimson M et al (2007) Rhizobacterial volatile emissions regulate auxin homeostasis and cell expansion in *Arabidopsis*. Planta 226:839–851

Zhang H, Kim MS, Sun Y, Dowd SE, Shi H, Paré PW (2008a) Soil bacteria confer plant salt tolerance by tissue specific regulation of the sodium transporter HKT1. Mol Plant Microbe Interact 21:737–744

Zhang H, Xie X, Kim MS, Kornyeyev DA, Holaday S, Paré PW (2008b) Soil bacteria augment *Arabidopsis* photosynthesis by decreasing glucose sensing and abscisic acid levels. Plant J 56:264–273

Zhang H, Sun Y, Xie X, Kim M-S, Dowd SE, Paré PW (2009) A soil bacterium regulates plant acquisition of iron via deficiency-inducible mechanisms. Plant J 58:568–577

Chapter 16
Signal and Nutrient Exchange in the Interactions Between Soil Algae and Bacteria

Max Teplitski and Sathish Rajamani

16.1 Introduction

In aquatic and terrestrial environments, microorganisms are typically found in multicellular consortia, which often include prokaryotic and eukaryotic organisms. Such consortia have important roles in aquatic and terrestrial ecosystems. For example, in many arid environments consortia of cyanobacteria, algae, and lichens (as well as biological polymers and small molecules released by them) form biological soil crusts (BSC). The crusts contribute to preventing erosion, improving soil structure and texture; they serve as sources of fixed carbon and nitrogen in ecosystems that otherwise have limited productivity. Aquatic microbial consortia ("biofilms") have similarly important functions in nutrient cycling, surface conditioning, etc. Precise gene regulation and exchange of chemical cues between the members of the consortium contribute to structuring and function of these communities. The goal of this chapter is to discuss signal and nutrient exchange that may contribute to the interactions between bacteria and soil algae within multispecies communities.

Even though soil is sometimes considered an inhospitable environment for photosynthetic microorganisms (like cyanobacteria and algae), they have been identified in samples collected from terrestrial ecosystems on all continents (Flechtner 1998; Otsuka et al. 2008; Langhans et al. 2009). Furthermore, over one hundred species of photosynthetic organisms representing 70 different genera have been recovered

M. Teplitski (✉)
Soil and Water Science Department, School of Natural Resources and Environment, University of Florida-Institute of Food and Agricultural Sciences, 2159 McCarty Hall, Gainesville, FL 32618, USA
e-mail: maxtep@ufl.edu

S. Rajamani
Department of Microbiology and Immunology, Dartmouth Medical School, 208 Vail building, Hanover, NH 03755, USA

G. Witzany (ed.), *Biocommunication in Soil Microorganisms*, Soil Biology 23,
DOI 10.1007/978-3-642-14512-4_16,

from surfaces or interior of rocks (van Thielen and Garbary 1998 and references therein). These observations indicate that soils and rocks can sustain diverse populations of eukaryotic and prokaryotic photosynthetic microorganisms. The interpretation of earlier studies on the diversity of soil algae may be complicated by the recent changes in the classification of algae and related organisms. For example, cyanobacteria were traditionally grouped with algae under the name "blue-green algae." However, cytological and genetic evidence places cyanobacteria within Negibacteria of the kingdom Bacteria (Cavalier-Smith 2004). Within the Six Kingdoms of Life, organisms that have been traditionally referred to as eukaryotic algae are now placed into two kingdoms, Plantae and Chromista (Cavalier-Smith 2004). The latter also contains oomycetes (including plant pathogens like *Pythium* and *Phytophtora*), which have been earlier classified as fungi. In this review, we will mostly consider interactions of proteobacteria and posibacteria with algae from the phyla Glaucophyta, Rhodophyta, Chlorophyta (kingdom Plantae), and members of the kindgom Chromysta. Even though our main goal is to focus on terrestrial algae, this review will be incomplete without comparing terrestrial prokaryote–algal interactions with the interactions that take place in aquatic environments. As appropriate, such comparisons will be introduced and discussed in this chapter.

16.2 Phylogenetic Diversity of Algal-Associated Prokaryotic Microbiota

Soil surfaces and subsurface environments not only harbor but also support growth of an impressive diversity of algae, their close relatives oomycetes, as well as cyanobacteria. For example, BSC contain up to 69 different chlorolichens, 68 representatives of algae, 62 bryophytes, 35 different species of cyanobacteria, and 13 cyanolichens [(Langhans et al. 2009) and references therein]. A sample of BCS harvested in sand dunes of the upper Rhine Valley (Germany) contained 26 algal species (including *Bracteacoccus* cf. *minor*, cf. *chlorosarcinopsis*, *Chlamydomonas*, *Chlorella*, *Chlorococcum* cf. *infusionum*, *Cocomyxa* cf. *confluens*, *Cylindrocystis*) and 13 species of cyanobacteria (including *Nostoc*, *Lyngbya*, *Microcoleus*) (Langhans et al. 2009). Interestingly, studies in other arid and semiarid areas similarly identified microalgae *Chlorella vulgaris* and *Bracteacoccus* (as well as *Stichococcus* and *Diplosphaera*) as dominant Chlorophytes consistently found in seven different locations in North American deserts (Flechtner 1998). *Chlorella* and *Stichococcus* were also commonly isolated from surfaces or in the interior of rocks (van Thielen and Garbary 1998). In addition to these common algae, soil samples collected in California, Arizona, and New Mexico contained 14–35 different species of Chlorophytes, 1–8 species of Xanthophytes and some samples also contained Eustigmatophytes (Flechtner 1998). It should be noted, however, that the dominant algal species identified in different studies may vary (Flechtner 1998), and some of this uncertainty could be due to the techniques used to recover and enumerate algae (e.g., direct counts vs. culture-based techniques) (Langhans et al. 2009).

In soil as well as aquatic environments, algae are found in association with bacteria. To characterize bacteria associated with soil isolates of *Chlorella* spp., Otsuka et al. carried out denaturing gradient gel electrophoresis (DGGE) profiling of DNA isolated from seven independent algal cultures and also sequenced 16S ribosomal RNA (rRNA) genes of the bacterial associates of the alga. Prior to DNA extraction, bacteria and algae were cocultured in a liquid medium. PCR-DGGE profiles revealed banding patterns that were unique to each *Chlorella* isolate, although several common bands were also present (Otsuka et al. 2008). Sequencing of the 16S rRNA genes of the culturable bacteria isolated from *Chlorella* revealed that previously unculturable planctomycetes and flavobacteria, as well as *Sphingomonas melonis*, were associated with multiple *Chlorella* cultures. At least six novel[1] bacteria were common to multiple cultures of *Chlorella* (Otsuka et al. 2008). A comparison of 16S rRNA gene sequence profiles of bacteria associated with the alga after 1 month and 1 year of nonauxenic *Chlorella* cultures revealed that temporally separated samples contained flavobacteria, sphingobacteria, and α-proteobacteria (Otsuka et al. 2008). In a 1-year-old nonauxenic culture of *Chlorella*, Otsuka et al. also identified 16S rRNA gene sequences belonging to α-proteobacteria (*Afipia massiliensis*, *Caulobacter vibriodes*, *Phyllobacterium leguminum*, *Azospirillum* spp.), β−proteobacteria (*Ralstonia* spp.), γ-proteobacteria (*Lysobacter koreensis*, *Pseudomonas fragi*, *P. migulae*), and actinobacteria (Otsuka et al. 2008). Interestingly, *Sphingomonas* spp. and *Ralstonia* spp. were isolated from nonauxenic laboratory cultures of *Chlorella* propagated by another group (Watanabe et al. 2005). Even though bacterial profiles of soils from which the alga were harvested were not determined, and despite the fact that less than only 50 16S rRNA sequences were characterized in the two studies, it is still tempting to suggest that some soil bacteria may have evolved to interact with the soil algae.

A hypothesis that the phycosphere microbiota is specific to a particular species of microalgae was suggested by studies of bacteria associated with phytoplankton blooms (Hasegawa et al. 2007; Sapp et al. 2007). Ribosomal Intergenic Spacer Analysis (RISA) fingerprints of bacterial communities associated with six phytoplankton species in Helgoland Roads (Germany) were clearly distinct (Sapp et al. 2007). Sequencing of the most prominent DGGE bands suggested α- and γ-proteobacteria, as well *Flavobacteria-Sphingobacteria*, were most commonly found associated with the six algae. The majority (89%) of α-proteobacteria were either *Roseobacter* or *Sulfitobacter*; approximately 6% were *Sphingomonads*. Alteromonads and oceanosprilliae were most common γ-proteobacteria isolated from phytoplankton (Sapp et al. 2007). Phylogenetically similar bacteria were isolated from a toxic dinoflagellate *Alexandrium fundyense* in Canada (Hasegawa et al. 2007).

[1]"Novel" was defined by the authors as having less than 94% similarity in the V3 region of the 16S rRNA gene to the closest known relative (Otsuka et al. 2008).

Microbiota associated with *A. fundyense* was distinct from free-living bacteria in the water column and those isolated from particles; however, there was also a significant overlap in the microbial species composition in these three habitats (Hasegawa et al. 2007), which makes it difficult to establish unequivocally that a particular bacterium is an obligate symbiont associated with phytoplankton.

Collectively, the results of these studies suggest that some bacterial species enter into commensal or mutualistic interactions with algae in soil and aquatic environments. It is far from clear, however, whether these interactions are truly coevolved. Studies in other eukaryote-bacterial symbioses have revealed intricate signal and nutrient exchange between the partners, their ability to alter gene expression and effect organogenesis (Hirsch et al. 2003; Gil et al. 2004; Nyholm and Mcfall-Ngai 2004). Below, we will analyze recent discoveries of the signal and nutrient exchange between bacteria and algae to test the hypothesis that it contributes to the establishment of algae-associated microbial communities.

16.3 Nutrient and Signal Exchange in Algal–Bacterial Interactions

16.3.1 Carbon and Nitrogen Exchange in the Phycosphere

Mucus released by the algae is the main source of fixed carbon in the phycosphere. The mucus sheath of *Chlorella sorokiana*, for example, consists of carbohydrates (3.6 mg g^{-1} of dry cell weight), proteins (0.8 mg g^{-1} of dry cell weight), and metals (mostly Mg^{2+}, Fe^{2+}, Mn^{2+}, at 1168, 4.7, and 3.3 mg g^{-1} of dry cell weight), respectively (Watanabe et al. 2006). Sucrose and ribose were the most abundant sugars in mucus of *C. sorokiana* (1718 mg g^{-1} of dry cell weight and 216 mg g^{-1} of dry cell weight, respectively); it also contained galacturonic acid (750 mg g^{-1} of dry cell weight), xylitol (435 mg g^{-1}), innositol (317 mg g^{-1}), as well as smaller amounts of mannose, galactose, arabinose, rhamnose, and fructose (Watanabe et al. 2006). The composition of *Chlorella* mucus is clearly distinct from plant root mucus (mucilage) secreted by the vascular plants in their rhizosphere; the latter mostly consists of arabinose and galactose, with smaller amounts of uronic acids and other sugars [(Knee et al. 2001) and references therein].

The ability to efficiently utilize mucus polymers from the host is usually an important trait of coevolved symbionts. For example, supplementation of a mineral salts medium with high molecular weight mucilage from pea promoted growth of the plant symbiotic bacterium *Rhizobium leguminosarum* by ~50–100-fold. The utilization of pea mucus by *R. leguminosarum* was further increased by the addition of naringenin, a plant flavonoid symbiotic signal (Knee et al. 2001). Other soil bacteria were also capable of utilizing pea mucus, although to a lesser degree: the supplementation of mineral medium with pea mucilage increased their growth by 10–50-fold (compared with mineral salts). The addition of naringenin did not increase pea mucilage utilization by nonsymbiotic bacteria (Knee et al. 2001). A similar study with bacteria isolated from surfaces of microalgae revealed that the addition of *Chlorella* extracellular organic carbon increased growth of its bacterial commensals in a pure culture by at least fivefold (Watanabe et al. 2005). However, there is no published evidence that mucus secreted by *Chlorella* is utilized differently by its commensals or free-living soil bacteria.

In addition to using organic carbon released by the algae, bacteria isolated from surfaces of microalgae can in turn promote growth of algae. A coculture of *Chlorella sorokiana* with bacteria increased growth of the alga by 10–20% (Watanabe et al. 2005). Auxenic cultures of *C. sorokiana* that were maintained in light for 5 months on agar slants lost ~40% of their chlorophyll, compared with cocultures with a bacterial consortium under the same conditions (Watanabe et al. 2005). Similarly, growth (and/or chlorophyll content) of pure cultures of *Chlamydomonas reinhardtii* was increased by ~sixfold in the presence of native bacteria (Nikolaev et al. 2008). The supplementation of *C. reinhardtii* cultures with *Bacillus* spp. or *Rhodococcus terrea* had a more modest effect on growth and/or chlorophyll production (Nikolaev et al. 2008). These results clearly indicate that both the microalgae and their bacterial commensals derive benefits from the association.

To better understand the mechanisms of algal growth promotion by bacteria, artificial "symbioses" between *Chlorella*, *Chlamydomonas*, and various well-characterized plant growth promoting bacteria were started under laboratory conditions. In earlier studies, cultures of *Chlamydomonas reinhardtii* were mixed with nitrogen-fixing *Azotobacter* on agar plates lacking carbon and/or nitrogen (Gyurjan et al. 1984, 1986). Under these conditions, a monoculture of *C. reinhardtii* lost chlorophyll and died within the first 2 months of the study, while *Azotobacter–Chlamydomonas* consortia persisted for at least 2 years (Gyurjan et al. 1986). The growth promoting effects were at least in part due to the ability of the bacteria to fix nitrogen, as suggested by nitrogenase activity (measured as acetylene reduction) (Gyurjan et al. 1984, 1986). The ability of nitrogen-fixing *Bacillus pumilus* to promote growth of *Chlorella vulgaris* was recently demonstrated using coimmobilization of the two organisms in alginate beads (De-Bashan and Bashan 2008; Hernandez et al. 2009). In the presence of *B. pumilus* (originally isolated from arid soils), cell numbers of *C. vulgaris* increased by 3×10^6 compared with a culture that was not supplemented with the bacilli, reaching population densities that were essentially the same as in cultures supplemented with ammonium chloride (Hernandez et al. 2009). The growth promoting effects of *B. pumilus* on algae were abolished in the presence of ammonia, suggesting that nitrogen fixed by the bacilli is most likely responsible for the growth promoting effects (Hernandez et al. 2009).

Results of these studies demonstrate that microalgae and their associated microbiota can benefit from the interaction. Similar loose interactions between plants and free-living nitrogen-fixing bacteria (*Azotobacter* spp., *Azosprillum* spp among others) are now well documented (Baldani and Baldani 2005).

16.3.2 Bacterially Produced Plant Hormones Stimulate Algal Growth

Production of plant hormones (or their analogs) by bacteria plays an important role in many plant–bacterial interactions. Plant symbiotic bacteria (e.g., *Rhizobia* spp, *Azospirillum*) and plant pathogens (*Agrobacterium* spp., *Erwinia herbicola*) produce plant hormones, which are thought to contribute to the development of new plant organs occupied by the microorganisms (Lambrecht et al. 2000). Thus, it appears that the ability to manipulate gene expression and relevant physiological changes by the production of plant hormones is an important, coevolved trait in plant-associated bacterial symbionts and pathogens.

A hypothesis that plant hormones produced by bacteria would also stimulate growth of microalgae was tested (de-Bashan et al. 2008; De-Bashan and Bashan 2008) in the medium that already contained soluble nitrogen (25 mg/L NH_4Cl). In the presence of *A. brasiliense* and increasing concentrations of tryptophan (a precursor for an auxin plant hormone IAA), growth of *Chlorella vulgaris* was increased fourfold (de-Bashan et al. 2008; De-Bashan and Bashan 2008). A coculture of *C. vulgaris* and IAA-deficient mutants of *A. brasiliense* had either a reduced effect on algal growth or had no growth promoting effect at all (de-Bashan et al. 2008). Growth of the alga was promoted by the culture filtrate of the bacterial IAA mutant only if the culture filtrates were also supplemented with IAA. In control experiments, supplementation of the growth medium with 10 mg mL^{-1} IAA promoted growth of *C. vulgaris* (de-Bashan et al. 2008).

16.3.3 Bidirectional Vitamin Exchange and Vitamin-Mediated Signaling in Algal–Bacterial Interactions

Many algae require vitamin B_{12} (cobalamin), vitamin B_1 (thiamine), and vitamin H (biotin) for growth, although not all algae require all three of these vitamins (Croft et al. 2005; Grossman et al. 2007). Recent genomic sequencing of *Chlamydomonas reinhardtii* and parallel physiological studies indicate that this microalga does not require these vitamins for growth (Grossman et al. 2007). At least 171 species of algae (of 326 tested), however, required external cobalamin for methionine synthesis and growth, suggesting that many algae rely on their commensal or symbiotic bacteria for the supply of cobalamin (Croft et al. 2005). An isolate of *Halomonas* sp. was shown to provide cobalamin to a marine red alga *Porphyridium purpureum* (Croft et al. 2005). Interestingly, in a legume-*Sinorhizobium* symbiosis, a bacterial mutant that was unable to synthesize cobalamin was defective in forming symbiotic nodules (at least on some plant hosts) (Medina et al. 2009). However, because cobalamin is required for methionine synthesis, a cobalamin mutant is also a methionine auxotroph (Medina et al. 2009). It is not yet clear whether the symbiotic defect of the cobalamin-deficient *Sinorhizobium* was due to the defect in the vitamin exchange with the plant host or whether it was a result of the inability to synthesize methionine.

In addition to their role as enzyme cofactors, vitamins appear to play important signaling roles. For example, vitamin riboflavin (vitamin B_2) and its derivative lumichrome (Fig. 16.1) were shown to affect plant growth and physiology. Treatment of plant seedlings with riboflavin promoted their resistance to viral, bacterial, and fungal pathogens (Dong and Beer 2000). The addition of lumichrome at nanomolar level increased plant shoot and root growth (Phillips et al. 1999; Matiru and Dakora 2005). Treatment of plant seeds and seedlings with lumichrome increased growth and stimulated stomatal conductance (Phillips et al. 1999). These studies suggested that both riboflavin and lumichrome, both self-produced and supplied by the commensal bacteria, have important roles in eukaryote–bacterial interactions. Intriguingly, lumichrome and riboflavin produced and secreted by *Chlamydomonas reinhardtii* were shown to alter population density-dependent gene expression in bacteria (Rajamani et al. 2008).

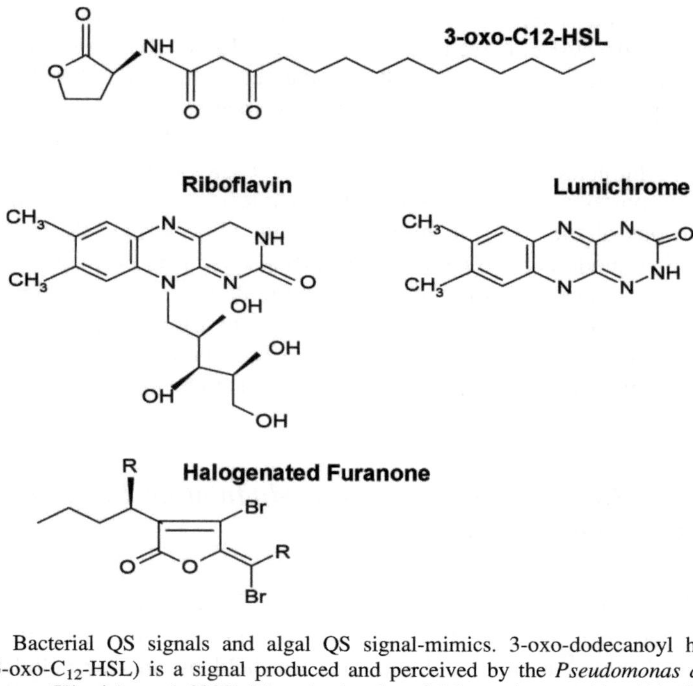

Fig. 16.1 Bacterial QS signals and algal QS signal-mimics. 3-oxo-dodecanoyl homoserine lactone (3-oxo-C_{12}-HSL) is a signal produced and perceived by the *Pseudomonas aeruginosa* Las QS system (Kiratisin et al. 2002). Vitamin signals and QS agonists capable of binding to LasR and stimulating LasR-mediate gene expression were identified in culture filtrates of a soil microalga *C. reinhardtii* (Rajamani et al. 2008), although these compounds are known to be produced by bacteria and by plants (Treadwell and Metzler 1972; Phillips et al. 1999; Joseph and Phillips 2003). Halogenated furanone produced by a red alga *Delisea pulchra* are the best characterized QS antagonists (Givskov et al. 1996). These compounds block bacterial QS by binding to the AHL receptor polypeptides and targeting them for degradation (Manefield et al. 2002; Koch et al. 2005)

Lumichrome and riboflavin were recently identified in a search for microalgal compounds capable of affecting cell-to-cell signaling in bacteria (Rajamani et al. 2008). Many bacteria rely on small diffusible signal molecules to effect changes in gene expression that parallel increases in bacterial population densities within diffusion-limited environments [rev. Dobretsov et al. (2009)]. This type of cell-to-cell signaling is known as "quorum sensing" (QS). Acyl homoserine lactones (AHLs, Fig. 16.1) are one of the best characterized bacterial QS signals. Inside bacterial cells, AHLs are bound by LuxR-like regulators, and the LuxR–AHL complex then binds within promoters of the genes subject to QS control (Zhang et al. 2002; Koch et al. 2005). Compounds that bind to LuxR proteins and thus inhibit bacterial QS have previously been characterized (see below); however, lumichrome and riboflavin are the first characterized biologically derived agonists that are structurally distinct from AHLs yet are capable of interacting with AHL receptors.

The ability of lumichrome and riboflavin to interact with AHL receptors was first detected using semisynthetic bacterial QS reporters (Rajamani et al. 2008). These reporters consist of a gene encoding an AHL receptor (*lasR*, a *luxR* homologue from *Pseudomonas aeruginosa*), a promoter controlled by LasR and a downstream promoterless *luxCDABE* cassette (Winson et al. 1998). To rigorously test the hypothesis that lumichrome and riboflavin interact with the LasR AHL receptor, additional reporters were constructed and their responses to synthetic lumichrome and riboflavin were tested. A study of Rajamani et al. (2008) demonstrated that the effect of lumichrome and riboflavin on the LasR-based tandem dimer RFP (tdTomato) reporter required the same amino acid residues that are also involved in the interactions of the receptor with the cognate AHL signals (Rajamani et al. 2008). In silico modeling and gel mobility shift assays using purified LasR further supported the hypothesis that lumichrome and riboflavin are the first characterized vitamin signals produced by microalga and capable of affecting QS in soil bacteria (Rajamani et al. 2008). The function of these compounds in structuring of the bacterial communities associated with algae remains to be elucidated.

16.3.4 The Role of Algal Signals in Modulating Bacterial Quorum Sensing

In terrestrial and aquatic environments, microorganisms are found within multicellular consortia. Microphotographs reveal that bacteria colonize surface of *Chlorella* and *Chlamydomonas* as microcolonies that are held by an extracellular matrix (Gyurjan et al. 1984, 1986; Watanabe et al. 2005; Imase et al. 2008). This is not uncommon: bacteria that colonize surfaces of plants are also found as micro-colonies or biofilms encased in the extracellular matrix of plant and microbial origin [rev. Danhorn and Fuqua (2007)]. To form multicellular communities, to interact with other organisms within these communities, and to colonize biotic substrata, bacteria rely on a variety of self-produced signals and chemical cues.

Quorum Sensing is one of bacterial gene regulatory systems that contributes to structuring of biofilms [rev. (Pasmore and Costerton (2003); Wolfe et al. (2003); Stanley and Lazazzera (2004); Dobretsov et al. (2009)]. The presence of QS was not tested in the phycosphere of soil algae. However, bacterial AHL signal production and perception associated with QS is well-documented in the rhizosphere of plants (Pierson and Pierson 1996; Ramos et al. 2001; Gao and Teplitski 2008); therefore, it is reasonable to hypothesize that bacteria similarly use QS to control gene expression within their colonies on surfaces of algae.

Algae and vascular plants produce compounds that alter QS in the associated bacterial communities (Givskov et al. 1996; Teplitski et al. 2000, 2004; Manefield et al. 2001; Gao et al. 2003, 2007; Bjarnsholt et al. 2005; Koch et al. 2005; Skindersoe et al. 2008). Halogenated furanones produced by a marine red alga *Delisea pulchra* are the best characterized eukaryotic inhibitors of bacterial QS [(Givskov et al. 1996) and Fig. 16.1]. In vivo, these compounds bind to the nascent AHL receptor polypeptide and prevent its correct folding, thus targeting the misfolded peptide for degradation by proteases (Manefield et al. 2001; Koch et al. 2005). Under laboratory conditions, halogenated furanones inhibit QS-mediated behaviors in gram-negative bacteria (Givskov et al. 1996; Manefield et al. 2001; Arevalo-Ferro et al. 2003; Hentzer and Givskov 2003; Hentzer et al. 2003). In situ, vesicle-mediated release of halogenated furanones on the surfaces of algal thalli shifts population of associated bacteria from gram negative (common marine microorganisms) to gram positive, which are typically under-represented in marine environments (Dworjanyn et al. 1999; Dworjanyn et al. 2006).

The ability to produce inhibitors of QS was tested in four soil algae: *Chlamydomonas reinhardtii*, *Chlorella vulgaris*, *C. fusca*, and *C. mutabilis* (Teplitski et al. 2004). As shown in Fig. 16.2, overlays of algal colonies with bacteria in which QS contributes to the production of light suggest that some soil algae are capable of modulating bacterial light production, either by affecting AHL-mediated signaling or the cell-to-cell signal transduction cascade controlled by an AI-2 signal. In addition to affecting QS-mediated light production, colonies of *C. reinhardtii* secreted compounds that reduced QS-controlled production of antibiotic pigments violacein and phenazine in two soil bacteria (Fig. 16.3).

In addition to the compounds that inhibit bacterial QS, *C. reinhardtii* was shown to produce chemically separable activities that stimulate QS in the semisynthetic reporters and also in the wild-type soil bacteria (Teplitski et al. 2004). Further bioassay-guided purification of bioactive compounds identified lumichrome as a QS agonist (Rajamani et al. 2008) and also revealed at least two peaks of activity separable with reverse phase Si HPLC (Teplitski et al. 2004). Treatment of prequorate cultures of a wild-type soil bacterium *Sinorhizobium meliloti* with a purified QS signal mimic from *C. reinhardtii* affected accumulation of 25 polypeptides. Sixteen of the 25 polypeptides responsive to the algal mimic were also subject to regulation by bacterial AHLs (Teplitski et al. 2004). These results indicate that both aquatic and soil algae are capable of QS in the associated bacteria and thus alter bacterial behaviors that may be relevant to the algal–bacterial interactions.

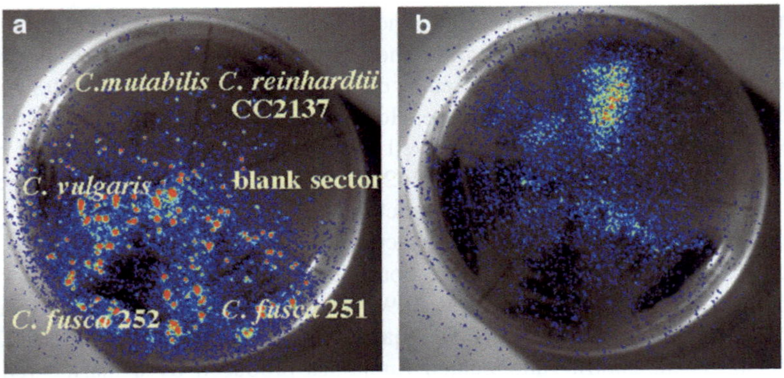

Fig. 16.2 Soil algae affect luminescence in *Vibrio harveyi*. *Chlorella mutabilis*, *Chlamydomonas reinhardtii*, *Chlorella vulgaris*, and two strains of *Chlorella fusca* were grown on TAP agar. Plates with algal streaks were then overlaid with a soft agar suspension of the wild-type *V. harveyi* 404 (PMH 2193 SK) (**a**) or *V. harveyi* BB170 (a reporter in which luminescence largely depends on the production of the AI-2 signal) (**b**). Luminescence was measured with a Hamamtsu C2400 intensified CCD camera. The false-color image of luminescence intensity was superimposed onto the black and white image of the plates

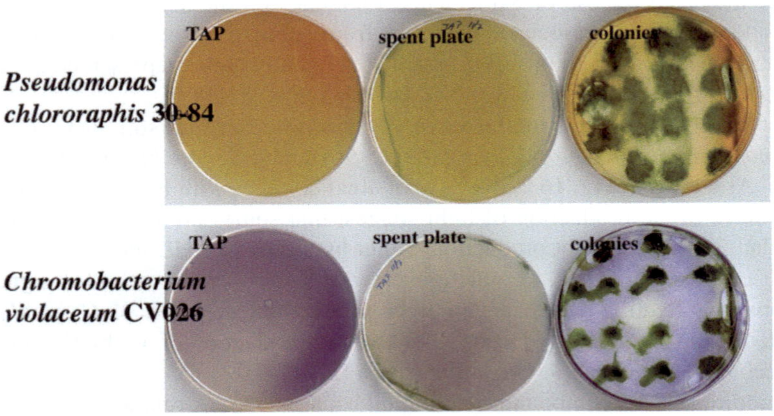

Fig. 16.3 QS-dependent pigment production in soil bacteria is affected by *Chlamydomonas reinhardtii*. In *Pseudomonas chlororaphis* 30–84 and in *Chromobacterium violaceum*, production of the antibiotic pigments requires functional QS circuitry (Pierson and Pierson 1996; McClean et al. 1997). As shown in the top panel, wild-type *P. chlororaphis* 30–84 produces bright *orange* pigment phenazine when seeded in LB agar overlaid on the TAP medium (Tris-acetate agar used to culture *Chlamydomonas*). Less phenazine was produced when bacterial suspension was overlaid on TAP agar, where *C. reinhardtii* was previously cultured (middle panel). Production of phenazine was further reduced in the bacterial lawn seeded on top of colonies of *C. reinhardtii* CC2137 (*right panel*). Similarly, less violacein was produced by *Chomobacterium violaceium* CV026 reporter when seeded onto spent plates or on top of algal colonies. Because CV026 does not produce own AHLs, top agar overlays were supplemented with C_4-HSL as in (Teplitski et al. 2000). *C. reinhardtii* CC2137 was grown on cellulose Whatman #1 filters, which were placed on top of TAP agar. For the assays, filter paper with algal colonies was lifted off the plates. Spent plates and filter paper with algal colonies were overlaid with suspensions of the bacteria in LB agar (0.3% wt/v)

In addition to manipulating bacterial QS, some algae appear to detect AHL signals produced by bacteria. For example, C_4-HSL (one of seven AHLs tested) promoted release and settlement of spores produced by a rhodophyte *Acrochaetium* (Weinberger et al. 2007). In these assays, C_4-HSL was active at 100 mM (Weinberger et al. 2007). Such high concentrations of AHLs are usually found within biofilms (Charlton et al. 2000; Dobretsov et al. 2009). Preferential settlement of spores from *Ulva intestinalis* on AHL-producing biofilms was also demonstrated (Joint et al. 2002). *Ulva* spores also exhibited chemokinesis along a gradient of AHLs (Wheeler et al. 2005). The ability to detect and respond to bacterial AHLs is not uncommon in eukaryotes and was reported in plants and animals (Smith et al. 2002; Joseph and Phillips 2003; Mathesius et al. 2003). However, the mechanism(s) by which eukaryotes detect these bacterial signals are not yet known.

16.4 Conclusions and Future Directions

Studies of the interactions of soil and aquatic algae with their associated microbiota suggest that some species of bacteria may be more commonly isolated from phycosphere of specific algae. This conclusion, however, is based on a limited number of studies. Further surveys are needed to rigorously establish bacterial diversity and species richness in phycosphere.

Several studies have demonstrated that algae may benefit from the association with bacteria. Algae may derive nitrogen, vitamins, and also plant hormones from their bacterial associates. Using defined and random mutants of bacteria, it will be important to learn whether other bacterial behaviors or metabolites are capable of modulating growth of the algae and contribute to structuring of algal-associated bacterial communities.

References

Arevalo-Ferro C et al (2003) Identification of quorum-sensing regulated proteins in the opportunistic pathogen *Pseudomonas aeruginosa* by proteomics. Environ Microbiol 5:1350–1369

Baldani JI, Baldani VL (2005) History on the biological nitrogen fixation research in graminaceous plants: special emphasis on the Brazilian experience. An Acad Bras Cienc 77:549–579

Bjarnsholt T et al (2005) Garlic blocks quorum sensing and promotes rapid clearing of pulmonary *Pseudomonas aeruginosa* infections. Microbiology 151:3873–3880

Cavalier-Smith T (2004) Only six kingdoms of life. Proc Biol Sci 271:1251–1262

Charlton TS et al (2000) A novel and sensitive method for the quantification of N-3-oxoacyl homoserine lactones using gas chromatography-mass spectrometry: application to a model bacterial biofilm. Environ Microbiol 2:530–541

Croft MT, Lawrence AD, Raux-Deery E, Warren MJ, Smith AG (2005) Algae acquire vitamin B12 through a symbiotic relationship with bacteria. Nature 438:90–93

Danhorn T, Fuqua C (2007) Biofilm formation by plant-associated bacteria. Annu Rev Microbiol 61:401–422 de-Bashan LE, Antoun H, Bashan Y (2008) Involvement of indole-3-acetic acid produced by the growth-promoting bacterium *Azospirillum* spp. in promoting growth of *Chlorella vulgaris*. J Phycol 44:938–947

De-Bashan LE, Bashan Y (2008) Joint immobilization of plant growth-promoting bacteria and green microalgae in alginate beads as an experimental model for studying plant-bacterium interactions. Appl Environ Microbiol 74:6797–6802

Dobretsov S, Teplitski M, Paul V (2009) Mini-review: quorum sensing in the marine environment and its relationship to biofouling. Biofouling 25:413–427

Dong H, Beer SV (2000) Riboflavin induces disease resistance in plants by activating a novel signal transduction pathway. Phytopathology 90:801–811

Dworjanyn S, de Nys R, Steinberg P (1999) Localisation and surface quantification of secondary metabolites in the red alga *Delisea pulchra*. Mar Biol 133:727–736

Dworjanyn SA, de Nys R, Steinberg PD (2006) Chemically mediated antifouling in the red alga *Delisea pulchra*. Mar Ecol Prog Ser 318:153–163

Flechtner VR (1998) Enigmatic desert soil algae. In: Seckbach J (ed) Enigmatic microorganisms and life in extreme environments. Kluwer Academic Publishers, Dordecht/Boston/London, pp 233–241

Gao M et al (2007) Effects of AiiA-mediated quorum quenching in *Sinorhizobium meliloti* on quorum-sensing signals, proteome patterns, and symbiotic interactions. Mol Plant Microbe Interact 20:843–856

Gao M, Teplitski M (2008) RIVET-a tool for *in vivo* analysis of symbiotically relevant gene expression in *Sinorhizobium meliloti*. Mol Plant Microbe Interact 21:162–170

Gao M, Teplitski M, Robinson JB, Bauer WD (2003) Production of substances by *Medicago truncatula* that affect bacterial quorum sensing. Mol Plant Microbe Interact 16:827–834

Gil R, Latorre A, Moya A (2004) Bacterial endosymbionts of insects: insights from comparative genomics. Environ Microbiol 6:1109–1122

Givskov M et al (1996) Eukaryotic interference with homoserine lactone-mediated prokaryotic signalling. J Bacteriol 178:6618–6622

Grossman AR et al (2007) Novel metabolism in *Chlamydomonas* through the lens of genomics. Curr Opin Plant Biol 10:190–198

Gyurjan I, Nghia NH, Toth G, Turtoczky I, Stefanovits P (1986) Photosynthesis, Nitrogen fixation and enzyme activities in *Chlamydomonas–Azotobacter* symbioses. Biochem Physiol Pflanz 181:147–153

Gyurjan I, Turtoczky I, Toth G, Paless G, Nghia NH (1984) Intercellular symbiosis of nitrogen-fixing bacteria and green algae. Acta Bot Hung 30:249–256

Hasegawa Y, Martin JL, Giewat MW, Rooney-Varga JN (2007) Microbial community diversity in the phycosphere of natural populations of the toxic alga, *Alexandrium fundyense*. Environ Microbiol 9:3108–3121

Hentzer M, Givskov M (2003) Pharmacological inhibition of quorum sensing for the treatment of chronic bacterial infections. J Clin Invest 112:1300–1307

Hentzer M et al (2003) Attenuation of *Pseudomonas aeruginosa* virulence by quorum sensing inhibitors. EMBO J 22:3803–3815

Hernandez JP, De-Bashan LE, Rodriguez DJ, Rodriguez Y, Bashan Y (2009) Growth promotion of the freshwater microalga *Chlorella vulgaris* by the nitrogen-fixing, plant growth-promoting bacterium *Bacillus pumilus* from arid zone soils. Eur J Soil Biol 45:88–93

Hirsch AM, Bauer WD, Bird DM, Cullimore J, Tyler B, Yoder JI (2003) Molecular signals and receptors: controlling rhizosphere interactions between plants and other organisms. Ecology 84:858–868

Imase M, Watanabe K, Aoyagi H, Tanaka H (2008) Construction of an artificial symbiotic community using a *Chlorella*-symbiont association as a model. FEMS Microbiol Ecol 63:273–282

Joint I et al (2002) Cell-to-cell communication across the prokaryote–eukaryote boundary. Science 298:1207

Joseph CM, Phillips DA (2003) Metabolites from soil bacteria affect plant water relations. Plant Physiol Biochem 41:189–192

Kiratisin P, Tucker KD, Passador L (2002) LasR, a transcriptional activator of *Pseudomonas aeruginosa* virulence genes, functions as a multimer. J Bacteriol 184:4912–4919

Knee EM et al (2001) Root mucilage from pea and its utilization by rhizosphere bacteria as a sole carbon source. Mol Plant Microbe Interact 14:775–784

Koch B, Liljefors T, Persson T, Nielsen J, Kjelleberg S, Givskov M (2005) The LuxR receptor: the sites of interaction with quorum-sensing signals and inhibitors. Microbiology 151: 3589–3602

Lambrecht M, Okon Y, Vande Broek A, Vanderleyden J (2000) Indole-3-acetic acid: a reciprocal signalling molecule in bacteria-plant interactions. Trends Microbiol 8:298–300

Langhans TM, Storm C, Schwabe A (2009) Community assembly of biological soil crusts of different successional stages in a temperate sand ecosystem, as assessed by direct determination and enrichment techniques. Microb Ecol 58(2):394–407

Manefield M et al (2002) Halogenated furanones inhibit quorum sensing through accelerated LuxR turnover. Microbiology 148:1119–1127

Manefield M, Welch M, Givskov M, Salmond GP, Kjelleberg S (2001) Halogenated furanones from the red alga, *Delisea pulchra*, inhibit carbapenem antibiotic synthesis and exoenzyme virulence factor production in the phytopathogen *Erwinia carotovora*. FEMS Microbiol Lett 205:131–138

Mathesius U et al (2003) Extensive and specific responses of a eukaryote to bacterial quorum-sensing signals. Proc Natl Acad Sci USA 100:1444–1449

Matiru VN, Dakora FD (2005) The rhizosphere signal molecule lumichrome alters seedling development in both legumes and cereals. New Phytol 166:439–444

McClean KH et al (1997) Quorum sensing and *Chromobacterium violaceum*: exploitation of violacein production and inhibition for the detection of N-acylhomoserine lactones. Microbiology 143(Pt 12):3703–3711

Medina C, Crespo-Rivas JC, Moreno J, Espuny MR, Cubo MT (2009) Mutation in the *cobO* gene generates auxotrophy for cobalamin and methionine and impairs the symbiotic properties of *Sinorhizobium fredii* HH103 with soybean and other legumes. Arch Microbiol 191:11–21

Nikolaev YA et al (2008) Effect of bacterial satellites on *Chlamydomonas reinhardtii* growth in an algo-bacterial community. Microbiology 77:78–83

Nyholm SV, Mcfall-Ngai MJ (2004) The winnowing: establishing the squid-Vibrio symbiosis. Nat Rev Microbiol 2:632–642

Otsuka S, Abe Y, Fukui R, Nishiyama M, Sendoo K (2008) Presence of previously undescribed bacterial taxa in non-axenic *Chlorella* cultures. J Gen Appl Microbiol 54:187–193

Pasmore M, Costerton JW (2003) Biofilms, bacterial signaling, and their ties to marine biology. J Ind Microbiol Biotechnol 30:407–413

Phillips DA, Joseph CM, Yang GP, Martinez-Romero E, Sanborn JR, Volpin H (1999) Identification of lumichrome as a *Sinorhizobium* enhancer of alfalfa root respiration and shoot growth. Proc Natl Acad Sci USA 96:12275–12280

Pierson LS 3rd, Pierson EA (1996) Phenazine antibiotic production in *Pseudomonas aureofaciens*: role in rhizosphere ecology and pathogen suppresion. FEMS Microbiol Lett 136: 101–108

Rajamani S et al (2008) The vitamin riboflavin and its derivative lumichrome activate the LasR bacterial Quorum-Sensing receptor. Mol Plant Microbe Interact 21:1184–1192

Ramos C, Licht TR, Sternberg C, Krogfelt KA, Molin S (2001) Monitoring bacterial growth activity in biofilms from laboratory flow chambers, plant rhizosphere, and animal intestine. Meth Enzymol 337:21–42

Sapp M, Schwaderer AS, Wiltshire KH, Hoppe HG, Gerdts G, Wichels A (2007) Species-specific bacterial communities in the phycosphere of microalgae? Microb Ecol 53:683–699

Skindersoe ME, Ettinger-Epstein P, Rasmussen TB, Bjarnsholt T, de Nys R, Givskov M (2008) Quorum sensing antagonism from marine organisms. Mar Biotechnol (NY) 10:56–63

Smith RS, Kelly R, Iglewski BH, Phipps RP (2002) The *Pseudomonas autoinducer* N-(3-oxododecanoyl) homoserine lactone induces cyclooxygenase-2 and prostaglandin E2 production in human lung fibroblasts: implications for inflammation. J Immunol 169:2636–2642

Stanley NR, Lazazzera BA (2004) Environmental signals and regulatory pathways that influence biofilm formation. Mol Microbiol 52:917–924

Teplitski M et al (2004) *Chlamydomonas reinhardtii* secretes compounds that mimic bacterial signals and interfere with quorum sensing regulation in bacteria. Plant Physiol 134:1–10

Teplitski M, Robinson JB, Bauer WD (2000) Plants secrete substances that mimic bacterial N-acyl homoserine lactone signal activities and affect population density-dependent behaviors in associated bacteria. Mol Plant Microbe Interact 13:637–648

Treadwell GE, Metzler DE (1972) Photoconversion of riboflavin to lumichrome in plant tissues. Plant Physiol 49:991–993

van Thielen N, Garbary DJ (1998) Life in the rocks – endolithic algae. In: Seckbach J (ed) Enigmatic microorganisms and life in extreme environments. Kluwer Academic Publishers, Dordrecht/Boston/London, pp 245–253

Watanabe K, Imase M, Sasaki K, Ohmura N, Saiki H, Tanaka H (2006) Composition of the sheath produced by the green alga *Chlorella sorokiniana*. Lett Appl Microbiol 42:538–543

Watanabe K et al (2005) Symbiotic association in *Chlorella* culture. FEMS Microbiol Ecol 51:187–196

Weinberger F et al (2007) Spore release in *Acrochaetium* sp (Rhodophyta) is bacterially controlled. J Phycol 43:235–241

Wheeler GL, Tait K, Taylor A, Browlee C, Joint I (2005) Acyl-homoserine lactones modulate the settlement rate of zoospores of the marine alga Ulva intestinalis via a novel chemokinetic mechanism. Plant Cell Environ 29:608–618

Winson MK et al (1998) Construction and analysis of *luxCDABE*-based plasmid sensors for investigating N-acyl homoserine lactone-mediated quorum sensing. FEMS Microbiol Lett 163:185–192

Wolfe AJ et al (2003) Evidence that acetyl phosphate functions as a global signal during biofilm development. Mol Microbiol 48:977–988

Zhang RG et al (2002) Structure of a bacterial quorum-sensing transcription factor complexed with pheromone and DNA. Nature 417:971–974

Chapter 17
Communication Among Soil Bacteria and Fungi

Ilona Pfeiffer

17.1 Introduction

It is difficult to reconstruct the evolutionary history of colonisation of the land by the ancestors of terrestrial fungi (phylum: Zygomycota, Glomeromycota, Ascomycota, Basidiomycota) in the lack of fossils. It is hypothesised that they appeared before the Precambrium area (Retallack 1994) but the first fossilised fungi originated from Devon about 400 million years ago (Remy et al. 1994). This roughly coincides with the appearance of plants in the earth.

Whenever the colonisation happened, fungi had to share the same habitat with bacteria already existed in the soil. Fungi are heterotrophic organisms, their existence is based on dead (saprobes) or live (parasites) organic nutrient resource. Many of the soil bacteria are also heterotrophic and as like fungi they can also be saprobes or parasites. The co-existence in the soil and sharing the same nutrients led to different kinds of interactions between them. The interactions can be symbiotic (or "cooperative"), what is beneficial for both groups or antagonistic ("non-cooperative") when one group makes unfavourable conditions for the other resulting the suppression of growth or death of the organisms.

Communication is established by different kinds of signals produced by the microorganisms and eliciting answers from the signal receivers. Although not all the elements of the whole communication process are fully understood, some examples detailed below will evidence the cross-talking between the kingdom of Bacteria and Fungi.

I. Pfeiffer
Department of Microbiology, University of Szeged, Közép fasor 52. H-6726, Szeged, Hungary
e-mail: pfeiffer@bio.u-szeged.hu

G. Witzany (ed.), *Biocommunication in Soil Microorganisms*, Soil Biology 23,
DOI 10.1007/978-3-642-14512-4_17, © Springer-Verlag Berlin Heidelberg 2011

17.2 Antagonistic Interactions

17.2.1 Alteration of Abiotic Environmental Factors

The terrestrial environment is usually poor in nutrients, may the cell density be rather high in soil samples; therefore, microorganisms are in competition for substrate and developed antagonistic strategies.

Natural soils often inhibit fungal or bacterial spore germination and hyphal growth or bacterial cell division (called fungistasis; Dobbs and Hinson 1953 and bacteriostasis; Brown 1973). Previously both phenomenons were explained as a consequence of competition for nutrients. In a recent publication, De Boer and his co-workers (2003) pointed out that microbial community composition rather than low availability of nutrients is responsible for this phenomenon. They revealed that the presence and activity of certain bacteria especially pseudomonads in the soil is essential to suppress fungal growth.

A possible explanation for antagonistic interaction can be that microorganisms are able to change the physical and chemical characteristics of the environments (e.g. pH); in this way, they control the growth of other microorganisms. Fungi prefer acidic environments to growth. They produce and secrete organic acids (oxalic acid, malic acid, citric acid, etc.) to make their surrounding favourable for them (Cromack and Caldwell 1981) what is inhibitory for the most of the soil bacteria which prefer neutral or alkaline pH.

Microorganisms require iron for growth. Under iron-limited conditions, low molecular weight compounds called siderophores are synthesised by microbes (Varma and Chincholkar 2007). Siderophores chelate ferric ions and transport it to the cell via receptor-mediated way. As iron is an indispensable growth factor, microbes producing siderophores can capture it from the others. Iron-deprivation causes decreasing metabolic activity and stop cell growth. *Pseudomonas* species synthesise numerous iron-chelators, e.g. pyoverdin and pyochelin (Cox et al. 1981; Cox and Adams 1985). These compounds play an important role in restricting the growth of plant-pathogen fungi, e.g. *Pythium* and *Fusarium* species (Matthijs et al. 2007).

17.2.2 Antibiotic Production

Several microorganisms (both fungi and bacteria) produce different kinds of antibiotics. Antibiotics are low molecular compounds produced by microorganisms that inhibit the growth of other microbes. They are well-known since A. Fleming discovered their first representative, penicillin in 1929. Their therapeutic use for curing infectious diseases became wide-spread after the Second World War. Since that time several studies were published in association with their clinical application. Recent papers revealed that some antibiotics applied in sub-inhibitory concentration can function as signal mediators influencing the transcription of different target genes (Goh et al. 2002; Linares et al. 2006). Increasing number of results demonstrate that low concentration of antibiotics stimulates transcription even of genes involved in pathogenesis. These results evidence that antibiotics exhibit different properties applied in low or in high concentration. The phenomenon is called hormesis. Considering the fact that the concentration of the antibiotics in the nature probably never reaches the inhibitory concentration used in therapy makes more plausible that they are produced as intermicrobial signals, rather than inhibitors (Davies et al. 2006).

One of the striking and complex example for communication between bacteria and fungi using an antibiotic as signal is the case of 2,4-diacetylphloroglucinol (DAPG). 2,4-diacetylphloroglucinol is a well-known bacterial metabolite produced by *Pseudomonas fluorescens* strains (Shanahan et al. 1992). This compound is active against different plant-pathogenic fungi (e.g. *Gaeumannomyces graminis* var. *tritici*, *Pythium ultimum*, *Thielaviopsis basicola*). Several positive and negative inter- and intraspecific communication signals were identified influencing DAPG synthesis. DAPG biosynthesis is regulated by PhlF transcription repressor protein, which binds to a specific operator sequence phO. DAPG destabilises the binding between PhlF and phO, therefore, suppresses the repression of its own transcription (Maurhofer et al. 2004). The mycotoxin fusaric acid produced by plant-pathogen fungi *Fusarium* species acts also as signal for DAPG biosynthesis. Fusaric acid suppresses the expression of DAPG via enhancing the expression of the PhlF repressor (Notz et al. 2001). Moreover, another plant-pathogen fungus *Pythium ultimum* is also able to repress the gene expression in *Pseudomonas fluorescens* via an as-yet unidentified diffusible factor (Fedi et al. 1997). The response was heat-labile and protease-sensitive, which indicates that a protein or a peptide acts as signal in this process. Extracellular metabolites salicylate and the phenolic polyketide derivative pyoluteorin produced by *Pseudomonas* species also regulate the expression of DAPG via PhlF protein (Schnider-Keel et al. 2000). Salicylate interacts with PhlF protein and enhances its binding to the phO operator (Abbas et al. 2002) in this way repress the DAPG transcription.

17.2.3 Trehalose

Trehalose is a widely distributed storage disaccharide accumulated in unfavourable growth conditions in fungi. Gaballa et al. (1997) demonstrated that trehalose can serve as interspecies signal and it induces antagonism in the bacteria *Pseudomonas fluorescens* against the fungus *Pythium debaryanum*. Trehalose released by *P. debaryanum* enters into the *Pseudomonas* cells (Matthijs et al. 2000) via a specific transport mechanism (trehalose phosphotranspherase, treP) and induces the transcription of trehalase (treA) and probably other genes which leads to restriction of the fungal growth.

17.2.4 Excretion of Extracellular Enzymes

Extracellular enzymes produced by bacteria can affect the development of conidia and hyphae of different fungal species (Table 17.1). They promote ultrastructural changes, e.g. swelling and vacuolation of conidia, discolouration and malformation of hyphae resulting in the degradation of the cell wall and out-flow of the cytoplasm leading to complete lysis and death of the cell. Several soil bacteria produce chitinase, β-1, 3-glucanase, cellulase (de Boer et al. 1998; Sindhu and Dadarwal 2001; Hong and Meng 2003; El-Tarabily et al. 1996) that act directly on fungal cell wall components. Sometimes one bacterial species excretes different types of extracellular enzyme each of them acts on different fungal species. Serine protease of *Stenotrophomonas maltophilia* suppresses the growth of *Pythium ultimum* (Dunne et al. 2000), while its chitinase is active against *Bipolaris sorokiniana* (Zhang and Yuen 2000).

The enzymes can act indirectly on the partner. The germination of *Pythium ultimum* sporangia is stimulated by long chain unsaturated fatty acids released by seeds of different plants. *Enterobacter cloacea* produces fatty acid degrading enzymes in this way prevents the germination of the sporangia and therefore protects the plant from the disease (van Dijk and Nelson 2000).

Fungi produce and secrete enormous amount of extracellular enzymes but only few of them are known to affect bacteria directly. A haloperoxidase enzyme of *Curvularia* sp. (Hansen et al. 2004) can change the expression of certain genes in *Escherichia coli* and arrest cell growth temporarily. Extracellular enzymes of certain wood-decaying Basidiomycetes can attack bacteria causing complete cell lysis (Thorn and Tsuneda 1992).

The components of dead plant materials (e.g. cellulose, lignin) are abundant recycling organic carbon source in the soil; however, because of their recalcitrant crystalline form they are relatively inaccessible. Extracellular cellulolytic enzymes of bacteria and fungi convert cellulose to an easily degradable carbon source, glucose. While under anaerobic conditions cellulose is mainly hydrolysed by

Table 17.1 Examples of extracellular enzyme producing bacteria and parasitised fungi

Bacteria	Parasitised fungi	References
Enterobacter cloacae	*Phytium ultimum*	van Dijk and Nelson (2000)
Arthrobacter sp.	*Pythium debaryanum*	Whipps (2001)
Micromonospora carbonacea	*Phytophtora cinnamomi*	El-Tarabily et al. (1996)
Actinomycetes sp.	*Phytophtora fragariae var. rubi*	Valois et al. (1996)
Bacillus cereus	*Rhizoctonia solani*	Gavini et al. (1989)
Pantoea agglomerans	*Rhizoctonia solani*	Chernin et al. (1995)
Stenotrophomonas maltophila	*Pythium ultimum*	Dunne et al. (2000)
Bacillus subtilis	*Fusarium oxysporum*, *Cylindrocarpon* sp.	Schelkle and Peterson (1996)

cellulosomes of bacteria, in aerobic circumstances both fungal and bacterial cellulases take part in its degradation (de Boer et al. 2005). Another rigid component of plant cell wall is lignin. Its breakdown is resulted by the function of laccases and peroxidases produced by species of Basidiomycota (Cullen and Kersten 1992) and by actinomycetes in the case of grasses. As both lignin and cellulose is a good substrate for bacteria and fungi, competitive interactions take place for it (Lang et al. 1997).

17.3 Symbiotic Interactions

The interactions are not necessarily detrimental even in the case of limited amounts of nutrients. Microbes can co-operate to access the substrate.

As it was mentioned above, fungi excrete low molecular weight organic acids (citric, fumaric, malic, oxalic acid) to make their environment favourable for themselves, therefore, suppress the growth of bacteria. These compounds can also be useful making possible for fungi to mobilise nutrients directly from minerals and make them available for not only the producer themselves but other organisms including bacteria (Landeweert et al. 2001). Moreover, formation of stabile insoluble complexes with toxic metal ions (e.g. metal oxalates) promotes the survival of the cells in contaminated soils (Gadd 1999).

Bacterial metabolic substances can enhance fungal growth and stimulate spore germination. For example, volatile 2-methylisoborneol produced by *Streptomyces orientalis* stimulates spore germination of the arbuscular mycorrhizal fungus *Gigaspora margarita* (Carpenter-Boggs et al. 1995).

Bacteria and fungi are usually in competition for utilisation of lignocellulose, nevertheless breakdown of the end products of the cellulose and lignin degradation by bacteria can stimulate fungal extracellular enzyme production and hyphal growth.

17.3.1 Mycorrhization Helper Bacteria

The majority of terrestrial plants possess mycorrhiza. This symbiosis is useful for both partners as plants provide carbon source, while fungi supply the plants with water and minerals; furthermore, they can protect plants from pathogens and environmental stress. Mycorrhiza is divided into different types. Arbuscular mycorrhiza (AM) is the most frequent one, when the hyphae of the obligate symbiotic fungi (species belong to phylum Glomeromycota) enter the root tissue. In the case of the ectomycorrhiza, the hyphae form mantle around the short roots and form the Hartig network.

Several bacterial strains stimulate hyphal growth consequently promote the formation of both ecto- and arbuscular mycorrhiza (Poole et al. 2001; Artursson et al. 2006). These bacteria are known as mycorrhization helper bacteria (MHB; Frey-Klett et al. 2007). Up to now it is not clear that bacteria prefer to live in association with fungi rather than in the bulk soil or fungi select bacteria around themselves. The helper effect presumes close contact between bacteria and fungi. Many of the bacteria are able to attach directly to the fungal hyphae. Those bacteria what are able to produce extracellular polysaccharides colonise fungal hyphae better than others (Toljander et al. 2006). Probably no specific receptors are involved in the colonisation process, but some authors conclude that polar flagella play a role in the attachment (Sen et al. 1986). The cell wall composition of the hyphae can also influence the attachment from the fungal side. A recent study indicates that probably trehalose has important role in the bacterial–fungal interaction (Duponnois and Kisa 2006).

The helper bacteria associated with mycelia of plant-pathogenic fungi can influence the pathogenicity with production of different kinds of extracellular enzymes. Lipases produced by helper bacteria of *Stagonospora nodorum* (Dewey et al. 2002), pectolytic enzyme of *Erwinia carotovora* ssp. *atroseptica* and *Bacillus lentus* increase pathogenicity of *Septoria tritici* (Newton and Toth 2002) and *Botrytis fabae* (Harrison 1983), respectively. In these cases, the enzymes do not act on the fungus rather on the permeability of the plant cell membrane, therefore, promote the colonisation of the plant tissue by the fungus.

17.3.2 Endosymbionts

The most intimate interaction between bacteria and fungi supposing continuous communication between them is endosymbiosis. Occurrence of bacteria inside fungal cells is known both in arbuscular- and ectomycorrhizal fungi (Table 17.2). Most AM-forming species of the familia Gigasporaceae harbour symbiotic endobacteria (Bianciotto et al. 2000). The bacteria are obligate endosymbionts as they can not be cultured in cell-free media and they propagate only across vertical

Table 17.2 Endosymbiotic bacteria of fungal species

Endobacteria	Host fungal species	References
Burkholderia endofungorum, *Burkholderia rhizoxinica*	*Rhizopus microsporus*	Partida-Martinez et al. (2007)
Candidatus Glomeribacter gigasporarum	*Gigaspora margarita*	Bianciotto et al. (2003)
Cytophaga-Flexibacter-Bacteroides related species	*Tuber borchii*	Barbieri et al. (2000)
Nostoc punctiforme	*Geosiphon pyriforme*	Gehrig et al. (1996)
Paenibacillus spp.	*Laccaria bicolor*	Bertaux et al. (2003)
Unidentified	*Morchella elata*	Buscot (1994)

transmission via the spores of the fungi (Bianciotto et al. 2004). How bacteria colonise fungi is unknown, but the identification of vacB gene (coding exoribonuclease RNase R) in the genome of the endobacteria suggests that they probably use the same system as pathogenic bacteria (Ruiz-Lozano and Bonfante 2000). The presence of the elements of nitrogen fixation (Minerdi et al. 2001) and P transporter genes (Ruiz-Lozano and Bonfante 1999) in the bacterial genome demonstrate that the bacteria actively take part in the metabolism of the fungus. Ectomycorrhizal fungi (species of *Russula*, *Suillus*, *Laccaria*, *Morchella*, *Tuber* genus) can also harbour endobacteria belonging to the genus *Pseudomonas*, *Burkholderia*, and *Bacillus* (Izumi et al. 2006).

The occurrence of endobacteria within the mycelia of non-mycorrhizal fungi belonging to Zygomycetes is common. The plant-pathogenic species *Rhizopus microsporus* harbours endosymbiontic bacteria identified as *Burkholderia rhizoxinica* and *Burkholderia endofungorum* (Partida-Martinez et al. 2007). Bacteria contribute to the pathogenesis caused by the fungus as they are responsible for mycotoxin – rhizoxin and its derivatives – production (Partida-Martinez and Hertweck 2005) which causes the rice seedling blight disease.

The interaction of *Geosiphon pyriformis* and *Nostoc punctiforme* is a symbiosis between a cyanobacteria and a fungus (Kluge 2002). The symbiotic process starts with touching *Nostoc* primordia to the apical tip of the fungal cell in an as-yet unidentified way, after the attachment of the bacterial cell fungus traps it inside by the mechanism of endocytosis. Mannose, which is present in primordia of *Nostoc* cells (in other cell types, i.e. hormoginia and hetocyst not), probably can be involved in the host recognition process. The volume of the *Nostoc* cells existing inside the fungal cells is ten times more than outside and contains more photosynthetic pigments as well. Heterocysts are also present evidencing that bacteria are able to fix nitrogen, but their main function is photosynthesis. *G. pyriforme* consume sugar produced by the photosynthesis of *Nostoc*, while *Nostoc* cells get minerals and CO_2 from the fungus.

17.4 Concluding Remarks

Our knowledge of the communication process between bacteria and fungi is rather poor. Several chemical signals produced either by bacteria or fungi are revealed. The phenotypic changes given as response can be detected but little is known about the intervening steps, how microorganisms perceive and interpret the signals.

The communication is the bases of different kinds of interactions, which has inestimable influence on the terrestrial ecosystem and sustainable agriculture. Antagonistic interactions can be applied in biological control of plant diseases by avoiding the use of pesticides. Understanding the mechanism of antagonism could help to identify the favourable circumstances to improve the efficacy of the inhibitory effect. Research targeted to reveal the molecular mechanism of the action of antibiotics at different concentrations can improve their therapeutic application in the future.

The degradation of recalcitrant organic polymers by bacteria and fungi, furthermore, helper effect of bacteria in mycorrhization process promotes the proper nutrient-supply of plants without fertilisation and pollution of the environment.

References

Abbas A, Morrissey JP, Marquez PC, Sheehan MM, Delany IR, O'Gara F (2002) Characterization of interactions between the transcriptional repressor PhlF and its binding site at the phlA promoter in *Pseudomonas fluorescens* F113. J Bacteriol 184:3008–3016

Artursson V, Finlay RD, Jansson JK (2006) Interactions between arbuscular mycorrhizal fungi and bacteria and their potential for stimulating plant growth. Environ Microbiol 8(1):1–10

Barbieri E, Potenza L, Rosshi I, Sisti D, Giomaro G, Rossetti S, Beimfohr C, Stocchi V (2000) Phylogenetic characterization and in situ detection of a *Cytophaga-Flexibacter-Bacteroides* phylogroup bacterium in *Tuber borchii* Vittad. ectomycorrhizal mycelium. Appl Environ Microbiol 66:5035–5042

Bertaux J, Schmid M, Prevost-Boure NC, Churin JL, Hartmann A, Garbaye J, Frey-Klett P (2003) In situ identification of intracellular bacteria related to *Paenibacillus* spp. in the mycelium of the ectomycorrhizal fungus *Laccaria bicolor* S238N. Appl Environ Microbiol 69(7): 4243–4248

Bianciotto V, Lumini E, Lanfranco L, Minerdi D, Bonfante P, Perotto S (2000) Detection and identification of bacterial endosymbionts in arbuscular mycorrhizal fungi belonging to the family Gigasporaceae. Appl Environ Microbiol 66(10):4503–4509

Bianciotto V, Lumini E, Bonfante P, Vandamme P (2003) '*Candidatus* Glomeribacter gigasporarum' *gen. nov., sp nov.*, an endosymbiont of arbuscular mycorrhizal fungi. Int J Syst Evol Microbiol 53:121–124

Bianciotto V, Genre A, Jargeat P, Lumini E, Bécard G, Bonfante P (2004) Vertical transmission of endobacteria in the arbuscular mycorrhizal fungus *Gigaspora margarita* through generation of vegetative spores. Appl Environ Microbiol 70(6):3600–3608

Brown ME (1973) Soil bacteriostasis limitation in growth of soil and rhizosphere bacteria. Can J Microbiol 19:195–199

Buscot F (1994) Ectomycorrhizal types and endobacteria associated with ectomycorrhizas of *Morchella elata* (Fr.) Boudier with *Picea abies* (L.) Karst. Mycorrhiza 4(5):223–232

Carpenter-Boggs I, Loynachan't TE, Stahl PD (1995) Spore germination of *Gigaspora margarita* stimulated by volatiles of soil-isolated Actinomycetes. Soil Biol Biochem 27(11):1445–1451

Chernin L, Ismailov Z, Haran S, Chet I (1995) Chitinolytic *Enterobacter agglomerans* antagonistic to fungal plant pathogens. Appl Environ Microbiol 61(5):1720–1726

Cox CD, Rinehart KL, Moore ML, Cook JC (1981) Pyochelin: novel structure of an iron-chelating growth promoter from *Pseudomonas aeruginosa*. Proc Natl Acad Sci USA 78:302–308

Cox CD, Adams P (1985) Siderophore activity of pyoverdin for *Pseudomonas aeruginosa*. Infect Immun 48:130–138

Cromack K, Caldwell BA (1981) The role of fungi in litter decomposition and nutrient cycling. In: Carroll GC, Wicklow DT (eds) The Fungal community: its organization and role in the ecosystem. Marcel Dekker, New York, pp 653–668

Cullen D, Kersten P (1992) Fungal enzymes for lignocellulose degradation. In: Kinghorn JR, Turner G (eds) Applied molecular genetics of filamentous fungi. Chapman and Hall, London, pp 100–131

Davies J, Spiegalman GB, Yim G (2006) The world of subinhibitory antibiotic concentration. Curr Opinion Microbiol 9:445–453

De Boer W, Klein Gunnewiek PJA, Lafeber P, Janse JD, Spit BE, Woldendorp JW (1998) Antifungal properties of chitinolytic dune soil bacteria. Soil Biol Biochem 30(2):193–203

de Boer W, Verheggen P, Klein Gunnewiek PJA, Kowalchuk GA, van Veen JA (2003) Microbial community composition affects soil fungistasis. Appl Environ Microbiol 69(2):835–844

de Boer W, Folman LB, Summerbell RC, Boddy L (2005) Living in a fungal word: impact of fungi on soil bacterial niche development. FEMS Microbiol Rev 29:795–811

Dewey FM, Li Wong Y, Seery R, Hollins TW, Gurr SJ (2002) Bacteria associated with *Stagonospora (Septoria) nodorum* increase pathogenicity of the fungus. New Phytol 144(3):489–497

Dobbs CG, Hinson WH (1953) A widespread fungistasis in soil. Nature 172:197–199

Dunne C, Moënne-Loccoz Y, de Bruijn FJ, O'Gara F (2000) Overproduction of an inducible extracellular serine protease improves biological control of *Pythium ultimum* by *Stenotrophomonas maltophilia* strain W81. Microbiology 146:2069–2078

Duponnois R, Kisa M (2006) The possible role of trehalose in mycorrhiza helper bacterium effect. Can J Bot 84:1005–1008

El-Tarabily KA, Sykes ML, Kurtböke ID, GEStJ H, Barbosa AM, Dekker RFH (1996) Synergistic effects of a cellulase-producing *Micromonospora carbonacea* and an antibiotic-producing *Streptomyces violascens* on the suppression of *Phytophthora cinnamomi* root rot of *Banksia grandis*. Can J Bot 74:618–624

Fedi S, Tola E, Moenne-Loccoz Y, Dowling DN, Smith LM, O'Gara F (1997) Evidence for signaling between the phytopathogenic fungus *Pythium ultimum* and *Pseudomonas fluorescens* F113: *P. ultimum* represses the expression of genes in *P. fluorescens* F113, resulting in altered ecological fitness. Appl Environ Microbiol 63(11):4261–4266

Frey-Klett P, Garbaye J, Tarkka M (2007) The mycorrhiza helper bacteria revisited. New Phytol 176(1):22–36

Gaballa A, Abeysinghe PD, Urich G, Matthijs S, De Greve H, Cornelis P, Koedam N (1997) Trehalose induces antagonism towards *Pythium debaryanum* in *Pseudomonas fluorescens* ATCC 17400. Appl Environ Microbiol 63(11):4340–4345

Gadd GM (1999) Fungal production of citric and oxalic acid: importance in metal speciation, physiology and biogeochemical processes. Adv Microb Physiol 41:47–92

Gavini F, Mergaert J, Beji A, Mielcarek C, Izard D, Kersters K, De Ley J (1989) Transfer of *Enterobacter agglomerans* (Beijerinck 1988) Ewing and Fife, 1972 to *Pantoea* gen. nov. as *Pantoea agglomerans* comb. nov. and description of *Pantoea dispersa* sp. nov. Int J Syst Bacteriol 39:337–345

Gehrig H, Schüßler A, Kluge M (1996) *Geosiphon pyriformis*, a fungus forming endocytobiosis with *Nostoc* (cyanobacteria), is an ancestral member of the Glomales: evidence by SSU rRNA analysis. J Mol Evol 43:71–81

Goh E-B, Yim G, Tsui W, McClure JA, Surette MG, Davies J (2002) Transcriptional modulation of bacterial gene expression by subinhibitory concentrations of antibiotics. Proc Natl Acad Sci USA 99(26):17025–17030

Hansen EH, Schembri MA, Klemm P, Schäfer T, Molin S, Gram L (2004) Elucidation of the antibacterial mechanism of the Curvularia haloperoxidase system by DNA microarray profiling. Appl Environ Microbiol 70(3):1749–1757

Harrison JG (1983) Growth of lesions caused by *Botrytis fabae* on field bean leaves in relation to foliar bacteria, non-enzymic phytotoxins, pectic enzymes and osmotica. Ann Botany 52:823–838

Hong T-Y, Meng M (2003) Biochemical characterization and antifungal activity of an endo-1, 3-b-glucanase of Paenibacillus sp. isolated from garden soil. Appl Microbiol Biotechnol 61:472–478

Hughes DT, Sperandio V (2008) Inter-kingdom signalling: communication between bacteria and their hosts. Nat Rev Microbiol 6:111–120

Izumi H, Anderson IC, Alexander IJ, Killham K, Moore ERB (2006) Endobacteria in some ectomycorrhiza of Scots pine (*Pinus sylvestris*). FEMS Microbiol Ecol 56(1):34–43

Kluge M (2002) A fungus eats a cyanobacterium: the story of the *Geosiphon pyriformis* endocyanosis. Biol Environ: Proc R Ir Acad 102B(1):11–14

Landeweert R, Hoffland E, Finlay RD, Kuyper TW, van Breemen N (2001) Linking plants to rocks: ectomycorrhizal fungi mobilize nutrients from minerals. Trends Ecol Evol 16 (5):248–254

Lang E, Kleeberg I, Zadrazil F (1997) Competition of *Pleurotus* sp. and *Dichomitus squalens* with soil microorganisms during lignocellulose decomposition. Bioresour Technol 60(2):95–99

Linares JF, Gustafsson I, Baquero F, Martinez JL (2006) Antibiotics as intermicrobial signaling agents instead of weapons. Proc Natl Acad Sci USA 103(51):19484–19489

Matthijs S, Koedam N, Cornelis P, De Greve H (2000) The trehalose operon of *Pseudomonas fluorescens* ATCC 17400. Res Microbiol 151(10):845–851

Matthijs S, Tehrani KA, Laus G, Jackson RW, Cooper RM, Cornelis P (2007) Thioquinolobactin, a Pseudomonas siderophore with antifungal and anti-Pythium activity. Environ Microbiol 9(2):425–434

Maurhofer M, Baehler E, Notz R, Martinez V, Keel C (2004) Cross talk between 2,4-diacetylphloroglucinol-producing biocontrol pseudomonads on wheat roots. Appl Environ Microbiol 70:1990–1998

Minerdi D, Fani R, Gallo R, Boarino A, Bonfante P (2001) Nitrogen fixation genes in an endosymbiotic *Burkholderia* strain. Appl Environ Microbiol 67(2):725–732

Newton AC, Toth IK (2002) Helper bacteria and pathogenicity assessments. New Phytol 144(3):385–386

Notz R, Maurhofer M, Schnider-Keel U, Duffy B, Haas D, Défago G (2001) Biotic factors affecting expression of the 2,4-diacetylphloroglucinol biosynthesis gene phlA in Pseudomonas fluorescens biocontrol strain CHA0 in the rhizosphere. Phytopathology 91:873–881

Partida-Martinez LP, Hertweck C (2005) Pathogenic fungus harbours endosymbiotic bacteria for toxin production. Nature 437:884–888

Partida-Martinez LP, Groth I, Schmitt I, Richter W, Roth M, Hertweck C (2007) *Burkholderia rhizoxinica* sp. nov. and *Burkholderia endofungorum* sp. nov., bacterial endosymbionts of the plant-pathogenic fungus *Rhizopus microsporus*. Int J Syst Evol Microbiol 57:2583–2590

Poole EJ, Bending GD, Whipps JM, Read DJ (2001) Bacteria associated with *Pinus sylvestris–Lactarius rufus* ectomycorrhizas and their effects on mycorrhiza formation in vitro. New Phytol 151:743–751

Remy WT, Taylor N, Hass H, Kerp H (1994) Four hundred-million-year-old vesicular arbuscular mycorrhiza. Proc Nat Acad Sci USA 9:11841–11843

Retallack GJ (1994) Were the Ediacaran fossils lichens? Paleobiology 20(4):523–544

Ruiz-Lozano JM, Bonfante P (1999) Identification of a putative P-transporter operon in the genome of a *Burkholderia* strain living inside the arbuscular mycorrhizal fungus *Gigaspora margarita*. J Bacteriol 181:4106–4109

Ruiz-Lozano JM, Bonfante P (2000) Intracellular *Burkholderia* of the arbuscular mycorrhizal fungus *Gigaspora margarita* possesses the *vacB* gene, which is involved in host cell colonization by bacteria. Microb Ecol 39:137–144

Schnider-Keel U, Seematter A, Maurhofer M, Blumer C, Duffy B, Gigot-Bonnefoy C, Reimmann C, Notz R, Défago G, Haas D, Keel C (2000) Autoinduction of 2,4-diacetylphloroglucinol biosynthesis in the biocontrol agent Pseudomonas fluorescens CHA0 and repression by the bacterial metabolites salicylate and pyoluteorin. J Bacteriol 182:1215–1225

Schelkle M, Peterson RL (1996) Suppression of common root pathogens by helper bacteria and ectomycorrhizal fungi in vitro. Mycorrhiza 6(6):481–485

Sen P, Chatterjee G, Kumar PM, Sen SK (1986) Enhancement of attachment of *Agrobacterium tumefaciens* to plant cell surface results in increase in genetic transformation. J Exp Biol 24:153–155

Shanahan P, O'Sullivan DJ, Simpson P, Glennon JD, O'Gara F (1992) Isolation of 2,4-diacetylphloroglucinol from a fluorescent pseudomonad and investigation of physiological parameters influencing its production. Appl Environ Microbiol 58:353–358

Sindhu SS, Dadarwal KR (2001) Chitinolytic and cellulolytic *Pseudomonas* sp. antagonistic to fungal pathogens enhances nodulation by *Mesorhizobium* sp. Cicer in chickpea. Microbiol Res 156(4):353–358

Thorn RG, Tsuneda A (1992) Interactions between various wood-dacaying fungi and bacteria: antibiosis, attack, lysis, and inhibition. Rept Tottori Mycol Inst 30:13–20

Toljander JF, Artursson V, Paul LR, Jansson JK, Finlay RD (2006) Attachment of different soil bacteria to arbuscular mycorrhizal fungal extraradical hyphae is determined by hyphal vitality and fungal species. FEMS Microbiol Lett 254:34–40

van Dijk K, Nelson EB (2000) Fatty acid competition as a mechanism by which *Enterobacter cloacae* suppresses *Pythium ultimum* sporangium germination and damping-off. Appl Environ Microbiol 66:5340–5347

Varma A, Chincholkar S (2007) Microbial siderophores. Springer, Berlin

Valois D, Fayad K, Barasubiye T, Garon M, Dery C, Brzezinski R, Beaulieu C (1996) Glucanolytic actinomycetes antagonistic to *Phytophthora fragariae var. rubi*, the causal agent of raspberry root rot. Appl Environ Microbiol 62(5):1630–1635

Whipps JM (2001) Microbial interactions and biocontrol in the rhizosphere. J Exp Bot 52:487–511

Zhang Z, Yuen GY (2000) The role of chitinase production by *Stenotrophomonas maltophilia* strain C3 in biological control of *Bipolaris sorokiniana*. Phytopathology 90(4):384–389

Chapter 18
Microbe–Microbe, Microbe–Plant Biocommunication

Ralf Oelmüller, Neeraj Shrivastava, Meghna Pohani, Kailash C. Upadhyaya, Irena Sherameti, Paul Murugan, Shashibala Singh, and Ajit Varma

18.1 Introduction

For a long time, it was believed that unicellular organisms exist as individual cells which primarily search for nutrients and multiply. Unicellular bacteria exist for almost 4 billion years and still survive in spite of dramatic changes in evolutionary history. These simple organisms reproduce by binary fission, divide and produce more of their kind, compete with neighbors, thus recognition and cooperation between cells may appear unlikely. The widely held view was that these unicellular organisms are solitary and respond to external stimuli in isolation. This concept has undergone a radical change in recent times. The discovery of intercellular communication among bacteria has led to the realization that bacteria are capable of coordinated activity that was believed to be restricted to multicellular organisms (Witzany 2008; Miller and Bassler 2001; Waters and Bassler 2005; Crespi 2001). Initially, it is considered as a rare phenomenon, restricted to few microbes, but now it is clear that a broad range of microorganisms have the ability to communicate,

R. Oelmüller, and I. Sherameti
Institute of General Botany and Plant Physiology, Friedrich-Schiller University of Jena, Dornburger Street 159, 07743 Jena, Germany
e-mail: b7oera@hotmail.de; irenasherameti@yahoo.de

N. Shrivastava, M. Pohani, and A. Varma (✉)
Amity Institute of Microbial Technology, Amity University Uttar Pradesh, Sector-125, Noida, Uttar Pradesh, India
e-mail: nsrivastava1@amity.edu; meghanapohani@gmail.com; ajitvarma@aihmr.amity.edu

K.C. Upadhyaya
School of Life Sciences, Jawaharlal Nehru University, New Delhi, India
e-mail: kcupadhyaya@hotmail.com

P. Murugan, and S. Singh
Defence Institute of High Altitude Research (DIHAR), Defence R & D Organisation, Leh, Ladakh 194101, India
e-mail: drpalmurugan@gmail.com; drshashisingh@gmail.com

G. Witzany (ed.), *Biocommunication in Soil Microorganisms*, Soil Biology 23,
DOI 10.1007/978-3-642-14512-4_18, © Springer-Verlag Berlin Heidelberg 2011

develop, organize, and coordinate social life with a great variety of behavioral patterns in which they organize themselves like multicellular organisms. It became clear that bacteria are highly interactive and exhibit a number of social behaviors, such as swarming motility, conjugal plasmid transfer, antibiotic resistance, biofilm maturation, and virulence. The capacity to behave collectively as a group has obvious advantages, for example, the ability to migrate to a more suitable environment/better nutrient supply and to adopt new modes of growth, such as sporulation or biofilm formation, which may afford protection from deleterious environments (Witzany 2010; Diggle et al. 2003, 2007; Schwarz et al. 2008).

Generally, bacteria release a wide variety of small molecules including secondary metabolites such as antibiotics and siderophores (iron chelators), metabolic end products, and cell-to-cell signaling molecules which function as pheromones and are sometimes termed "autoinducers" (where they function in part to stimulate their own synthesis) (Williams et al. 2007). These autoinducers diffuse through the bacterium membrane into the environment. For the communication purpose, bacteria send these specific, self-generated signal molecules into the environment. These sigalling molecules are used as a "language"; by sensing concentration of these signal molecules a bacterium senses the population density. This process is known as "quorum sensing" (QS). The ability to communicate within the microbial population gives the advantage to coordinate a group's behavior leading to a higher fitness in the environment. QS is the term of description for sign-mediated interactions in which chemical molecules are produced and secreted by bacteria (Manefield et al. 2002; Greenberg 2003). The term "QS" was first used in a review by Fuqua et al. (1994), which essentially reflected the minimum threshold level of individual cell mass required to initiate a concerted population response (Gera and Srivastava 2006). The principle of QS phenomenon, or cell-to-cell communication, is based on the fact that a single bacterium releases signal molecules into the environment (Waters and Bassler 2005). The accumulation of it in the environment takes place if there is a high concentration of bacteria in that space and or ecological niche. If this occurs, a special gene will be activated and an increase in the production/activation of signaling molecules is observed. However, when sufficient bacteria are present, signal molecule concentrations reach a threshold level that allows the bacteria to sense a critical cell mass and, in response, to switch on/off specific genes in the bacterial cells, leading to a coordinated population response. Thus, once a threshold concentration of the molecule (and consequently a specific population density) is achieved, a coordinated change in bacterial behavior is initiated. Most of the bacteria thus far identified that utilize QS systems are associated in some way with plants or animals. Generally, QS signals are derived from small molecules or peptides. Some of them are given in Table 18.1 (Williams 2007; Witzany 2010; Lal 2009; Williams et al. 2007; Schwarz et al. 2008; Diggle et al. 2002).

Today several classes of microbial-derived signaling molecules have been discovered, these can be divided into two main categories on the basis of utilization by bacteria. The first group are amino acids and short peptide derivatives, utilized by gram-positive bacteria (Diggle et al. 2007; Lazazzera and Grossman 1998; Shapiro 1998) and the second group are fatty acid derivatives, named homoserine

Table 18.1 Structure of some QS signal molecules (Lal 2009)

Signal	Structure	Organisms
C4-HSL (an AHL)	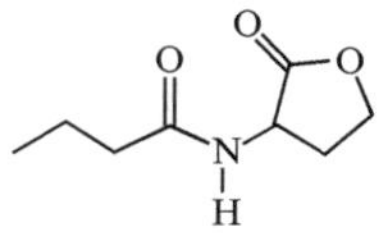	*Aeromonas hydrophila, Pseudomonas aeruginosa*
C6-HSL	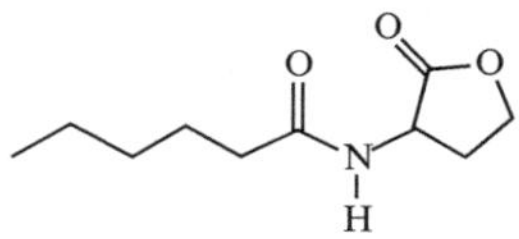	*Erwinia carotovora, Pseudomonas aureofaciens, Yersinia enterocolitica*
3-Oxo-C6-HSL	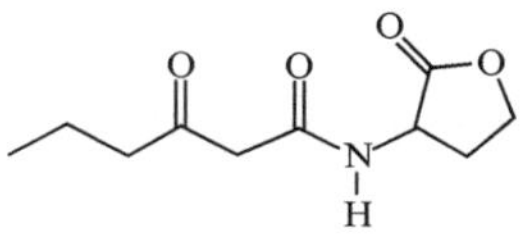	*E. carotovora, Vibrio fischeri, Y. enterocolitica*
3-Oxo-C8-HSL	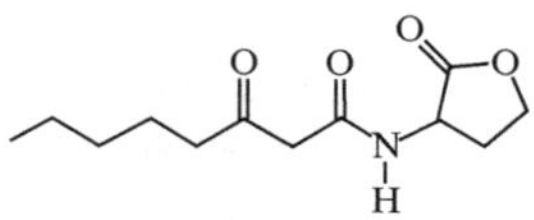	*Agrobacterium tumefaciens*
Autoinducing peptide (AIP)-I	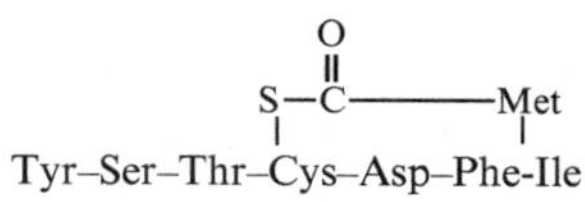	*Straphylococcus aureus* Group I strains
AI-2 (*S*-THMF-borate)	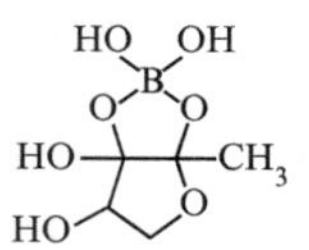	*Vibrio harveyi*
Competence and sporulation stimulating factor (CSF)	Glu – Arg – Gly – Met – Thr	*Bacillus subtilis*
Farnesol	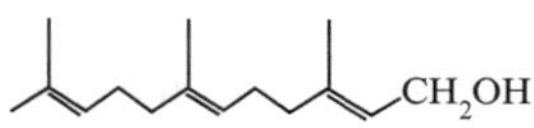	*Candida albicans*

lactones (HSLs) frequently utilized by gram-negative members (Dunny and Winans 1999; Whitehead et al. 2001). The main function of these signal molecules is to reenter into the network system of a cell by either simple diffusion or active transport. The signaling mechanism involves subsequent interaction with an intracellular effector that will induce the pathway for the concerned phenotype (Gera and Srivastava 2006). For any signal molecule to be classed as a QS signal, the following criteria should be fulfilled:

1. Stimulation or production of the signal molecule should take place during environment changes or specific stages of growth.
2. The signal should accumulate in the extra-cellular environment and should be recognized by specific bacterial receptors.
3. The concentration of signal molecules should be greater than a particular threshold value so that it can stimulate the cellular response.
4. These cellular responses could lead to physiological changes required for a coordinated population response (Diggle et al. 2007; Williams 2007; Witzany 2010; Miller and Bassler 2001).

18.2 Rhizosphere and Biocommunication

18.2.1 What Is Rhizosphere/Mycorhizosphere?

Some of the most complex chemical, physical, and biological interactions experienced by terrestrial plants are those that occur between the roots and their surrounding soil environment. Rhizosphere interactions include root–root, root–insect, and root–microbe associations. Plant roots exude an enormous range of potentially valuable small molecular weight compounds and signaling molecules into the rhizosphere. Many microbes grow and interact in the rhizosphere by utilizing nutrients directly or indirectly originating from plants.

The term "rhizosphere" was first brought into light by Hiltner (1904) who defined rhizosphere as a zone of soil immediately adjacent to legume roots that supports high levels of bacterial activity. The term is derived from the Greek word "rhiza" (root) and "sphere" (field of influence). However, more recently the term has been broadened to include both the volume of soil influenced by the root and the root tissues colonized by microorganisms (Pinton et al. 2001). Microbes and their exudates interact with plant roots in a range of positive, negative, and neutral mechanisms. Almost all plants live in symbiotic interactions with microbes, where they express different lifestyles ranging from mutualism (beneficial for both partners) through commensalism (beneficial for the microbe, while the host species is neither positively nor negatively affected) to parasitism (the host is noticeably harmed or deprived at the expense of the microbe).

The rhizosphere can be divided into several distinct parts. These include the ectorhizosphere (the soil immediately adjacent to the root), the rhizoplane (the root surface with the epidermis and mucilaginous polysaccharide layer), and the endorhizosphere (root tissue including the endodermis and cortical layers). Plants that are colonized by mycorrhizal fungi have an additional zone termed the mycorrhizosphere (Linderman 1988). As mycorrhizal fungi can extend out from the plant root for some distance, this zone can be significant. Nitrogen fixing bacteria in the soil clearly have the potential to manipulate the mycorrhizal fungi. The presence of genes for N_2 fixation has been shown for many bacteria including the endosymbiotic *Burkholderia* sp. (Minerdi et al. 2001), but expression of this activity at levels significantly influencing the growth of the mycorrhizal association has yet to be demonstrated. *Rhizobium* spp. may act synergistically with arbuscular mycorrhizal (AM) fungi on their plant hosts (Gryndler 2000) (Fig. 18.1). However, since rhizobial bacteria and mycorrhizal fungi utilize partially overlapping signaling pathways in Legumes for the establishment of their symbiosis (Oldroyd et al. 2005), they could also compete with each other which might have an important influence on the microbial community in the rhizophere.

18.2.2 Rhizoplane Organisms

Notwithstanding the diverse effects of soil chemistry, environment, and other nonbiological factors, in the rhizosphere, plants, fungi, and bacteria can all develop highly structured symbiotic interactions. Most of these obey a simple rule: plants are better at obtaining carbon, and fungi and bacteria are better at acquiring phosphorus, nitrogen, potassium, and other microelements when these nutrients or ions are inadequate. Traditional techniques are insufficient to study the actual interaction of the over 10,000 distinct species of microbes found in cohabitation

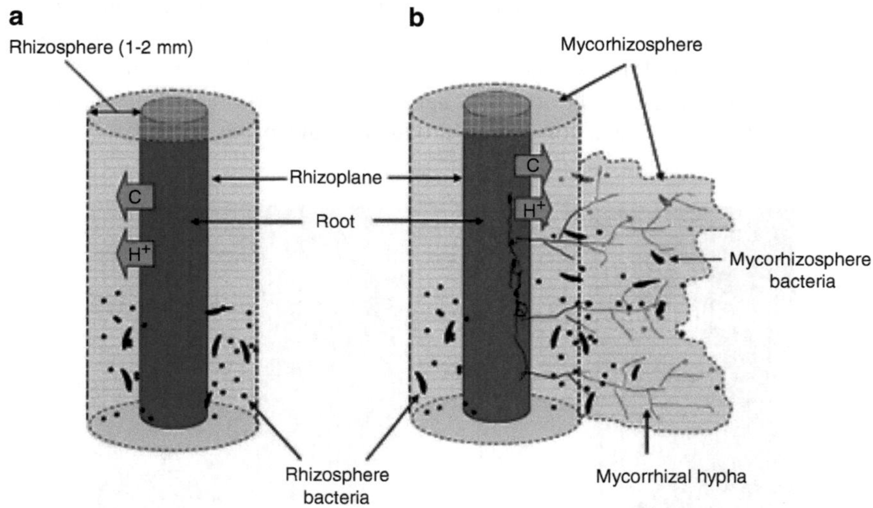

Fig. 18.1 A typical functional anatomy of plant root rhizosphere. Note subterranean root is surrounded by a large number of microorganisms including fungi, better termed as "Mycorhizosphere" (c.f. http://www.scielo.org.co/img/revistas/rfnam/v60n1/a01fig01.gif); Hiltner 1904; Varma 2009)

with them, particularly since most of the rhizosphere organisms are unculturable (Kent and Triplett 2002; Walker et al. 2003).

The presence of bacteria that are directly concerned in mycorrhiza formation was first indicated by the studies of Bowen and Theodorou (1979), which showed that some bacterial isolates promoted and others inhibited the colonization of *Pinus radiate* roots by *Rhizopogon luteolus*. In subsequent work, the presence of bacteria able to promote mycorrhiza formation was confirmed in ectomycorrhiza (Garbaye and Bowen 1987) and in AM (Meyer and Linderman 1986; Ames 1989). Several bacteria are also associated with fungi and promote mycorrhizal development. These bacteria were collectively named "Mycorrhiza Helper Bacteria" (MHB; Duponnois and Garbaye 1991; Garbaye 1994).

Varma and his collaborators have described *Piriformospora indica* in 1998 as an axenically cultivable plant growth-promoting root endophytic fungus. The fungus was originally found in soil samples from the rhizosphere of the woody shrubs *Prosopsis juliflora* and *Zizyphus nummularia* growing in the western part of Rajasthan. In contrast to other AM fungi, *P. indica* can be cultured axenically on various synthetic simple and complex media at 25–35°C (Varma et al. 1999). It was named according to its characteristic pear-shaped chlamydospores (Verma et al. 1998, Fig. 18.2). Exploitation of *P. indica* may not only complement crop-growing strategies but also serve as a model system to study molecular traits that affect disease resistance and grain yields in many commercially valuable plants. The fungus (1) functions as a plant promoter and biofertilizer in nutrient-deficient soils, (2) as a bioprotector against biotic and abiotic stresses including root and leaf fungus pathogens and insect invaders, (3) as a bioregulator for plant growth development, early flowering, enhanced seed production, and stimulation of active ingredients in medicinal plants (4) as well as a bio-agent for the hardening of tissue-culture-raised plants. Positive interaction has been established for many plants of economic importance in arboriculture, agro-forestry, flori-horticulture including Orchids, and those utilized for energy production and paper industry. *P. indica*

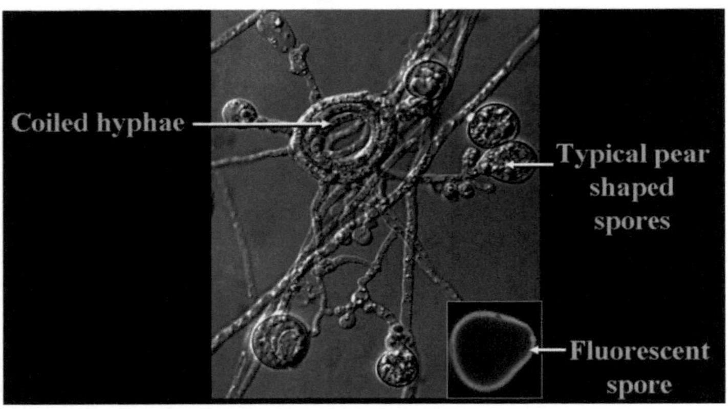

Fig. 18.2 Model organism *Piriformospora indica* (Varma et al. 1999)

also interacts with members of bryophyte, *Aneura pinguis*, pteridophyte, *Pteris ensiormis*, gymnosperms (*Pinus halepensis*), and a large number of angiosperms including the model plant *Arabidopsis thaliana* and other members of the mustard family. Similar to AM fungi, *P. indica* stimulates nutrient uptake in the roots and solubilizes insoluble phosphate and sulfur components in the soil. The interaction of *P. indica* with the model plants *A. thaliana* and barley (*Hordeum vulgare* L.) is being used to understand the molecular basis of this beneficial plant/microbe interaction (Oelmüller et al. 2009). The interaction of *P. indica* with the roots of many plant species allows also to study the influence of beneficial symbioses on microbial communities in the rhizophere.

The important role of *P. indica* on plant growth can be nicely shown under stress conditions. The fungus has considerably reduced the dormancy period in large number of leafy and horticultural crops at low temperature at (4–8°C) as shown by the Defence Research and Development Organization Laboratory located in Leh (Laddakh, India). The valley of Leh is situated at quite high altitude; mountains up to 3,500 m dominate the landscape around the Leh. Leh has a cold, arid climate with long, harsh winters from October to early March, with temperatures below freezing for most of the winter (temperature range from −28°C in winter to 33°C in summer). Recent results on plant/*P. indica* interactions for vegetable and horticultural plants are summarized in Table 18.2 and Fig. 18.3.

18.2.3 Root Exudates

In the complex ecosystem of the rhizosphere, plants release great varieties of various organic substances from the roots into the rhizosphere. These exudates directly or indirectly influence the physiological surface activity, which are

Table 18.2 Seed germination with fungal biomass as compared with control in polyench green house

Hosts	Days after treatment	% Increase over the control
Cabbage	25	>100
Endive	25	100
Swisschord (palak)	25	>100
Swisschord (red)	25	100
Radish – 100% (25 days)	25	100
Onion	25	>100
Carrot	21	84
Cauliflower	21	84
Beetroot	20	80
Peas	15	60
Snowpea	12	48

Experiment was performed at −18°C whereas the temperature inside the polyench green house was maintained between 4 and 8°C. The site of the experiment was at high altitude (Leh, Laddakh)

Cabbage Endive

Fig. 18.3 Seed germination in polyench green-house in the presence of *P. indica* at high altitude (Leh, Laddakh). Outside temperature was −18°C and inside 4–8°C

responsible for intense microbial activity in rhizosphere regions and have a strong influence on the root–microbe communication. From a great variety of organic compounds, few of them play an important role in root–microbe interactions such as flavonoids which, e.g., activate *Rhizobium meliloti* genes responsible for the nodulation process (Walker et al. 2003; Peters et al. 1986). Furthermore, AM fungi show extensive branching in the vicinity of host roots before formation of the appressorium, a structure used to penetrate the plant root. Host roots release signaling molecules that trigger hyphal branching. Akiyama et al. (2005) have isolated a branching factor from the root exudates of *Lotus japonicas* and identified it as a strigolactone, 5-deoxy-strigol. Strigolactones are a group of sesquiterpene lactones, previously isolated as seed-germination stimulants for the parasitic weeds *Striga* and *Orobanche.* The authors showed that natural strigolactones 5-deoxy-strigol, sorgolactone and strigol, and a synthetic analog, GR24, induced extensive hyphal branching in germinating spores of the AM fungus *Gigaspora margarita* at very low concentrations.

Interestingly, the soil-bacteria have also evolved complicated signal transduction mechanisms to perceive sensory information in accordance with the changing environmental conditions including changes in pH, temperature, osmotic potential, and nutrient availability. Such mechanisms have been normally seen to incorporate a two-component sensory transduction system consisting of a sensor protein that detects the environmental stimulus and a second component that acts as a regulator controlling the expression of particular genes, thus facilitating an adaptive response (e.g., Senadheera and Cvitkovitch 2008). Phosphorylation events via regulatory two-component systems followed by alterations in gene expression are now recognized as one of the major global regulatory networks in bacteria. However, not all types of sensor-regulator circuits relay on information through phosphoryl-transfer reactions. Small diffusible molecules termed auto-inducers or pheromones that mediate alternative signaling systems have also been identified to be involved in the control of gene expression. Numerous signaling molecule-mediated sensing and response pathways have now been recognized and most of them can be associated with QS (Walker et al. 2003; Witzany 2008).

18.3 Rhizobia–Legumes Symbiosis

The best-studied example of signal exchange is the *Rhizobium*–legume symbiosis, in which the plant releases flavonoid compounds that act as signals for the bacterium to secrete Nod factors (Martínez-Romero 2009; Soto et al. 2006; Fauvart and Michiels 2008). Nod factors are perceived by plant root hairs and function in a hormone-like fashion to induce root nodules in which the *Rhizobium* bacterium can fix atmospheric nitrogen (Brencic and Winans 2005; Gray and Smith 2005). This symbiosis is a prime example of an intimate relationship between a soil bacterium and its host plant and illustrates the concept behind the term "plant growth promoting rhizobacteria" (PGPRs): in nitrogen-poor environments, the *Rhizobium* promotes legume growth by providing a limiting nutrient, nitrogen (Van Loon 2007; González and Marketon 2003).

18.4 QS in *Rhizobium*

QS signals are found in many species of legume-nodulating rhizobia. In the well-characterized strain *R. leguminosarum* biovar *viciae*, a variety of autoinducers are synthesized, and all have been identified as *N*-acyl-HSLs. One of these *N*-acyl-HSLs is *N*-(3-hydroxy-7-cis-tetradecenoyl)-L-homoserine lactone, previously known as small bacteriocin, which inhibits the growth of several other *R. leguminosarum* strains. The *cinRI* locus is responsible for the production of small bacteriocin. CinR induces *cinl* in response to the acylhomoserine lactone (AHL) and establishes a positive autoregulatory induction loop (Wisniewski-Dyé and Downie 2002; Blosser-Middleton and Gray 2001). A complex cascade of QS loops was characterized, in which the *cinMR* locus appears to be the master control for three other AHL-dependent QS control systems. Other rhizobial strains share some but not all of the QS loci. Small bacteriocin and other *N*-acyl-HSLs regulate growth inhibition of sensitive strains, transfer of a symbiotic plasmid, and expression of rhizosphere-specific genes that influence nodulation. Although much of the regulatory circuits of one of the most complex QS cascade are already understood, several of the functions regulated by the QS genes remain to be elucidated (Wisniewski-Dyé and Downie 2002; Brewin et al. 1980; Bassler et al. 1994).

18.5 QS in *Pseudomonas aeruginosa*

The QS mechanism is also exploited by the opportunistic bacterium *Pseudomonas aeruginosa* to coordinate biofilm formation, motility, exopolysaccharide synthesis, and cell aggregation. The bacterial virulence depends on a large panel of cell-associated and extracellular factors. QS facilitates cell-to-cell communication by sensing environmental changes and responding with appropriate changes in the expression of various genes within the bacterial population. QS is based on an interaction between a small diffusible AHL molecule and a transcriptional activator. Two QS systems, the *las* and *rhl* systems, have been identified in *P. aeruginosa* (Chugani et al. 2001; Pearson et al. 1995). The *las* system associates the transcriptional activator proteins LasR and LasI responsible for the synthesis of the AHLC12-HSL. Similarly, the *rhl* system associates the transcriptional activator proteins RhlR with RhlI, which is responsible for the synthesis of another AHL: C4-HSL. Synthesis and secretion of a number of virulence factors are controlled by QS. Different animal model systems showed the crucial role of QS in the pathogenesis of *P. aeruginosa* infections. The discovery of QS has given a new opportunity to treat bacterial infection by other means than the usual mechanism of growth inhibition. New drugs inhibiting QS were recently discovered: furanone compounds can repress a large number of QS-regulated genes, including numerous *P. aeruginosa* virulence factor genes (De Kievit and Iglewski 2000; Gambello and Iglewski 1991) (Fig. 18.4).

18.6 QS in *Azotobacter*

Many types of signaling molecules, which regulate diverse phenotypes across distant genera, have been described from *Azotobacter*. The most common signaling molecules found in these gram-negative bacteria are *N*-acyl derivatives of HSLs (acyl HSLs). Modulation of the physiological processes controlled by acyl HSLs occurs in a cell density- and growth phase-dependent manner. Therefore, these bacteria also use QS to monitor cell density before expressing a phenotype.

The interaction between the symbiotic fungus *P. indica* and several nitrogen fixing bacteria such as *Azospirillum Azotobacter chroococcum*, *Bradyrhizobium*, and *Burkholderia* result in enhanced fungal biomass production. Tables 18.3 and 18.4 and Fig. 18.5 show the enhancement of the fungal biomass by ~31% and of the number of spores. In contrast, growth of *P. indica* was severely inhibited by *Ps. fluorescence* (see Fig. 18.6).

In cocultivation experiments of *P. indica*, *Az. chroococcum* promoted hyphae proliferation whereas *Ps. fluorescence* lysed them. This provides another example for the different modes of interactions between microbes, which has a strong impact of the microbial community in the rhizophere.

18.7 QS in *Agrobacterium tumefaciens*

The crown gall tumor caused by the gram-negative soil bacterium *A. tumefaciens* results from the transfer of oncogenic genetic material to the nucleus of its host. Virulence is brought about *via* different factors that may act individually or

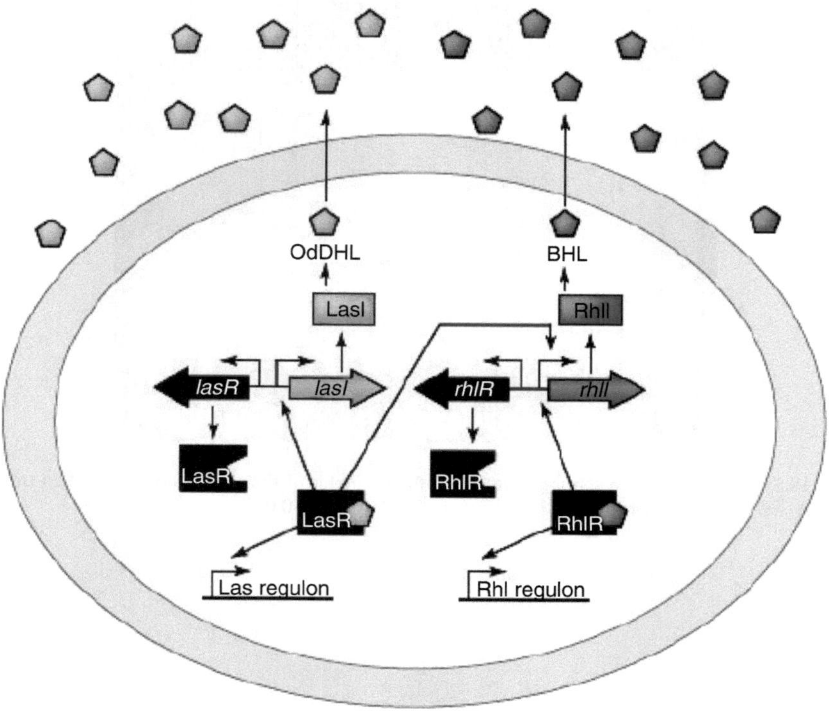

Fig. 18.4 The main QS system in *P. aeruginosa*. LasI and RhlI synthetases manufacture the autoinducer signaling molecules OdDHL and BHL, respectively. These signaling molecules diffuse out into the environment and, upon reaching a threshold concentration, activate the receptors LarS and RhlR. These receptors control a regulon of genes including virulence factors (Adonizio 2008 after modifying Schuster and Greenberg 2006)

Table 18.3 Fungal biomass with and without cocultivation of *P. indica* cultivated for 7 days in Hill and Käfer medium

Average fungal biomass of *P. indica* (g)	
Control (*P. indica*)	Treated (*P. indica* + *A. chroococcum*)
11.0	14.4

Table 18.4 Number of fungal spores in control and cocultured medium

Average cells/ml	
Control (*P. indica*)	Treated (*P. indica* + *A. chroococcum*)
1.86×10^5	1.8×10^6

simultaneously. The major virulence determinant of *A. tumefaciens* is its tumor inducing Ti plasmid (De Kievit and Iglewski 2000; Haudecoeur and Faure 2010; Miller and Bassler 2001).

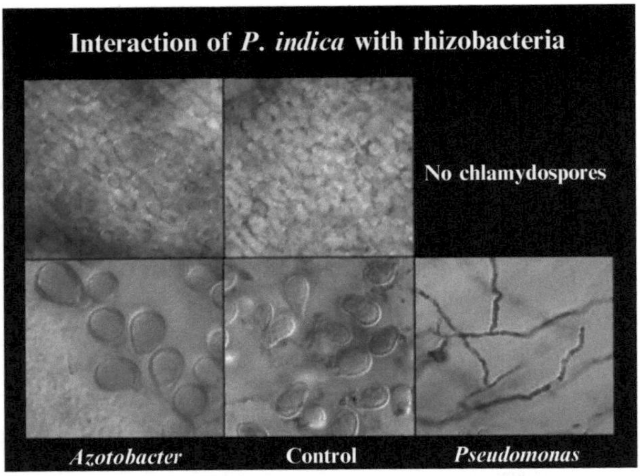

Fig. 18.5 Histograph representing increase in the diameter of the chlamydospores when treated with *A. chroococcum*. *Error bars* represent standard error (± 0.04)

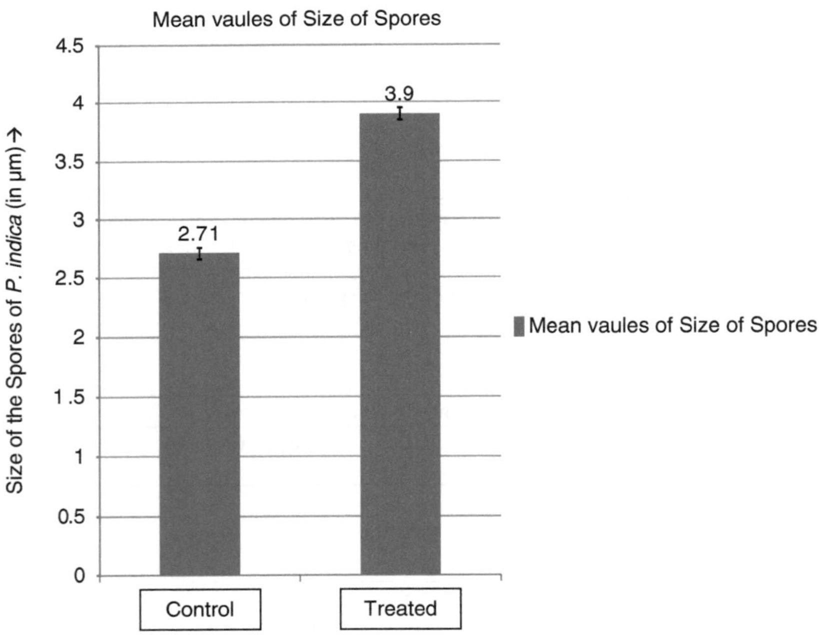

Fig. 18.6 Interaction of *P. indica* with the rhizobacteria *Azotobacter* and *Psedomonas*. Note dense and enlarged *P. indica* chlymydospores when cocultured with *Az. chroococcum*. In contrast, *Ps. fluorescense* completely blocked the production of chlamydospores

Upon infection, a region of the Ti plasmid, the T-DNA, is transferred from *A. tumefaciens* to the plant cell where it is integrated into the nuclear genome (Fuqua et al. 1994). The *tra* genes of the Ti plasmid are positively regulated by the

QS proteins TraR and TraI, homologues of LuxI/LuxR, in conjunction with the diffusible compound *N*-3-(oxooctanoyl)-L-homoserine lactone (AAI). The second regulatory determinant is TraM, which inhibits the activation of *tra* genes by TraR and AAI (Fuqua et al. 1994). This is thought to be a mechanism by which TraM sequesters TraR, preventing TraR-mediated AAI induction at low cell density, until the appropriate environmental conditions arise. Additionally, the T-DNA holds the gene for biosynthesis of opines which can be utilized by *A. tumefaciens* as sole nitrogen source enabling a competitive advantage over other soil bacteria. Opines also function as sensory molecules to induce conjugation. However, different opines induce different genes and it is worth noting that only when the respective opines are present does expression of the *tra* genes take place (De Kievit and Iglewski 2000; White and Finan 2009; Fuqua et al. 1994).

18.8 Microbe–Plant Communication

The microbial diversity of QS bacteria in the plant root rhizosphere contains more than 20 different groups of molecules with up to 100,000 different substances; these are known as secondary metabolites and are actively engaged in communicatory function at the root zone (Witzany 2010; Bais et al. 2003). In complex environments of the rhizosphere, many different QS bacteria establish signaling networks to maintain ecological balance (Lal 2009). The bacteria associated with the rhizospheric zone of plants must interact with other microorganisms in order to colonize and persist on the hosts.

Organisms of these regions develop the ability to survive in this competitive environment and they sense the environment via expressing different traits in the form of signals. These signals initiate many different responses in plants including multicellular differentiation, defense responses (cf. below), fruit body development and sporulation, as well as physiological communication processes such as bioluminescence (Sharma et al. 2003; Pierson et al. 1998).

Signal-mediated interactions with organisms belonging to other species, genera, families, and organismic kingdoms are vital for plants. They are highly coordinated and organized in parallel to establish symbiotic, parasitic, or mutualistic symbioses. The different forms of symbiotic communication represent different pragmatic contexts and require a great variety and highly coordinated behaviors from the participating partners. This involves a large numbers of complementary direct and indirect defenses, but also mutual beneficial behaviors. There are some indications that especially plant derived extra cellular *gamma*-amino butyrate (GABA) serves for sign-mediated communications between plants and animals, fungi, bacteria, and even among plants (Shelp et al. 2006; Waters and Bassler 2005; Witzany 2010). Intercellular communication processes are crucial in coordinating growth and development, shape, and dynamics. In general, QS systems facilitate the coordination of population behavior to enhance access to nutrients or specific environmental niches, collective defense against other competitor organisms or community escape where survival of the population is threatened (Williams 2007).

A wide variety of soil- and plant-associated bacteria with different prokaryotic and eukaryotic origin are involved in regulation of important phenotypes like virulence, motility, production of various enzymes, and antibiotic synthesis by QS. A variety of plant-associated bacteria produce Acyl HSLs (AHLs), 2-alkyl-4-quinolones (AQs), long-chain fatty acids, and fatty acid methyl esters as well as autoinducer-2 (AI-2), a collective term for a group of interconvertible furanones (derived from dihydroxyl pentanedione and various linear oligopeptides), as a chemical vocabulary used as signs in diverse processes (Williams 2007; Witzany 2008; Givskov et al. 1996).

A recent study has suggested that AHL production is quite common in plant-associated pseudomonads. After analyzing the pattern of AHLs produced by numerous plants, it appears that several species have one or more AHLs in common. For example, *Erwinia carotovora* pv. atroseptica, several xanthomonads, *A. tumefaciens*, and several rhizobia all have an AHL with a mobility on HPLC, which is identical to 3-oxo-C8-HSL (Cha et al. 1998; González and Marketon 2003). Additionally, most rhizobia had strong nonpolar AHL. It has been recommended that various chemical signals present in soil serve as a bacterial "Esperanto"; this is helpful in interactive or avoidable behavior of microorganisms in their quest to interrelate with their plant hosts (Cha et al. 1998; González and Marketon 2003; Givskov et al. 1996).

The symbiotic relationships formed between the nitrogen-fixing rhizobia and their legume hosts are the result of a complex signaling network. The process of nodulation is a series of biochemical interactions between the bacterium and its host. The *nod* genes are responsible for the various processes of nodule formation in plant roots for colonizing bacterial population in the inner cortex of the root. The plant-produced flavonoids are generally necessary, but not essential, for the expression of the *nod* genes. Control of *nod* gene expression in rhizobia varies from strain to strain but is usually mediated by NodD, a transcriptional regulator; these flavanoids bind with this NodD and activate *nod* gene expression and subsequent Nod factor production (González and Marketon 2003) (Fig. 18.7). Different *Rhizobium* species contain one or more symbiotic (*sym*) plasmids, which contain the genes for nodulation, nitrogen fixation, and exo-polysaccharide biosynthesis. In contrast, some species of bacteria i.e., *Bradyrhizobium*, *Azorhizobium*, and *Mesorhizobium* carry symbiotic information in clusters or islands on the chromosome.

Interestingly, Legumes form interactions not only with nitrogen-fixing bacteria, but also with mycorrhizal fungi. Although the structures formed to support these interactions are different (nodules compared with arbuscules), there is conservation in early signaling between these two symbioses. In the nodulation signaling pathway, calcium plays an essential role as a secondary messenger in plants, and the component that probably transduces the calcium signal is a unique calcium-activated kinase that is required for both mycorrhization and nodulation. The nodulation signaling pathway contains transcriptional regulators downstream of the calcium-activated kinase that are not required for the mycorrhizal symbiosis. This suggests that different symbiosis-specific signaling pathways are activated downstream of the calcium-activated kinase, and raises the question of how specificity of gene induction can be achieved in two pathways that are both likely to use

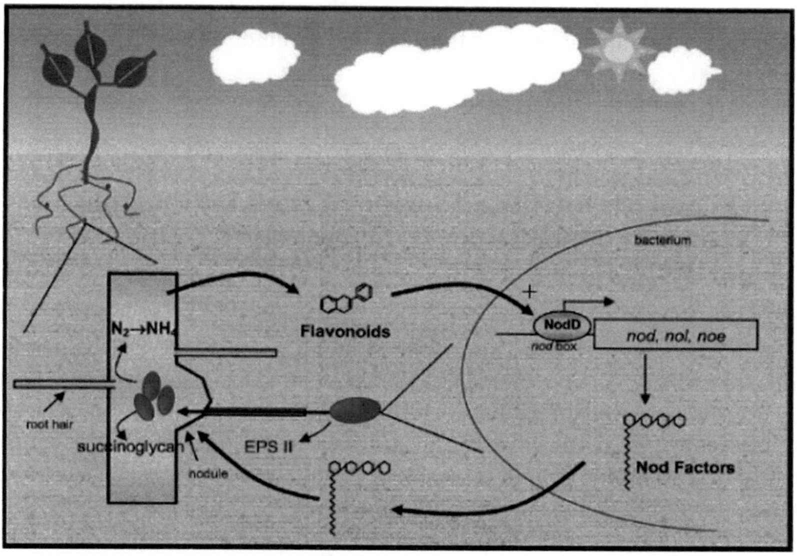

Fig. 18.7 *Rhizobium*-legume symbiosis model (González and Marketon 2003)

calcium and a unique calcium-activated kinase to induce different downstream events (Oldroyd and Downie 2006). It is likely that rhizobacteria and mycorrhiza forming fungi compete with the roots of the avaibable hosts, thereby influencing the population of Legume-interacting microbes in the rhizophere.

Gram-negative bacteria use various chemical compounds as small signal molecules, such as the acyl-homoserine lactone (acyl-HSL) system. This type of bacterial cell-to-cell communication was discovered in the context of microbial ecology, but it is now evident that acyl-HSL signaling is important in plant diseases. The property of this communication system is now exploited for enticing target for the development of antipathogenic therapies. Acyl-HSL signals are generated by the activity of a single enzyme S-adenosyl methionine, an intermediate of fatty acid biosynthesis and an acyl-acyl carrier protein. The cellular concentration of an acyl-HSL is defined by the environmental concentration, and environmental concentrations can rise only when there is a sufficient population of the signal-producing bacterium. The specific receptors for acyl-HSL signals are members of the LuxR family of transcriptional regulators. Acyl-HSL QS is most frequently observed in Gram-negative bacteria that interact with plant and animal hosts. Acyl-HSL signaling is critical for virulence of the plant pathogen *P. aeruginosa* in plants. Thus, acyl-HSL QS appears as a common theme in the interaction of several different bacterial species with eukaryotic hosts (Parsek and Greenberg 2000).

There could be more such mechanisms that are utilized by eukaryotic hosts to interfere with the QS behavior of associated bacteria. It is a task for the future to identify new communication systems, since they could prove to be a potential solution for the manipulation of pathogenic and symbiotic bacteria (González and Marketon 2003).

18.9 QS and Mycorrhizal Fungi or Beneficial Endophytes: Tripartite Associations Between Plants, Fungi, and Bacteria

The interactions among plants, mycorrhizal fungi, and bacteria can be defined as tripartite associations. The interaction between mycorrhizal fungi and bacteria range from loose associations between the microbes to endobacteria. Release of active molecules, including volatiles, and physical contact among the partners seem important for the establishment of the bacteria/mycorrhizal fungus/plant network. Although the nature of the complex interspecies/interphylum interactions remains unclear, QS might be important for establishing and maintaining the complex network (summarized in Bonfante and Anca 2009). Here, we focus on some aspects related to the endophyte *P. indica*.

P. indica, a facultative symbiont and a member of Basidiomycotina, has enormous potential for growth promotion of plants by colonizing their roots. In particular since the fungus has a broad host spectrum, the underlying growth-promoting events should be very general and not host-specific. *P. indica* mobilizes insoluble phosphates and translocates the phosphorus to the host in an energy-dependent process. In the presence of roots, the fungus releases so-far unidentified factors, which are recognized by the root cells and induce intracellular signaling events leading to better plant performance. Our present knowledge is mainly based on genetic studies with the model plant *A. thaliana*. Fungus-derived signals are secreted into the medium or rhizosphere and activate a plant receptor at the root plasma membrane. Early signaling events in the root cells include phospholipids, phosphorylation events, and an increase in cytoplasmic and nuclear calcium levels (summarized in Oelmüller et al. 2009; Vadassery et al. 2009a). Besides having access to reduced carbon, it appears that the plant provides a shelter for the fungus. Considering that *P. indica* was isolated from a desert region, living in harmony with a host and strengthening its performance under stress conditions also ensure better propagation of the fungus.

It is well established that interaction of symbiotic microbes like strains of *Azotobacter*, *Azospirillum*, *Bradyrhizobium*, Mycorrhizae, and members of Sebacinales enhance the growth and impart value addition to plant. One of the reasons is an enhanced transport of essential nutrients and unknown trace elements into the host plant. However, the terrestrial roots are not only colonized by useful fungi but also a wide spectrum of pathogenic and nonsymbiotic bacteria. The ecological niches of these bacteria are rhizosphere, rhizoplane, and mycorrhizosphere (Fig. 18.8). In nature, it is not possible to see any root without being surrounded

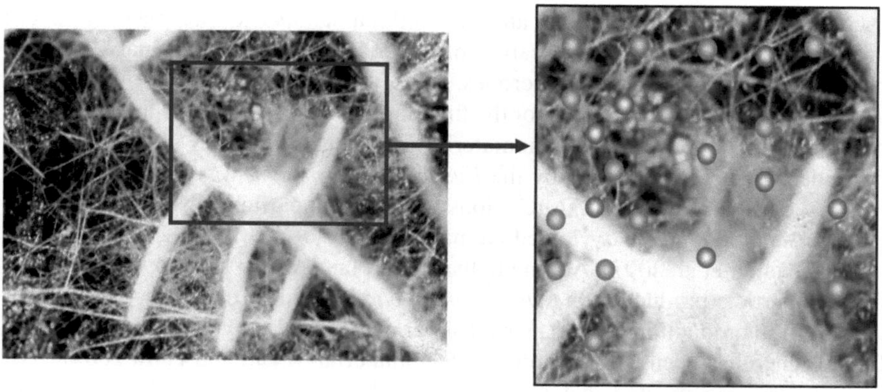

Fig. 18.8 Root and root hairs are surrounded by intense growth of fungal mycelia. Embedded in the matrix are high populations of known and unknown bacteria. Unknown nanoparticles are present in the mycorhizospheric matrix (see *golden particles*) (Patent application number 1632/DEL/2009, Reference Number: E-2/179/2010-DEL)

by beneficial and nonbeneficial microorganisms. Root and root hairs are colonized by highly branched fungal hyphae, which provide habitats for a wide spectrum of microbes. Only in tissue-culture raised plantlets (where individual parameters can be changes), beneficial and nonbeneficial traits in the complex symbioses can be studied which is far away from the situation in nature, in particular if tripartite interactions between plants, fungi, and bacteria are considered (Bonfante and Anca 2009). However, identification of factors/genes/proteins which determine the interactions is required for the understanding of the situation in nature.

Many microbes from bacteria to fungi and also plants produce nanomaterial. *Pseudomonas stutzeri* AG259 isolated from silver mines produces silver nanoparticles (Joerger et al. 2000). Nair and Pradeep (2002) reported that common *Lactobacillus* strains found in buttermilk assisted the growth of microscopic gold, silver, and gold-silver alloy crystals of well-defined morphology. Sastry et al. (2003) observed that the extremophilic actinomycete, *Thermomonospora* sp., when exposed to gold ions reduced the metal ions extracellularly, yielding gold nanoparticles with a much improved polydispersity. Ahmad et al. (2003) further tried to elucidate the mechanism and conditions favoring the formation of nanoparticles with desired features. They carried out the reduction of Au ions by using an extremophilic *Thermomonospora* sp. biomass that has resulted in efficient synthesis of monodisperse gold nanoparticles.

Candida glabrata and *Schizosaccharomyces pombe* were used for the first time for the biosynthesis of cadmium sulfide (CdS) nanocrystals. These nanocrystals were produced using cadmium salts and are now used in quantum semiconductor crystallites (Dameron et al. 1989). The *Verticillium* sp. fungal biomass when exposed to aqueous $AgNO_3$ solution resulted in the intracellular formation of silver nanoparticles, while *Fusarium oxysporum* biomass resulted in the formation of extracellular silver nanoparticles (Senapati et al. 2004). Many more fungi have been used to produce both intracellular and extracellular nanoparticles. Birla et al. (2009) used *Phoma glomerata* in fabrication of silver nanoparticles and analyzed these nanoparticles against selected microbes.

Mukherjee et al. (2002) have for the first time opened up a novel fungal/enzyme-based in vitro approach for nanomaterial synthesis. A species-specific NADH-dependent reductase, released by the *Fusarium oxysporum*, was successfully used to carry out the reduction of $AuCl_4$ ions to gold nanoparticles. Nitrate reductases from *Fusarium oxysporum* yielded Ag nanoparticles of nearly 10–25 nm extracellularly (Kumar et al. 2007). Adding to the list of plants which are showing potential for nanoparticle production, a *Cinnamomum camphora* leaf extract has been identified for the production of gold as well silver nanoparticles by Huang et al. (2007). Even Lichen has been used to produce Nanomaterial (Shahi and Patra 2003).

It is also documented that nanomaterial kills the microbes including viruses (Durán et al. 2007). However, a positive influence of nanomaterials on fungal and bacterial biomass production has also been observed (Varma, unpublished). The nanomaterials used are gold, silver, activated TNP (Titanium Nanoparticles), and CNT (Carbon Nanotubes). Activated and nonactivated charcoal did not play any significant role on the morphology and the functioning of *P. indica*. The nanomaterial-embedded materials also enhanced plant productivity (Varma et al., unpublished). Nonmaterial-containing microbial culture filtrate also influence breaking the seed dormancy and accelerate the growth processes.

Based on the available literature data, we assume that interaction among microbes and interaction between microbes (*P. indica*) and plants leads to the production of certain unidentified, uncharacterized nanomaterials. Nanomaterials are produced by microbes when they are provided with the specific requirements like gold, silver, titanium, and carbon nanotubes (CNT). It is clearly depicted in literature that nanomaterials influence the growth of plant and also microbes.

Since many fungi if given appropriate conditions can develop nanomaterials, their use for nanomaterial synthesis is sustainable and an efficient method for biotechnological applications. For instance, *Usnea* produces usnic acid as a nanomaterial (Shashi and Patra 2003), but there are no reports in the literature that mycorrhiza is producing nanomaterial. Again, *P. indica*, while interacting with various plant species and microbes, provides a good study object for nanomaterial production. It remains to be determined how nonoparticles influence plant/microbe interactions and the microbial community.

Since *P. indica* interacts with many different plant species, this endophyte, like others, has emerged as a model organism to understand how a plant (root), its fitness, and metabolic state influence the microbial communities in the rhizophere. It appears that genetic parameters of the *P. indica*-interacting plants and environmental cues influencing plant performance are crucial for growth and propagation of individual members of the soil microbial community. Although the interaction of *P. indica* with all tested plant species is beneficial for both partners, the plant activates defense processes which restrict fungal growth in the root and rhizophere and thereby has a strong influence on the propagation of individual microbes in the rhizophere and the community in general. In sum, a large battery of defense components is synthesized in the roots in response to pathogenic and beneficial fungi, with highly specific or broader fungal targets. Recently, Knecht et al. (2010) have shown that growth of pathogenic fungi can be limited by specific germin-like proteins expressed in the roots, with little or no influence on the mode of interaction of the root with *P. indica*. On the other hand, genetic inactivation of specific defense pathways in the plants, such as the ethylene/jasmonic acid or salicylic acid signaling pathway (Camehl et al. 2010), the oxidative stress-inducible pathway utilizing phosphatidic acid-dependent kinases and oxidative stress-inducible kinases, the synthesis of glucosinolates (Sherameti et al. 2008), or other defense compounds weakens the plants and allows both pathogenic and beneficial microbes to propagate faster in the root and root environment. This has a strong influence on the microbial community, and both the population density and composition can change in the rhizophere. An example for an *environmental* factor crucial for the plant performance that influences the microbial society in the vicinity of roots is sulfur. Sulfur limitations force Brassicaceae to redirect their metabolism toward the synthesis of primary sulfur-containing compounds on the expanse of secondary sulfur-containing compounds such as glucosinolates which are crucial for defense. Since glucosinolates can be specifically directed against individual soil microbes, the sulfur metabolic state of a plant participates in shaping the root community. Another, so far little investigated environmental factor crucial for many plant metabolic processes is the availability of iron. Iron limitation or availability for the plant prevent many primary metabolic pathways and defense processes in roots resulting in severe changes in the microbial communities which interact with them (Tripathy et al. 2010). Therefore, it is not surprising that many plant genes involved in iron uptake and metabolism are targets of microbes in both beneficial and pathogenic root/microbe interactions. Nutrients (Sherameti et al. 2005), phytohormones (Vadassery et al. 2008), the redox state in the plant cell (Vadassery et al. 2009b), the efficiency of the photosynthetic electron transport in the leaves (Johnson and Oelmüller, unpublished), and many other plant parameters influence the microbial communities, the information flow between them and their hosts, and only very minor changes (such as a single mutation in a plant defense gene) have a strong influence on the community. Barely any of the underlying mechanisms is understood yet.

18.10 Parallel Communication of Plant Roots with Bacteria and Fungi

In order to conduct neuronal-like activities and establish symbiotic relationships with bacteria (Denison and Kiers 2004), plants use their respective plant-specific synapses (Baluska et al. 2005). Similar mutually profitable relationships are established with mycorrhizal fungi. A special type of plant "synapses" resembles the immunological synapses of animal cells and allows plants to respond to pathogen and parasite attacks as well as to establish stable symbiotic interactions with rhizobia bacteria and fungal mycorrhiza (Baluska et al. 2005; Witzany 2008, 2010). Electrical signals can reinforce chemical signals or overcome short-distance responses of fungal mycelia that can be present on root surfaces (West et al. 2006). Interestingly, rhizobia are taken up into plant cells via phagocytosis during symbiotic interactions with roots of leguminous plants (Samaj et al. 2004).

Legumes and rhizobial bacteria share symbiotic liaison that leads to the creation of nitrogen-binding nodules in the root zone. Nod factor signaling and thigmotrophic responses of root hairs have common characteristics here as well. This shows context dependency, i.e., how the same signaling pathways are used for different content transfer (Guerts et al. 2005). Today, several hundred species of fungi colonize more than 100,000 different plant species. This type of cohabitation requires elaborated symbiotic signaling (Lammers 2004). We are only at the very beginning to understand the information flow in complex multiorganismic systems, such as the rhizosphere. Roots develop from rhizomes to provide better conditions for mycorrhizal fungi, which in turn supply plants with better nutrients (Brundrett 2002). For the fungus, the relationship is either balanced or predatory. Many endophytic fungi, however, live in plants without triggering disease symptoms (Brundrett 2002). Plants and microbes share a particular repertoire of signals. Interestingly, some are therefore also employed strategically.

18.11 Conclusion

Microbes and plants fundamentally depend on successful communication. The behavior in the specific interaction can be misinterpreted. A plant can feign mutualism, for example, in order to gain a one-sided advantage from the interaction and to damage or kill the partner. However, this cannot be the representative form of communication because no individuals would survive if all plants behaved in this manner. The majority of interactions must be successful for several participants. Communication processes are successful when the rules are correctly followed (Witzany 2008; Schwarz et al. 2008).

The whole QS circuit relies on the intracellular production and export of low-molecular mass signaling molecules, and their extracellular concentration grows with the population density of the producing organism. The signaling molecule can be sensed and imported into the cells of the responding organism, thus allowing the whole population or society to respond to changing environment (Gera and Srivastava 2006). Since the list of bacteria that utilize QS systems is growing, our knowledge is by far not complete. It is therefore certain that our current comprehension of QS is severely limited and that the true extent of cell–cell communication in the environment and its significance await further and exciting discoveries. Interestingly, the soil-bacteria have also evolved complicated signal transduction mechanisms to perceive sensory information in accordance with the changing environmental conditions including changes in pH, temperature, osmotic potential, and nutrient availability.

18.12 Future Perspectives

QS mechanisms have been exploited by diverse group of bacteria for regulation of various phenotypes as a part of their pathogenic or symbiotic lifestyles. The ability to restrict or promote these systems provides a powerful tool to solve many environmental and agricultural problems and enhance productivity. Genetically modified plants can be used to produce AHLs to maneuvre plant–bacterium associations. QS may also provide the much needed approach to control multiple-drug-resistant strains of bacteria that employ their transduction signaling systems for virulence and pathogenicity and for control of host-cell responses. Apart from this, QS and related mechanisms can be investigated for novel antimicrobial targets and immunomodulatory agents. Therefore, it is important to take a look at the methods employed, which can enhance/degrade this signaling cascade (Gera and Srivastava 2006; Schwarz et al. 2008).

The continuing research is directed at refining the molecular basis of our understanding of QS and in characterizing new signaling "languages." Future research in this growing area will clearly benefit from refined approaches. Established concepts and definitions developed in the fields of ecology and evolution should be taken into account. Understanding of the cell-to-cell communication which operates in complex bacterial consortia, e.g., within the human intestinal tract, is an important task for the future (Williams et al. 2007; Waters and Bassler 2005).

With the increase in the number of microorganisms that employ QS systems, the possibilities for the analysis of the underlying regulatory mechanisms increase. It is evident that many important animal and plant pathogens use QS to regulate pathogenicity; hence, strategies designed to interfere with these signaling systems will likely have broad applicability for biological control of the virulent organisms. It will be fascinating to see in the future, if additional human pathogens utilize QS as part of their pathogenic lifestyle and, if so, whether synthesis of the signal molecules can be exploited to control infections. The recent production of AHLs in plants represents an exciting new approach to control crop diseases as well as to manipulate plant–microbe interactions for improved crop production in the future (De Kievit and Iglewski 2000; Schwarz et al. 2008). The key questions to be addressed are: delineating morphological, physiological, and developmental parameters involved in fungal–bacterial interactions; comparative profiling of protein patterns of bacteria enhancing and inhibiting growth and differentiation of fungi such as *P. indica*; visualization of factors specifically expressed in bacterial–fungal cocultivation; analysis of potential molecular modulators involved in the interaction and a comparative profile of differentially expressed fungal and bacterial genes and proteins in the rhizosphere. Finally, a comparative analysis of microbial communities under laboratory conditions (which allow manipulations of individual factors) and in nature will help to understand the mechanisms and to identify the factors that determine the interactions between the organisms from three kingdoms.

References

Adonizio AL (2008) Anti-quorum sensing agents from South Florida medicinal plants and their attenuation of *Pseudomonas Aeruginosa* pathogenicity. Electronic Theses and Dissertations, Florida International University

Ahmad A, Senapati S, Khan MI, Kumar R, Sastry M (2003) Extracellular biosynthesis of monodisperse gold nanoparticles by a novel extremophilic actinomycete, *Thermomonospora* sp. Langmuir 19:3550–3553

Akiyama K, Matsuzaki K, Hayashi H (2005) Plant sesquiterpenes induce hyphal branching in arbuscular mycorrhizal fungi. Nature 435:824–827

Ames BN (1989) Mycorrhiza development in onion in response to chitin-decomposing actinomycetes. New Phytol 112:423–427

Bais HP, Park SW, Weir TL, Callaway RM, Vivanco JM (2003) How plants communicate using the underground information superhighway. Trends Plant Sci 9:26–32

Baluska F, Volkmann D, Menzel D (2005) Plant synapses: actin-based domains for cell-to-cell communication. Trends Plant Sci 10:106–111

Bassler BL, Wright M, Silverman MR (1994) Multiple signaling systems controlling expression of luminescence in *Vibrio harveyi*: sequence and function of genes encoding a second sensory pathway. Mol Microbiol 13:273–286

Birla SS, Tiwari VV, Gade A, Ingle AP, Yadav A, Rai MK (2009) Fabrication of silver nanoparticles by *Phoma glomerata* and its combined effect against *Escherichia coli*, *Pseudomonas aeruginosa* and *Staphylococcus aureus*. Lett Appl Microbiol 48:173–179

Blosser-Middleton RS, Gray KM (2001) Multiple N-acyl homoserine lactone signals of *Rhizobium leguminosarum* are synthesized in a distinct temporal pattern. J Bacteriol 183:6771–6777

Bonfante P, Anca IA (2009) Plants, mycorrhizal fungi, and bacteria: a network of interactions. Annu Rev Microbiol 63:363–383

Bowen GD, Theodorou C (1979) Interactions between bacteria and ectomycorrhizal fungi. Soil Biol Biochem 11:119–126

Brencic A, Winans SC (2005) Detection of and response to signals involved in host-microbe interactions by plant-associated bacteria. Microbiol Mol Biol Rev 69:155–194

Brewin NJ, Beringer JE, Buchanan-Wollaston AV, Johnston AWB, Hirsch PR (1980) Transfer of symbiotic genes with bacteriocinogenic plasmids in *Rhizobium leguminosarum*. J Gen Microbiol 116:261–270

Brundrett MC (2002) Coevolution of roots and mycorrhizas of land plants. New Phytol 154:275–304

Camehl I, Sherameti I, Venus Y, Bethke G, Varma A, Lee J, Oelmüller R (2010) Ethylene signalling and ethylene-targeted transcription factors are required to balance beneficial and nonbeneficial traits in the symbiosis between the endophytic fungus *Piriformospora indica* and *Arabidopsis thaliana*. New Phytol 185:1062–1073

Cha C, Gao P, Chen YC, Shaw PD, Farrand SK (1998) Production of acyl-homoserine lactone quorum-sensing signals by gram-negative plant-associated bacteria. Mol Plant Microbe Interact 11:1119–1129

Chugani SA, Whiteley M, Lee KM, D'Argenio D, Manoil C, Greenberg EP (2001) QscR, a modulator of quorum-sensing signal synthesis and virulence in *Pseudomonas aeruginosa*. Proc Natl Acad Sci USA 98:2752–2757

Crespi BJ (2001) The evolution of social behaviour in microorganisms. Trends Ecol Evol 16:178–183

Dameron CT, Reeser RN, Mehra RK, Kortan AR, Carroll PJ, Steigerwaldm ML, Brus LE, Winge DR (1989) Biosynthesis of cadmium sulphide quantum semiconductor crystallites. Nature 338:596–597

De Kievit TR, Iglewski BH (2000) Bacterial quorum sensing in pathogenic relationships. Infect Immun 68(9):4839–4849

Denison FR, Kiers TE (2004) Why are most rhizobia beneficial to their plants, rather than parasitic? Microbes Infect 6:1235–1239

Diggle SP, Winzer K, Lazdunski A, Williams P, Ca'mara M (2002) Advancing the quorum in *Pseudomonas aeruginosa*: the regulation of N-acylhomoserine lactone production and virulence gene expression. J Bacteriol 184:2576–2586

Diggle SP, Crusz SA, Cámara M (2003) Quorum sensing. Curr Biol 17:R908–R910

Diggle SP, Gardner A, West SA, Griffin AS (2007) Evolutionary theory of bacterial quorum sensing: when is a signal not a signal? Philos Trans R Soc Lond B Biol Sci 362:1241–1249

Dunny GM, Winans SC (1999) Cell–cell signalling in bacteria. ASM Press, Washington, DC, pp 1–5

Duponnois R, Garbaye J (1991) Mycorrhization helper bacteria associated with the Douglas fir *Laccarialaccata* symbiosis: effect *in vitro* and in glasshouse conditions. Ann Sci For 48:239–251

Durán N, Marcato PD, De Souza G, Alves OL, Esposito E (2007) Antibacterial effect of silver nanoparticles produced by fungal process on textile fabrics and their effluent treatment. J Biomed Nanotechnol 3:203–208

Fauvart M, Michiels J (2008) Rhizobial secreted proteins as determinants of host specificity in the *Rhizobium*-legume symbiosis. FEMS Microbiol Lett 285:1–9

Fuqua WC, Winans SC, Greenberg EP (1994) Quorum sensing in bacteria: the LuxR-LuxI family of cell density-responsive transcriptional regulators. J Bacteriol 176:269–275

Gambello MJ, Iglewski BH (1991) Cloning and characterization of the *Pseudomonas aeruginosa lasR* gene, a transcriptional activator of elastase expression. J Bacteriol 173:3000–3009

Garbaye J (1994) Helper bacteria – a new dimension to the mycorrhizal symbiosis. New Phytol 128:197–210

Garbaye J, Bowen GD (1987) Effect of different microflora on the success of ectomycorrhizal inoculation of *Pinus radiata*. Planta 17:941–943

Gera C, Srivastava S (2006) Quorum-sensing: the phenomenon of microbial communication. Curr Sci 90:666–676

Givskov M, de Nys R, Manefield M, Gram L, Maximilien R, Eberl L, Molin S, Steinberg PD, Kjelleberg S (1996) Eukaryotic interference with homoserine lactone-mediated prokaryotic signalling. J Bacteriol 178:6618–6622

González JE, Marketon MM (2003) Quorum sensing in nitrogen-fixing rhizobia. Microbiol Mol Biol Rev 67:574–592

Gray EJ, Smith DL (2005) Intracellular and extracellular PGPR: commonalities and distinctions in the plant bacterium signaling process. Soil Biol Biochem 37:395–412

Greenberg PE (2003) Bacterial communication: tiny teamwork. Nature 424:134

Gryndler M (2000) Interactions of arbuscular mycorrhizal fungi with other soil organisms. In: Kapulnik Y, Douds DDJ (eds) Arbuscular mycorrhizas: physiology and function. Kluwer Academic Publishers, Dordrecht, The Netherlands, pp 239–262

Guerts R, Fedorova E, Bisseling T (2005) Nod factor signalling genes and their function in the early stages of *Rhizobium* infection. Curr Opin Plant Biol 8:346–352

Haudecoeur E, Faure D (2010) A fine control of quorum-sensing communication in *Agrobacterium tumefaciens*. Commun Integr Biol 3:1–5

Hiltner L (1904) Über neuere Erfahrungen und Probleme auf dem Gebiete der Bodenbakteriologie unter besonderer Berücksichtigung der Begrünung und Brache. Arbeiten der Deutschen Landwirtschaftlichen Gesellschaft 98:59–78

Huang J, Li Q, Sun D, Lu Y, Su Y, Yang X, Wang H, Wang Y, Shao W, He N, Hong J, Chen C (2007) Biosynthesis of silver and gold nanoparticles by novel sun dried *Cinnamomum camphora* leaf. Nanotechnology 18:105104–105114

Joerger R, Klaus T, Granqvist CG (2000) Biologically produced silver-carbon composite materials for optically functional thin film coatings. Adv Mater 12:407–409

Kent AD, Triplett EW (2002) Microbial communities and their interactions in soil and rhizosphere ecosystems. Annu Rev Microbiol 56:211–236

Knecht K, Seyffarth M, Desel C, Thurau T, Sherameti I, Lou B, Oelmüller R, Cai D (2010) Expression of BvGLP-1 encoding a germin-like protein from sugar beet in *Arabidopsis thaliana* leads to resistance against phytopathogenic fungi. Mol Plant Microbe Interact 23:446–457

Kumar SA, Abyaneh MK, Gosavi SW, Kulkarni SK, Pasricha R, Ahmad A, Khan MI (2007) Nitrate reductase-mediated synthesis of silver nanoparticles from $AgNO_3$. Biotechnol Lett 29:439–445

Lal A (2009) Quorum sensing. Resonance 14:866–871

Lammers PJ (2004) Symbiotic signalling: new functions for familiar proteins. New Phytol 16:324–326

Lazazzera BA, Grossman AD (1998) The ins and outs of peptide signalling. Trends Microbiol 7:288–294

Linderman RG (1988) Mycorrhizal interactions with the rhizosphere microflora: the mycorrhizosphere effect. Phytopathology 78:366–371

Manefield M, Rasmussen TB, Henzter M, Andersen JB, Steinberg P, Kjelleberg S, Givskov M (2002) Halogenated furanones inhibit quorum sensing through accelerated LuxR turnover. Microbiology 148:1119–1127

Martínez-Romero E (2009) Coevolution in *Rhizobium*-legume symbiosis? DNA Cell Biol 28:361–370

Meyer JR, Linderman RG (1986) Response of subterranean clover to dual inoculation with vesicular arbuscular fungi and a plant growth-promoting bacterium, *Pseudomonas putida*. Soil Biol Biochem 18:185–190

Miller MB, Bassler BL (2001) Quorum sensing in bacteria. Annu Rev Microbiol 55:165–199

Minerdi D, Fani R, Gallo R, Boarino A, Bonfante P (2001) Nitrogen fixation genes in an endosymbiotic *Burkholderia* strain. Appl Environ Microbiol 67:725–732

Mukherjee P, Senapati S, Mandal D, Ahmad A, Khan MI, Kumar R, Sastry M (2002) Extracellular synthesis of gold nanoparticles by the fungus Fusarium oxysporum. Chembiochem 3:461–463

Nair B, Pradeep T (2002) Coalescence of nanoclusters and formation of submicron crystallites assisted by *Lactobacillus* strains. Cryst Growth Des 4:295–298

Oelmüller R, Sherameti I, Tripathi S, Varma A (2009) *Piriformospora indica*, a cultivable root endophyte with multiple biotechnological applications. Symbiosis 49:1–17

Oldroyd GE, Downie JA (2006) Nuclear calcium changes at the core of symbiosis signalling. Curr Opin Plant Biol 9:351–357

Oldroyd GE, Harrison MJ, Udvardi M (2005) Peace talks and trade deals. Keys to long-term harmony in legume-microbe symbioses. Plant Physiol 137:1205–1210

Parsek MR, Greenberg EP (2000) Acyl-homoserine lactone quorum sensing in gram-negative bacteria: a signalling mechanism involved in associations with higher organisms. Proc Natl Acad Sci USA 97:8789–8793

Pearson JP, Passador L, Iglewski BH, Greenberg EP (1995) A second N-acylhomoserine lactone signal produced by *Pseudomonas aeruginosa*. Proc Natl Acad Sci USA 92:1490–1494

Peters NK, Frost JW, Long SR (1986) A plant flavone, luteolin, induces expression of *Rhizobium meliloti* nodulation genes. Science 233:977–980

Pierson EA, Wood DW, Cannon JA, Blachere FM, Piers LS (1998) Interpopulation signalling *via* N-acyl-homoserine lactones among bacteria in the wheat rhizosphere. Mol Plant Microbe Interact 11:1078–1084

Pinton R, Varanini Z, Nannipieri P (2001) The rhizosphere as a site of biochemical interactions among soil components, plants and microorganisms. In: Pinton R, Varanini Z, Nannipieri P (eds) The rhizosphere: biochemistry and organic substances in the soil–plant interface. Dekker, New York, NY, pp 1–17

Samaj J, Baluska F, Voigt B, Schlicht M, Volkmann D, Menzel D (2004) Endocytosis, actin cytoskeleton and signaling. Plant Physiol 135:1150–1161

Sastry M, Ahmad A, Khan MI, Kumar R (2003) Biosynthesis of metal nanoparticles using fungi and actinomycete. Curr Sci 85:162–170

Schuster M, Greenberg EP (2006) A network of networks: quorum-sensing gene regulation in *Pseudomonas aeruginosa*. Int J Med Microbiol 296:73–81

Schwarz M, Romano D, Gheorghe M (2008) Visualizing bacteria quorum sensing swarm intelligence algorithms and applications symposium, AISB 2008 convention communication, interaction and social intelligence, Aberdeen, Scotland, UK, 1–4 April 2008

Senadheera D, Cvitkovitch DG (2008) Quorum sensing and biofilm formation by *Streptococcus mutans*. Adv Exp Med Biol 631:178–188

Senapati S, Mandal D, Ahmad A, Khan MI, Sastry M, Kumar R (2004) Fungus mediated synthesis of silver nanoparticles: a novel biological approach. Indian J Phys 78A:101–105

Shahi SK, Patra P (2003) Microbially synthesized bioactive nanoparticles and their formulation active against human pathogenic fungi. Rev Adv Mater Sci 5:501–509

Shapiro JA (1998) Thinking about bacterial populations as multicellular organisms. Annu Rev Microbiol 52:81–104

Sharma A, Sahgal M, Johri BN (2003) Microbial communication in the rhizosphere: operation of quorum sensing. Curr Sci 85:1164–1172

Shashi SK, Patra M (2003) Microbially synthesized bioactive nanoparticles and their formulation active against human pathogenic fungi. Rev Adv Mater Sci 5:501–509

Shelp BJ, Bown AW, Faure D (2006) Extracellular g-aminobutyrate mediates communication between plants and other organisms. Plant Physiol 142:1350–1352

Sherameti I, Shahollari B, Venus Y, Altschmied L, Varma A, Oelmüller R (2005) The endophytic fungus *Piriformospora indica* stimulates the expression of nitrate reductase and the starch-degrading enzyme glucan-water dikinase in tobacco and *Arabidopsis* roots through a home-odomain transcription factor that binds to a conserved motif in their promoters. J Biol Chem 280:26241–26247

Sherameti I, Venus Y, Drzewiecki C, Tripathi S, Dan VM, Nitz I, Varma A, Grundler FM, Oelmüller R (2008) PYK10, a beta-glucosidase located in the endoplasmatic reticulum, is crucial for the beneficial interaction between *Arabidopsis thaliana* and the endophytic fungus *Piriformospora indica*. Plant J 54:428–439

Soto MJ, Sanjuán J, Olivares J (2006) Rhizobia and plant-pathogenic bacteria: common infection weapons. Microbiology 152:3167–3174

Tripathy BC, Sherameti I, Oelmüller R (2010) Siroheme: an essential component for life on earth. Plant Signal Behav 5:1–5

Vadassery J, Ritter C, Venus Y, Camehl I, Varma A, Shahollari B, Novák O, Strnad M, Ludwig-Müller J, Oelmüller R (2008) The role of auxins and cytokinins in the mutualistic interaction between *Arabidopsis* and *Piriformospora indica*. Mol Plant Microbe Interact 21:1371–1383

Vadassery J, Ranf S, Drzewiecki C, Mithöfer A, Mazars C, Scheel D, Lee J, Oelmüller R (2009a) A cell wall extract from the endophytic fungus *Piriformospora indica* promotes growth of Arabidopsis seedlings and induces intracellular calcium elevation in roots. Plant J 59:193–206

Vadassery J, Tripathi S, Prasad R, Varma A, Oelmüller R (2009b) Monodehydroascorbate reductase 2 and dehydroascorbate reductase 5 are crucial for a mutualistic interaction between *Piriformospora indica* and *Arabidopsis*. J Plant Physiol 166:1263–1274

Van Loon LC (2007) Plant responses to plant growth-promoting rhizobacteria. Eur J Plant Pathol 119:243–254

Varma A, Verma S, Sudha SN, Butehorn B, Franken P (1999) *Piriformospora indica*, a cultivable plant-growth-promoting root endophyte. Appl Environ Microbiol 65:2741–2744

Varma A(2009) In: Ajit Varma (Ed) Mycorrhiza Springer-Verlag, Germany, pp 1–783

Verma S, Varma A, Karl-Heinz R, Hassel A, Kost G, Sarbhoy A, Bisen P, Bütehorn P, Franken P (1998) *Piriformospora indica* gen. nov., a new root-colonizing fungus. Mycologia 90:895–909

Walker TS, Bais HP, Grotewold E, Vivanco JM (2003) Root exudation and rhizosphere biology. Plant Physiol 132:44–51

Waters CM, Bassler BL (2005) Quorum sensing: cell-to-cell communication in bacteria. Annu Rev Cell Dev Biol 21:319–346

West SA, Griffin AS, Gardner A, Diggle SP (2006) Social evolution theory for microorganisms. Nat Rev Microbiol 4:597–607

White CE, Finan TM (2009) Quorum quenching in *Agrobacterium tumefaciens*: chance or necessity? J Bacteriol 191:1123–1125

Whitehead NA, Barnard AM, Slater H, Simpson NJL, Salmond GPC (2001) Quorum-sensing in gram-negative bacteria. FEMS Microbiol Rev 25:365–404

Williams P (2007) Quorum sensing, communication and cross-kingdom signalling in the bacterial world. Microbiology 153:3923–3938

Williams P, Winzer K, Chan WC, Camara M (2007) Look who's talking: communication and quorum sensing in the bacterial world. Philos Trans R Soc Lond B Biol Sci 362:1119–1134

Wisniewski-Dyé F, Downie JA (2002) Quorum-sensing in *Rhizobium*. Antonie Van Leeuwenhoek 81:397–407

Witzany G (2008) Biocommunication of unicellular and multicellular organisms. TripleC 6:24–53

Witzany G (2010) Plant communication. In: Witzany G (ed) Biocommunication and natural genome editing. Springer, Dordrecht, The Netherlands, pp 27–52

Index

G. Witzany (ed.), *Biocommunication in Soil Microorganisms*, Soil Biology 23,
DOI 10.1007/978-3-642-14512-4, © Springer-Verlag Berlin Heidelberg 2011

N

O

P

MIX
Papier aus verantwortungsvollen Quellen
Paper from responsible sources
FSC® C105338

If you have any concerns about our products,
you can contact us on
ProductSafety@springernature.com

In case Publisher is established outside the EU,
the EU authorized representative is:
Springer Nature Customer Service Center GmbH
Europaplatz 3, 69115 Heidelberg, Germany

Printed by Libri Plureos GmbH
in Hamburg, Germany